Theoretical and Computational Polymers Science: Physics, Chemistry and Biology

Guest Editor
Hector Eduardo Roman
Department of Physics
University of Milano-Bicocca
Milan
Italy

Editorial Office
MDPI AG
Grosspeteranlage 5
4052 Basel, Switzerland

This is a reprint of the Special Issue, published open access by the journal *Polymers* (ISSN 2073-4360), freely accessible at: https://www.mdpi.com/journal/polymers/special_issues/998227Q5WT.

For citation purposes, cite each article independently as indicated on the article page online and as indicated below:

Lastname, A.A.; Lastname, B.B. Article Title. *Journal Name* **Year**, *Volume Number*, Page Range.

ISBN 978-3-7258-5115-7 (Hbk)
ISBN 978-3-7258-5116-4 (PDF)
https://doi.org/10.3390/books978-3-7258-5116-4

Cover image courtesy of Hector Eduardo Roman

Contents

About the Editor

Hector Eduardo Roman

Hector Eduardo Roman is currently working as scientific collaborator at the Department of Physics and Department of Earth and Environmental Sciences of the University of Milano-Bicocca. He has been awarded a visiting professorship at Pohang University of Science and Technology (POSTECH) working at the Division of IT Convergence Engineering (Pohang, Korea). He has been guest scientist at the Max Planck Institute for the Physics of Complex Systems (MPIPKS, Germany) and research fellow at the Department of Physics of the University of Milan. He has been nominated Research Assistant (Privatdozent) at the University of Hamburg and University of Giessen (Germany), and has been awarded a Feodor Lynen Fellowship from the Alexander von Humboldt Foundation (Germany). His interests cover different areas of complex systems such as polymers, disordered systems and fractals, biomolecules, sensor networks and analysis of acoustic environmental phenomena, atomic-soft matter-plasma physics, and finance. He is co-editor/co-author of 10 books and has over 300 publications in International Journals.

theories [93,94], orientational relaxation [95–97], large-scale molecular simulations [98–100], actin filaments [101–103], conformational dynamics of GLP-2 peptide [104–106], soft vesicle–polymer chain conformations [107,108], entangled polymer networks [109–111], polyelectrolyte pore translocation [112–115], protein adsorption dynamics [116–119], biomedical applications of polymers [120–122], and polymer hydrophilic networks for energy storage [123,124].

Conflicts of Interest: The author declares no conflicts of interest.

References

1. Qiao, L.; Giannakou, M.; Schmid, F. An Efficient and Accurate SCF Algorithm for Block Copolymer Films and Brushes Using Adaptive Discretizations. *Polymers* **2024**, *16*, 1228. [CrossRef]
2. Mian, S.H.; Umer, U.; Moiduddin, K.; Alkhalefah, H. Predicting Mechanical Properties of Polymer Materials Using Rate-Dependent Material Models: Finite Element Analysis of Bespoke Upper Limb Orthoses. *Polymers* **2024**, *16*, 1220. [CrossRef]
3. Jung, J.K.; Lee, J.H.; Park, J.Y.; Jeon, S.K. Modeling of the time-dependent H_2 emission and equilibrium time in H_2-enriched polymers with cylindrical, spherical and sheet shapes and comparisons with experimental investigations. *Polymers* **2025**, *16*, 2158. [CrossRef]
4. Xu, X.; Zhang, B.; Shi, F.; Liu, K.; Peng, G.; Gao, L.; Gao, J.; Du, Y. Study on the Influence of Hygrothermal Aging on the Mechanical Properties of Carbon Fabric/Polyetheretherketone Composites. *Polymers* **2025**, *17*, 724. [CrossRef] [PubMed]
5. Beer, G.F. Complex Networks in Phase-Separating Gels: A Computer Simulation Study. *Polymers* **2025**, *17*, 880. [CrossRef] [PubMed]
6. Li, C.; Wang, J.; Li, M.; Wang, H.; Song, Y.; Meng, X.; Liu, R. Simulation of the Design Performance of Carbon Fiber/Glass Fiber Hybrid-Reinforced Resin Matrix Composite Rotors. *Polymers* **2025**, *17*, 1668. [CrossRef]
7. Jing, Y.; Liu, G. Systematic Investigation on the Swelling Response and Oil Resistance of NBR Using the Prediction Models Determined by the Modified Flory–Huggins Interaction Parameter. *Polymers* **2025**, *16*, 2696. [CrossRef] [PubMed]
8. Du, Y.; Zhang, H.; Han, W.; Du, X.; Shang, Y.; Yang, H.; Wang, X.; Chen, Q.; Li, Z. The network construction of a new byproduct-free XLPE-based insulation using a click chemistry-type reaction and a theoretical study of the reaction mechanism. *Polymers* **2024**, *16*, 3536. [CrossRef]
9. Yang, K.; Wang, S.; Chen, Y.; Dong, H.; Wang, Q.; Cheng, Q. Dynamic Mechanical Properties and Energy Absorption Capabilities of Polyureas Through Experiments and Molecular Dynamic Simulation. *Polymers* **2025**, *17*, 107. [CrossRef]
10. Du, Y.; Zhang, H.; Deng, C.; Du, X.; Shang, Y.; Wang, X.; Chen, Q.; Li, Z. The Structural Design of a New Graftable Antioxidant and the Theoretical Study of Its Role in the Cross-Linking Reaction Process of Polyethylene. *Polymers* **2025**, *17*, 546. [CrossRef]
11. Tang, B.; Liu, X.; Deng, S.; Zhong, W.; Shao, J. Dielectric Constant Estimation of Spherical Particle-Filled Nanocomposites Based on the Poon and Shin Model, Considering Interphase Properties. *Polymers* **2025**, *17*, 1035. [CrossRef]
12. Škrbić, T.; Giacometti, A.; Hoang, T.X.; Maritan, A.; Banavar, J.R. A Tale of Two Chains: Geometries of a Chain Model and Protein Native State Structures. *Polymers* **2024**, *16*, 502. [CrossRef]
13. Anhesini, D.A.; Caetano, D.L.Z.; Caruso, I.P.; Cherstvy, A.G.; de Carvalho, S.J. Critical adsorption of polyelectrolytes onto patchy particles with a low-dielectric interior. *Polymers* **2025**, *17*, 2205. [CrossRef]
14. Roman, H.E. Polymers in Physics, Chemistry and Biology: Behavior of Linear Polymers in Fractal Structures. *Polymers* **2024**, *16*, 3400. [CrossRef]
15. Schiemann, R.; Bachmann, M.; Janke, W. Exact enumeration of three-dimensional lattice proteins. *Comput. Phys. Commun.* **2005**, *166*, 8–16. [CrossRef]
16. Tsolou, G.; Stratikis, N.; Baig, C.; Stephanou, P.S.; Mavrantzas, V.G. Melt structure and dynamics of unentangled polyethylene rings: Rouse theory, atomistic molecular dynamics simulation, and comparison with the linear analogues. *Macromolecules* **2010**, *43*, 10692–10713. [CrossRef]
17. Araki, T.; Tanaka, H. Three-dimensional numerical simulations of viscoelastic phase separation: Morphological characteristics. *Macromolecules* **2001**, *34*, 1953–1963. [CrossRef]
18. Tu, S.; Choudhury, C.K.; Luzinov, I.; Kuksenok, O. Recent advances towards applications of molecular bottlebrushes and their conjugates. *Curr. Opin. Solid State Mater. Sci.* **2019**, *23*, 50–61. [CrossRef]
19. Shah, S.; Leon, L. Structural dynamics, phase behavior, and applications of polyelectrolyte complex micelles. *Curr. Opin. Colloid Interface Sci.* **2021**, *53*, 101424. [CrossRef]
20. Tang, X.; Xiao, S.; Lei, Q.; Yuan, L.; Peng, B.; He, L.; Luo, J.; Pei, Y. Molecular dynamics simulation of surfactant flooding driven oil-detachment in nano-silica channels. *J. Phys. Chem. B* **2018**, *123*, 277–288. [CrossRef] [PubMed]
21. Manoharan, V.N. Colloidal matter: Packing, geometry, and entropy. *Science* **2015**, *349*, 1253751. science.1253751. [CrossRef]

22. Espinosa, J.R.; Vega, C.; Valeriani, C.; Frenkel, D.; Sanz, E. Heterogeneous versus homogeneous crystal nucleation of hard spheres. *Soft Matter* **2019**, *15*, 9625–9631. [CrossRef]
23. Luu, H.T.; Perrot, C.; Monchiet, V.; Panneton, R. Three-dimensional reconstruction of a random fibrous medium: Geometry, transport, and sound absorbing properties. *J. Acoust. Soc. Am.* **2017**, *141*, 4768–4780. [CrossRef]
24. Mostaço-Guidolin, L.B.; Smith, M.S.; Hewko, M.; Schattka, B.; Sowa, M.G.; Major, A.; Ko, A.C.T. Fractal dimension and directional analysis of elastic and collagen fiber arrangement in unsectioned arterial tissues affected by atherosclerosis and aging. *J. Appl. Physiol.* **2019**, *126*, 638–646. [CrossRef]
25. Yethiraj, A.; Schweizer, K.S. Self-consistent polymer integral equation theory: Comparisons with Monte Carlo simulations and alternative closure approximations. *J. Chem. Phys.* **1992**, *97*, 1455–1464. [CrossRef]
26. Jaramillo, E.; Wu, D.T.; Grest, G.S.; Curro, J.G. Anomalous mixing behavior of polyisobutylene/polypropylene blends: Molecular dynamics simulation study. *J. Chem. Phys.* **2004**, *120*, 8883–8886. [CrossRef] [PubMed]
27. Sofos, F.; Karakasidis, T.E.; Spetsiotis, D. Molecular Dynamics simulations of ion separation in nano-channel water flows using an electric field. *Mol. Simul.* **2019**, *45*, 1395–1402. [CrossRef]
28. Ansong, O.E.; Jansen, S.; Wei, Y.; Pomrink, G.; Lu, H.; Patel, A.; Li, S. Accelerated controlled radical polymerization of methacrylates. *Polym. Int.* **2009**, *58*, 54–65. [CrossRef]
29. Wang, X.J.; Zhao, X.Y.; Li, Q.G.; Chan, T.W.; Wu, S.Z. Artificial neural network modeling and mechanism study for relaxation of deformed rubber. *Ind. Eng. Chem. Res.* **2016**, *55*, 4059–4070. [CrossRef]
30. John, M.J.; Thomas, S. Biofibres and biocomposites. *Carbohydr. Polym.* **2008**, *71*, 343–364. [CrossRef]
31. Faruk, O.; Bledzki, A.K.; Fink, H.P.; Sain, M. Biocomposites reinforced with natural fibers: 2000–2010. *Prog. Polym. Sci.* **2012**, *37*, 1552–1596. [CrossRef]
32. Tian, W.; Qiu, T.; Shi, Y.; He, L.; Tuo, X. The facile preparation of aramid insulation paper from the bottom-up nanofiber synthesis. *Mater. Lett.* **2017**, *202*, 158–161. [CrossRef]
33. Frischknecht, A.L.; Winey, K.I. The evolution of acidic and ionic aggregates in ionomers during microsecond simulations. *J. Chem. Phys.* **2019**, *150*, 064901. [CrossRef]
34. Schwarz, A.E.; Ligthart, T.N.; Boukris, E.; Van Harmelen, T. Sources, transport, and accumulation of different types of plastic litter in aquatic environments: A review study. *Mar. Pollut. Bull.* **2019**, *143*, 92–100. [CrossRef]
35. Hong, S.-H.; Gorce, J.B.; Punzmann, H.; Francois, N.; Xia, H. Surface waves control bacterial attachment and formation of biofilms in thin layers. *Sci. Adv.* **2020**, *6*, aaz9386. [CrossRef]
36. Schamboeck, V.; Iedema, P.D.; Kryven, I. Dynamic networks that drive the process of irreversible step-growth polymerization. *Sci. Rep.* **2019**, *9*, 2276. [CrossRef] [PubMed]
37. Zhang, H.P.; Cao, J.J.; Jiang, W.B.; Yang, Y.Q.; Zhu, B.Y.; Liu, X.Y.; Wu, Y.; Sun, X.; Essouma, A.F.B.E.; Liu, J.; et al. Synthesis and mechanical properties of polyacrylamide gel doped with graphene oxide. *Energies* **2022**, *15*, 5714. [CrossRef]
38. Ferrarese Lupi, F.; Giammaria, T.J.; Miti, A.; Zuccheri, G.; Carignano, S.; Sparnacci, K.; Seguini, G.; De Leo, N.; Boarino, L.; Perego, M.; et al. Hierarchical order in dewetted block copolymer thin films on chemically patterned surfaces. *ACS Nano* **2018**, *12*, 7076–7085. [CrossRef]
39. Wang, A.; Ye, J.; Humphrey, M.G.; Zhang, C. Graphene and Carbon-Nanotube Nanohybrids Covalently Functionalized by Porphyrins and Phthalocyanines for Optoelectronic Properties. *Adv. Mater.* **2018**, *30*, 1705704. [CrossRef] [PubMed]
40. Escárcega-Bobadilla, M.V.; Zelada-Guillén, G.A.; Pyrlin, S.V.; Wegrzyn, M.; Ramos, M.M.; Giménez, E.; Stewart, A.; Maier, G.; Kleij, A.W. Nanorings and rods interconnected by self-assembly mimicking an artificial network of neurons. *Nat. Commun.* **2013**, *4*, 2648. [CrossRef]
41. Zhang, X.; Nguyen, H.; Daly, M.; Nguyen, S.T.; Espinosa, H.D. Nanoscale toughening of ultrathin graphene oxide-polymer composites: Mechanochemical insights into hydrogen-bonding/van der Waals interactions, polymer chain alignment, and steric parameters. *Nanoscale* **2019**, *11*, 12305–12316. [CrossRef]
42. Guo, J.; Li, X.; Liu, Y.; Liang, H. Flow-induced translocation of polymers through a fluidic channel: A dissipative particle dynamics simulation study. *J. Chem. Phys.* **2011**, *134*, 134906. [CrossRef] [PubMed]
43. Chelakkot, R.; Winkler, R.G.; Gompper, G. Flow-induced helical coiling of semiflexible polymers in structured microchannels. *Phys. Rev. Lett.* **2012**, *109*, 178101. [CrossRef] [PubMed]
44. Min, C.K.; Wu, T.B.; Yang, W.T.; Chen, C.L. Functionalized mesoporous silica/polyimide nanocomposite thin films with improved mechanical properties and low dielectric constant. *Compos. Sci. Technol.* **2008**, *68*, 1570–1578. [CrossRef]
45. Lee, T.; Park, S.S.; Jung, Y.; Han, S.; Han, D.; Kim, I.; Ha, C.S. Preparation and characterization of polyimide/mesoporous silica-hybrid nanocomposites based on water-soluble poly (amic acid) ammonium salt. *Eur. Polym. J.* **2009**, *45*, 19–29. [CrossRef]
46. Cai, W.; Shalaev, V. *Optical Metamaterials: Fundamentals and Applications*; Springer: New York, NY, USA, 2010. [CrossRef]
47. Cummer, S.A.; Christensen, J.; Alù, A. Controlling sound with acoustic metamaterials. *Nat. Rev. Mater.* **2016**, *1*, 16001. [CrossRef]
48. Li, Y.; Li, W.; Han, T.; Zheng, X.; Li, J.; Li, B.; Fan, S.; Qiu, C.W. Transforming heat transfer with thermal metamaterials and devices. *Nat. Rev. Mater.* **2021**, *6*, 488–507. [CrossRef]

103. Tassieri, M. Dynamics of semiflexible polymer solutions in the tightly entangled concentration regime. *Macromolecules* **2017**, *50*, 5611–5618. [CrossRef]
104. Venneti, K.C.; Hewage, C.M. Conformational and molecular interaction studies of glucagon-like peptide-2 with its N-terminal extracellular receptor domain. *FEBS Lett.* **2011**, *585*, 346–352. [CrossRef]
105. Li, Y.; Sun, J.; Li, D.; Lin, J. Activation and conformational dynamics of a class B G-protein-coupled glucagon receptor. *Phys. Chem. Chem. Phys.* **2016**, *18*, 12642–12650. [CrossRef]
106. Song, G.; Yang, D.; Wang, Y.; De Graaf, C.; Zhou, Q.; Jiang, S.; Liu, K.; Cai, X.; Dai, A.; Lin, G.; et al. Human GLP-1 receptor transmembrane domain structure in complex with allosteric modulators. *Nature* **2017**, *546*, 312–315. [CrossRef] [PubMed]
107. Antonietti, M.; Förster, S. Vesicles and liposomes: A self-assembly principle beyond lipids. *Adv. Mater.* **2003**, *15*, 1323–1333. [CrossRef]
108. Li, W.; Luo, T.; Yang, Y.; Tan, X.; Liu, L. Formation of controllable hydrophilic/hydrophobic drug delivery systems by electrospinning of vesicles. *Langmuir* **2015**, *31*, 5141–5146. [CrossRef]
109. Lang, P.; Frey, E. Disentangling entanglements in biopolymer solutions. *Nat. Commun.* **2018**, *9*, 494. s41467-018-02837-5. [CrossRef] [PubMed]
110. Händler, T.; Tutmarc, C.; Glaser, M.; Freitag, J.S.; Smith, D.M.; Schnauß, J. Measuring structural parameters of crosslinked and entangled semiflexible polymer networks with single-filament tracing. *Phys. Rev. E* **2021**, *103*, 062501. [CrossRef]
111. Elbalasy, I.; Mollenkopf, P.; Tutmarc, C.; Herrmann, H.; Schnauß, J. Keratins determine network stress responsiveness in reconstituted actin-keratin filament systems. *Soft Matter* **2021**, *17*, 3954–3962. [CrossRef]
112. Palyulin, V.V.; Ala-Nissila, T.; Metzler, R. Polymer translocation: The first two decades and the recent diversification. *Soft Matter* **2014**, *10*, 9016–9037. [CrossRef]
113. Payet, L.; Martinho, M.; Merstorf, C.; Pastoriza-Gallego, M.; Pelta, J.; Viasnoff, V.; Auvray, L.; Muthukumar, M.; Mathé, J. Temperature effect on ionic current and ssDNA transport through nanopores. *Biophys. J.* **2015**, *109*, 1600–1607. [CrossRef]
114. Lee, N.K.; Jung, Y.; Johner, A.; Joanny, J.F. Globular polyampholytes: Structure and translocation. *Macromolecules* **2021**, *54*, 2394–2411. [CrossRef]
115. Malkin, A.; Ilyin, S.; Vasilyev, G.; Arinina, M.; Kulichikhin, V. Pressure losses in flow of viscoelastic polymeric fluids through short channels. *J. Rheol.* **2014**, *58*, 433–448. [CrossRef]
116. Senaratne, W.; Andruzzi, L.; Ober, C.K. Self-assembled monolayers and polymer brushes in biotechnology: Current applications and future perspectives. *Biomacromolecules* **2005**, *6*, 2427–2448. [CrossRef] [PubMed]
117. Barbey, R.; Lavanant, L.; Paripovic, D.; Schuwer, N.; Sugnaux, C.; Tugulu, S.; Klok, H.A. Polymer brushes via surface-initiated controlled radical polymerization: Synthesis, characterization, properties, and applications. *Chem. Rev.* **2009**, *109*, 5437–5527. [CrossRef]
118. Wei, S.; Knotts, T.A. A coarse grain model for protein-surface interactions. *J. Chem. Phys.* **2013**, *139*, 095102. 10.1063/1.4819131. [CrossRef]
119. Han, Y.; Cui, J.; Jin, J.; Jiang, W. Hydrogen bonding induced protein adsorption on polymer brushes: A Monte Carlo study. *J. Mater. Chem. B* **2017**, *5*, 8479–8486. [CrossRef]
120. Ramakrishna, S.; Mayer, J.; Wintermantel, E.; Leong, K.W. Biomedical applications of polymer-composite materials: A review. *Compos. Sci. Technol.* **2001**, *61*, 1189–1224. [CrossRef]
121. Senra, M.R.; Marques, M.D.F.V. Synthetic polymeric materials for bone replacement. *J. Compos. Sci.* **2020**, *4*, 191. [CrossRef]
122. Agrawal, R.; Kumar, A.; Mohammed, M.K.; Singh, S. Biomaterial types, properties, medical applications, and other factors: A recent review. *J. Zhejiang-Univ.-Sci. A* **2023**, *24*, 1027–1042. [CrossRef]
123. Wang, Z.; Li, H.; Tang, Z.; Liu, Z.; Ruan, Z.; Ma, L.; Yang, Q.; Wang, D.; Zhi, C. Hydrogel electrolytes for flexible aqueous energy storage devices. *Adv. Funct. Mater.* **2018**, *28*, 1804560. [CrossRef]
124. Chan, C.Y.; Wang, Z.; Jia, H.; Ng, P.F.; Chow, L.; Fei, B. Recent advances of hydrogel electrolytes in flexible energy storage devices. *J. Mater. Chem. A* **2021**, *9*, 2043–2069. [CrossRef]

polymers

MDPI

Article

An Efficient and Accurate SCF Algorithm for Block Copolymer Films and Brushes Using Adaptive Discretizations

Le Qiao *, Marios Giannakou and Friederike Schmid *

Institut für Physik, Johannes Gutenberg-Universität Mainz, D55099 Mainz, Germany; mgiannak@uni-mainz.de
* Correspondence: le.qiao@uni-mainz.de (L.Q.); friederike.schmid@uni-mainz.de (F.S.)

Abstract: Self-consistent field (SCF) theory serves as a robust tool for unraveling the intricate behavior exhibited by soft polymeric materials. However, the accuracy and efficiency of SCF calculations are crucially dependent on the numerical methods employed for system discretization and equation-solving. Here, we introduce a simple three dimensional SCF algorithm that uses real-space methods and adaptive discretization, offering improved accuracy and efficiency for simulating polymeric systems at surfaces. Our algorithm's efficacy is demonstrated through simulations of two distinct polymeric systems, namely, block copolymer (BCP) films and polymer brushes. By enhancing spatial resolution in regions influenced by external forces and employing finer contour discretization at grafting chain ends, we achieve significantly more accurate results at very little additional cost, enabling the study of 3D polymeric systems that were previously computationally challenging. To facilitate the widespread use of the algorithm, we have made our 1D-3D SCF code publicly available.

Keywords: SCF; BCP; brush; adaptive; film; polymer

1. Introduction

Self-assembled copolymer materials have diverse applications in both industry and daily life due to the wide and complex spectrum of possible morphological patterns into which the copolymer molecules may assemble spontaneously. Self-consistent field (SCF) theory, which employs a simplified chain representation and a mean-field approximation to predict the spatial distribution of polymer segments, has proven to be a reliable and powerful tool for predicting the equilibrium morphology of many polymeric systems [1–3].

The SCF calculation iteratively computes space-dependent polymer density and associated potential fields from chain statistics propagators until the self-consistent condition is satisfied [4]. The chain propagators satisfy nonlinear modified diffusion equations in the variables space and chain contour ("time"), which have to be numerically solved using spectral or real-space methods. Pure spectral methods based on, e.g., Fourier series-based spectral solutions, have the advantage that they do not require a discretization of the chain contour. Matsen and coworkers have demonstrated their capability in accurately constructing morphology phase diagrams for periodic block-copolymer melts [3,5–7]. However, they demand prior knowledge of morphology, assuming symmetry in the considered phase, which limits their applicability for discovering new phases [8–11]. To address these limitations, the pseudo-spectral method was introduced, which does rely on a discretization of the contour and switches between Fourier and real-space representations of the system, utilizing the Fourier representation for the evaluation of gradient terms and the real-space representation for the evaluation of nonlinear self-consistent fields [10,12]. In contrast, pure real-space methods discretize the diffusion equation within a simulation box and solve for the solution using finite difference schemes [8,13–16]. This approach is particularly advantageous for complex polymer systems characterized by symmetry breaking or non-periodic boundary conditions, but can be computational expensive in three dimensions. A comprehensive exploration of the advantages and disadvantages of these methods can be found in Ceniceros and Fredrickson's detailed review [11].

Our primary objective in this paper is to enhance the efficiency and accuracy of real-space methods, which heavily rely on the discretization of space/chain contours. The solution of modified diffusion equations in the SCF context typically employs lower-order finite difference; however, unless the discretization is very fine, the accuracy and stability of SCF calculations can suffer, especially when dealing with polymer systems containing sharp interfaces [17,18]. For instance, in simulating polymer films, in which the substrate/air interface is often represented by a Dirichlet boundary condition and surface interactions with the polymer are introduced artificially through an external potential field, numerical inaccuracies can significantly affect the calculated free energy of the film [19]. Similarly, in the context of polymer brushes [14,20], ensuring proper attachment of the grafting end to the substrate involves fixing an end segment on one space grid using a Dirac delta function as the initial condition for the modified diffusion equations [2,14,21]. Achieving convergence to an accurate SCF solution necessitates much finer contour discretization compared to free chains, demanding additional computational effort.

The trade-off between spatial discretization and computational efficiency presents a critical challenge, especially for systems requiring higher-dimensional calculations such as cylinders and spheres in thin copolymer films [19,22,23] or particle-grafted chain polymer brushes with angular or radial-dependent morphologies [21,24,25]. To address this challenge, our paper introduces a simple scheme that adaptively increases discretization in the spatial domain where external forces are present and refines the discretization in the contour domain at the grafting point. This approach is similar in spirit to other more sophisticated adaptive methods that have recently been proposed in the literature, such as the use of Oc-Tree data structures [26], polygonal meshes [27], and finite element methods [16,28]. By optimizing the spatial resolution according to the system's composition, our approach achieves very high accuracy while keeping computational resources low compared to uniform finite-difference grid methods. In the following sections, we evaluate and demonstrate the effectiveness of our adaptive scheme with two test cases of polymeric systems, namely, block copolymer (BCP) films and polymer brushes.

2. Background: SCF Equations for Two Test Cases

We use the SCF theory for inhomogeneous systems of Gaussian polymers [29]. In the following, we list only the most important equations relevant for our test systems; readers may refer to the literature for the details of the derivations [30].

2.1. Test Case 1: Diblock Copolymer Film

We consider an incompressible melt of asymmetric AB diblock copolymer molecules with a degree of polymerization N, which is confined between two flat surfaces. We assume that the majority block A occupies a volume fraction f of each diblock copolymer chain and that both blocks share the same statistical segment length b. In the grand canonical ensemble, the free energy takes the following form [19]:

$$\begin{aligned}\frac{F}{k_bT} = &- e^{\mu}Q + \rho_c \int d\mathbf{r}\left[\chi N\phi_A(\mathbf{r})\phi_B(\mathbf{r}) + \frac{1}{2}\kappa N(\phi_A(\mathbf{r}) + \phi_B(\mathbf{r}) - 1)^2\right] \\ &- \rho_c \int d\mathbf{r}[\omega_A(\mathbf{r})\phi_A(\mathbf{r}) + \omega_B(\mathbf{r})\phi_B(\mathbf{r})] \\ &+ \rho_c \int d\mathbf{r}H(\mathbf{r})N[\Lambda_A\phi_A(\mathbf{r}) + \Lambda_B\phi_B(\mathbf{r})] \end{aligned} \quad (1)$$

where $\mu = \mu_0 + \ln G$ is the chemical potential, $G = \rho_c R_g^3$ is the rescaled dimensionless copolymer density in the bulk, $\rho_c = n/V$ is the average molecular number density (with n being the total number of copolymer molecules and V the volume of the film), $R_g = \sqrt{Nb^2/6}$ is the radius of gyration of the noninteracting copolymer chain, and serves as the spatial length unit throughout the paper, Q is the partition function of a single copolymer chain in the mean field of the other chains, $\phi_A(\mathbf{r})$ and $\phi_B(\mathbf{r})$ are the local concentrations of the A and B segments at a given point $\mathbf{r}$, and χ is the Flory–Huggins parameter specifying the incompatibility of the two segments. The incompressibility of the BCP melt is ensured by the inverse of the isothermal compressibility parameter κ. The last term on the right-hand side of Equation (1) describes the interaction energy with the substrate/interface, with $\Lambda_{A,B}H(\mathbf{r})$ being the surface field. We assume symmetric boundary wetting conditions; the surface interaction energies with A and B segments are Λ_A and Λ_B, respectively. The surface field

$$H(\mathbf{r}) = \begin{cases} (1+\cos(\pi z/\epsilon)) & 0 \leqslant z \leqslant \epsilon \\ 0 & \epsilon \leqslant z \leqslant h-\epsilon \\ (1+\cos(\pi(h-z)/\epsilon)) & h-\epsilon \leqslant z \leqslant h \end{cases} \quad (2)$$

is applied within a depth of ϵ from the two surfaces, where h is the distance between the two surfaces. By finding the extremum of the free energy in Equation (1) with respect to $\phi_{A,B}(\mathbf{r})$, we obtain the fields experienced by the A and B segments:

$$\frac{\omega_A(\mathbf{r})}{N} = \chi\phi_B(\mathbf{r}) + \kappa[\phi_A(\mathbf{r}) + \phi_B(\mathbf{r}) - 1] + \Lambda_A H(\mathbf{r}), \quad (3)$$

$$\frac{\omega_B(\mathbf{r})}{N} = \chi\phi_A(\mathbf{r}) + \kappa[\phi_A(\mathbf{r}) + \phi_B(\mathbf{r}) - 1] + \Lambda_B H(\mathbf{r}). \quad (4)$$

The partition function for a single chain is simply $Q = \int d\mathbf{r}\, q(\mathbf{r},s)q^\dagger(\mathbf{r},1-s)$, with $q(\mathbf{r},1-s)$ and $q^\dagger(\mathbf{r},s)$ being the partial partition functions for the first Ns and last $N(1-s)$ segments. Note that Q does not depend on the specific choice of s. The propagators $q(\mathbf{r},s)$ and $q^\dagger(\mathbf{r},1-s)$ obey the modified diffusion equation

$$\frac{\partial q(\mathbf{r},s)}{\partial s} = \Delta q(\mathbf{r},s) - \omega(\mathbf{r})q(\mathbf{r},s) \quad (5)$$

with initial conditions $q(\mathbf{r},0) = 1$ and $q^\dagger(\mathbf{r},1) = 1$. Here, Δ represents the Laplacian, $\mathbf{r}$ denotes the spatial coordinate in units of R_g and $0 \leqslant s \leqslant 1$ represents the chain coordinate of the coarse-grained chain segment in the units of the chain contour length L_c. The field $\omega(\mathbf{r}) \equiv \omega_A(\mathbf{r})$ for $0 \leqslant s \leqslant f$ and $\omega(\mathbf{r}) \equiv \omega_B(\mathbf{r})$ for $f < s \leqslant 1$. The local concentrations of the A and B segments are simply

$$\phi_A(\mathbf{r}) = \frac{1}{\rho_c}e^{\mu}\int_0^f ds q(\mathbf{r},s)q^\dagger(\mathbf{r},1-s), \quad (6)$$

$$\phi_B(\mathbf{r}) = \frac{1}{\rho_c}e^{\mu}\int_f^1 ds q(\mathbf{r},s)q^\dagger(\mathbf{r},1-s). \quad (7)$$

Starting with an initial guess of the field in Equation (5), we first solve the propagator $q(\mathbf{r},s)$ and $q^\dagger(\mathbf{r},1-s)$. Next, the new local concentrations $\phi_A(\mathbf{r})$ and $\phi_B(\mathbf{r})$ are obtained from Equations (6) and (7), respectively. Finally, these concentrations are used in Equations (3) and (4) to solve for the new fields and the old and new fields are mixed according to a prescription of choice [31]. This is repeated until convergence to either a metastable or an equilibrium state.

2.2. Test Case 2: Homopolymer Brush

We consider a monodisperse brush solution in which n linear homopolymer chains with a degree of polymerization N are grafted onto a flat substrate at one end. In the canonical ensemble, the free energy of such a system is provided by

$$F/k_BT = \rho_o \int d\mathbf{r}\left[\frac{1}{2}v\phi(\mathbf{r})^2 + \Lambda H(\mathbf{r})\phi(\mathbf{r}) - \phi(\mathbf{r})\omega(\mathbf{r})/N\right] - n(\ln(\rho_o Q/n) + 1), \quad (8)$$

where ν is the excluded-volume parameter. Similar to the BCP film system, the interactions between the segments and the substrate where the chains are grafted are imposed by an external field $H(r)$, as in Equation (2). The self-consistent equations are determined by the extremum of the free energy, leading to

$$\omega(\mathbf{r})/N = \nu\phi(\mathbf{r}) + \Lambda H(\mathbf{r}), \tag{9}$$

$$\phi(\mathbf{r}) = \frac{nN}{Q\rho_o}\int_0^1 ds\, q(\mathbf{r},s)q^\dagger(\mathbf{r},s), \tag{10}$$

$$Q = \int d\mathbf{r}\, q(\mathbf{r},1). \tag{11}$$

The initial conditions for such a system are provided by

$$q(\mathbf{r},0) = 1, \tag{12}$$

$$q^\dagger(\mathbf{r},0) = \delta(z - g_p), \tag{13}$$

where g_p is the grafting point and δ is the Dirac delta function.

3. Adaptive Discretization

We solve the diffusion equation using the semi-implicit Crank–Nicolson discretization scheme combined with the alternating direction implicit method (ADI) [32]. For simplicity, we illustrate the method for the one-dimensional diffusion equation. Provided that the discretization points are arranged on a grid, extension of the ADI to three dimensions is straightforward. In one dimension, the diffusion equation reads

$$\frac{\partial q(z,s)}{\partial s} = \frac{\partial^2}{\partial z^2}q(z,s) - \omega(z,s)q(z,s), \tag{14}$$

which can be discretized in s and z as

$$\frac{q_i^{m+1} - q_i^m}{\delta s_{m+1}} = (\Delta_z - \omega_i)\frac{q_i^m + q_i^{m+1}}{2}, \tag{15}$$

where $q_i^m \equiv q(z_i, s_m)$, with $m = \{1, N\}$ being the contour steps in s with contour step size $\delta s_m = s_m - s_{m-1}$ and $i = \{1, n_z\}$ being the spatial steps in z with spatially varying discretization $\delta z_i = z_i - z_{i-1}$. Taking the second-order approximation for the Laplacian Δ_z [33], we have $\Delta_z f_i/2 \approx a_i f_{i-1} - b_i f_i + c_i f_{i+1}$, with $a_i = \frac{1}{\delta z_i(\delta z_i + \delta z_{i+1})}$, $b_i = \frac{1}{\delta z_i \delta z_{i+1}}$, and $c_i = \frac{1}{\delta z_{i+1}(\delta z_i + \delta z_{i+1})}$. Rearranging Equation (15), we obtain

$$\begin{aligned} &-a_i\delta s_{m+1}q_{i-1}^{m+1} + (1 + b_i\delta s_{m+1} + \frac{\omega_i\delta s_{m+1}}{2})q_i^{m+1} - c_i\delta s_{m+1}q_{i+1}^{m+1} \\ &= a_i\delta s_{m+1}q_{i-1}^m + (1 - b_i\delta s_{m+1} - \frac{\omega_i\delta s_{m+1}}{2})q_i^m + c_i\delta s_{m+1}q_{i+1}^m. \end{aligned} \tag{16}$$

We can calculate the propagator q_i^{m+1} at step $m+1$ from the propagator q_i^m at step m. The key lies in choosing proper adaptive discretization tailored to the system, which enables gains in both computational time and accuracy.

In the thin film case, where the loss of accuracy mainly arises from the external potential added to mimic interactions with the substrate or air interface, we employ finer discretization only in the z direction where external force is present. We keep a uniform contour discretization where $\delta s/L_c = 1/N$. Our approach involves adaptive discretization with a total of n_z grids, achieved through either a cosine function [16] or a step function. The former has a form of $\delta z_i = \frac{L_z}{2}\left[\cos(\frac{i-1}{n_z}\pi) - \cos(\frac{i}{n_z}\pi)\right]$ for $i = \{1, n_z\}$, providing continuous discretization. The latter uses a finer $\delta z_s = \frac{\epsilon}{\alpha_s n_z}$ near the surface ($z \leqslant \epsilon$ or $L_z - \epsilon < z \leqslant L_z$)

and a coarser $\delta z_b = \frac{L_z - 2\epsilon}{(1-2\alpha_s)n_z}$ for the rest. The latter provides finer discretization; here, α_s represents the fraction of grids allocated to the surface region ϵ. Throughout this work, we refer to these adaptive schemes as "cos" and "step", respectively.

For polymer brushes, additional adaptive contour discretization turns out to be crucial. We utilize $\delta s_m = \left[\cos(\frac{m-1}{2N}\pi) - \cos(\frac{m}{2N}\pi)\right]$, $m = \{1, N\}$ to discretize the contour. Similar to the approach for thin films, we employ two Dirichlet wall boundaries in the z direction. The separation L_z is chosen to be greater than the brush height to ensure that the brush's free end is maintained. The grafting point of the polymer is located at a distance of $g_p = 0.05$ from the substrate. Initially, we employ a uniform grid spacing of $\delta z_g = 0.001$ for $z \leqslant 2g_p$, utilizing $n_g = 100$ grids to ensure fine discretization at the grafting site. Subsequently, we gradually increase the discretization by $\delta z_i = \delta z_g + \frac{L_z - 2g_p}{2}\left[\cos(\frac{i-1}{2(n_z-n_g)}\pi) - \cos(\frac{i}{2(n_z-n_g)}\pi)\right]$, $i = \{1, n_z - n_g\}$.

4. Results

4.1. Block Copolymer Film

We first demonstrate the effectiveness of our adaptive discretization scheme in computing the free energy of BCP films, specifically focusing on lamellar-forming BCP films (with $f = 0.5$) as an example. Figure 1a–c illustrates the calculated free energy as a function of the film thickness using three different discretization schemes.

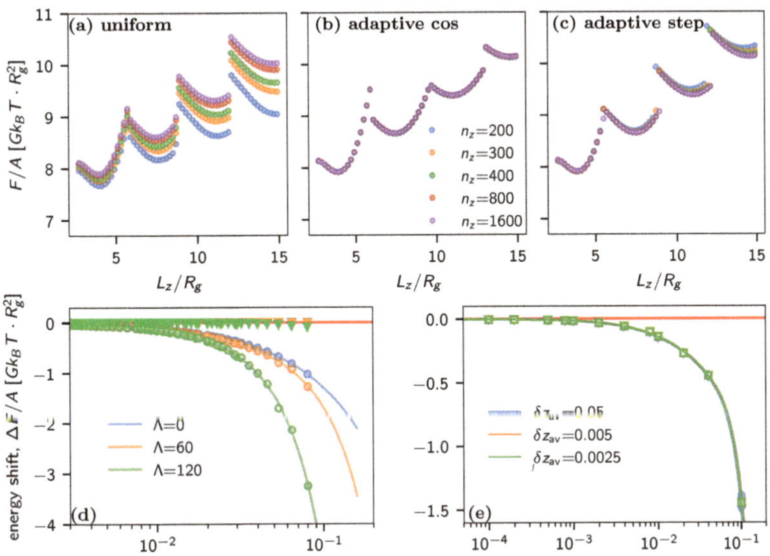

Figure 1. (**a–c**) SCF-calculated free energy F/A versus the thickness of the film L_z/R_g when using the three different discretization schemes described in the text. The data are generated using $f = 0.5$, $\mu_0 = 2.55$, $\chi N = 20$, $\kappa N = 25$, and $\Lambda_A = \Lambda_B = 60$. (**d**) Energy shift per area ($\Delta F/A$) as a function of the averaged space discretizations δz_{av} for three different strengths of surface interactions $\Lambda_A = \Lambda_B = 0$, 60 and 120, where $\Delta F = F(\Delta z) - F(0)$, with $F(0)$ being the extrapolated value for $N_z = \infty$. The empty circle (○) and solid triangle (▾) symbols represent uniform discretization and adaptive "cos" discretization, respectively, while the red horizontal line indicates $\Delta F = 0$. (**e**) Energy shift per area ($\Delta F/A$) as a function of contour discretizations $\delta s/L_c$ with $\Lambda_A = \Lambda_B = 60$. The empty circle (○), square (□), and triangle (▽) indicate uniform, cos-adaptive, and step-adaptive discretization, respectively. The solid lines of different colors in (**d**,**e**) are polynomial fits of ΔF using $f(x) = ax + bx^2 + cx^3$ for different surface interaction strengths and δz_{av}.

In the case of uniform discretization, the energy curves diverge for varying discretizations, denoted by different grid numbers. These discrepancies become more pronounced for larger thicknesses due to the coarser discretization. In contrast, our adaptive scheme consistently generates the same free energy curves regardless of the number of discretization grids used. Although a minor discrepancy arises for larger thicknesses in the step-adaptive case, the overall performance remains superior to that of the uniform case.

Figure 1d illustrates the inaccuracies versus discretization, quantified by the shift in energy from the extrapolated energy at spatial discretization $\delta z \to 0$. To facilitate comparison with the uniform case, the energy shift in the adaptive case is plotted versus the averaged discretization, defined by $\delta z_{\rm av} \equiv L_z/n_z$. In the uniform case, the energy shift can be fitted by a cubic polynomial, which converges to zero with increased discretization. As expected, the accuracy in the energy deteriorates more rapidly with coarser discretization when stronger surface interactions are employed. In contrast, the cos-adaptive scheme consistently produces highly accurate energy values for all tested discretizations, as is evident in the data points overlapping on the zero-error baseline. While surface interaction slightly impacts the shift in energy for larger discretizations, it is inconspicuous compared to the uniform case.

Figure 1e plots the energy shift versus $\delta s/L_c$ for three different discretization schemes and various $\delta z_{\rm av}$. The fact that all the curves collapse regardless of the discretization scheme and that $\delta z_{\rm av}$ indicates that adaptive discretization of the space does not require special treatment of δs. The numerical error caused by δs converges to zero when $\delta s/L_c < 10^{-3}$ for all three cases.

Our SCF calculations for lamellar BCP films show a significant improvement in accuracy with adaptive discretization, especially when using the cos-adaptive scheme. This allows for coarser discretization, making SCF simulations more efficient and enabling the investigation of multi-layered structures in thicker films.

As a further example, we studied thin films of sphere-forming BCPs. In the bulk, prior SCF studies revealed tiny free energy differences between Hexagonally Close-Packed (HCP) and Face-Centered Cubic (FCC) packings, showing that the HCP phase is the true stable phase [7]. Here, we investigate whether this remains true for thin films. Unlike in the case of thin films with lamellar or cylindrical order, the study of sphere packings in thin films requires three-dimensional calculations and large systems, which presents a substantial computational challenge.

Using the cos-adaptive scheme, it is possible to compute the free energy of films containing three layers of spheres (Figure 1a). The close packings of FCC and HCP correspond to ABC and ABA stackings of three layers, as illustrated by the inset drawing in Figure 2a and the SCF-calculated density plot in Figure 1b,c. The free energy curves demonstrate that the HCP phase remains the stable phase in films, i.e., it has the lower free energy. Furthermore, they show that the HCP film has a smaller equilibrium thickness (free energy minimum at $h^* = 9.57\ R_g$) than the FCC film ($h^* = 9.60\ R_g$).

4.2. Homopolymer Brush

Next, we consider polymer brushes. In this case, inaccuracies in SCF calculations continue to arise due to spatial discretization errors; in addition, the SCF results turn out to suffer from contour discretization errors as well. Specifically, implementing the delta function to graft the chain onto the substrate demands a much smaller δs compared to free polymer chains. Additionally, as the system approaches the strong stretching limit, which is characterized by significant chain interactions and brush thicknesses much larger than the radius of gyration of free chains, R_g, SCF calculations typically face convergence issues unless a very fine contour discretization is chosen.

Figure 3a plots the energy shift as a function of the averaged spatial discretization $\delta z_{\rm av}/R_g$ for both uniform and adaptive spatial discretizations and a fixed contour discretization of $\delta s/L_c = 0.0001$. Similar to the previous example of the thin film, the calculations perform badly in the case of uniform discretization as δz increases. This is particularly evident in scenarios with attractive walls, where more segments are attracted to regions with external potential (the blue curve). Conversely, cos-adaptive δz_{av} consistently yields much more accurate energy calculations, as illustrated by the fact that all filled symbols align with the zero-error baseline.

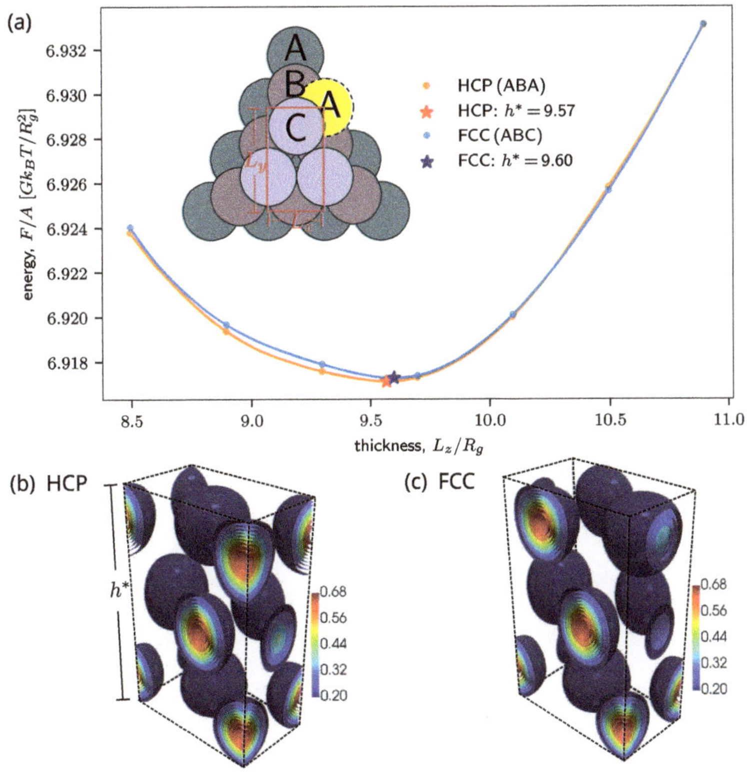

Figure 2. Spherical BCP film consisting of three layers. (**a**) SCF free energy per area F/A of the film as a function of film thickness L_z. The data are generated using $f = 0.76$, $\mu_0 = 2.25$, $\chi N = 20$, $\kappa N = 25$, $\Lambda_A = 60$, and $\Lambda_B = 5$. The optimum thickness, which corresponds to the minimum free energy, is marked by $\star$. The inset drawing shows the grid of the FCC (ABC) and HCP (ABA) packing. The green rectangle shows the selected periodic cell for the SCF calculation. (**b**,**c**): Three-dimensional contour plot of the density of the B-block calculated with the SCF for HCP and FCC packings.

In Figure 3b, we compare the energy shift relative to the average contour discretization δs_{av} for both uniform and cos-adaptive methods; we consider various wall interactions while maintaining a small uniform $\delta z/R_g = 0.002$ throughout the analysis. For the case of uniform δs, the energy shift converges to zero at around $\delta s/L_c = 10^{-5}$, contrasting with 10^{-3} in the free chain melt case. Moreover, the SCF calculation fails to converge beyond $\delta s/L_c \gtrapprox 10^{-3}$ under uniform conditions. However, the adaptive δs scheme is able to accommodate larger discretization values while consistently maintaining significantly higher accuracy for different types of wall interactions, as shown by the empty triangles.

We further extend our calculations to the strong stretching limit. Strong Stretching Theory (SST) predicts a parabolic density profile for the polymer [34,35]. The normalized concentration can be written as

$$\phi_{sst}(z) = \frac{3}{2}\phi_o\left(1 - \left(\frac{z}{L}\right)^2\right)\frac{R_g}{L}, \tag{17}$$

where $\phi_0 = \frac{\sqrt{6}N^{1/2}n}{b\rho_0 A}$ is the renormalization factor for the concentration and $\frac{L}{R_g} = \left(\frac{24N\nu}{\pi^2}\phi_0\right)^{1/3}$ is the absolute thickness of the brush.

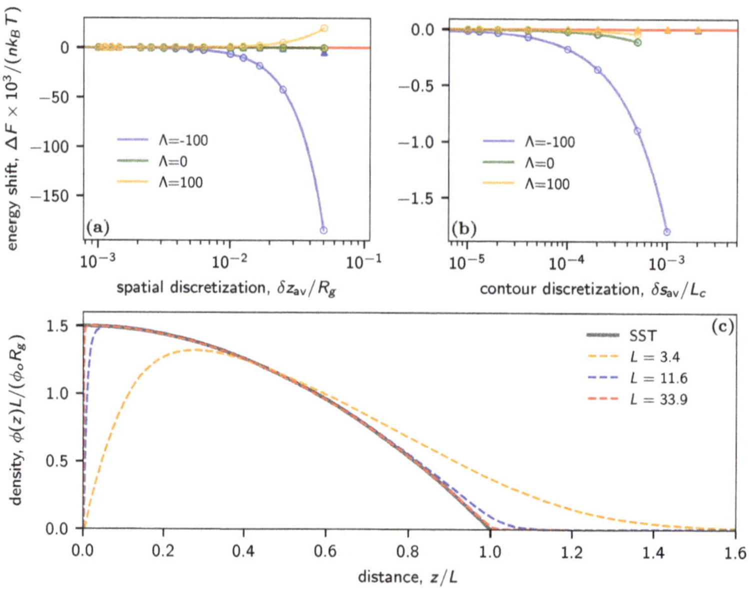

Figure 3. (**a**) SCF-calculated energy shift per polymer as a function of the averaged spatial discretization with attractive ($\Lambda = -100$), repulsive ($\Lambda = 100$), and neutral ($\Lambda = 0$) interactions with the substrate. Here, $\phi_0 = 2$ and $N\nu = 1$. (**b**) SCF-calculated free energy per polymer as a function of the contour discretization. In (**a**,**b**), the empty circles correspond to uniform discretization in space (**a**) and contour (**b**), while the filled triangles correspond to their counterparts using adaptive discretization. The red horizontal line indicates zero error $\Delta F = 0$. The solid curves in (**a**,**b**) are fits to the function $f(x) = ax + bx^2 + cx^3$. (**c**) Normalized segment density $\phi(z)L/\phi_0 R_g$ as a function of z/L (see text for definitions) at $\Lambda = 0$. The distance is rescaled by the thickness of the brush estimated by the Strong Stretching Theory (SST).

The SCF-calculated densities are showcased in Figure 3c. In the dilute regime, which is characterized by small L, the SCF captures both the depletion region near the wall and the tail at the top of the brush. As the graft density escalates, the monomer density profile in the brush approaches the prediction of the strong stretching limit. Here, the depletion width narrows, accompanied by a sharp concentration increase from zero. Despite nearing the strong stretching limit, our SCF calculations using the adaptive scheme continue to deliver accurate results. This is evident from the close alignment between the SST theory and the red dashed curve, particularly notable at $L = 33.9$.

5. Conclusions

We have proposed a method for incorporating adaptive discretization schemes in real-space SCF calculations to improve both their accuracy and computational efficiency. By implementing finer discretization near surfaces, where strong polymer–substrate interactions occur, the proposed method effectively reduces numerical errors in the calculated free energy. Notably, our study shows that the cos-adaptive scheme consistently achieves superior accuracy, even with spatial discretizations larger than $\delta z_{av} > 0.1$, outperforming the uniform scheme by requiring discretizations that are at least ten times smaller.

To illustrate the potential of our method, we studied the morphologies of sphere-forming BCP films, focusing on the question of whether HCP or FCC stacking is more favorable. Our analysis indicates that HCP stacking is the thermodynamically stable state in thin films, at least in the example studied by us (confinement between two attractive surfaces); however, the free energy differences between HCP and FCC stacked films are very small.

When looking at polymer brushes, we again found that adaptive spatial discretization significantly enhances the accuracy of SCF calculations. In addition, adaptive discretization of the contour length parameter turns out to be essential for obtaining accurate results as well as for obtaining converged SCF solutions to begin with when the total number of discretization points is low. Choosing finer values of δs_{av} enables more subtle features to be captured, such as the depletion region near the grafting point, which are easily overlooked when using uniform discretization schemes, especially if δz is comparable to the depletion width. Importantly, our adaptive scheme alleviates the computational burden associated with brush simulations by allowing larger values of the averaged discretization parameter δs_{av} to be chosen, similar to the case of ungrafted chains in the bulk, without sacrificing accuracy. This enhancement in efficiency is particularly noteworthy in view of the fact that the costs of SCF calculations are dominated by the costs of repeatedly solving modified diffusion equations in order to obtain the propagators q and $q^{\dagger}$. Furthermore, our adaptive scheme remains robust in the strong stretching limit, accommodating scenarios where inter-chain interactions exert substantial influence, resulting in brush heights $L \gg R_g$. While polymer brush SCF calculations in one dimension remain feasible with manageable computational costs even when using uniform discretization schemes, the integration of adaptive contour discretization and spatial discretization extends the computational capabilities, facilitating SCFT calculations in two or three dimensions for intricate morphologies [21,24,25,36,37].

In summary, our study highlights the crucial role of adaptive discretization schemes in advancing SCF calculations, delivering significant enhancements in accuracy and computational efficiency for various polymeric systems. The versatility of our simple approach is demonstrated by two illustrative examples showcasing its applicability to problems involving interfaces and external potentials. In general, the required accuracy of SCF calculations depends on the specific quantity of interest. In the present paper, we have mainly focused on the free energy, which helps to identify the true equilibrium phase from a set of competing candidate structures. In other applications, accurate prediction of the density profiles of specific chain segments may be more important.

We have introduced two types of adaptive discretization schemes, namely, adaptive spatial discretization and adaptive contour discretization. Adaptive spatial discretization is useful in all situations where density or composition profiles vary strongly within the selected regions in space. On the other hand, adaptive contour discretization is useful in situations where particularly strong variations of the propagator function $q(\mathbf{r}, s)$ are expected for well-defined values of s as is the case, for instance, with polymer brushes or copolymers close to junction points that connect different blocks. In general, it would be desirable to couple spatial discretization and contour discretization. This is because the basic equation of any SCF iteration scheme, Equation (5), has the form of a modified diffusion equation. In such cases, the contour step should not be chosen independent of the spatial discretization [38,39]. In fact, an upper bound for δs should typically scale with δz, as $\delta s \sim \delta z^{1+\epsilon}$ with $\epsilon = 2$ [39]. Here, we have chosen a sufficiently small δs, i.e., smaller than the upper bound. In future work, we will explore possibilities for coupling the spatial and contour length discretizations such that the contour discretization is adjusted and becomes finer in those regions of space where the spatial discretization is finer.

It should be noted that the term "adaptive" in the present work refers to situations in which a specific inhomogeneous discretization scheme is chosen at the beginning of an SCF calculation and is not changed thereafter. In fact, we would strongly recommend not proceeding otherwise, as changing the discretization in the middle of an SCF iteration loop is likely to result in convergence problems. However, in algorithms that involve many successive SCF calculations, such as dynamic density functional (DDF) simulations, our scheme could be used to set up dynamically adaptive grids that adjust to the current state of the polymer system.

To foster broader adoption and further refinement, we have made our SCF code publicly available (see below). The code has a modular structure; it provides both adaptive real-space methods and uniform discretization pseudospectral methods, as well as, in the latter case, options to perform dynamic density functional (DDF) simulations following the DDF models used in [40–42]. It can be applied to arbitrary mixtures of linear multiblock copolymers, and can easily be extended to other polymer architectures as well.

Author Contributions: Conceptualization, F.S. and L.Q.; software, F.S., L.Q. and M.G.; methodology, investigation and data curation, L.Q. and M.G.; writing—original draft preparation, L.Q. and M.G.; writing—review and editing, F.S.; supervision, F.S. All authors have read and agreed to the published version of the manuscript.

Funding: This research was funded by funded by the Deutsche Forschungsgemeinschaft (DFG, Germany), grant numbers 248882694 (L.Q.) and 446008821 (M.G.). F.S. and L.Q. are members of and acknowledge partial funding from CRC 1552, grant number 465145163 (project C02). Parts of this research were conducted using the MOGON 2 supercomputer, which is a part of the AHRP (Alliance for High Performance Computing in Rhineland Palatinate, https://www.ahrp.info/ (accessed on 24 April 2024)) and the Gauss Alliance e.V. The authors gratefully acknowledge the computing time granted on the MOGON 2 supercomputer at Johannes Gutenberg University Mainz (https://hpc.uni-mainz.de/) (accessed on 24 April 2024).

Data Availability Statement: The code used to produce the data in the manuscript is available at https://github.com/leqiao/ADSCF.git (accessed on 24 April 2024).

Conflicts of Interest: The authors declare no conflicts of interest.

References

1. Vigil, D.L.; Quah, T.; Sun, D.; Delaney, K.T.; Fredrickson, G.H. Self-Consistent Field Theory Predicts Universal Phase Behavior for Linear, Comb, and Bottlebrush Diblock Copolymers. *Macromolecules* **2022**, *55*, 4237–4244. [CrossRef]
2. Müller, M. Phase Diagram of a Mixed Polymer Brush. *Phys. Rev. E* **2002**, *65*, 030802. [CrossRef]
3. Matsen, M.W.; Schick, M. Stable and Unstable Phases of a Diblock Copolymer Melt. *Phys. Rev. Lett.* **1994**, *72*, 2660–2663. [CrossRef]
4. Matsen, M.W. Self-Consistent Field Theory and Its Applications. In *Soft Matter*; John Wiley & Sons, Ltd.: Hoboken, NJ, USA, 2005; Chapter 2, pp. 87–178. [CrossRef]
5. Matsen, M.W.; Bates, F.S. Unifying Weak- and Strong-Segregation Block Copolymer Theories. *Macromolecules* **1996**, *29*, 1091–1098. [CrossRef]
6. Matsen, M.W.; Whitmore, M.D. Accurate Diblock Copolymer Phase Boundaries at Strong Segregations. *J. Chem. Phys.* **1996**, *105*, 9698–9701. [CrossRef]
7. Matsen, M.W. Fast and Accurate SCFT Calculations for Periodic Block-Copolymer Morphologies Using the Spectral Method with Anderson Mixing. *Eur. Phys. J. E* **2009**, *30*, 361. [CrossRef] [PubMed]
8. Fredrickson, G.H.; Ganesan, V.; Drolet, F. Field-Theoretic Computer Simulation Methods for Polymers and Complex Fluids. *Macromolecules* **2002**, *35*, 16–39. [CrossRef]
9. Drolet, F.; Fredrickson, G.H. Combinatorial Screening of Complex Block Copolymer Assembly with Self-Consistent Field Theory. *Phys. Rev. Lett.* **1999**, *83*, 4317–4320. [CrossRef]
10. Rasmussen, K.Ø.; Kalosakas, G. Improved Numerical Algorithm for Exploring Block Copolymer Mesophases. *J. Polym. Sci. Part B Polym. Phys.* **2002**, *40*, 1777–1783. [CrossRef]
11. Ceniceros, H.D.; Fredrickson, G.H. Numerical Solution of Polymer Self-Consistent Field Theory. *Multiscale Model. Simul.* **2004**, *2*, 452–474. [CrossRef]
12. Tzeremes, G.; Rasmussen, K.Ø.; Lookman, T.; Saxena, A. Efficient Computation of the Structural Phase Behavior of Block Copolymers. *Phys. Rev. E* **2002**, *65*, 041806. [CrossRef]
13. Fraaije, J.G.E.M.; Van Vlimmeren, B.A.C.; Maurits, N.M.; Postma, M.; Evers, O.A.; Hoffmann, C.; Altevogt, P.; Goldbeck-Wood, G. The Dynamic Mean-Field Density Functional Method and Its Application to the Mesoscopic Dynamics of Quenched Block Copolymer Melts. *J. Chem. Phys.* **1997**, *106*, 4260–4269. [CrossRef]
14. Netz, R.R.; Schick, M. Polymer Brushes: From Self-Consistent Field Theory to Classical Theory. *Macromolecules* **1998**, *31*, 5105–5122. [CrossRef]
15. Daoulas, K.C.; Müller, M. Exploring Thermodynamic Stability of the Stalk Fusion-Intermediate with Three-Dimensional Self-Consistent Field Theory Calculations. *Soft Matter* **2013**, *9*, 4097–4102. [CrossRef]

16. Revelas, C.J.; Sgouros, A.P.; Lakkas, A.T.; Theodorou, D.N. RuSseL: A Self-Consistent Field Theory Code for Inhomogeneous Polymer Interphases. *Computation* **2021**, *9*, 57. [CrossRef]
17. Cochran, E.W.; Garcia-Cervera, C.J.; Fredrickson, G.H. Stability of the Gyroid Phase in Diblock Copolymers at Strong Segregation. *Macromolecules* **2006**, *39*, 2449–2451. [CrossRef]
18. Vigil, D.L.; García-Cervera, C.J.; Delaney, K.T.; Fredrickson, G.H. Linear Scaling Self-Consistent Field Theory with Spectral Contour Accuracy. *ACS Macro Lett.* **2019**, *8*, 1402–1406. [CrossRef] [PubMed]
19. Vu, G.T.; Abate, A.A.; Gómez, L.R.; Pezzutti, A.D.; Register, R.A.; Vega, D.A.; Schmid, F. Curvature as a Guiding Field for Patterns in Thin Block Copolymer Films. *Phys. Rev. Lett.* **2018**, *121*, 087801. [CrossRef]
20. Milner, S.T. Polymer Brushes. *Science* **1991**, *251*, 905–914. [CrossRef] [PubMed]
21. Vorselaars, B.; Kim, J.U.; Chantawansri, T.L.; Fredrickson, G.H.; Matsen, M.W. Self-Consistent Field Theory for Diblock Copolymers Grafted to a Sphere. *Soft Matter* **2011**, *7*, 5128. [CrossRef]
22. Li, W.; Liu, M.; Qiu, F.; Shi, A.C. Phase Diagram of Diblock Copolymers Confined in Thin Films. *J. Phys. Chem. B* **2013**, *117*, 5280–5288. [CrossRef] [PubMed]
23. Yang, Y.B.; Jeon, Y.M.; Kim, J.U.; Cho, J. Diblock and Triblock Copolymer Thin Films on a Substrate with Controlled Selectivity. *Eur. Phys. J. E* **2012**, *35*, 86. [CrossRef] [PubMed]
24. Roan, J.R.; Kawakatsu, T. Self-Consistent-Field Theory for Interacting Polymeric Assemblies. I. Formulation, Implementation, and Benchmark Tests. *J. Chem. Phys.* **2002**, *116*, 7283–7294. [CrossRef]
25. Chantawansri, T.L.; Bosse, A.W.; Hexemer, A.; Ceniceros, H.D.; García-Cervera, C.J.; Kramer, E.J.; Fredrickson, G.H. Self-Consistent Field Theory Simulations of Block Copolymer Assembly on a Sphere. *Phys. Rev. E* **2007**, *75*, 031802. [CrossRef]
26. Ouaknin, G.; Laachi, N.; Delaney, K.; Fredrickson, G.H.; Gibou, F. Self-consistent field theory simulations of polymers on arbitrary domains. *J. Comput. Phys.* **2016**, *327*, 168–185. [CrossRef]
27. Wei, H.; Wang, X.; Chen, C.; Jiang, K. An adaptive virtual element method for the polymeric self-consistent field theory. *Comput. Math. Appl.* **2023**, *141*, 242–254. [CrossRef]
28. Revelas, C.J.; Sgouros, A.P.; Lakkas, A.T.; Theodorou, D.N. Addressing Nanocomposite Systems via 3D-SCFT: Assessment of Smearing Approximation and Irregular Grafting Distributions. *Macromolecules* **2023**, *56*, 1731–1746. [CrossRef]
29. Helfand, E. Theory of inhomogeneous polymers—Fundamentals of Gaussian Random-walk model. *J. Chem. Phys.* **1975**, *62*, 999–1005. [CrossRef]
30. Matsen, M. The standard Gaussian model for block copolymer melts. *J. Phys. Cond. Matter* **2002**, *14*, R21–R47. [CrossRef]
31. Müller, M.; Schmid, F. Incorporating fluctuations and dynamics in self-consistent field theories for polymer blends. In *Advanced Computer Simulation Approaches for Soft Matter Sciences II*; Holm, C., Kremer, K., Eds.; Advances in Polymer Science Series; Springer: Berlin/Heidelberg, Germany, 2005; Volume 185, pp. 1–58. [CrossRef]
32. Douglas, J.; Rachford, H.H. On the numerical solution of heat conduction problems in two and three space variables. *Trans. Am. Math. Soc.* **1956**, *82*, 421–439. [CrossRef]
33. Chong, T.H. A Variable Mesh Finite Difference Method for Solving a Class of Parabolic Differential Equations in One Space Variable. *SIAM J. Numer. Anal.* **1978**, *15*, 835–857. [CrossRef]
34. Milner, S.T.; Witten, T.A.; Cates, M.E. Theory of the Grafted Polymer Brush. *Macromolecules* **1988**, *21*, 2610–2619. [CrossRef]
35. Zhulina, E.B.; Borisov, O.V.; Priamitsyn, V.A. Theory of Steric Stabilization of Colloid Dispersions by Grafted Polymers. *J. Colloid Interface Sci.* **1990**, *137*, 495–511. [CrossRef]
36. Wang, R.; Wei, Q.; Sheng, W.; Yu, B.; Zhou, F.; Li, B. Driving Polymer Brushes from Synthesis to Functioning. *Angew. Chem. Int. Ed. Engl.* **2023**, *62*, e202219312. [CrossRef] [PubMed]
37. Zhou, X.; Liu, X.; Xie, Z.; Zheng, Z. 3D-patterned Polymer Brush Surfaces. *Nanoscale* **2011**, *3*, 4929–4939. [CrossRef]
38. Press, W.H. *Numerical Recipes: The Art of Scientific Computing*, 3rd ed.; Cambridge University Press: Cambridge, UK, 2007.
39. Rezzolla, L.; Zanotti, O. *Relativistic Hydrodynamics*; Oxford University Press: Oxford, UK, 2013.
40. Qi, S.; Schmid, F. Dynamic density functional theories for inhomogeneous polymer systems compared to Brownian dynamics simulations. *Macromolecules* **2017**, *50*, 9831–9845. [CrossRef]
41. Mantha, S.; Qi, S.; Schmid, F. Bottom-up construction of dynamic density functional theories for inhomogeneous polymer systems from microscropic simulations. *Macromolecules* **2020**, *53*, 3409–3423. [CrossRef]
42. Schmid, F.; Li, B. Dynamic self-consistent field approach for studying kinetic processes in multiblock copolymer melts. *Polymers* **2020**, *12*, 2205. [CrossRef]

polymers

MDPI

Article

Predicting Mechanical Properties of Polymer Materials Using Rate-Dependent Material Models: Finite Element Analysis of Bespoke Upper Limb Orthoses

Syed Hammad Mian [1,2], Usama Umer [1,2,*], Khaja Moiduddin [1,2] and Hisham Alkhalefah [1,2]

1 Advanced Manufacturing Institute, King Saud University, Riyadh 11421, Saudi Arabia
2 King Salman Center for Disability Research, Riyadh 11614, Saudi Arabia
* Correspondence: uumer@ksu.edu.sa

Abstract: Three-dimensional printing—especially with fused deposition modeling (FDM)—is widely used in the medical field as it enables customization. FDM is versatile owing to the availability of various materials, but selecting the appropriate material for a certain application can be challenging. Understanding materials' mechanical behaviors, particularly those of polymeric materials, is vital to determining their suitability for a given application. Physical testing with universal testing machines is the most used method for determining the mechanical behaviors of polymers. This method is resource-intensive and requires cylinders for compression testing and unique dumbbell-shaped specimens for tensile testing. Thus, a specialized fixture must be designed to conduct mechanical testing for the customized orthosis, which is costly and time-consuming. Finite element (FE) analysis using an appropriate material model must be performed to identify the mechanical behaviors of a customized shape (e.g., an orthosis). This study analyzed three material models, namely the Bergström–Boyce (BB), three-network (TN), and three-network viscoplastic (TNV) models, to determine the mechanical behaviors of polymer materials for personalized upper limb orthoses and examined three polymer materials: PLA, ABS, and PETG. The models were first calibrated for each material using experimental data. Once the models were calibrated and found to fit the data appropriately, they were employed to examine the customized orthosis's mechanical behaviors through FE analysis. This approach is innovative in that it predicts the mechanical characteristics of a personalized orthosis by combining theoretical and experimental investigations.

Keywords: finite element analysis; Bergström–Boyce model; three network viscoplastic model; upper limb orthosis; customization; 3D printing; mechanical testing

1. Introduction

In the current highly competitive business environment, there is an increasing need to enhance, simplify, and expedite design and fabrication processes. Additive manufacturing (AM) or 3D printing is a cost-effective method used to fabricate complex shapes with less waste compared with traditional subtractive manufacturing methods [1]. Among the fabrication techniques, fused filament fabrication (FFF), also known as fused deposition modeling (FDM), is widely used. In this process, a solid material, typically a raw filament, is forced through a heated nozzle at temperatures below the melting point but above the glass transition temperature. Then, this quasi-melted filament is gradually applied to a bed layer by layer to construct the component from the bottom to the top [2]. FDM is widely used owing to its ease of use, accessibility, and cost-effectiveness [3]. This tendency to create the highest quality product while minimizing resource usage is crucial for success across various industries. Thus, computational tools are increasingly being employed at the start of design processes to minimize reliance on trial-and-error methods and physical trials. Conducting physical tests to investigate mechanical properties directly can be expensive and time-consuming, particularly when numerous parameters are involved [4]. Alternative methods involve employing mathematical simulations that utilize numerical techniques. The finite element (FE) method is a well-established technique that enables accurate forecasts of deformations and stresses in products under both normal and accelerated loading conditions [5]. As a result of ongoing advancements in FE processes, multiple commercially accessible FE packages can effectively conduct complex simulations.

Polymers demonstrate both elastic and viscous material properties [6]. The dynamic behavior of 3D-printing-based elastic polymers has garnered significant interest from researchers in the fields of applied physics, material science, and engineering. Given polymeric materials's key role in the 3D printing industry, advanced material models should be incorporated into simulation tools to achieve accurate analysis and design.

The polymeric materials acrylonitrile butadiene styrene (ABS), polylactic acid (PLA), and polyethylene terephthalate glycol (PETG) are widely employed in medical settings [7,8]. When these polymers are utilized in biomedical devices, such as orthoses' structural components, it is necessary to include additional factors to accurately describe their mechanical behavior. Furthermore, the mechanical behaviors of polymers are correlated with their thermal properties, becoming particularly significant under a high strain rate [9]. When polymers undergo substantial deformations, they generate heat due to the loss of energy caused by inelastic viscosity. This leads to an increased material temperature, resulting in thermal softening [10]. Thermal effects significantly impact polymers' mechanical properties under a high strain rate [11]. In addition, material nonlinearity caused by plastic behavior poses a challenge because it introduces a highly complex mathematical formulation in the analysis. Moreover, experimental studies for polymer materials require a substantial amount of time and financial resources [12]. To overcome these issues, numerical analysis is often employed. Moreover, advanced constitutive models in simulation tools should be used for accurate analysis and design.

Mathematical constitutive models provide a framework to elucidate the mechanical behaviors of materials in any given context [13]. These models can vary in complexity and may rely on various factors that typically represent a material's mechanical properties [14]. To have reliable numerical simulations, it is essential to accurately quantify the parameters (i.e., constants) employed in the governing equations of the chosen constitutive model. Model calibration entails the mechanical characterization of materials. Various constitutive methodologies have been developed to examine polymers' mechanical characteristics. Initial methodologies, such as those used by Hughes [15], have elucidated the mechanical behaviors of polymers through the application of conventional isotropic, rate-independent plasticity utilizing the Mises yield criterion. Numerous studies have reported the utilization of FE simulations for the analysis of polymer materials for orthosis applications [16–18].

Elastomers are essential components in the study of medical devices, including orthoses. Some of these applications subject elastomers to significant deformations, small strain, and dynamic loads to test the material's nonlinear viscoelastic behavior. Various methodologies are necessary to comprehensively investigate polymeric materials in medical equipment applications. The Bergström–Boyce (BB) model [19,20], three-network (TN) [21,22], and three-network viscoplastic (TNV) models [23,24] are commonly used in Abaqus as a build-in model and are known for their accuracy and utility.

This study examined and evaluated three constitutive mathematical models, namely the BB, TN, and TNV models, to accurately describe the mechanical behaviors of PLA, ABS, and PETG used in individualized upper-limb orthoses. These models have been calibrated for diverse 3D printing materials and could be employed in the field of biomedical engineering. These models can be used to examine orthosis devices, such as upper limb orthosis that are capable of being forced to undergo dynamic loading. Furthermore, these models provide additional opportunities for the examination of deformations and their implications.

2. Materials and Methods

Three constitutive models were compared to evaluate the behavior of thermoplastic materials in unpredictable deformation scenarios. Three material models, namely the BB, TN, and TNV implemented in ABAQUS/Standard from Dassault Systems (Dassault Systems, Vélizy-Villacoublay, France), were used to determine their suitability. These models were independently simulated for three polymer materials, namely PLA, ABS, and PETG, to assess their mechanical behavior. The objective was to develop an accurate and validated constitutive material model to select the most appropriate polymer material for hand orthosis.

Two main components of the proposed approach were used to calibrate the models using data obtained through tensile and compression testings for specific materials, as well as using the calibrated material models, to forecast the mechanical behaviors of the three polymers for hand orthosis. The study methodology is illustrated in Figure 1.

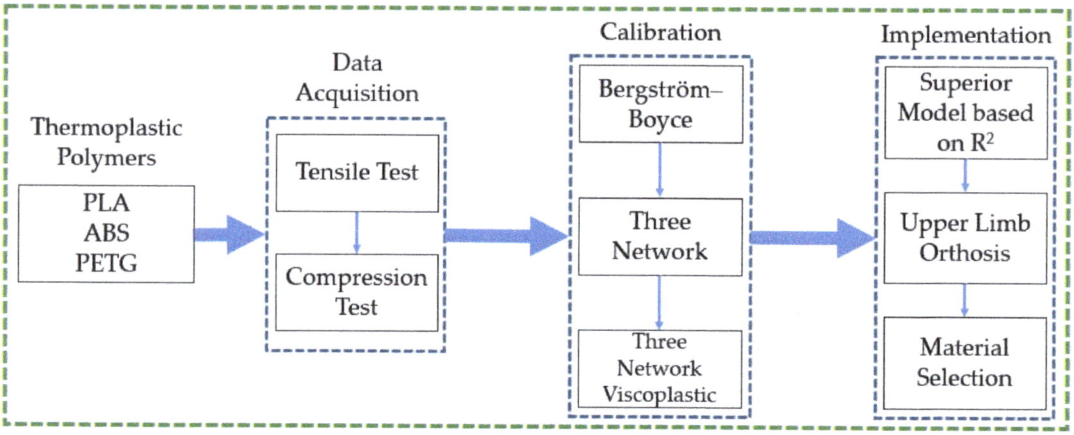

Figure 1. Methodology adopted to identify suitable material for upper-limb orthosis.

2.1. Data Acquisition

PLA, ABS, and PETG were examined in this investigation. In the data-collection phase, the true mechanical behaviors of 3D-printed specimens were explored based on these materials. Three constitutive material models were calibrated using information gathered from tensile and compression tests. These models are covered in detail in the subsequent section. For data acquisition, the first step was to fabricate standard specimens (dog-bone and cylindrical-shaped specimens) using a 3D printer.

As illustrated in Figure 2a, the specimens employed in this work were produced utilizing an open-filament 3D printer called the Raise 3D Pro 3 (Raise 3D Technologies Inc., Irvine, CA, USA). The specimens were produced using the ideal printing parameters suggested by Raise 3D, the material filament supplier. As demonstrated in Figure 2b, the test specimens were constructed flatwise along the X-direction. The raster angle of the grid infill pattern was altered by 90° between two succeeding layers. For instance, one layer was printed at a raster angle of 45°, followed by a subsequent layer printed at 135°, and so on, until the entire component was produced (refer to Figure 2b). The same layer height of 0.1 mm and infill percentage of 100% were used to produce all of the specimens. All materials have ideal temperature settings and optimum speed, which ensures seamless and precise 3D printing. Consequently, PLA, ABS, and PETG were processed using the Raise-3D-recommended speed and default temperature settings (bed and nozzle temperatures) [25–27]. Table 1 lists he 3D printing parameters applied for producing standard specimens using various materials.

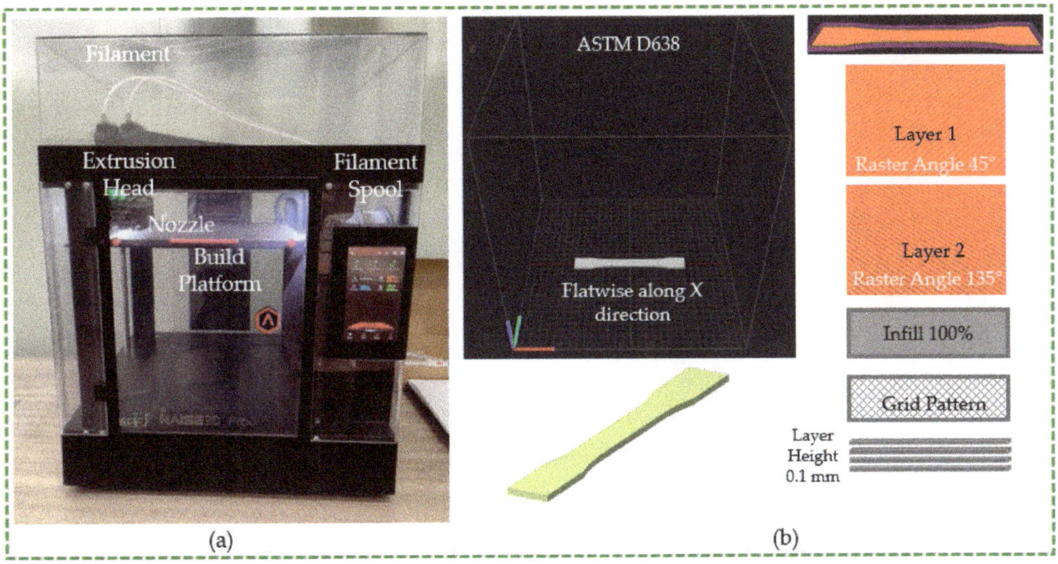

Figure 2. (**a**) Open-filament 3D printer (**b**) Graphical representation of process parameters.

Table 1. Three-dimensional printing parameters utilized for producing standard specimens.

Material	Variables					
	Bed Temperature (°C)	Nozzle Temperature (°C)	Speed (mm/s)	Layer Thickness (mm)	Infill (%)	Extrusion Width (mm)
PLA	55	205	70	0.1	100	0.4
ABS	100	250				
PETG	60	245				

Tensile and compression test results were deployed to calibrate the constitutive material models. The mechanical behaviors were examined using tensile and compression tests. Printed samples were subjected to compression and tensile testing based on ASTM standards. The standards used were ASTM D638 [28] and ASTM D695 [29] for tensile and compression tests, respectively.

As shown in Figure 3a, tensile and compression tests were performed using a Zwick Z100 electromechanical universal testing machine (ZwickRoell, Ulm, Germany) fitted with a cell with a 100 kN load. In the elastic range, the modulus of elasticity was experimentally measured using an extensometer. To measure elongation, the extensometer was fastened to the tensile test sample (Figure 3b). This enabled an accurate calculation of Young's modulus [30]. Identifying the yield point, which denotes changes in the material's behavior from elastic to plastic, is difficult. If yield restriction is crossed, the material is permanently plastically deformed. The 0.2% strain offset approach is useful for determining yield strength [31]. The 0.2% offset (yield strength) is the stress value that equates to 0.2% plastic strain. Thus, this technique was used to calculate yield strength [32].

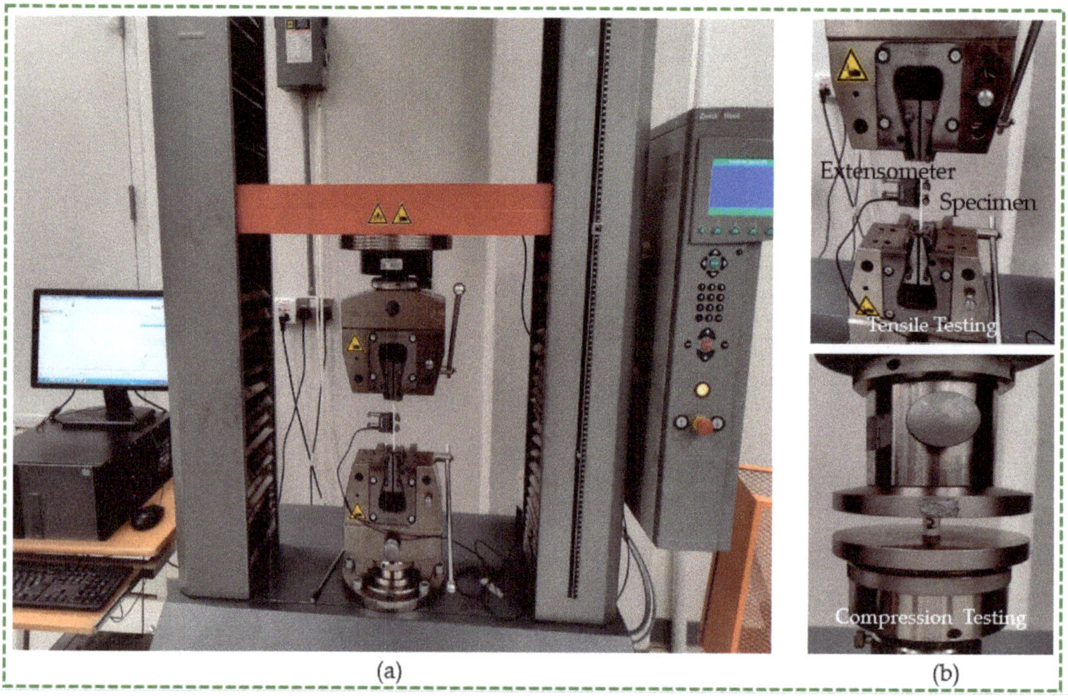

Figure 3. (**a**) Universal testing machine and (**b**) tensile and compression testing setups.

It is vital to use data for various strain rates when calibrating constitutive material models. Thus, tension and compression samples were deformed with varying rates of displacement control. These tests were conducted at various strain rates. The strain rates for tensile tests were 0.0005 s^{-1} and 0.005 s^{-1}, and those for compression tests were 0.018 s^{-1}, 0.011 s^{-1}, and 0.0011 s^{-1}. The stress–strain experimental curves were recorded for each test conducted at room temperature.

2.2. Material Models

Material responses were captured using the BB, TN, and TNV constitutive material models. The details of these models are as follows.

2.2.1. BB

The BB is a comprehensive material model used to forecast the large-strain, time-dependent behavior of elastomers [19,33–35]. BB as a material model was used to replicate the mechanical behaviors of thermoplastic materials.

The imposed deformation gradient in the BB model acts on F_A and F_B, two parallel macromolecular networks [33]. The deformation gradient operating on network B is broken down into viscoelastic and elastic constituents as $F = F_B{}^e\, F_B{}^v$. The response of network A is provided by an eight-chain model using Equation (1).

$$\sigma_A = \frac{\mu}{J\overline{\lambda}^*} \frac{\mathcal{L}^{-1}(\overline{\lambda}^*/\lambda_L)}{\mathcal{L}^{-1}(1/\lambda_L)}\, dev\,[\mathrm{b}^*] + \kappa\,(J-1)\,\mathrm{I} \tag{1}$$

Stress on network B is estimated using the eight-chain model (see Equation (2)), although the effective shear modulus differs.

$$\sigma_B = \frac{s\mu}{J_B^e \overline{\lambda_B^{e*}}} \frac{\mathcal{L}^{-1(\overline{\lambda_B^{e*}}/\lambda_L)}}{\mathcal{L}^{-1(1/\lambda_L)}}\, dev\,[b_B^{e*}] + \kappa\,(J_B^e - 1)\mathrm{I} \tag{2}$$

where λ_B^{e*} refers to the chain stretch in the elastic portion of Network B and s is a dimensionless material property indicating the shear modulus of Network B in relation to Network A. Equation (3) can be employed to determine the total Cauchy stress.

$$\sigma = \sigma_A + \sigma_B \tag{3}$$

The velocity gradient on network B L_B can be separated into elastic and viscous elements (refer to Equation (4)).

$$L_B = L_B^e + \tilde{L}_B^v \tag{4}$$

where $L_B^e = \dot{F}_B^e\,(F_B^e)^{-1}$; $\tilde{L}_B^v = \tilde{D}_B^v + \tilde{W}_B^v$; $L_B^v = \dot{F}_B^v\,(F_B^v)^{-1} = D_B^v + W_B^v$

$\tilde{W}_B^v \equiv 0$ must be specified to make the unloading distinct. Equation (5) is used to constitutively define the viscous deformation rate of network B.

$$\tilde{D}_B^v = \dot{Y}_B\left(\sigma_{B,\, b_B^{e*}}\right)\mathrm{N}_B^v \tag{5}$$

where $\mathrm{N}_B^v = \frac{dev[\sigma_B]}{\tau} = \frac{dev[\sigma_B]}{\|dev[\sigma_B]\|_F}$

τ is the effective stress driving the viscous flow. The time derivative of $\dot{F}_B^v$ can be expressed using Equation (6).

$$\dot{F}_B^v = \dot{Y}_B^v (F_B^e)^{-1} = \frac{dev\,[\sigma_B]}{\|dev\,[\sigma_B]\|_F}\,\mathrm{F}_B^e\,\mathrm{F}_B^V \tag{6}$$

Equation (7) provides the rate equation for viscous flow.

$$\dot{Y}_B^v = \dot{Y}_o\left(\overline{Y}_B^v - 1 + \xi\right)^C\left[R\left(\frac{\tau}{\tau_{base}} - \hat{\tau_{cut}}\right)\right]^m \tag{7}$$

R (x) = $\frac{(x+|x|)}{2}$ is the ramp function, $\hat{\tau_{cut}}$ represents the cutoff stress beyond which no flow will occur; a constant, $\dot{Y}_o \equiv 1/s$, is included to ensure dimensional continuity, and $\overline{Y}_B^v = \sqrt{\frac{tr[b_B^v]}{3}}$ denotes the viscoelastic chain stretch. The expression for effective stress causing the viscous flow is defined by Equation (8).

$$\tau = \|dev\,[\sigma_B]\|_F = \sqrt{[\sigma_B' \sigma_B']} \tag{8}$$

2.2.2. TN

Three components, or parallel-acting molecule networks, constitute the TN model, a material model designed especially for thermoplastic materials [22]. Entropic resistance governs the massive strain response in this model, whereas two distinct energy-activation methods equivalent to amorphous and semi-crystalline domains characterize the early viscoplastic response [36].

In this model structure, the deformation gradient operating on network A is broken down into elastic and viscoplastic elements: F = $F_A{}^e\ F_A{}^V$. An eight-chain model with temperature dependence provides the Cauchy stress applying on network A (refer to Equation (9) [20,37].

$$\sigma_A = \frac{\mu_A}{J_A^e \overline{\lambda}_A^{e*}} \left[1 + \frac{\theta - \theta_0}{\hat{\theta}}\right] \frac{\mathcal{L}^{-1}(\overline{\lambda}_A^{e*}/\lambda_L)}{\mathcal{L}^{-1}(1/\lambda_L)} \, dev\,[b_A^{e*}] + \kappa\,(J_A^e - 1)\,I \tag{9}$$

where $J_A^e = \det[F_A^e]$, with μ_A representing the initial shear modulus and λ_L denoting chain locking stretch. The current temperature is represented by θ, while θ_0 is the reference temperature, and $\hat{\theta}$ is a material parameter that indicates the stiffness's temperature dependence. The Cauchy–Green deformation tensor is expressed as $b_A^{e*} = (J_A^e)^{-2/3}\, F_A^e\, (F_A^e)^T$, and $\overline{\lambda}_A^{e*} = \sqrt{\frac{tr[b_A^{e*}]}{3}}$ indicates the effective chain stretch dependent on the eight-chain topology hypothesis [37], $\mathcal{L}^{-1}(x)$ represents the inverse Langevin function, where $\mathcal{L}(x) = \frac{coth\,(x) - 1}{x}$ and κ indicate the bulk modulus.

The viscoelastic deformation gradient for network B can be split into elastic and viscoplastic elements: $F = F_B^e\, F_B^v$. By adopting the eight-chain network model as in network A, network B's Cauchy stress can be calculated using Equation (10).

$$\sigma_B = \frac{\mu_B}{J_B^e \overline{\lambda}_B^{e*}} \left[1 + \frac{\theta - \theta_0}{\hat{\theta}}\right] \frac{\mathcal{L}^{-1}(\overline{\lambda}_B^{e*}/\lambda_L)}{\mathcal{L}^{-1}(1/\lambda_L)} \, dev\,[b_B^{e*}] + \kappa\,(J_B^e - 1)\,I \tag{10}$$

where $J_B^e = \det[F_B^e]$, μ_B is the initial shear modulus and $b_B^{e*} = (J_B^e)^{-2/3}\, F_B^e\, (F_B^e)^T$ defines the Cauchy–Green deformation tensor. The effective chain stretch is represented as $\overline{\lambda}_B^{e*} = \sqrt{\frac{tr[b_B^{e*}]}{3}}$. With the introduction of plastic strain, the effective shear modulus changes from an original value of μ_{Bi} to a final value of μ_{Bf}, as per Equation (11).

$$\dot{\mu_B} = -\beta\left[\mu_B - \mu_{Bf}\right] \cdot \dot{Y}_A \tag{11}$$

where $\dot{Y}_A$ represents the viscoplastic flow rate defined in Equation (14). The model captures the transition from the initial yield to large-scale flow by modulating stiffness with plastic strain.

Equation (12) determines network C's Cauchy stress, adopting the eight-chain model with I_2-dependence of the first order.

$$\sigma_C = \frac{1}{1+q}\left\{\frac{\mu_C}{J\lambda_{chain}}\left[1 + \frac{\theta - \theta_0}{\hat{\theta}}\right]\frac{\mathcal{L}^{-1}(\lambda_{chain}/\lambda_L)}{\mathcal{L}^{-1}(1/\lambda_L)} dev[b^*] + \kappa(J-1)I + q\frac{\mu_C}{J}\left[I_1^* b^* - \frac{2I_2^*}{3} I - (b^*)^2\right]\right\} \tag{12}$$

where $J = \det[F]$, μ_C is the initial shear modulus, $b^* = (J)^{-2/3}\, F\, (F)^T$ is the Cauchy–Green deformation tensor, $\lambda_{chain} = \sqrt{\frac{tr[b^*]}{3}}$ represents the effective chain stretch, and q modulates the I_2 dependence.

The system's total Cauchy stress is expressed by Equation (13).

$$\sigma = \sigma_A + \sigma_B + \sigma_C \tag{13}$$

To complete the description of this model framework, the rate kinematics should be determined. The rate kinematics for network A's total velocity gradient, $L = \dot{F}F^{-1}$, can be split into elastic and viscous components: $L = L_A^e + \left(F_A^e L_A^v F_A^e\right)^{-1} = L_A^e + \tilde{L}_A^v$, where $L_A^v = \dot{F}_A^v \left(F_A^v\right)^{-1} = D_A^v + W_A^v$ and $\tilde{L}_A^v = \tilde{D}_A^v + \tilde{W}_A^v$.

An undefined rigid body rotation of an intermediary state maintains the stress-free state, resulting in the nondistinct characterization of unloading processes coupling the deformed state with the intermediary state. Several techniques can be employed to make the intermediary state distinctive [38], one of which is to specify $\tilde{W}_A^v \equiv 0$. Elastic and inelastic deformation gradients incorporating rotation will generally arise from this. After this, network A's viscoplastic flow rates can be determined utilizing Equation (14).

$$\tilde{D}_A^v = \dot{Y}_A N_A \tag{14}$$

The term $\dot{Y}_A$ indicates the true deviatoric flow rate, whereas the tensor N_A specifies the driving deviatoric stress's direction, which is transferred from the relaxed to the current configuration.

Given that σ_A is calculated in a loaded configuration, the driving deviatoric stress transferred from the relaxed to the current configuration is given as $\sigma'_A = dev[\sigma_A]$. The effective stress, denoted by τ_A, is defined using the Frobenius norm $\tau_A = \|\sigma'_A\|_F = \sqrt{tr[\sigma'_A \sigma'_A]}$. Thus, the direction of the driving deviatoric stress N_A is $\frac{\sigma'_A}{\tau_A}$. The effective deviatoric flow rate is then determined by the power law described in Equation (15).

$$\dot{Y}_A = \dot{Y}_o \left(\frac{\tau_A}{\hat{\tau}_A + aR(p_A)} \right)^{mA} \cdot \frac{\theta}{\theta_o}^{n} \tag{15}$$

where $\dot{Y}_o \equiv 1/s$ is included to guarantee dimensional continuity, $p_A = -[(\sigma_A)_{11} + (\sigma_A)_{22} + (\sigma_A)_{33}]/3$ indicates hydrostatic pressure, $R(x) = (x + |x|)/2$ represents ramp function, and $\hat{\tau}_A$, a, mA, and n are material parameters. This paradigm assumes that the flow rate's temperature dependence takes the form of power law. To summarize, the viscoelastic flow of the velocity gradient of network A can be determined by applying Equation (16).

$$\dot{F}_A^v = \dot{Y}_A^v (F_A^e)^{-1} = \frac{dev\,[\sigma_A]}{\tau_A} \text{F} \tag{16}$$

Network B's total velocity gradient can be determined as that of network A. For example, $L = \dot{F}F^{-1}$, can be split into elastic and viscous components: $L = L_B^e + (F_B^e L_B^v F_B^e)^{-1} = L_B^e + \tilde{L}_B^v$, where $L_B^v = \dot{F}_B^v\,(F_B^v)^{-1} = D_B^v + W_B^v$ and $\tilde{L}_B^v = \tilde{D}_B^v + \tilde{W}_B^v$. The intermediary state is defined by $\tilde{W}_B^v \equiv 0$. Network B's viscoplastic flow rate is determined utilizing Equation (17).

$$\tilde{D}_B^v = \dot{Y}_B\, N_B = \frac{dev[\sigma_B]}{\tau_B} \tag{17}$$

where $\tau_B = \|\sigma'_B\|_F \equiv = \sqrt{tr[\sigma'_B \sigma'_B]}$

The deviatoric flow rate can be estimated using Equation (18).

$$\dot{Y}_B = \dot{Y}_o \left(\frac{\tau_B}{\hat{\tau}_B + aR(p_B)} \right)^{mB} \cdot \frac{\theta}{\theta_o}^{n} \tag{18}$$

where $\dot{Y}_o \equiv 1/s$ is included to ensure dimensional continuity, $p_B = -[(\sigma_B)_{11} + (\sigma_B)_{22} + (\sigma_B)_{33}]/3$ indicates hydrostatic pressure, and $\hat{\tau}_B$, a, mA, and n are material parameters. The viscoelastic flow of the velocity gradient of network B can be measured using Equation (19).

$$\dot{F}_B^v = \dot{Y}_B^v\,(F_B^e)^{-1} = \frac{dev\,[\sigma_B]}{\tau_B} \text{F} \tag{19}$$

2.2.3. TNV

The TNV is a versatile viscoplastic material model that can accurately represent experimentally observed behaviors of numerous thermoplastics, including volumetric plastic flow, damage accumulation, time- and pressure-dependent plastic flow, pressure-dependent bulk modulus, and triaxiality-dependent failure [39]. One-, two-, or three-parallel networks and one additional failure model are used in the TNV model. The material model's fourth parameter indicates the failure model type, whereas the initial three parameters identify the type of network for each network.

The TNV model is a substitute for the TN one. In many cases, TNV is more accurate and more extensive than TN. Moreover, volumetric plastic flow can be predicted using TNV.

2.3. *Abaqus Implementation*

FE models for tensile testing have been developed using ABAQUS/CAE. Besides ABAQUS, there are other effective numerical techniques such as the finite difference method [40], the bezier multi-step method [41], and the differential quadrature method [42] which can be utilized for numerical analysis. These approaches are unique because they can handle complex geometries, are robust, and can be applied to large-scale engineering simulations. This paper employed ABAQUS for numerical analysis because of the authors' expertise and the program's well-established capabilities.

The FE model had been discretized using 8,322 C3D8R elements (i.e., eight-node brick elements with decreased integration and hourglass control) in ABAQUS. One end of the specimen was fixed, i.e., all structural degrees of freedom were constrained, whereas the other end had a displacement boundary condition (BC) for the required strain (refer to Figure 4). The Mesh and BCs are shown in Figure 4. The BB, TN, and TNV models were implemented using the PolyUmod plugin for ABAQUS developed by PolymerFEM LLC, for the calculation of stresses at each integration point [43]. Visco analysis was conducted using ABAQUS/STANDARD using direct method and asymmetric matrix storage.

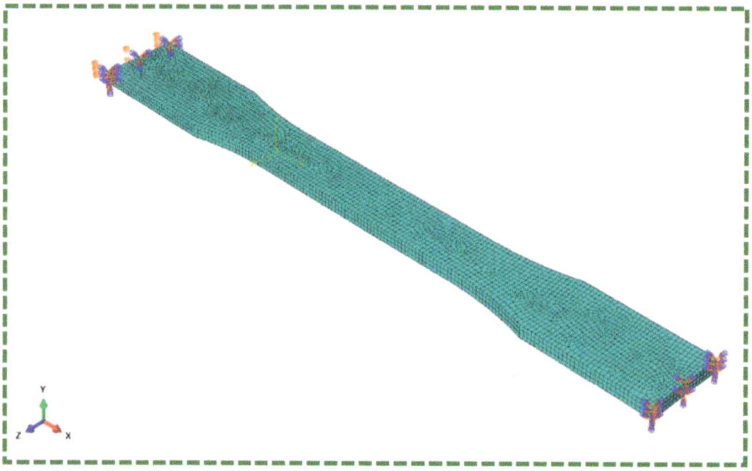

Figure 4. Boundary conditions for simulation in ABAQUS.

The implementation of the BB, TN, and TNV models in ABAQUS is computationally more challenging than that of other viscoelastic and viscoplastic models. The complete material behavior is decomposed into two or three networks to accurately capture the material nonlinear response at different strain and strain rates. This requires the determination of the deformation gradient and Cauchy stress tensor for parallel and series networks. For example, with the BB model, the true behavior of a certain polymer can be decomposed into two parallel networks (e.g., A and B) with non-linear hyperelastic behavior. In addition, network B also contains non-linear viscoelastic behavior elements connected in series. This requires additional steps for the calculation of a consolidated deformation gradient considering both elastic and viscoelastic components. The accuracy of the models depends on model parameters selection and the number of experimental data available for calibration and testing. For example, using a TN model to predict a typical thermoplastic without considering the temperature effects at least requires both uniaxial tension and compression data at two different strain rates.

3. Results and Discussion

The experimental findings of uniaxial tensile testing and the numerical results obtained using the BB model for PLA, ABS, and PETG are illustrated in Figure 5a, Figure 8a and Figure 11a, respectively. The data indicated that all three materials exhibited nearly linear elastic behavior initially before beginning to soften and yield until ultimate breakage. This indicates that these materials depend substantially on strain rate. The flow stress was found to be highly sensitive to strain rate. The flow stress increased as the strain rate increased. In the present investigation, it was observed that the tensile strength, yield strength, Young's modulus, etc., all increased as the strain rate increased for all three polymers. Similar outcomes of increased mechanical characteristics at a greater strain rate have also been reported in the literature [44–47]. The identical results were observed in the current study, where an increase in strain rates led to an increase in all mechanical properties. This improvement in the mechanical characteristics at greater strain rates can be attributed to the reduced time available for polymer relaxation and rearrangement processes. The shorter test duration inhibits the formation of flaws or cracks and reduces the likelihood of failures due to in-layer or inter-layer fracture modes [46]. There is another justification for greater mechanical properties at higher strain rates. When strain rate rises, strain hardening also increases because the stress-relaxation time decreases. This leads to the formation of additional dislocations and defects in the material, which lowers molecular mobility [48]. In addition to enhanced strength at higher strain rates, a slightly higher elongation at break was also noted at higher strain rates for all three materials. This increase in elongation at break can be a result of thermal softening. At higher strain rates, friction inside the polymer chains causes localized heating, which increases the temperature. This temperature rise causes the polymer to become softer, which increases the elongation before failure. Similar outcomes of increased elasticity modulus, ultimate strength, and total elongation at higher strain rates were reported by Cui et al. [49] and Wang et al. [46] for glass-fiber-reinforced thermoplastic composites. In summary, the self-heating in the current investigation was minimal at the strain rates that were investigated, and the stress–strain curves showed no significant indications of thermal softening. Nonetheless, at strain rates considerably higher than those explored here, the consequences of thermal softening might be substantial.

For PLA, plastic deformation occurred at about 2% strain, resulting in a Young's modulus of 2.28 GPa, tensile strength of 30.84 MPa, and a breaking strain of 3.56%. Under tensile stress, the ABS curve displayed a region of elastic and plastic deformation that displayed a stress–strain response that was initially linear, similar to PLA. Furthermore, ABS's elongation at break (3.86%) was only slightly longer than that of PLA. Furthermore, it was found that ABS was 20% weaker than PLA, with a tensile strength of 24.58 MPa. Additionally, it was observed that, up until the ABS sample fractures, the resistive stresses remained nearly constant (following the yield point) with an increase in strain. PETG exhibited the lowest elastic modulus—that is, the gradient of the stress–strain curve in an elastic region—compared with PLA and ABS. The elastic modulus of PETG was measured to be 1.64 GPa. Furthermore, PETG was as strong as PLA, but it became considerably more brittle and did not extend considerably when it reached its yielding point. There is a high chance of failure because PETG does not deform plastically.

A linear zone exists for all the materials (uniaxial compression testing) wherein the material experiences elastic deformation and restores to its initial size upon stress removal (Figure 5b, Figure 8b and Figure 11b). This region ends when a material reaches its yield point and begins to exhibit plastic behavior. In this situation, when stress is released, it no longer reverts to its initial size. However, the linear zone displays varying elongation based on material, indicating that some materials have higher elastic regimes than others. Thus, different compressive properties are anticipated. For instance, at a strain rate of 0.0011/s,

PLA exhibited the highest yield compressive strength at 51.93 MPa, followed by ABS at 50.03 MPa and PETG at 44.94 MPa.

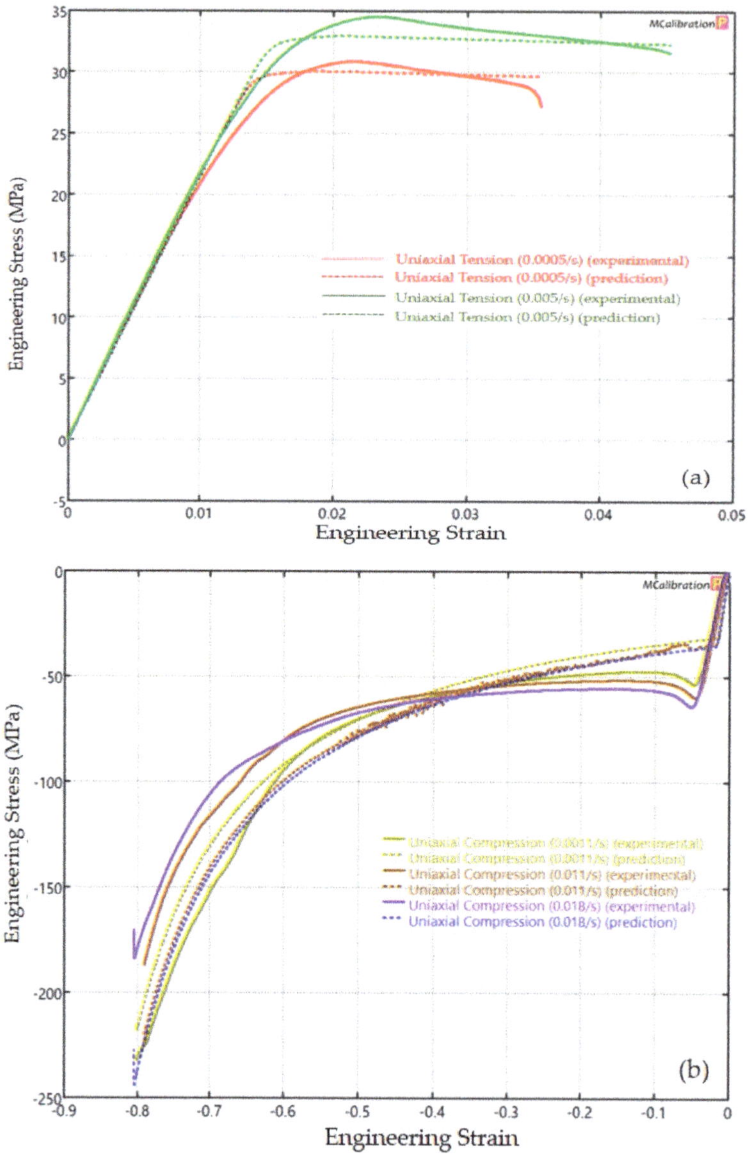

Figure 5. Experimental and BB predicted results for PLA: (**a**) uniaxial tension; (**b**) uniaxial compression.

The initial step in a material model's execution is its calibration to the uniaxial experimental data. The best estimate of the material's properties must be made using the entire set of uniaxial data. These models can be utilized for prediction once the proper material parameters are determined. For the most precise calibration, the automatic extensive procedure was selected. The automatic optimization process continued until the solution could no longer be improved. This approach toggles between various optimization methods to identify the suitable combination of material parameters. This technique involves various algorithms, including random search, the Levenberg–Marquardt algorithm, and NEWUOA search. Tables 2–4 for BB, TN, and TNV, respectively, provide the final material properties.

Table 2. Material parameters used by the Bergström–Boyce.

Symbol	Value	Description
μ	2.11919	Shear modulus of network A
λ_L	6.27995	Locking stretch
κ	42616.8	Bulk modulus
s	340.526	Relative stiffness of network B
x_i	0.83272	Strain adjustment factor
c	−0.0189804	Strain exponential
τ	32.418	Flow Resistance
m	24.999	Strain exponential
τ_{cut}	0.01	Normalized cut-off stress for flow

Table 3. Material parameters used by the three-network model.

Symbol	Value	Description
μ_A	355.991	Shear modulus of network *A*
λ_L	4.21835	Locking stretch
κ	4261.68	Bulk modulus
$\hat{\tau}_A$	12.4453	Flow resistance of network *A*
a	0.00596287	Pressure dependence of flow
$m_A = m_B$	22.2823	Stress exponential of network *A*
μ_{Bi}	395.969	Initial shear modulus of network *B*
μ_{Bf}	273.751	Final shear modulus of network *B*
β	9.42152	Evolution rate of μ_B
$\hat{\tau}_B$	21.5989	Flow resistance of network B
μ_C	0.805924	Shear modulus of network C
q	0	Relative contribution of I_2 of network C

Table 4. Material parameters used by the Three Network Viscoplastic model.

Symbol	Value	Description
C10	3.34032	Yeoh parameter 1 of Network A
C20	−0.210461	Yeoh parameter 2 of Network A
C30	0.00479607	Yeoh parameter 3 of Network A
κ_1	2130.84	Bulk modulus 1 of Network A
κ_2	0	Bulk modulus 2 of Network A
κ_3	0	Bulk modulus 3 of Network A
C10	596.03	Yeoh parameter 1 of Network B
C20	0	Yeoh parameter 2 of Network B
C30	0	Yeoh parameter 3 of Network B
κ_1	2130.84	Bulk modulus 1 of Network B
κ_2	0	Bulk modulus 2 of Network B
κ_3	0	Bulk modulus 3 of Network B
$\hat{\tau}_B$	9.93545	Flow resistance of network B
mm	19.6805	Stress exponent of Network B
bb	0	Volumetric flow coefficient of Network B
p_o	−0.194728	Pressure dependence of flow of Network B
fff	1.40761	Yield evaluation of $\hat{\tau}_B$
epsF	0.489528	Yield evaluation of characteristics strain of Network B
ceps	0.1	Flow damage strain of Network B
fss	1	Flow damage final state of Network B
C10	239.134	Yeoh parameter 1 of Network C

Table 4. *Cont.*

Symbol	Value	Description
C20	0	Yeoh parameter 2 of Network C
C30	0	Yeoh parameter 3 of Network C
κ_1	2130.84	Bulk modulus 1 of Network C
κ_2	0	Bulk modulus 2 of Network C
κ_3	0	Bulk modulus 3 of Network C
$\hat{\tau}_C$	51.8926	Flow resistance of network C
mm	8.59769	Stress exponent of Network C
bb	0	Volumetric flow coefficient of Network C
p_o	0.77947	Pressure dependence of flow of Network C
fff	0.25083	Yield evaluation of $\hat{\tau}_c$ of Network C
epsF	0.236314	Yield evaluation of characteristics strain of Network C
ceps	0.1	Flow damage strain of Network C
fss	1	Flow damage final state of Network C

Figures 5–13 present a comparison of three models for each material between experimental results and model predictions. Both the strain-rate dependency of the material and the yield evolution were reproduced by the calibrated models; this finding is consistent with those of mechanical tests. Although good agreement was observed for certain models, the same was not true for other materials. The coefficient of determination (R^2) and average fitness error were used to quantify this. The R^2 measures a model's ability to explain and forecast prospective results. Its value ranges from 0 to 1, with a greater coefficient typically denoting a model with a superior fit. Moreover, the average fitness error was employed to assess the performance of different models. This indicates the accuracy of a model in making predictions. An accurate prediction model with minimal error has a low average fitness error value.

Figures 5–7 present a comparison of PLA predictions made by the BB, TN, and TNV models with experimental results. These figures demonstrate how the calibrated model can accurately capture and forecast stress–strain response. For the calibrated BB model, the average fitness error was 11.588 and the coefficient of prediction (R^2) was 0.827. Similarly, for the calibrated TN model, the R^2 and average fitness error were 0.884 and 10.896, respectively. For the calibrated TNV model, these values were 0.886 and 12.100, respectively.

In Figures 8–10, the outcomes of the calibrated BB, TN, and TNV models for ABS are displayed. For the calibrated BB model, the R^2 was 0.756 and the average fitness error was 13.763. Likewise, the calibrated TN model had an R^2 of 0.508 and an average fitness error of 17.028, respectively. For the calibrated TNV model, these values were 0.743 and 18.455, respectively.

Figures 11–13 display the findings of calibrated BB, TN, and TNV models for PETG. For the calibrated BB model, the R^2 and average fitness error were 0.974 and 7.343, respectively. These values were 0.973 and 6.991, respectively, for the calibrated TN model and 0.972 and 6.923 for the calibrated TNV model.

According to the R^2 metric—which is the most favored in the literature—the best model for a given material was chosen. Furthermore, none of the models had considerably high average fitness errors. While the average fitness error is another option, the R^2 metric was used to compare various models in this investigation. Due to its ease of interpretation and comprehension, R^2 has been the most frequently used statistic to compare the different prediction models in the literature [50–54]. Table 5 lists the R^2 values of all three models. The BB was the best material model for ABS and PETG, whereas the TNV was the best model for PLA. Therefore, to examine upper limb orthoses, the appropriate chosen model was employed for a given material. For example, the TNV material model and corresponding material constants were used to forecast deformation and stresses in design while employing PLA material for upper-limb orthoses. Likewise, the BB material model with optimal material constants was applied to predict the stresses and deformation in orthoses based on ABS and PETG. The TNV model has been chosen as the material model for the numerical analysis of the PLA-based orthosis because of its higher R^2 value (=0.886) and comprehensiveness. This suggests that the TNV model yielded a superior fit and increased performance for PLA in comparison to the other models. In the current instance for PLA, 88.6% of the data fit the TNV model, as opposed to 82.7% and 88.4% for the BB and TN models. Generally speaking, a higher coefficient denotes a better fit for the model. Moreover, the BB model was chosen for the ABS and PETG since it yielded the highest coefficients, 0.756 and 0.974, respectively.

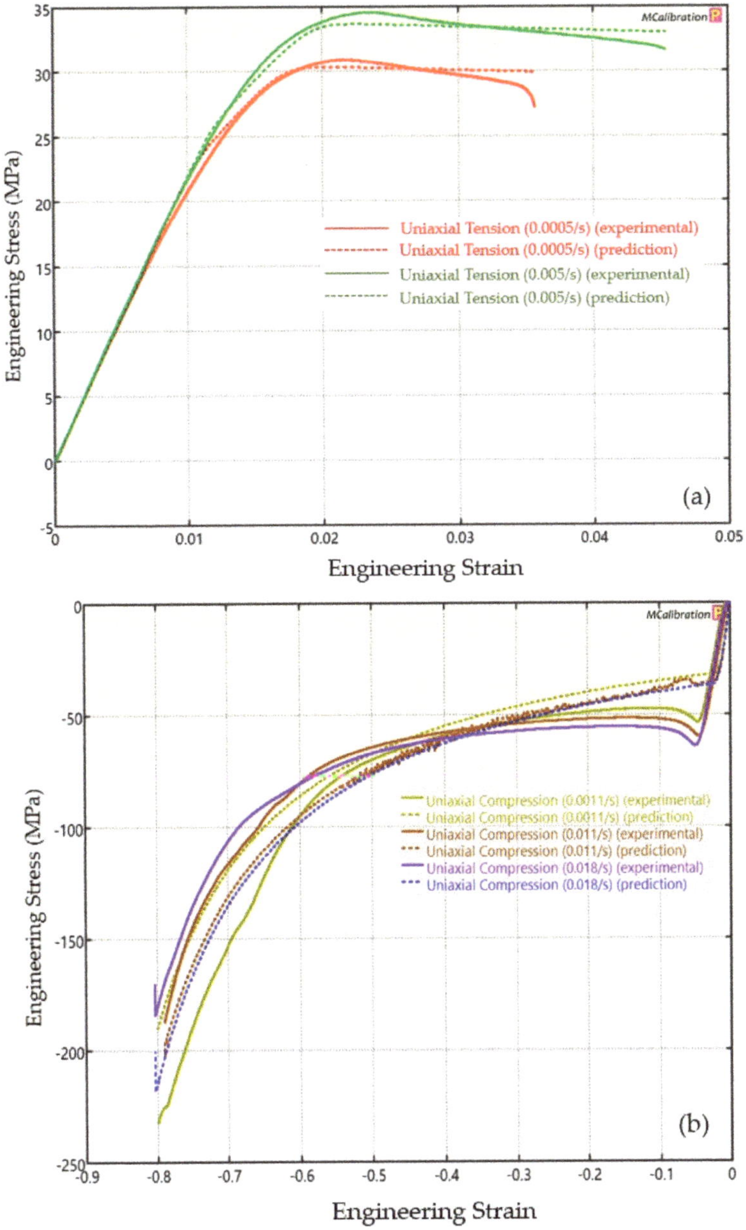

Figure 6. Experimental and TN predicted results for PLA: (**a**) uniaxial tension; (**b**) uniaxial compression.

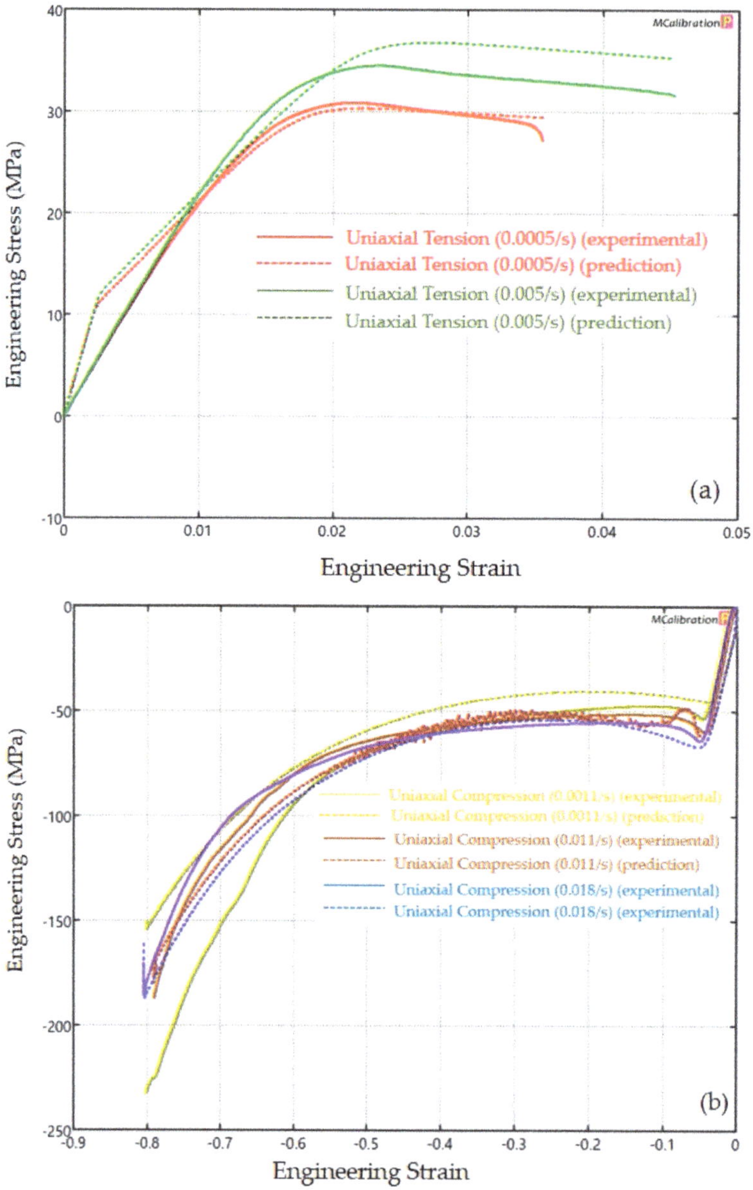

Figure 7. Experimental and TNV predicted results for PLA: (**a**) uniaxial tension; (**b**) uniaxial compression.

Each material model has its limitations and capabilities and was developed to capture certain behaviors not available with other material models. The BB model was developed for hyperelastic polymers and to incorporate large-strain dynamic behavior. TN model is more suitable for most of the thermo polymers, and it can also predict temperature-dependent phenomena, but it is unable to include failure/damage effects. In contrast, the TNV model is more general, and due to its open configuration, it can be tailored for different polymers by selecting suitable constitutive laws for each network. For example, for isothermal plastics, the following constitutive laws are recommended for each network:

Network A: Yeoh hyperelascticity; Network B and C: Yeoh hyperelascticity with the power-law. With regard to the results for materials used in the study, these models are unable to accurately predict fluctuations in the strain hardening behavior of ABS during compression testing within the strain range from 0.6 to 0.8.

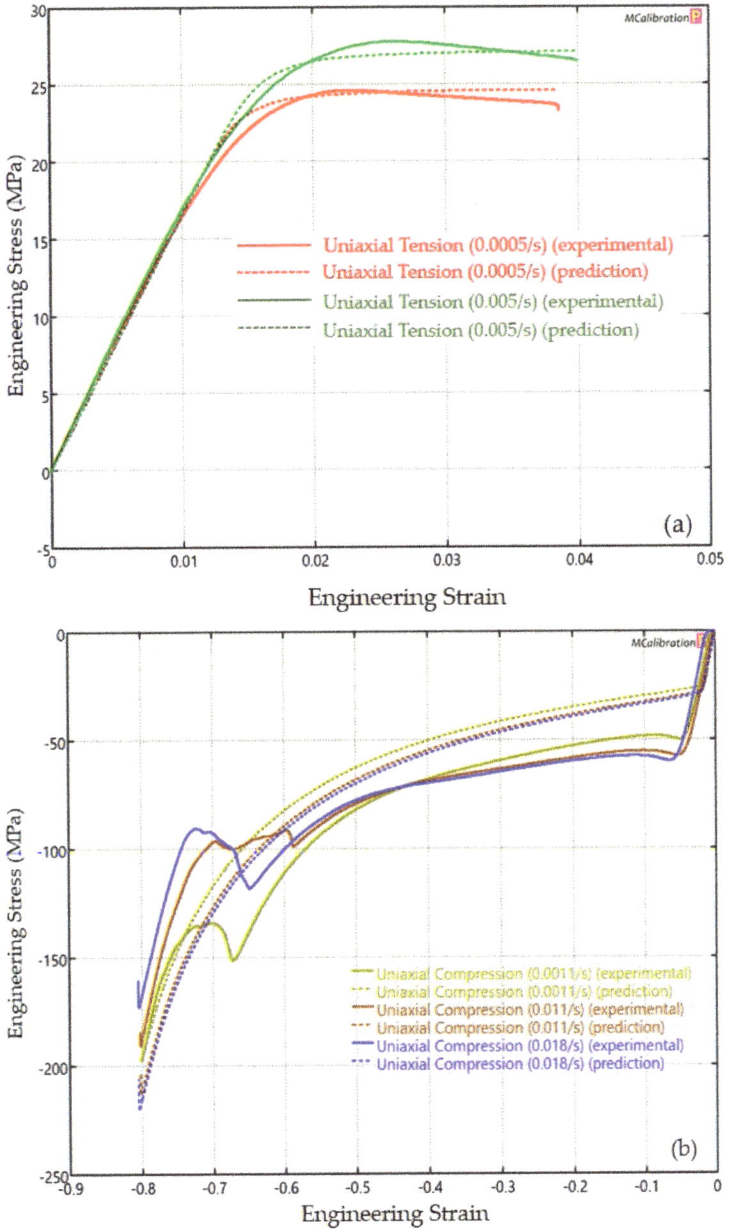

Figure 8. Experimental and BB predicted results for ABS: (**a**) uniaxial tension; (**b**) uniaxial compression.

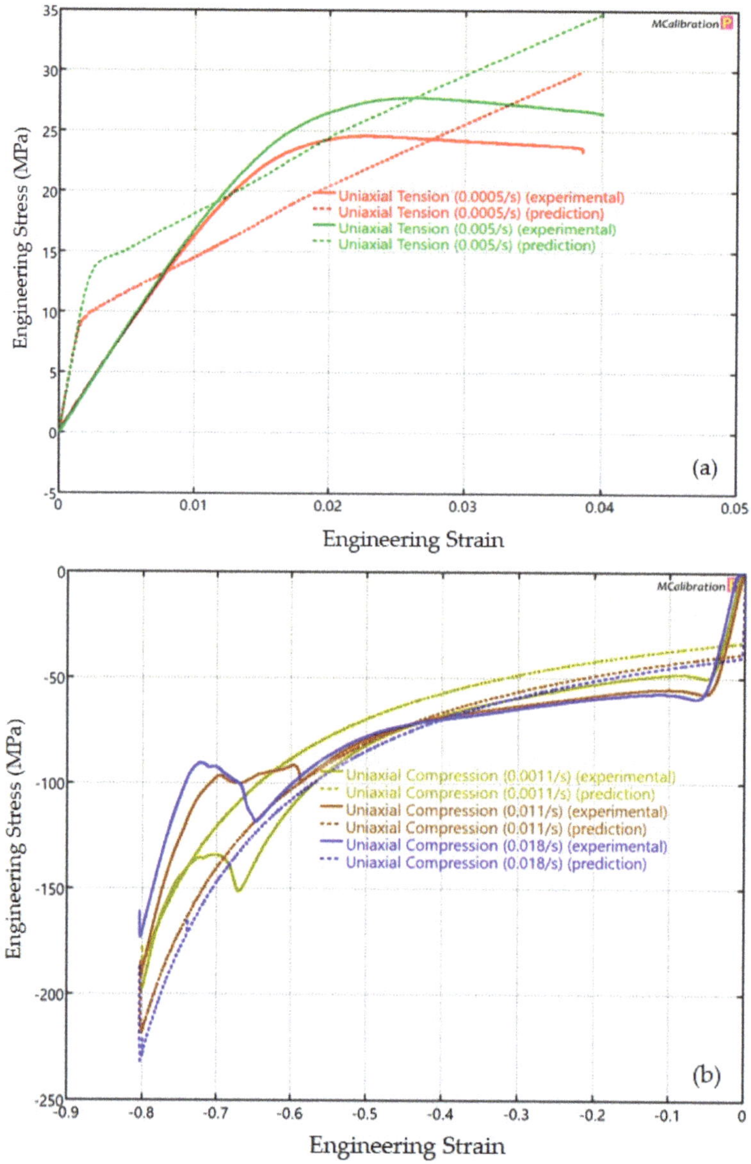

Figure 9. Experimental and TN predicted results for ABS: (**a**) uniaxial tension; (**b**) uniaxial compression.

A brief comparison of the three material models are also introduced here for the readers' understanding. The TN and TNV models can simulate strain rate-dependent thermal softening more accurately than the BB model owing to their ability to capture numerous deformation mechanisms [24,55]. The TNV model, which incorporates both viscoelastic and viscoplastic effects, provides the most thorough characterization and is particularly ideally suited for reproducing complicated phenomena at high strain rates [21,56,57]. Furthermore, TN and TNV models' higher computational complexity may render them less useful for some applications than the more straightforward BB model. In comparison to the BB model, the TNV and TN models involve a significantly higher number of material constants or parameters [58]. Consequently, the model selection process should consider the specific requirements of the application as well as the required degree of prediction accuracy.

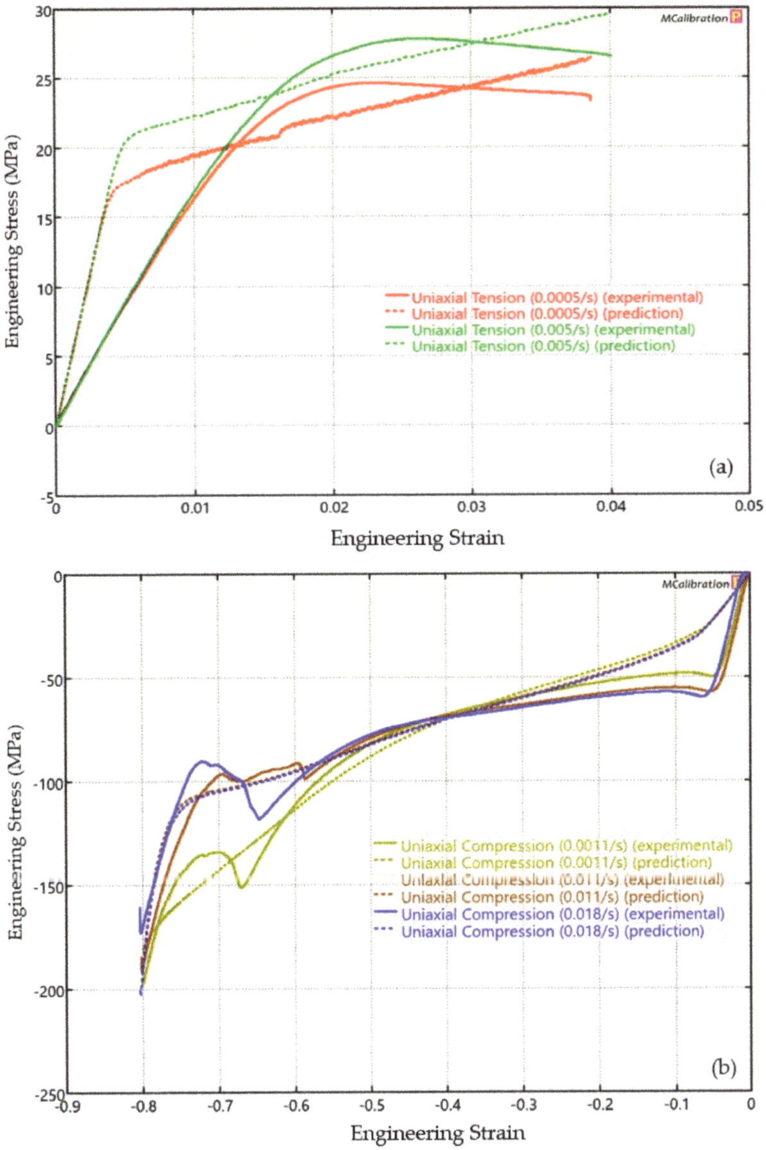

Figure 10. Experimental and TNV predicted results for ABS: (**a**) uniaxial tension; (**b**) uniaxial compression.

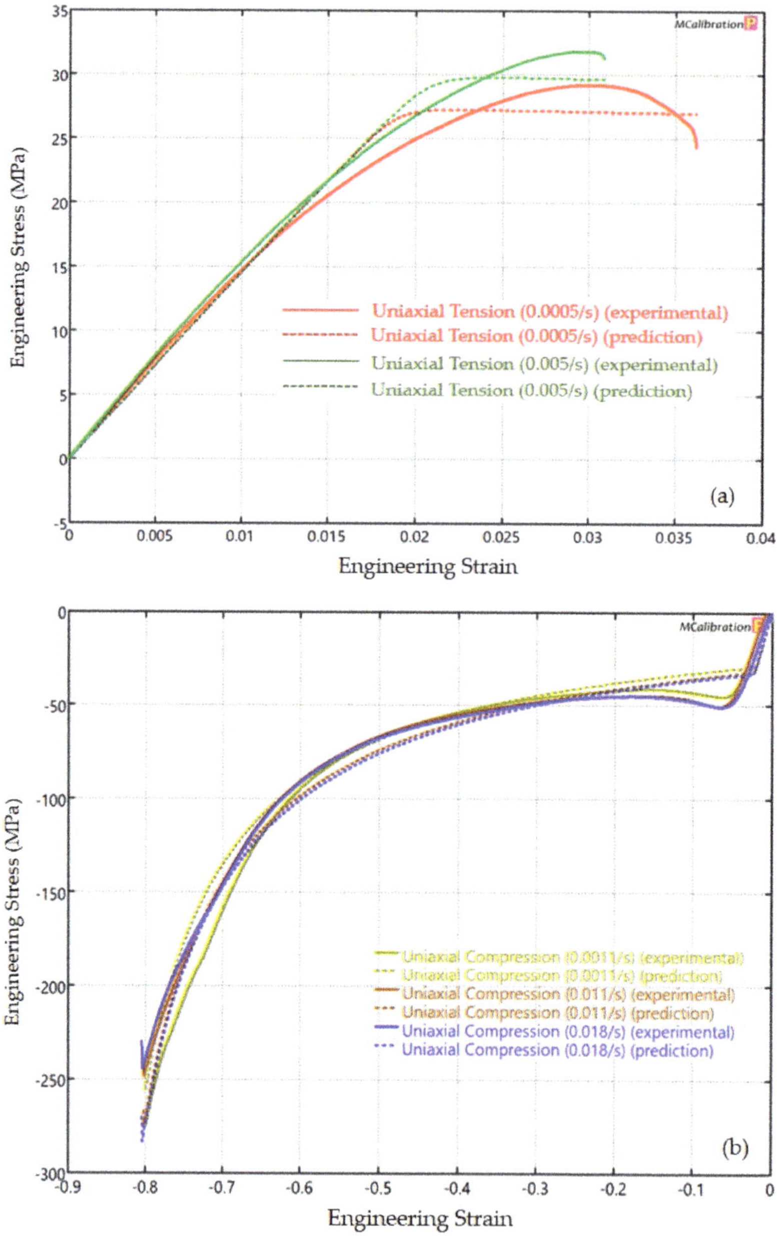

Figure 11. Experimental and BB predicted results for PETG: (**a**) uniaxial tension; (**b**) uniaxial compression.

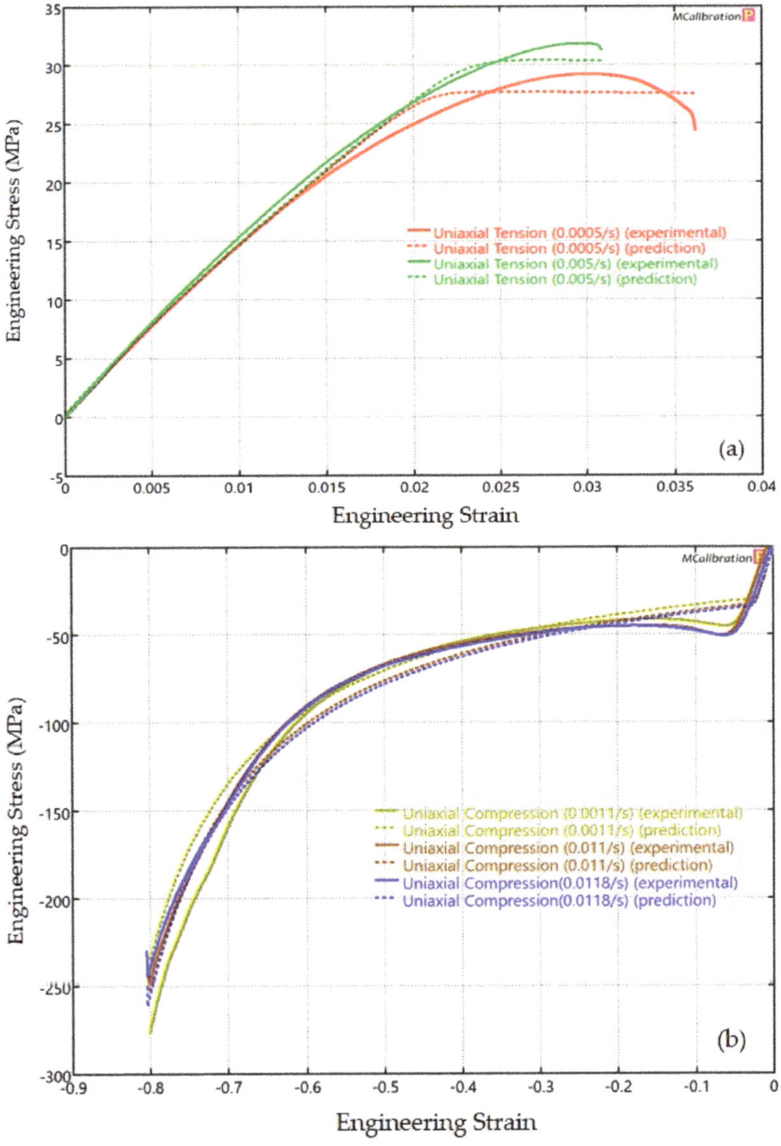

Figure 12. Experimental and TN predicted results for PETG: (**a**) uniaxial tension; (**b**) uniaxial compression.

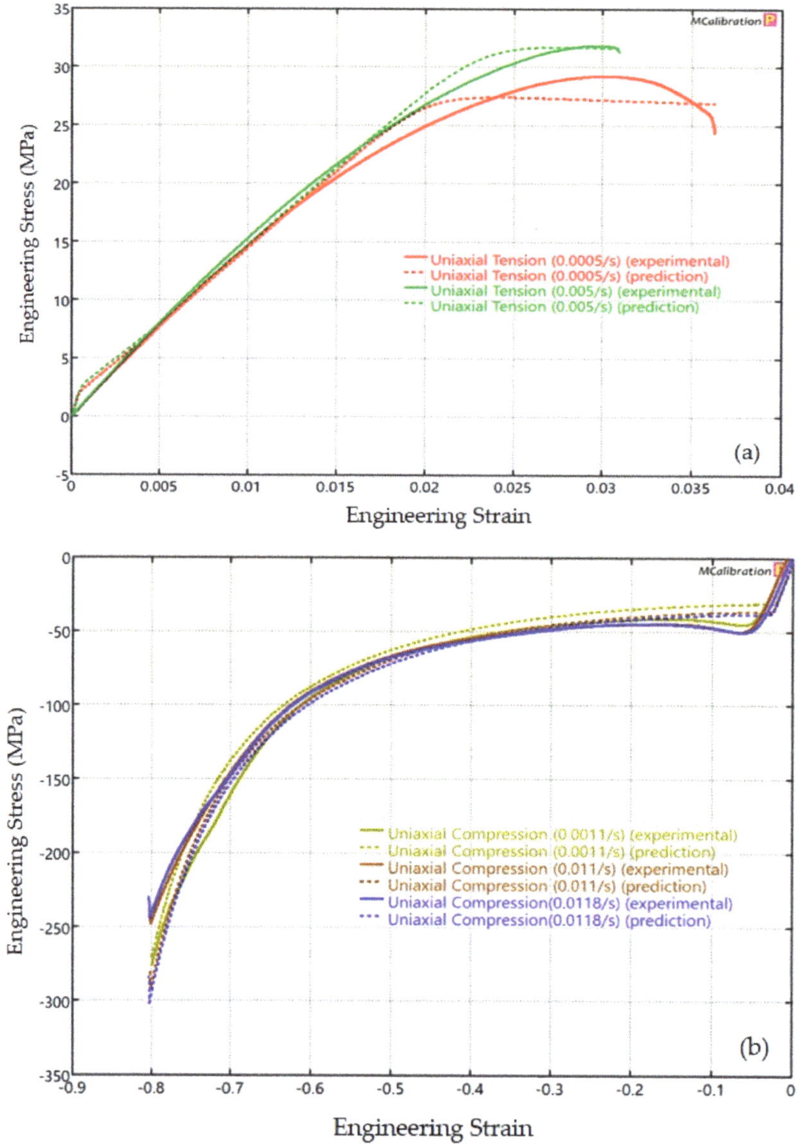

Figure 13. Experimental and TNV predicted results for PETG: (**a**) uniaxial tension; (**b**) uniaxial compression.

Table 5. Summary of R^2 values for different material models.

Material	Material Model		
	BB	TN	TNV
PLA	0.827	0.884	0.886
ABS	0.756	0.508	0.743
PETG	0.974	0.973	0.972

4. Case Study—Upper-Limb Orthosis

The upper-limb hand splint was examined in this study employing two designs, namely solid and porous (square perforated), through FE analysis. The perforated design had 22 levels with 24 perforations at a 15° angle at each level. The square perforation's side measured 4.43 mm. Figure 14a,b depict the solid and perforated hand splint designs, respectively. The durability and dependability of the orthosis during usage are largely dependent on its strength, especially in static orthoses. However, because of the lack of suitable fixtures for their tensile or compressive testing, it is challenging to establish their strength in physical trials. Thus, a new approach was used to provide actual data using standard specimens from universal testing machines, calibrate models for different polymer materials, and then choose the most favorable model for that particular material. The material chosen was employed to assess the mechanical properties of various orthosis designs.

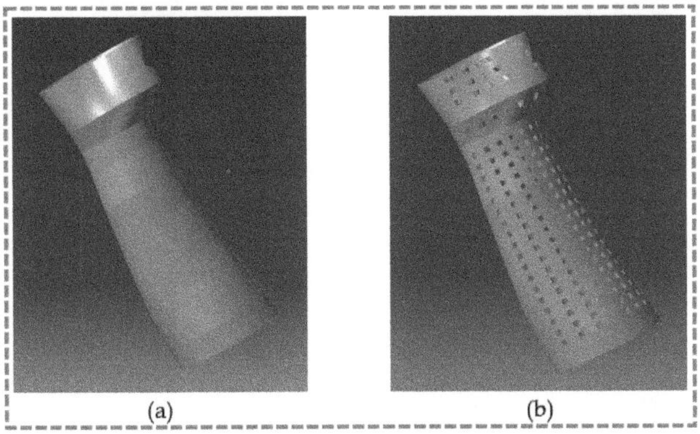

Figure 14. Design for hand splint (**a**) Solid (**b**) Perforated.

FE analysis was performed based on methods proposed by Cazon et al. [59]. The proposed designs comprised of 3-mm-thick solid and perforated (square-shaped) splints. For every design, ABS, PLA, and PETG were analyzed. This thickness was deemed to be suitable given the AM material's superior stiffness. These materials' attributes, which had been incorporated into the FE model (Table 6), were obtained through experiments and found in literature sources.

Table 6. Properties of the 3D printing materials.

Material	Properties		
	Young's Modulus (GPa)—Estimated	**Poisson Ratio**	**Yield Strength (MPa)—Estimated**
PLA	2.28	0.30 [60]	25.98
ABS	1.80	0.35 [61]	20.70
PETG	1.64	0.33 [62]	19.55

Dassault Systems' Abaqus software (Dassault Systems Headquarters, Vélizy-Villacoublay, France) was employed for the FE analysis of orthoses. The Initial Graphics Exchange Specification file format was used to import the designs into Abaqus. In the FE model, the splint exhibited 11,481 node points and 37,225 linear four-node tetrahedral elements (C3D4) (Figure 15).

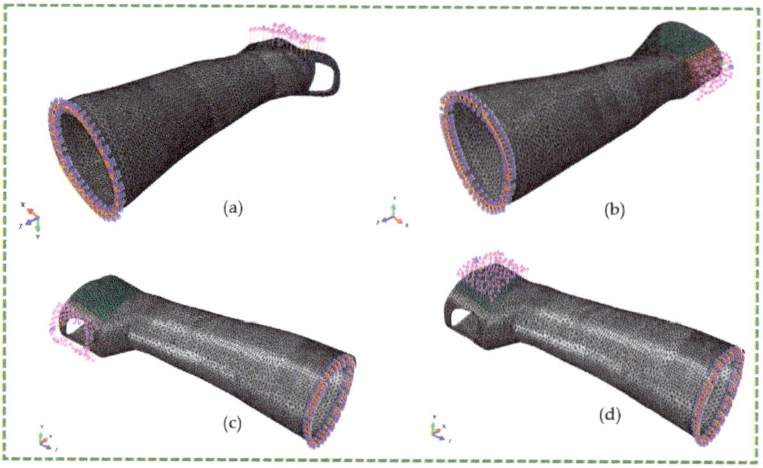

Figure 15. FE mesh, BCs, and load directions for four crucial wrist movements. (**a**) flexion, (**b**) extension, (**c**) radial and (**d**) ulnar.

The predominant biaxial mobility of the wrist at the carpus area results in four wrist movements: radial-ulnar deviation and flexion-extension [63,64]. The wrist's strength required for these movements is usually determined by employing subjects who have completed static (isometric) and dynamic (isokinetic) testing. This study examined the maximum flexion-extension torque and radial-ulnar deviation torque produced by healthy volunteers [65]. Under real-world circumstances, people with stroke would often be unable to apply the complete strength necessary to break the splint. Thus, analyses at 8% of the maximal isometric strength of a healthy individual were performed. The stresses and torques imparted on the splint for these main movements are listed in Table 7.

Table 7. Considered torque and loads applied to the splints.

	Vanwearingen Torque (N m)	Applied 8% Loads (N)	Load Direction
Flexion	14.8	11.8	−Y-axis
Extension	8.4	6.7	Y-axis
Radial deviators	11.4	9.1	−X-axis
Ulnar deviators	9.9	7.9	X-axis

The input force (N) required for simulation was estimated by dividing torque by the distance between the wrist and metacarpophalangeal joints [66]. Assuming that 50% of subjects (without hand-wrist medical conditions) have this measurement, the distance between the wrist and metacarpophalangeal joints was considered to be 0.1 m [67]. These force values were applied to splint surfaces that can carry these forces in real life (Table 6). It was essential to demonstrate a natural interaction between splints and hands/wrists and prevent stress resulting from point forces. The FE mesh, BCs, and load directions for the flexion, extension, radial, and ulnar are illustrated in Figure 15a–d. In addition to the most noticeable surfaces under stresses based on each movement, areas in touch with the thumb were considered for radial movement, whereas the palm section was only assumed for ulnar movement. For simulations, it was hypothesized that the patient's forearm was at rest or fixed and the hand could be free to move but restricted by the splint. The BC could securely connect splint surfaces in the forearm region. For all four movements, the splint's back surfaces were secured in six degrees of freedom. Each motion had a secondary limit on the central surface of the splint to establish a 3-point pressure system and matched the simulated and real-world forearm-splint relationships.

Twenty-four FE analysis simulations were conducted to examine displacements and stresses for the two designs in flexion-extension and radial-ulnar wrist movements. These simulations assessed the effectiveness of various materials and solid and perforated designs for splints. Because minor deformations were assumed in all the simulations, the materials remained in the elastic component of the stress–strain curve. The simulations were run using a static and comprehensive procedure that utilized ABAQUS/STANDARD's implicit solver. These displacements and Von Mises stress were estimated and compared in these analyses. Every simulation was performed on a PC running Windows 10, 64 bits, with an Octa-core 3.7 GHz Xeon processor and 16 GB RAM. About one minute was needed for each simulation.

Analysis was conducted to compare different materials used in hand splint designs. The TNV model was used for PLA and the BB model for ABS and PETG. The values of displacements and Von Mises stress for the 3-mm-thick hand splint obtained from FE simulations for the 8% load are listed in Tables 8 and 9, respectively. Figure 16 provides a graphical representation of Von Mises stress values and average displacements for PLA.

Table 8. Highest displacement (mm) attained in the major axes for 8% load.

Design	Material		Flexion (−Y) (mm)	Extension (Y) (mm)	Radial Deviators (−X) (mm)	Ulnar Deviators (X) (mm)
Solid	PLA	X	−0.25	0.09	−0.78	0.77
		Y	−2.30	1.17	−0.07	0.13
		Z	0.40	0.26	0.29	−0.25
	ABS	X	−0.22	0.08	−0.93	0.93
		Y	−2.68	1.40	−0.05	0.12
		Z	0.47	0.30	0.34	−0.29
	PETG	X	−0.32	0.12	−1.12	1.11
		Y	−3.28	1.68	−0.09	0.17
		Z	0.57	0.37	0.41	−0.36
Perforated (Square Perforations)	PLA	X	−0.20	0.08	−0.95	0.73
		Y	−3.38	1.56	−0.07	0.12
		Z	−0.59	0.33	0.35	−0.23
	ABS	X	−0.23	0.10	−1.18	0.91
		Y	−4.14	1.92	−0.07	0.13
		Z	−0.75	0.40	0.44	−0.29
	PETG	X	−0.28	0.12	−1.38	1.06
		Y	−4.90	2.26	−0.09	0.16
		Z	−0.86	0.48	0.51	−0.34

Table 9. Von Mises stress values (MPa) achieved for 8% load.

Design	Material	Flexors (−Y) (MPa)	Extensors (Y) (MPa)	Radial Deviators (−X) (MPa)	Ulnar Deviators (X) (MPa)
Solid	PLA	9.79	4.68	2.27	2.60
	ABS	5.74	2.71	2.08	2.06
	PETG	8.16	3.61	2.10	2.22
Perforated (Square (Perforations)	PLA	31.66	13.67	8.02	5.70
	ABS	33.05	15.10	9.25	6.55
	PETG	36.61	15.48	8.51	6.03

For average displacements, the PLA hand splint (both designs) was the stiffest material juxtaposed to ABS and PETG. The PLA-based solid hand splint was sturdier and more durable than the ABS-based orthosis, exhibiting approximately 11.52%, 14.61%, 12.50%, and 14.47% lower displacement in flexion, extension, radial, and ulnar directions, respectively. Likewise, the PLA-made perforated hand splint outperformed the ABS-made one, exhibiting around 18.55%, 18.60%, 17.28%, and 17.73% lower displacement in the flexion, extension, radial, and ulnar directions, respectively. The performance of PETG was poor for displacement. For example, the PETG-based solid hand splint displayed 24.69%, 21.91%, 25%, and 21.05% greater displacement than the ABS material in flexion, extension, radial, and ulnar directions, respectively. Compared with a perforated hand splint composed of ABS material, the PETG perforated hand splint demonstrated poorer performance. The PETG material exhibited approximately 17.97%, 18.18%, 18.52, and 17.33 higher displacement than ABS across flexion, extension, radial, and ulnar directions, respectively. Irrespective of these directions, splint regions with the greatest displacement were those with curvature variation, including the metacarpus. It was noted that the observed displacement was in line with the average displacement for all the materials in flexion, extension, radial, and ulnar movements. Moreover, in the perforated hand split, the overall increase in the displacements was observed for the three materials in all three directions. Undoubtedly, the existence of perforations that could serve as stress concentration sites and a smaller material volume could be the cause of the increased deformation in perforated orthosis compared to its solid variant.

Stress peaks were observed in and around the hand's palmer digital region for the solid splint. For all the materials, the peak stress was lower than their yield strength in the solid splint (Table 6). This finding indicates that stroke patients can safely use the 3-mm-thick solid hand splint (made of PLA, ABS, and PETG). The 3-mm-thick perforated hand splint collapsed in flexion because the von Mises stress was greater than the material's yield strength. Thus, square perforated hand splints should not be used unless significant design changes are made. Possible modifications could include thickening the splint, changing the pattern of holes, or decreasing the hole number. Moreover, the splint's curved regions—such as its sides and the region surrounding the thumb—both on palmer and dorsal sides generate greater stress levels.

It is crucial to acknowledge that the quality of experimental data has a substantial impact on the prediction accuracy of prediction models (in this case, the BB, TN, and TNV models) [68–70]. 3D printing has numerous advantages, like the ability to customize, and manufacture intricate parts without the need for tools or fixtures, etc. However, proper printing conditions are needed to get the best results and reap the benefits of 3D printing. Inadequate layer adhesion, inferior dimensional accuracy, inaccurate mechanical characteristics, flawed surface quality, component imperfections, invariable part qualities, fabrication issues, etc., are possible outcomes of improper parameters selection [71–76]. The standard parts produced will be erroneous and the tensile/compression test results obtained afterward will not be reliable if the components are built using the incorrect printing parameters. Consequently, the prediction model's calibration and the determination of material constants for a particular material will be inaccurate when based on the incorrect test results. The FE analysis of the final application (in this case, custom orthosis design) when carried out using these incorrectly calibrated material models, will produce deceptive results. Serious ramifications could arise from the possibility that a weak design could yield positive outcomes or that a well-designed orthosis could result in increased stresses or deformation. This will ultimately lead to increased expenses, material waste, user discomfort, ineffectual use of the orthosis, etc.

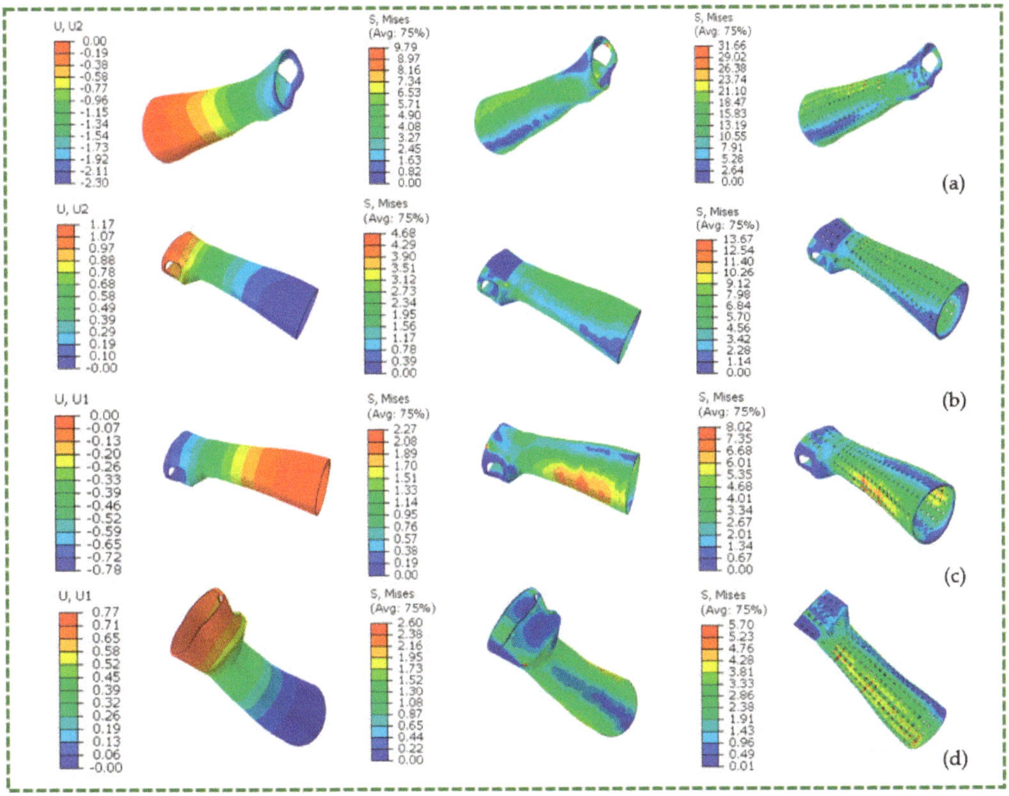

Figure 16. Displacement values (mm) and Von Mises stress values (MPa) for solid and perforated hand splint (**a**) flexion (**b**) extension (**c**) radial (**d**) ulnar movements in the 3-mm-thick PLA splint.

FE analysis and 3D printing contribute significantly to the domains of AM and biomedical engineering, especially in the development of personalized medical devices [77,78]. By using FE analysis, which can simulate different loading scenarios and other BCs, designs can be modified to meet the needs of each particular patient, increasing the medical device's fit, appearance, comfort, and efficiency. Once analysis is completed, only 3D printing will be able to effectively build intricately shaped medical devices that mimic the bodily part. It is also feasible and economical to efficiently iterate and improve the design for user satisfaction based on their input by using FE analysis and 3D printing. This method, which uses 3D printing and FE analysis, also allows for the rapid testing a wide range of materials and designs. Moreover, with the development of customized tools based on FE models, healthcare providers can visualize the performance of different designs and materials which would be helpful in selecting the optimal design according to the patient's specific problem and budget constraints.

Although, this study seeks to replicate the behaviors of a real patient's hand in a splint. However, this research overlooked or oversimplified a number of dynamics because there was not enough information available. It is challenging to identify the appropriate load intensity because wrist joints with stroke are not yet well-studied. Due to this lack of information, it was assumed that strength loss in a wrist impaired by stroke is comparable to that in a wrist affected by arthritis. Future research should incorporate the actual pressure that a stroke patient might consciously apply to a splint in a true deployment scenario. Nonetheless, this study used real experimental data to obtain the mechanical parameters needed for FE simulations. Because these data were gathered in actual settings, they may fairly depict real-world circumstances. Other materials with higher elasticity and solutions to thicken the splint and incorporate unique lattice architecture should be examined. The effects of other loads and BCs on FE simulations should also be considered.

5. Conclusions

To assess a material's suitability for any given application, its mechanical behavior, especially that of polymeric materials, should be examined. Standard specimens must be used for tensile and compression testing when using a universal testing machine, which makes physical testing resource-intensive. To perform mechanical testing for orthosis, a particular fixture must be designed; this is an expensive and time-consuming process. Thus, FE analysis is utilized to identify the mechanical behaviors of an upper-limb hand orthosis with a suitable material model.

Three constitutive mathematical models—BB, TN, and TNV for PLA, ABS, and PETG—were investigated. The optimal model for a material was identified based on R^2, having first been calibrated using data from physical tests. Appropriate material parameters, which were established using various optimization techniques, were necessary for the calibration of material models. This investigation revealed that TNV was the most appropriate model for PLA, whereas BB was the best model for ABS and PETG. Subsequently, the hand orthosis based on two designs and three materials was examined, each using its material model. The PLA hand splint (both designs) had the lowest average displacement, followed by ABS and PETG. Additionally, compared with solid hand splints, perforated splints exhibited overall increase in displacements in flexion, extension, radial, and ulnar directions for all three materials. In the solid splint, the peak stress was lower than yield strength. These findings indicate that stroke patients can safely use the 3-mm-thick solid hand splint. However, the 3-mm-thick perforated hand splint failed in flexion because the von Mises stress was greater than the material's yield strength. Further design exploration would be advantageous, even though this would require volunteers and financial resources. The cost-effectiveness and utility of diverse materials for 3D-printed splints will eventually need to be demonstrated through human research.

This work has used FE simulation to investigate the mechanical performance of the orthosis designs under different loading scenarios. Despite the fact that FE modeling is an invaluable tool, it is unable to accurately represent the complex and dynamic mechanical stresses that arise in real-world applications [79]. Thus, real-world clinical evaluation is required in addition to FE simulation to validate the effectiveness of orthoses composed of different materials. Similarly, FE simulation cannot be used to assess comfort, which is a subjective performance indicator. Characterization of comfort necessitates more research, such as clinical tests, user feedback, ergonomic assessment, etc., since it depends on fit, skin response, aesthetics, and other factors. Although FE models can provide valuable insights into the mechanical characteristics of orthosis designs. Yet, real-world clinical efficacy is still required, especially with regard to comfort, durability, and therapeutic efficacy. It is undoubtedly possible to achieve enhanced upper limb rehabilitation for increased patient satisfaction through optimal orthosis design by combining FE simulation with clinical assessments. The clinical assessment is not included in this work due to resource constraints; however, clinical trials will be conducted in the future to further improve the design and enhance the reliability of the findings. Furthermore, the future direction regarding material modeling is to develop more accurate and robust material models that can handle multiple scenarios that could occur during the use of 3D-printed orthotic devices. Another area is to develop and calibrate material models with temperature dependency for the simulation of distortion and residual stress development during 3D printing. FE models for simulating 3D printing of polymers for orthotic devices could save a lot of time and cost usually associated with trial experimental runs.

Author Contributions: Conceptualization, U.U. and S.H.M.; methodology, U.U., S.H.M. and K.M.; software, U.U. and S.H.M.; validation, K.M. and H.A.; resources, U.U., S.H.M. and H.A.; data curation,

H.A.; writing—original draft preparation, S.H.M. and K.M.; writing—review and editing, U.U., S.H.M. and H.A.; supervision, H.A.; project administration, U.U. and S.H.M.; funding acquisition, U.U., S.H.M. and H.A. All authors have read and agreed to the published version of the manuscript.

Funding: This research was funded by the King Salman center For Disability Research through Research Group no KSRG-2022-060.

Institutional Review Board Statement: Not applicable.

Informed Consent Statement: Not applicable.

Data Availability Statement: The data presented in this study are available in the article.

Acknowledgments: The authors extend their appreciation to the King Salman center For Disability Research for funding this work through Research Group no KSRG-2022-060.

Conflicts of Interest: The authors declare no conflict of interest.

Nomenclature

Symbol	Description
μ	Shear modulus of network A
λ_L	Locking Stretch
κ	Bulk Modulus
s	Relative stiffness of network B
x_i	Strain adjustment factor
c	Strain exponential
τ	Flow Resistance
m	Strain exponential
τ_{cut}	Normalized cut-off stress for flow
λ_B^{e*}	Chain stretch in the elastic portion of Network B
σ	Cauchy stress
L_B	Velocity gradient on network B
$\tilde{D}_B^v$	Viscous deformation rate of Network B
$\hat{\tau_{cut}}$	Cutoff stress
μ_A	Shear modulus of network A
λ_L	Locking stretch
κ	Bulk modulus
$\hat{\tau}_A$	Flow resistance of network A
a	Pressure dependence of flow
$m_A = m_B$	Stress exponential of network A
μ_{Bi}	Initial shear modulus of network B
μ_{Bf}	Final shear modulus of network B
β	Evolution rate of μ_B
$\hat{\tau}_B$	Flow resistance of network B
μ_C	Shear modulus of network C
q	Relative contribution of I_2 of network C
θ	Current temperature
θ_0	Reference temperature
$\hat{\theta}$	Stiffness's temperature dependence
$\mathcal{L}^{-1}(x)$	Inverse Langevin function
κ	Bulk modulus
$\dot{Y}_A$	Viscoplastic flow rate
p_A	Hydrostatic pressure
$R(x)$	Ramp function

References

1. Zhang, P.; Li, J.; Ashour, E.A.; Chung, S.; Wang, H.; Vemula, S.K.; Repka, M.A. Development of Multiple Structured Extended Release Tablets via Hot Melt Extrusion and Dual-Nozzle Fused Deposition Modeling 3D Printing. *Int. J. Pharm.* **2024**, *653*, 123905. [CrossRef] [PubMed]
2. Cano-Vicent, A.; Tambuwala, M.M.; Hassan, S.S.; Barh, D.; Aljabali, A.A.A.; Birkett, M.; Arjunan, A.; Serrano-Aroca, Á. Fused Deposition Modelling: Current Status, Methodology, Applications and Future Prospects. *Addit. Manuf.* **2021**, *47*, 102378. [CrossRef]
3. Jandyal, A.; Chaturvedi, I.; Wazir, I.; Raina, A.; Ul Haq, M.I. 3D Printing—A Review of Processes, Materials and Applications in Industry 4.0. *Sustain. Oper. Comput.* **2022**, *3*, 33–42. [CrossRef]
4. Raj, R.; Dixit, A.R.; Łukaszewski, K.; Wichniarek, R.; Rybarczyk, J.; Kuczko, W.; Górski, F. Numerical and Experimental Mechanical Analysis of Additively Manufactured Ankle–Foot Orthoses. *Materials* **2022**, *15*, 6130. [CrossRef]
5. Stopa, R.; Komarnicki, P.; Kuta, Ł.; Szyjewicz, D.; Słupska, M. Modeling with the Finite Element Method the Influence of Shaped Elements of Loading Components on the Surface Pressure Distribution of Carrot Roots. *Comput. Electron. Agric.* **2019**, *167*, 105046. [CrossRef]
6. Kang, K. Mechanical Properties of Metal and Polymeric Viscoelastic Materials and Their Applications. *J. Phys. Conf. Ser.* **2022**, *2152*, 012055. [CrossRef]
7. Durga Rajesh, K.V.; Ganesh, N.; Yaswanth Kalyan Reddy, S.; Mishra, H.; Teja Naidu, T.M.V.P.S. Experimental Research on the Mechanical Characteristics of Fused Deposition Modelled ABS, PLA and PETG Specimens Printed in 3D. *Mater. Today Proc.* **2023**. [CrossRef]
8. Sepahi, M.T.; Abusalma, H.; Jovanovic, V.; Eisazadeh, H. Mechanical Properties of 3D-Printed Parts Made of Polyethylene Terephthalate Glycol. *J. Mater. Eng. Perform.* **2021**, *30*, 6851–6861. [CrossRef]
9. Siviour, C.R.; Jordan, J.L. High Strain Rate Mechanics of Polymers: A Review. *J. Dyn. Behav. Mater.* **2016**, *2*, 15–32. [CrossRef]
10. Garcia-Gonzalez, D.; Garzon-Hernandez, S.; Barba, D.; Arias, A. Thermo-Mechanics of Polymers at Extreme and Failure Conditions: Influence of Strain Rate and Temperature. In *Handbook of Damage Mechanics: Nano to Macro Scale for Materials and Structures*; Voyiadjis, G.Z., Ed.; Springer: New York, NY, USA, 2020; pp. 1–28. ISBN 978-1-4614-8968-9.
11. Rohbeck, N.; Ramachandramoorthy, R.; Casari, D.; Schürch, P.; Edwards, T.E.J.; Schilinsky, L.; Philippe, L.; Schwiedrzik, J.; Michler, J. Effect of High Strain Rates and Temperature on the Micromechanical Properties of 3D-Printed Polymer Structures Made by Two-Photon Lithography. *Mater. Des.* **2020**, *195*, 108977. [CrossRef]
12. Esmaeili, H.; Rizvi, R. An Accelerated Strategy to Characterize Mechanical Properties of Polymer Composites Using the Ensemble Learning Approach. *Comput. Mater. Sci.* **2023**, *229*, 112432. [CrossRef]
13. Buljak, V.; Ranzi, G. Chapter 2—Calibration of Constitutive Models. In *Constitutive Modeling of Engineering Materials*; Buljak, V., Ranzi, G., Eds.; Academic Press: Cambridge, MA, USA, 2021; pp. 43–53. ISBN 978-0-12-814696-5.
14. Zhang, X.; Chen, Z.; Liu, Y. Chapter 6—Constitutive Models. In *The Material Point Method*; Zhang, X., Chen, Z., Liu, Y., Eds.; Academic Press: Oxford, UK, 2017; pp. 175–219. ISBN 978-0-12-407716-4.
15. Hughes, T.J.R. Numerical Implementation of Constitutive Models: Rate-Independent Deviatoric Plasticity. In *Theoretical Foundation for Large-Scale Computations for Nonlinear Material Behavior: Proceedings of the Workshop on the Theoretical Foundation for Large-Scale Computations of Nonlinear Material Behavior, Evanston, IL, USA, 24–26 October 1983*; Nemat-Nasser, S., Asaro, R.J., Hegemier, G.A., Eds.; Mechanics of Elastic and Inelastic Solids 6; Springer: Dordrecht, The Netherlands, 1984; pp. 29–63. ISBN 978-94-009-6213-2.
16. Ielapi, A.; Forward, M.; De Beule, M. Computational and Experimental Evaluation of the Mechanical Properties of Ankle Foot Orthoses: A Literature Review. *Prosthet. Orthot. Int.* **2019**, *43*, 339–348. [CrossRef] [PubMed]
17. Stefańczak, D.; Gajewski, J.; Rogala, M. Application of the Finite Element Method to the Design of an Ankle Orthosis. *J. Phys. Conf. Ser.* **2021**, *2130*, 012013. [CrossRef]
18. Ali, M.H.; Smagulov, Z.; Otepbergenov, T. Finite Element Analysis of the CFRP-Based 3D Printed Ankle-Foot Orthosis. *Procedia Comput. Sci.* **2021**, *179*, 55–62. [CrossRef]
19. Bergström, J.S.; Boyce, M.C. Constitutive Modeling of the Large Strain Time-Dependent Behavior of Elastomers. *J. Mech. Phys. Solids* **1998**, *46*, 931–954. [CrossRef]
20. Bergström, J.S.; Boyce, M.C. Large Strain Time-Dependent Behavior of Filled Elastomers. *Mech. Mater.* **2000**, *32*, 627–644. [CrossRef]
21. Almomani, A.; Deveci, S.; Mourad, A.-H.I.; Barsoum, I. Constitutive Model Calibration for the Thermal Viscoelastic-Viscoplastic Behavior of High Density Polyethylene under Monotonic and Cyclic Loading. *Polym. Test.* **2023**, *118*, 107911. [CrossRef]
22. Jorgen. PolyUMod Three Network (TN) Model. 2020. Available online: https://polymerfem.com/three-network-model/ (accessed on 31 March 2024).
23. Kumar, S.; Singh, S.S.; Rozycki, P. Numerical Simulation of Strain-Softening Behavior of Glass-Filled Polymer Composites: Comparison of Two-Dimensional and Three-Dimensional Analyses Using Arruda-Boyce and Three-Network Viscoplastic Models. *Mech. Mater.* **2022**, *175*, 104481. [CrossRef]
24. Kumar, M.; Ravi, K.; Singh, S.S. Predicting the Double-Yield Phenomenon in Low-Density Polyethylene Film Using Three-Network Viscoplastic Model. *Mech. Mater.* **2023**, *184*, 104736. [CrossRef]
25. Raise 3D. *Raise3D Premium PLA Technical Data Sheet*; Raise 3D: Lake Forest, CA, USA, 2019.
26. Raise 3D. *Raise3D Premium ABS Technical Data Sheet*; Raise 3D: Lake Forest, CA, USA, 2023.

27. Raise 3D. *Raise3D Premium PETG Technical Data Sheet*; Raise 3D: Lake Forest, CA, USA, 2017.
28. *ASTM D638−22*; Standard Test Method for Tensile Properties of Plastics. ASTM International: West Conshohocken, PA, USA, 2022.
29. *ASTM D695−23*; Standard Test Method for Compressive Properties of Rigid Plastics. ASTM International: West Conshohocken, PA, USA, 2023.
30. Dobos, J.; Hanon, M.M.; Oldal, I. Effect of Infill Density and Pattern on the Specific Load Capacity of FDM 3D-Printed PLA Multi-Layer Sandwich. *J. Polym. Eng.* **2022**, *42*, 118–128. [CrossRef]
31. Zhang, G.; Luo, J.; Zheng, G.; Bai, Z.; Cao, L.; Mao, H. Is the 0.2%-Strain-Offset Approach Appropriate for Calculating the Yield Stress of Cortical Bone? *Ann. Biomed. Eng.* **2021**, *49*, 1747–1760. [CrossRef] [PubMed]
32. Selvamani, S.K.; Rajan, K.; Samykano, M.; Kumar, R.R.; Kadirgama, K.; Mohan, R.V. Investigation of Tensile Properties of PLA–Brass Composite Using FDM. *Prog. Addit. Manuf.* **2022**, *7*, 839–851. [CrossRef]
33. Jorgen. Bergstrom-Boyce (BB) Model. 2020. Available online: https://polymerfem.com/bergstrom-boyce-bb-model/ (accessed on 31 March 2024).
34. Bergström, J.S.; Kurtz, S.M.; Rimnac, C.M.; Edidin, A.A. Constitutive Modeling of Ultra-High Molecular Weight Polyethylene under Large-Deformation and Cyclic Loading Conditions. *Biomaterials* **2002**, *23*, 2329–2343. [CrossRef] [PubMed]
35. Bergström, J.S.; Rimnac, C.M.; Kurtz, S.M. Prediction of Multiaxial Mechanical Behavior for Conventional and Highly Crosslinked UHMWPE Using a Hybrid Constitutive Model. *Biomaterials* **2003**, *24*, 1365–1380. [CrossRef] [PubMed]
36. Bergstrom, J.S.; Bischoff, J.E. An Advanced Thermomechanical Constitutive Model for UHMWPE. *Int. J. Struct. Chang. Solids* **2010**, *2*, 31–39.
37. Arruda, E.M.; Boyce, M.C. A Three-Dimensional Constitutive Model for the Large Stretch Behavior of Rubber Elastic Materials. *J. Mech. Phys. Solids* **1993**, *41*, 389–412. [CrossRef]
38. Boyce, M.C.; Weber, G.G.; Parks, D.M. On the Kinematics of Finite Strain Plasticity. *J. Mech. Phys. Solids* **1989**, *37*, 647–665. [CrossRef]
39. Jorgen. PolyUMod TNV Model. 2020. Available online: https://polymerfem.com/tnv-model/ (accessed on 31 March 2024).
40. Wang, Y.; Gu, Y.; Liu, J. A Domain-Decomposition Generalized Finite Difference Method for Stress Analysis in Three-Dimensional Composite Materials. *Appl. Math. Lett.* **2020**, *104*, 106226. [CrossRef]
41. Kabir, H.; Aghdam, M.M. A Generalized 2D Bézier-Based Solution for Stress Analysis of Notched Epoxy Resin Plates Reinforced with Graphene Nanoplatelets. *Thin-Walled Struct.* **2021**, *169*, 108484. [CrossRef]
42. Bert, C.W.; Malik, M. Differential Quadrature: A Powerful New Technique for Analysis of Composite Structures. *Compos. Struct.* **1997**, *39*, 179–189. [CrossRef]
43. PolymerFEM. MCalibration Documentation. 2021. Available online: https://polymerfem.com/introduction-to-mcalibration/ (accessed on 31 March 2024).
44. Swallowe, G.M. Strain Rate Effects. In *Mechanical Properties and Testing of Polymers: An A–Z Reference*; Swallowe, G.M., Ed.; Springer: Dordrecht, The Netherlands, 1999; pp. 214–218. ISBN 978-94-015-9231-4.
45. Richeton, J.; Ahzi, S.; Vecchio, K.S.; Jiang, F.C.; Adharapurapu, R.R. Influence of Temperature and Strain Rate on the Mechanical Behavior of Three Amorphous Polymers: Characterization and Modeling of the Compressive Yield Stress. *Int. J. Solids Struct.* **2006**, *43*, 2318–2335. [CrossRef]
46. Wang, Q.; Wang, J.; Wang, A.; Zhou, C.; Hu, J.; Pan, F. Effect of Strain Rate and Temperature on the Tensile Properties of Long Glass Fiber-Reinforced Polypropylene Composites. *Polymers* **2023**, *15*, 3260. [CrossRef]
47. Hameed, M.G.; Kadhim, B.J.; Jura, J.; Mehdi, A.H.; Al-Mutairi, N.H. Strain rate effect on the mechanical properties of thermoplastic polyolefin. *Constr. Optim. Energy Potential Bud. Zoptymalizowanym Potencjale Energetycznym* **2023**, *12*, 209–216. [CrossRef]
48. Amjadi, M.; Fatemi, A. Tensile Behavior of High-Density Polyethylene Including the Effects of Processing Technique, Thickness, Temperature, and Strain Rate. *Polymers* **2020**, *12*, 1857. [CrossRef]
49. Cui, J.; Wang, S.; Wang, S.; Li, G.; Wang, P.; Liang, C. The Effects of Strain Rates on Mechanical Properties and Failure Behavior of Long Glass Fiber Reinforced Thermoplastic Composites. *Polymers* **2019**, *11*, 2019. [CrossRef]
50. Gautam, V.K.; Pande, C.B.; Moharir, K.N.; Varade, A.M.; Rane, N.L.; Egbueri, J.C.; Alshehri, F. Prediction of Sodium Hazard of Irrigation Purpose Using Artificial Neural Network Modelling. *Sustainability* **2023**, *15*, 7593. [CrossRef]
51. Hagerty, M.R.; Srinivasan, V. Comparing the Predictive Powers of Alternative Multiple Regression Models. *Psychometrika* **1991**, *56*, 77–85. [CrossRef]
52. Panda, S.S.; Ames, D.P.; Panigrahi, S. Application of Vegetation Indices for Agricultural Crop Yield Prediction Using Neural Network Techniques. *Remote Sens.* **2010**, *2*, 673–696. [CrossRef]
53. Vasilyeva, N.; Fedorova, E.; Kolesnikov, A. Big Data as a Tool for Building a Predictive Model of Mill Roll Wear. *Symmetry* **2021**, *13*, 859. [CrossRef]
54. Krause, P.; Boyle, D.P.; Bäse, F. Comparison of Different Efficiency Criteria for Hydrological Model Assessment. *Adv. Geosci.* **2005**, *5*, 89–97. [CrossRef]
55. Tsai, J.-T.; Akin, S.; Bahr, D.F.; Jun, M.B.-G. A Predictive Modeling Approach for Cold Spray Metallization on Polymers. *Surf. Coat. Technol.* **2024**, *483*, 130711. [CrossRef]
56. Cumbicus, W.E.; Estrems, M.; Arizmendi, M.; Jiménez, A. Analysis of Self-Tapping Screw Joints in Fibre Glass Reinforced PEI Polymer Used in the Automotive Industry. *Int. J. Adv. Manuf. Technol.* **2023**, *126*, 5163–5178. [CrossRef]

57. Dalli, D.; Fanton, L.; Mallia, B.; Wismayer, P.S.; Buhagiar, J.; Mollicone, P. Polyethylene Wear Simulation Models Applied to a Prosthetic Hip Joint Based on Unidirectional Articulations. *J. Mech. Behav. Biomed. Mater.* **2023**, *142*, 105882. [CrossRef] [PubMed]
58. Schwerz, R.; Roellig, M. Non-Linear Viscoelastic Material Models of Polymers for Electronics Simulation-Measurement, Modelling, Validation. In Proceedings of the 2024 25th International Conference on Thermal, Mechanical and Multi-Physics Simulation and Experiments in Microelectronics and Microsystems (EuroSimE), Catania, Italy, 7–10 April 2024; pp. 1–7.
59. Cazon, A.; Kelly, S.; Paterson, A.M.; Bibb, R.J.; Campbell, R.I. Analysis and Comparison of Wrist Splint Designs Using the Finite Element Method: Multi-Material Three-Dimensional Printing Compared to Typical Existing Practice with Thermoplastics. *Proc. Inst. Mech. Eng. H* **2017**, *231*, 881–897. [CrossRef]
60. Oksiuta, Z.; Jalbrzykowski, M.; Mystkowska, J.; Romanczuk, E.; Osiecki, T. Mechanical and Thermal Properties of Polylactide (PLA) Composites Modified with Mg, Fe, and Polyethylene (PE) Additives. *Polymers* **2020**, *12*, 2939. [CrossRef]
61. Yavuz, G.A.; Kiral, B.G.; Katre, S.; Atilla, D. Effects of Topology and Material on Mechanical Properties of Structures Produced by the Additive Manufacturing Method. *Deufmd* **2021**, *23*, 755–765. [CrossRef]
62. Huysamen, H.W.; Kinnear, W.A.; Fonternel, T.E.; Turton, E.; Yadroitsava, I.; Yadroitsev, I. 3D Printed Laryngoscope for Endotracheal Intubation. *S. Afr. J. Ind. Eng.* **2020**, *31*, 209–217. [CrossRef]
63. Youm, Y.; McMurthy, R.Y.; Flatt, A.E.; Gillespie, T.E. Kinematics of the Wrist. I. An Experimental Study of Radial-Ulnar Deviation and Flexion-Extension. *J. Bone Jt. Surg. Am.* **1978**, *60*, 423–431. [CrossRef]
64. Volz, R.G.; Lieb, M.; Benjamin, J. Biomechanics of the Wrist. *Clin. Orthop. Relat. Res.* **1980**, *149*, 112–117. [CrossRef]
65. Vanswearingen, J. Measuring Wrist Muscle Strength. *J. Orthop. Sports Phys. Ther.* **1983**, *4*, 217–228. [CrossRef]
66. Glinsky, J.; Harvey, L.; Korten, M.; Drury, C.; Chee, S.; Gandevia, S.C. Short-Term Progressive Resistance Exercise May Not Be Effective at Increasing Wrist Strength in People with Tetraplegia: A Randomised Controlled Trial. *Aust. J. Physiother.* **2008**, *54*, 103–108. [CrossRef] [PubMed]
67. Garrett, J.W. The Adult Human Hand: Some Anthropometric and Biomechanical Considerations. *Hum. Factors* **1971**, *13*, 117–131. [CrossRef] [PubMed]
68. Korkmaz, M.E.; Gupta, M.K.; Kuntoğlu, M.; Patange, A.D.; Ross, N.S.; Yılmaz, H.; Chauhan, S.; Vashishtha, G. Prediction and Classification of Tool Wear and Its State in Sustainable Machining of Bohler Steel with Different Machine Learning Models. *Measurement* **2023**, *223*, 113825. [CrossRef]
69. Yang, J.; Jia, L.; Guo, Z.; Shen, Y.; Li, X.; Mou, Z.; Yu, K.; Lin, J.C.-W. Prediction and Control of Water Quality in Recirculating Aquaculture System Based on Hybrid Neural Network. *Eng. Appl. Artif. Intell.* **2023**, *121*, 106002. [CrossRef]
70. Tan, S.; Yang, J.; Ding, H. A Prediction and Compensation Method of Robot Tracking Error Considering Pose-Dependent Load Decomposition. *Robot. Comput.-Integr. Manuf.* **2023**, *80*, 102476. [CrossRef]
71. Baraheni, M.; Shabgard, M.R.; Tabatabaee, A.M. Effects of FDM 3D Printing Parameters on PLA Biomaterial Components Dimensional Accuracy and Surface Quality. *Proc. Inst. Mech. Eng. Part C J. Mech. Eng. Sci.* **2023**, 09544062231202142. [CrossRef]
72. Dhakal, N.; Wang, X.; Espejo, C.; Morina, A.; Emami, N. Impact of Processing Defects on Microstructure, Surface Quality, and Tribological Performance in 3D Printed Polymers. *J. Mater. Res. Technol.* **2023**, *23*, 1252–1272. [CrossRef]
73. Kechagias, J.; Zaoutsos, S. Effects of 3D-Printing Processing Parameters on FFF Parts' Porosity: Outlook and Trends. *Mater. Manuf. Process.* **2024**, *39*, 804–814. [CrossRef]
74. Pulipaka, A.; Gide, K.M.; Beheshti, A.; Bagheri, Z.S. Effect of 3D Printing Process Parameters on Surface and Mechanical Properties of FFF-Printed PEEK. *J. Manuf. Process.* **2023**, *85*, 368–386. [CrossRef]
75. Mwania, F.M.; Maringa, M.; Nsengimana, J. Investigating the Effect of Process Parameters on the Degree of Fusion of Two Adjacent Tracks Produced through Fused Deposition Modelling of Acrylonitrile Butadiene Styrene. *Polym. Test.* **2023**, *121*, 107981. [CrossRef]
76. Clarke, T.; Hosseini, A. Effects of Print Parameters on Tensile Characteristics of Additively Manufactured Polyethylene Terephthalate-Glycol (PETG). *Int. J. Adv. Manuf. Technol.* **2023**, *125*, 4953–4974. [CrossRef]
77. Pathak, K.; Saikia, R.; Das, A.; Das, D.; Islam, M.A.; Pramanik, P.; Parasar, A.; Borthakur, P.P.; Sarmah, P.; Saikia, M.; et al. 3D Printing in Biomedicine: Advancing Personalized Care through Additive Manufacturing. *Explor. Med.* **2023**, *4*, 1135–1167. [CrossRef]
78. Rossetos, I.; Gantes, C.J.; Kazakis, G.; Voulgaris, S.; Galanis, D.; Pliarchopoulou, F.; Soultanis, K.; Lagaros, N.D. Numerical Modeling and Nonlinear Finite Element Analysis of Conventional and 3D-Printed Spinal Braces. *Appl. Sci.* **2024**, *14*, 1735. [CrossRef]
79. Ereiz, S.; Duvnjak, I.; Fernando Jiménez-Alonso, J. Review of Finite Element Model Updating Methods for Structural Applications. *Structures* **2022**, *41*, 684–723. [CrossRef]

polymers

MDPI

Article

Modeling of the Time-Dependent H_2 Emission and Equilibrium Time in H_2-Enriched Polymers with Cylindrical, Spherical and Sheet Shapes and Comparisons with Experimental Investigations

Jae Kap Jung [1], Ji Hun Lee [1,2], Jae Yeong Park [1,*] and Sang Koo Jeon [1]

[1] Hydrogen Energy Group, Strategic Technology Research Institute, Korea Research Institute of Standards and Science, Daejeon 34113, Republic of Korea; jkjung@kriss.re.kr (J.K.J.); ljh93@kriss.re.kr (J.H.L.); sangku39@kriss.re.kr (S.K.J.)

[2] Department of Measurement Science, University of Science and Technology, 217 Gajeong-ro, Yuseong-gu, Daejeon 34113, Republic of Korea

* Correspondence: j.park@kriss.re.kr; Tel.: +82-42-868-5387

Abstract: Time-dependent emitted H_2 content modeling via a reliable diffusion analysis program was performed for H_2-enriched polymers under high pressure. Here, the emitted hydrogen concentration versus elapsed time was obtained at different diffusivities and volume dimensions for cylinder-, sphere- and sheet-shaped specimens. The desorption equilibrium time, defined as the time when the H_2 emission content is nearly saturated, was an essential factor for determining the periodic cyclic testing and high-pressure H_2 exposure effect. The equilibrium time in the desorption process was modeled. The equilibrium time revealed an exponential growth behavior with respect to the squared thickness and the squared diameter of the cylinder--shaped specimen, while it was proportional to the squared diameter for the sphere-shaped specimen and to the squared thickness for the sheet-shaped specimen. Linear relationships between the reciprocal equilibrium time and diffusivity were found for all shaped polymers. The modeling results were confirmed by analysis of the solutions using Fick's second diffusion law and were consistent with the experimental investigations. Numerical modeling provides a useful tool for predicting the time-dependent emitted H_2 behavior and desorption equilibrium time. With a known diffusivity, a complicated time-dependent emitted H_2 behavior with a multi-exponential form of an infinite series could also be predicted for the three shaped samples using a diffusion analysis program.

Keywords: enriched polymers; numerical modeling; hydrogen desorption equilibrium time; hydrogen uptake; diffusion; diffusion analysis program

1. Introduction

Gas sorption is a main physical process in polymer media, and desorption is the reversal process controlled by Fick's diffusion laws. H_2 molecule sorption and desorption are the predominant parameters used to determine the solubility, diffusivity and permeation of gas-sealed polymer in hydrogen infrastructure [1–6]. Moreover, the fracture phenomenon caused by H_2 embrittlement in O-ring seals is related to the residence time of the penetrated H_2 molecules in the polymer network during the sorption and desorption processes. Additionally, the H_2 transport characteristics are associated with the equilibrium features related to the diffusion mechanism of both processes under high-pressure environments [7,8].

The H_2 equilibrium time is described as the saturated attainment of nearly the maximum H_2 content in a polymer through the sorption and desorption processes and is affected predominantly by the sample shape, dimension and diffusion coefficient. Investigations of the saturated equilibrium time after high-pressure exposure in hydrogen permeation are essential for designing durable O-rings, reducing operating costs, gaining insights into adsorption and, finally, determining the appropriate exposure time of H_2 under high pressure through the cycling tests [9,10].

On the other hand, the H_2 diffusivity and desorbed (absorbed) content in gas-enriched polymers are mainly dependent on the filler spices and content [11–15]. For the measurement of the permeation parameters after hydrogen exposure in the polymers with different fillers, the same exposure conditions are needed. The same exposure conditions imply the equilibrium time in sorption and desorption processes. Specifically, the measurements of the hydrogen exposure effect need to be performed after the H_2 sorption equilibrium time is reached. Thus, we should determine the equilibrium time of the hydrogen sorption and desorption before starting the measurement. In a previous investigation, the equilibrium times of sorption and desorption were found to be the same, with a margin of nearly 10%. The polymer specimens normally have different H_2 diffusivity ranging from 10^{-9} m^2/s to 10^{-11} m^2/s.

In this study, we modeled only the time dependence of the emitted H_2 content and the equilibrium time in the desorption process for the cylinder-, sphere- and sheet-shaped polymer samples with different thicknesses, diameters and diffusivities of corresponding ranges without measurements. The desorption equilibrium time is important for determining the high-pressure exposure conditions in the periodic cycling test of rubbers and designing O-ring materials for high-pressure hydrogen gas sealing devices.

The time-dependent behavior of the emitted H_2 content and its equilibrium time were evaluated by performing numerical modeling based on the solutions from Fick's second law; this was conducted by employing a validated diffusion analysis program [16–19]. In particular, the relationships of the equilibrium time with respect to the specimen shape, dimension and diffusivity are investigated for the three shaped specimens, which are well consistent with the experimental results. Consequently, this research provides a useful tool for determining the equilibrium time of H_2 and the time evolution of desorption patterns after specimens are exposed to high pressure with different diffusivities. For the case of a known diffusivity, a complicated time-dependent emitted H_2 behavior with a multi-exponential form of an infinite series can be predicted for shaped samples with any dimension by employing a diffusion analysis program.

2. Modeling Background

Modeling Background for Determining the Sorption Equilibrium Time

In the presence of gas concentration gradients, H_2 is absorbed into a rubber specimen and diffuses across the specimen to permeate to the opposite side. The diffusion flux (J) is the flow of the specimen through a unit cross-sectional area in a unit time and is defined as Equation (1) [20–22].

$$J = -D\frac{dC}{dx} \tag{1}$$

where D is the H_2 diffusivity and C is the H_2 concentration. Fick's first law describing steady-state diffusion relates the diffusion flux to the change in the H_2 concentration.

The dissolution of H_2 into a specimen is described by Henry's law, $c = Sp$ [23], where S is the solubility and p is the pressure. Thus, Equation (1) can be expressed as follows:

$$J = -DS\frac{dp}{dx} = -P\frac{dp}{dx} \tag{2}$$

where P is the permeability expressed as $P = DS$.

Under unsteady conditions, the concentration of diffusing H_2 changes over time and is described by Fick's second law of diffusion as follows [21]:

$$\frac{\partial C}{\partial t} = \frac{\partial}{\partial x}\left(D\frac{\partial C}{\partial x}\right) \tag{3}$$

Assuming that the H_2 desorption is governed by a diffusion process, the emitted H_2 concentration $C_E(t)$ in the desorption process is expressed as Equation (4) [24,25]:

$$\frac{C_E(t)}{C_\infty} = 1 - \frac{32}{\pi^2} \times \left[\sum_{n=0}^{\infty} \frac{exp\left\{ \frac{-(2n+1)^2\pi^2 Dt}{l^2} \right\}}{(2n+1)^2} \right] \times \left[\sum_{n=1}^{\infty} \frac{exp\left\{ -\frac{D\beta_n^2 t}{\rho^2} \right\}}{\beta_n^2} \right] \quad (4)$$

Equation (4) shows the solutions to Fick's second diffusion law for a cylinder-shaped specimen, where the boundary condition for a constantly uniform H_2 concentration is initially maintained and the cylindrical surfaces are kept at a constant concentration. In Equation (4), l and ρ are the thickness and radius, respectively, of the cylindrical rubber sample, and β_n is the root of the zero-order Bessel function. In Equation (4), C_∞ is the saturated emitted hydrogen mass concentration at an infinitely long time, i.e., hydrogen uptake in the desorption process. D is the desorption diffusion coefficient.

The equilibrium time (t_{eq}) in the desorption process can be defined as the time at which the emitted H_2 content reaches 97% of the total sorption content, i.e., $C_E(t = t_{eq}) = 0.97\ C_\infty$. The dependence of the equilibrium time on the sample shape and diffusivity was investigated for cylinder-, sphere- and sheet-shaped specimens.

For cylindrical specimens with a fixed thickness (l_{fixed}), a fixed diameter (ρ_{fixed}) and $\frac{C_E(t_{eq})}{C_\infty} = 0.97$ at equilibrium, Equation (4) can be written as follows:

$$\frac{32}{\pi^2} \times \left[\sum_{n=0}^{\infty} \frac{exp\left\{ \frac{-(2n+1)^2\pi^2 D t_{eq}}{l_{fixed}^2} \right\}}{(2n+1)^2} \right] \times \left[\sum_{n=1}^{\infty} \frac{exp\left\{ -\frac{D\beta_n^2 t_{eq}}{\rho_{fixed}^2} \right\}}{\beta_n^2} \right] = 0.03 = \text{constant} \quad (5)$$

The dependence of the equilibrium time on the diffusivity is as follows: $D \cdot t_{eq}$ = constant. This relationship indicates that $\frac{1}{D}$ is linear with respect to t_{eq}.

For the cylindrical specimen with a fixed diffusivity (D_{fixed}), a fixed diameter (ρ_{fixed}) and $\frac{C_E(t_{eq})}{C_\infty} = 0.97$, Equation (4) can be modified as follows:

$$\frac{32}{\pi^2} \times \left[\sum_{n=0}^{\infty} \frac{exp\left\{ \frac{-(2n+1)^2\pi^2 D_{fixed} t_{eq}}{l^2} \right\}}{(2n+1)^2} \right] \times \left[\sum_{n=1}^{\infty} \frac{exp\left\{ -\frac{D_{fixed}\beta_n^2 t_{eq}}{\rho_{fixed}^2} \right\}}{\beta_n^2} \right] = 0.03 = \text{constant} \quad (6)$$

The relationship between the equilibrium time and the squared thickness is not linear but exponential because of the additive second exponential term in Equation (6). Similarly, for the cylindrical specimen, a fixed diffusivity (D_{fixed}), a fixed thickness (l_{fixed}) and $\frac{C_E(t_{eq})}{C_\infty}$ = 0.97, Equation (4) can be modified as follows:

$$\frac{32}{\pi^2} \times \left[\sum_{n=0}^{\infty} \frac{exp\left\{ \frac{-(2n+1)^2\pi^2 D_{fixed} t_{eq}}{l_{fixed}^2} \right\}}{(2n+1)^2} \right] \times \left[\sum_{n=1}^{\infty} \frac{exp\left\{ -\frac{D_{fixed}\beta_n^2 t_{eq}}{\rho^2} \right\}}{\beta_n^2} \right] = 0.03 = \text{constant} \quad (7)$$

The relationship between the equilibrium time and the squared diameter is not linear because of the additive first exponential term in Equation (7).

The emitted H_2 concentration $C_E(t)$ in the desorption process for a spherical sample is expressed as follows [24,26]:

$$\frac{C_E(t)}{C_\infty} = [1 - \frac{6}{\pi^2} \sum_{n=1}^{\infty} \frac{1}{n^2} \exp\left(-\frac{Dn^2\pi^2 t}{a^2} \right)] \quad (8)$$

Equation (8) shows the solutions to Fick's second diffusion law with an initially constant uniform hydrogen concentration and a constant concentration at the spherical surface. C_∞ is the hydrogen uptake measured in emission mode. a is the radius of the spherical rubber.

For the sphere-shaped specimen with a fixed diameter (a_{fixed}) and $\frac{C_E(t_{eq})}{C_\infty}$ = 0.97 at equilibrium, Equation (8) can be modified as follows:

$$\frac{6}{\pi^2}\sum_{n=1}^{\infty}\frac{1}{n^2}\exp\left(-\frac{Dn^2\pi^2 t_{eq}}{a_{fixed}}\right) = 0.03 = \text{constant} \tag{9}$$

Thus, $D{\cdot}t_{eq}$ = constant. This relationship indicates that $\frac{1}{D}$ is linear with respect to t_{eq}. Moreover, for the spherical specimen with a fixed diffusivity and $\frac{C_E(t_{eq})}{C_\infty}$ = 0.97, then $\frac{t_{eq}}{a^2}$ = constant. This relationship indicates that t_{eq} is linear with respect to a^2.

Equation (10) shows the solutions to Fick's second diffusion law for a plane sheet specimen for the emitted H_2 concentration $C_E(t)$ in the desorption process [24].

$$\frac{C_E(t)}{C_\infty} = [1 - \frac{8}{\pi^2}\sum_{n=0}^{\infty}\frac{1}{(2n+1)^2}\exp\left(-\frac{D(2n+1)^2\pi^2 t}{l^2}\right)] \tag{10}$$

where l is the thickness of the sheet-shaped rubber. Similarly, for a fixed-thickness sheet-shaped specimen and $\frac{C_E(t_{eq})}{C_\infty}$ = 0.97, then $D{\cdot}t_{eq}$ = constant. This relationship indicates that $\frac{1}{D}$ is linear with respect to t_{eq}. Moreover, for the sheet-shaped specimen with a fixed diffusivity and $\frac{C_E(t_{eq})}{C_\infty}$ = 0.97, then $\frac{t_{eq}}{l^2}$ = constant. This relationship indicates that t_{eq} is linear with respect to l^2.

3. Modeling for Three Shaped Specimens and Comparison with Experiment

To produce the time-varying H_2 concentration curve of multi-exponential functions with infinite series in Equations (4), (8) and (10), we developed a dedicated diffusion analysis program using Visual Studio 2019 written in c# language, based on a least-squares regression and the Nelder–Mead simplex optimization algorithm [18,27]. The program calculation included the summation of all the contributions up to n = 50–100 terms in infinite series.

Equations (4), (8) and (10) are dependent on the dimension and diffusivity of the specimen. Due to the modeling process of the analysis program, the H_2 emission behavior for the different diffusivities with different cylindrical thicknesses and radii, different spherical radii and different thicknesses of the sheet-shaped specimen, together with the desorption equilibrium times, can be predicted. The equilibrium times modeled for the remaining and emitted contents are the same. Thus, we modeled the desorption equilibrium time in emission mode with a validated diffusion analysis program [16–19]. The algorithm and code for the analysis program are described in Appendix A with an application example of the diffusion analysis program to obtain the diffusion coefficient and hydrogen uptake amount for NBR exposed to 8.9 MPa. In addition, the modeling results are compared with experimental results.

The measurement methods and principle are found in the previous literatures [16–18]. Figure 1 shows the diagram of manometric measurement sequence to measure the concentration and diffusivity of the emitted H_2. The system consists of high-pressure chamber for H_2 exposure in Figure 1A, cylinder-shaped specimen container with a USB-type data logger in Figure 1B. The commercial sensors used for the pressure/temperature measurements are data loggers to measure and record both the atmospheric pressure and temperature. Figure 1C is the results for measured pressure versus time together with analysis.

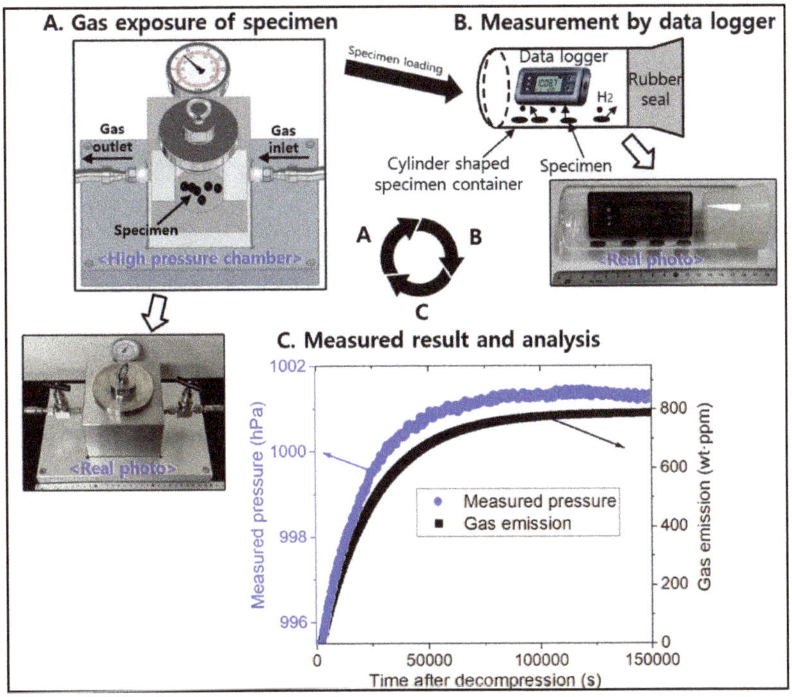

Figure 1. Diagram of manometric measurement sequence to measure the concentration and diffusivity of the emitted H_2. (**A**) High-pressure chamber for H_2 exposure. (**B**) Cylinder-shaped specimen container with a USB-type data logger. (**C**) Results for measured pressure versus time together with analysis.

After the exposure and the decompression from the chamber in Figure 1A, the specimen was loaded in the cylindrical container in Figure 1B. The H_2 released from the specimen increased the pressure in the specimen container with increasing time. Thus, the pressure (P) and temperature (T) of the gas inside the sample container varied with time. The gas in container was governed by ideal gas equation: $PV = nRT$, where R is the gas constant and n is mole number of emitted H_2 in the specimen container. The principle and measuring processes for obtaining H_2 concentration and diffusivity, including volumetric method, are detail described in the literature [16,18].

For experimental investigations, we employed the polymer specimens, such as nitrile butadiene rubber (NBR), ethylene propylene diene monomer (EPDM) and fluoroelastomer (FKM), which are used as sealing materials in the hydrogen refueling station. A low-density polyethylene (LDPE) sample was contained, which is employed as a plastic pipeline for H_2 transport. The polymer specimens are actually used as gas-sealing materials in the hydrogen infrastructure. Therefore, the investigations of the saturated equilibrium time and the permeation properties in H_2 infrastructure are essential for determining the appropriate exposure time/permeation of H_2 for the cycling test. Thus, we selected the polymers NBR, EPDM, FKM and LDPE, which are actually used in H_2 environments. In addition, the basic properties, application fields, chemical compositions and density of polymers are already described in the literatures [17].

3.1. Modeling for the Cylinder-Shaped Polymer Specimens and Comparison with Experimental Results

Figure 2 shows modeling and experimental results for the cylinder-shaped specimens. Modeling of cylinder-shaped specimens was performed using a dedicated diffusion analysis program with five different diffusivities of 1×10^{-8} m^2/s, 1×10^{-9} m^2/s, $1 \times$

10^{-10} m^2/s, 1×10^{-11} m^2/s and 1×10^{-12} m^2/s and thicknesses ranging from 1 mm to 45 mm at a fixed diameter of 20 mm, as shown in Figure 2a–c. Experimental results are represented in Figure 2d–g. Figure 2a shows the representative modeling result for the normalized hydrogen desorption content versus the elapsed time for a diffusivity of 1×10^{-10} m^2/s and five different thicknesses at a fixed diameter of 20 mm. The time-dependent normalized emission contents at the four different diffusivities were similar to those in Figure 2a. The normalized content was calculated as the emission content divided by the total uptake, $[C(t)/C_\infty]$. The black rectangle marked by blue arrows a, b, c, d and e in Figure 2a indicate the equilibrium times modeled at thicknesses of 1 mm, 3 mm, 7 mm, 15 mm and 45 mm, respectively. The equilibrium time obtained indicated the crossing point (black rectangle) on the x-axis between the normalized emission content line and 97% dotted line of the total sorption content shown with horizontal blue dotted line. Figure 2b depicts the desorption equilibrium time versus the squared thickness for five different diffusivities at a fixed diameter of 20 mm. The results were obtained using the numerical modeling and an exponential growth relationship was observed between the equilibrium time and the squared thickness with good squared correlation coefficients (R^2 = 0.99), as predicted by Equation (6).

(a)

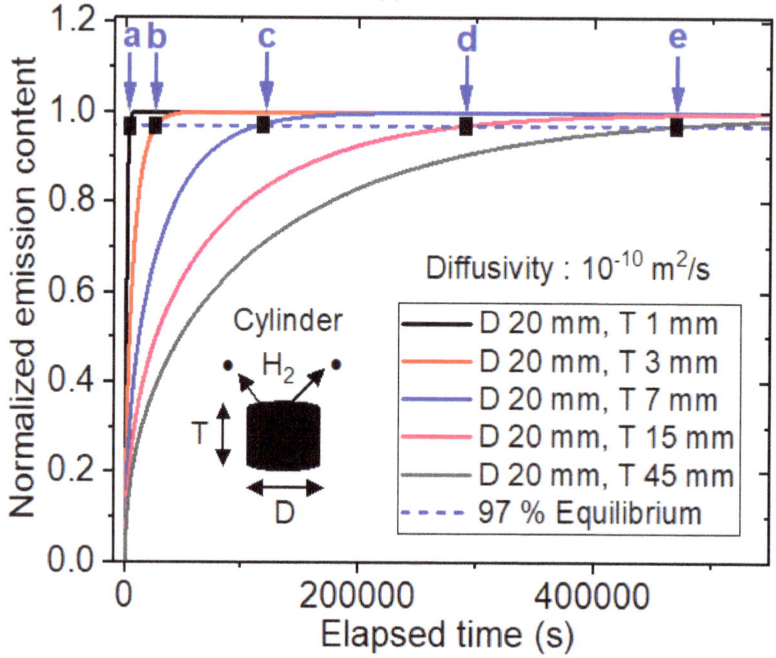

Figure 2. *Cont.*

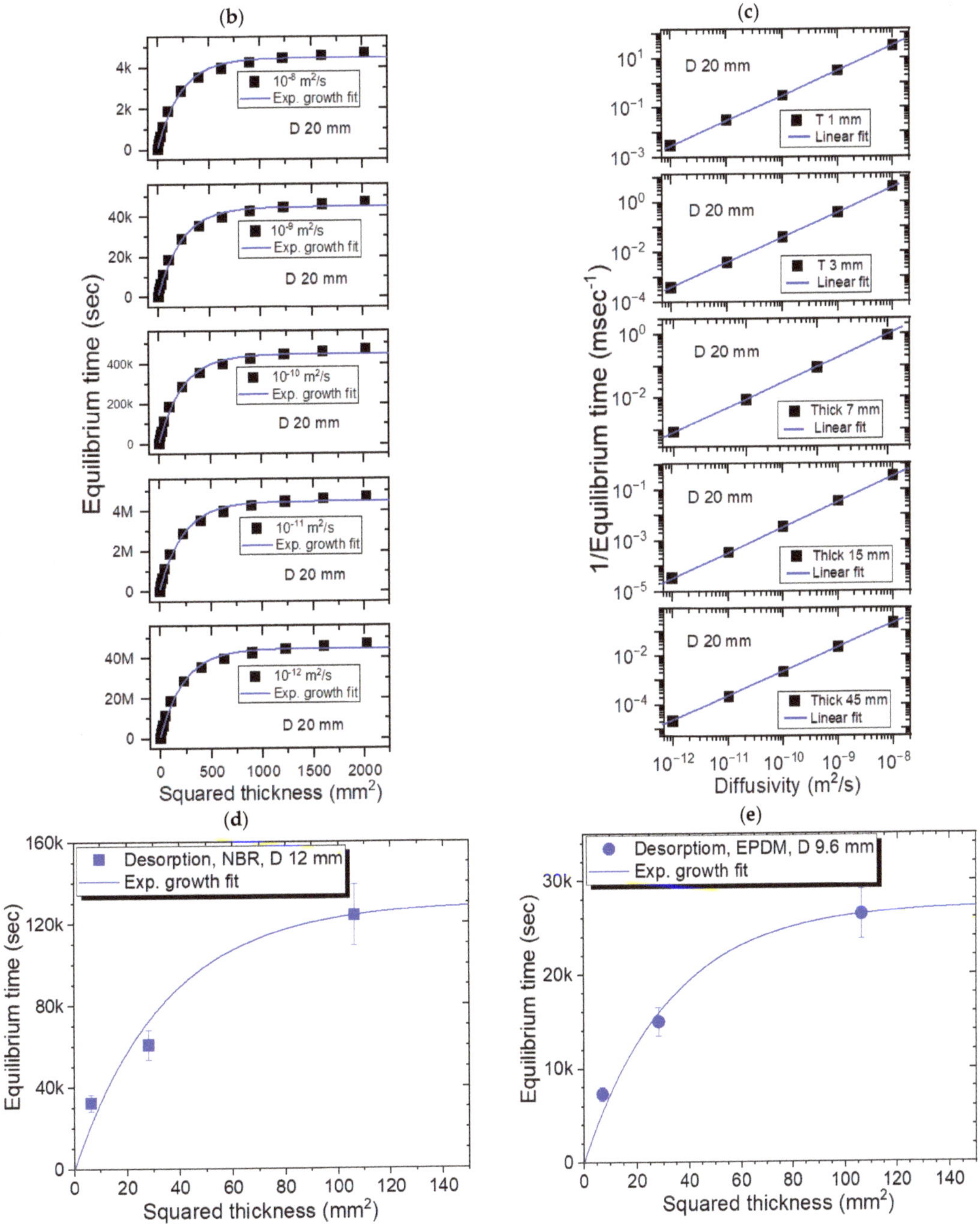

Figure 2. *Cont.*

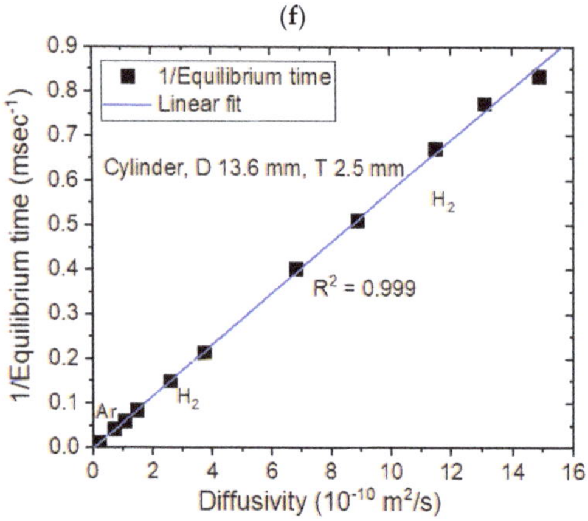

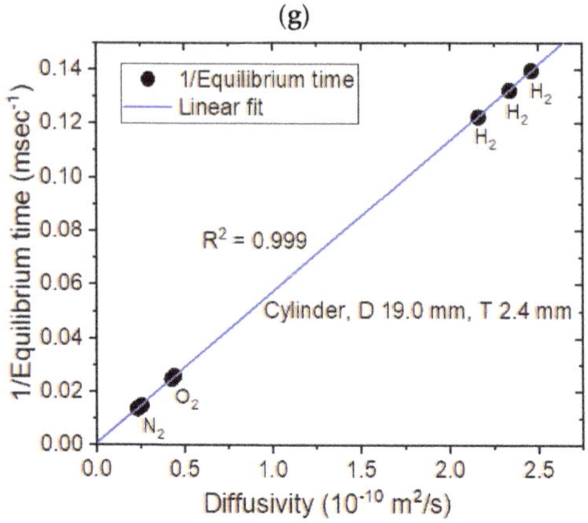

Figure 2. Modeling (**a**–**c**) and experimental (**d**–**g**) results for the cylinder-shaped specimen with a fixed diameter. (**a**) Numerical modeling result for the normalized hydrogen desorption content versus the elapsed time for a diffusivity of 1×10^{-10} m^2/s and five different thicknesses at a fixed diameter of 20 mm. (**b**) Numerical modeling result for the hydrogen equilibrium time versus the squared thickness with five different diffusivities at a fixed diameter of 20 mm. (**c**) Linear correlation between the reciprocal equilibrium time and diffusivity with five different thicknesses at a diameter of 20 mm. (**d**) Experimental results with the desorption equilibrium time versus the squared thickness at a fixed diameter of 12 mm in the NBR specimen. (**e**) Experimental results with the desorption equilibrium time versus the squared thickness at a fixed diameter of 9.6 mm in the EPDM specimen. (**f**) Experimental results with reciprocal equilibrium time versus the diffusivity at a fixed diameter of 13.6 mm and a fixed thickness of 2.5 mm. (**g**) Experimental results with reciprocal equilibrium time versus the diffusivity at a fixed diameter of 19.0 mm and a fixed thickness of 2.4 mm. D and T in Figure 2 indicate the diameter and thickness, respectively, of the cylindrical specimen.

Figure 2c displays the reciprocal equilibrium time versus the diffusivity for five different thicknesses at a fixed diameter of 20 mm. The results revealed a distinct linear relationship between the reciprocal equilibrium time and the diffusivity with perfect squared correlation coefficients (R^2 = 1.00), as predicted by Equation (5). Figure 2d,e displays the experimental results for desorption equilibrium time versus the squared thickness at a diameter of 12 mm for NBR and 9.6 mm for EPDM, respectively. The exponential growth behaviors in the experiment were consistent with the modeling result in Figure 2b. Figure 2f represents the experimental investigations revealing the linearity between the reciprocal equilibrium time versus diffusivity for a fixed diameter of 13.6 mm with a thickness of 2.5 mm, in which even though three points with low diffusivity are applied to Ar gas. Figure 2g also represents the experimental results showing the linearity between the reciprocal equilibrium time versus the diffusivity including different testing gases (N_2 and O_2). It means the linearity is valid regardless of species of testing gas. The gas independent linear dependence in Figure 2f,g supports the modeling results in Figure 2c.

On the other hand, Figure 3 shows the modeling results for the cylinder-shaped specimen at a fixed thickness and experimental results. The modeling of cylinder-shaped specimens was also performed using a dedicated diffusion analysis program with five different diffusivities of 1×10^{-8} m^2/s, 1×10^{-9} m^2/s, 1×10^{-10} m^2/s, 1×10^{-11} m^2/s and 1×10^{-12} m^2/s and diameters ranging from 1 mm to 60 mm at a fixed thickness of 2 mm. Figure 3a shows the representative modeling result for the normalized hydrogen desorption content versus the elapsed time for a diffusivity of 1×10^{-10} m^2/s and five different diameters at a fixed thickness of 2 mm. The black rectangle marked by blue arrows, a, b, c, d and e in Figure 3a indicate the equilibrium times of 97% of the total sorption content modeled at diameters of 1 mm, 2 mm, 5 mm, 15 mm and 50 mm, respectively.

The equilibrium time obtained indicated the crossing point (black rectangle) on the x-axis between the normalized emission content line and the 97% dotted line of the total sorption content shown by the horizontal blue dotted line. The time-dependent normalized emission contents at the four different diffusivities were similar to those in Figure 3a.

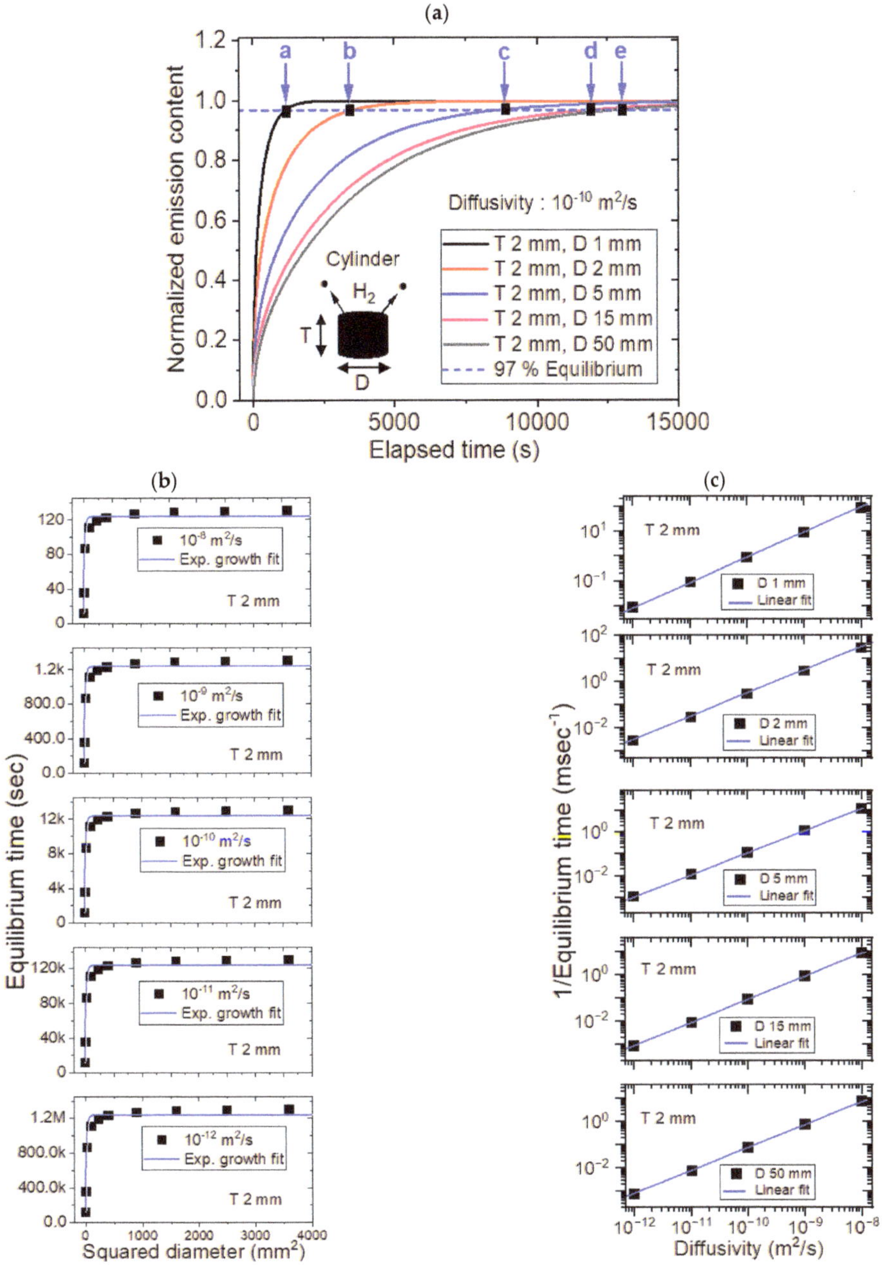

Figure 3. *Cont.*

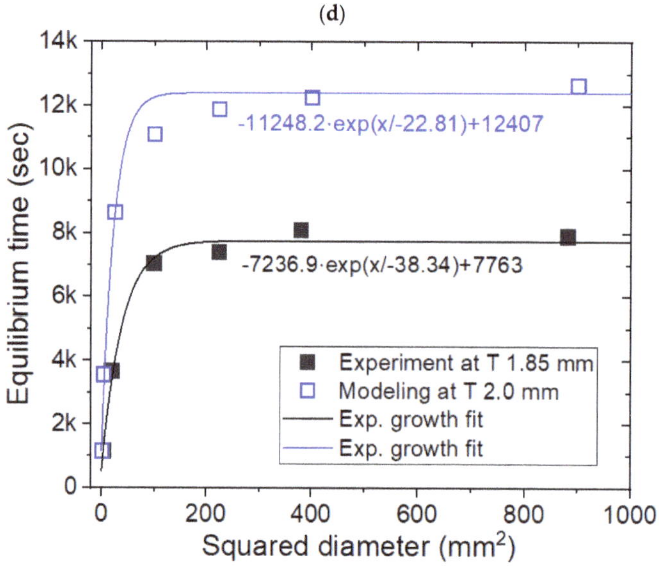

Figure 3. Modeling (**a–d**) and experimental (**d**) results for the cylinder-shaped specimen at a fixed thickness. (**a**) Numerical modeling for the normalized desorption hydrogen content versus the elapsed time for a diffusivity of 1×10^{-10} m^2/s with five different diameters at a thickness of 2 mm. (**b**) Numerical modeling results for the hydrogen equilibrium time versus the squared diameter for five different diffusivities at a fixed thickness of 2 mm. (**c**) Linear correlation between the reciprocal equilibrium time and the diffusivity with five different diameters at a thickness of 2 mm. (**d**) Comparison between experiment and modeling. D and T in Figure 3 indicate the diameter and thickness, respectively, of the cylindrical specimen.

Figure 3b depicts the desorption equilibrium time versus the squared diameter for five different diffusivities at a thickness of 2 mm. The results showed an exponential growth relationship between the squared diameter and the equilibrium time with good squared correlation coefficients (R^2 = 0.98), as predicted by Equation (7). Figure 3c displays the reciprocal equilibrium time versus the diffusivity for five different diameters at a fixed thickness of 2 mm. The results revealed a linear relationship between the reciprocal equilibrium time and the diffusivity with perfect squared correlation coefficients (R^2 = 1.00), as predicted by Equation (5). The overall behaviors with different diffusivities, diameters and thicknesses were similar to those in Figure 3. Figure 3d shows the comparison between experiment and modeling, which reveals same exponential growth behavior between the desorption equilibrium time versus the squared diameter. In Figure 3d, the difference in saturation value of equilibrium time is attributed to different thicknesses. Therefore, the experimental result supports again the modeling results in Figure 3b.

3.2. Modeling for the Sphere-Shaped Polymer Specimens and Comparison with Experimental Results

Figure 4 shows the modeling results for the spherical specimens, together with experimental results. Modeling of sphere-shaped polymer specimens was also performed using a dedicated diffusion analysis program with five different diffusivities of 1×10^{-8} m^2/s, 1×10^{-9} m^2/s, 1×10^{-10} m^2/s, 1×10^{-11} m^2/s and 1×10^{-12} m^2/s and diameters ranging from 10 mm to 40 mm. Figure 4a shows the representative modeling result for the normalized hydrogen desorption content versus the elapsed time for a fixed diffusivity of 1×10^{-10} m^2/s with five different diameters. The black rectangle marked by blue arrows a, b, c, d and e in Figure 4a indicate the equilibrium times modeled at diameters of 10 mm,

15 mm, 20 mm, 30 mm and 40 mm, respectively. Figure 4b shows the desorption equilibrium time versus squared diameter; here, a linear relationship was observed for the squared diameter for five diffusivities with perfect squared correlation coefficients (R^2 = 1), as predicted in the Section 2. Figure 4c displays the reciprocal equilibrium time versus diffusivity for the five different diameters. The results also showed a linear relationship between the reciprocal equilibrium time and the diffusivity with perfect squared correlation coefficients (R^2 = 1.00), as predicted by Equation (9). The overall behaviors with different diffusivities were similar to those in Figure 4b,c. Figure 4d shows the experimental investigations with the linearity between the desorption equilibrium time versus the squared diameter for three polymers (NBR, EPDM and FKM). The experimental linear behaviors coincided with the modeling results in Figure 4b. Figure 4e,f represents the experimental results with linear relationship between the reciprocal equilibrium time versus the diffusivity. The linear dependency was similar to the modeling results in Figure 4c.

(**a**)

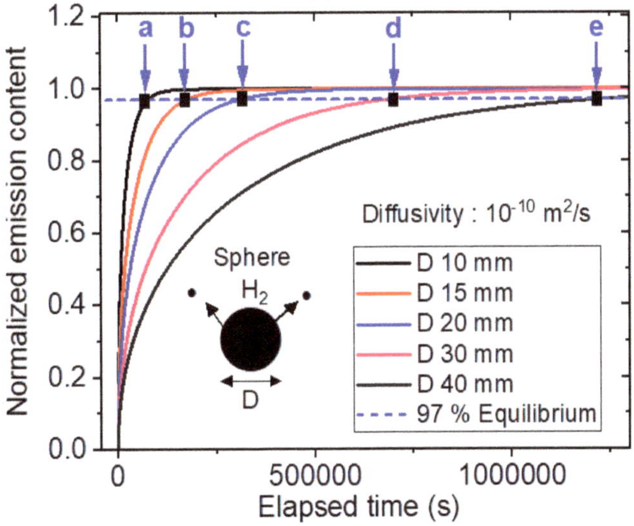

Figure 4. *Cont.*

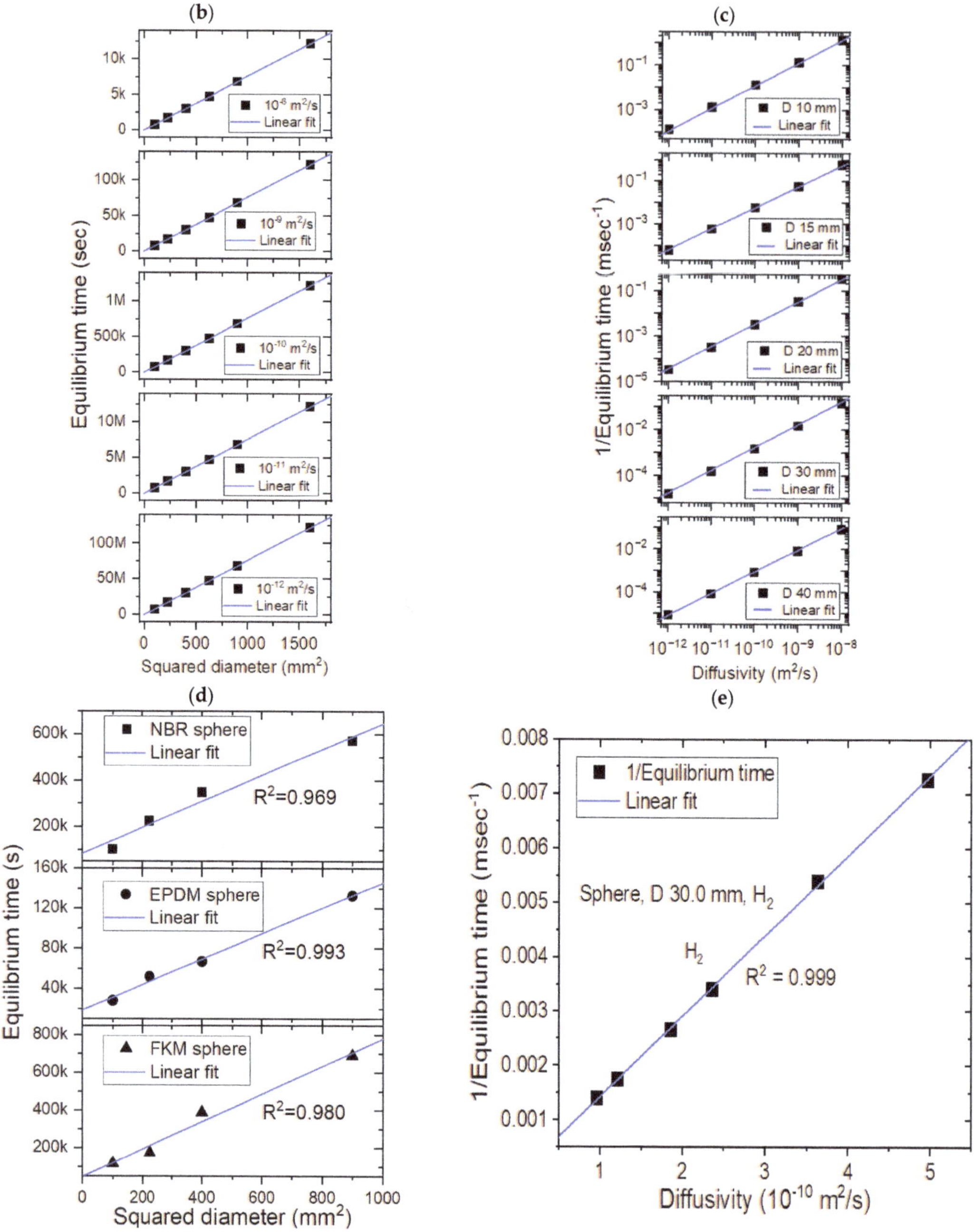

Figure 4. *Cont.*

(f)

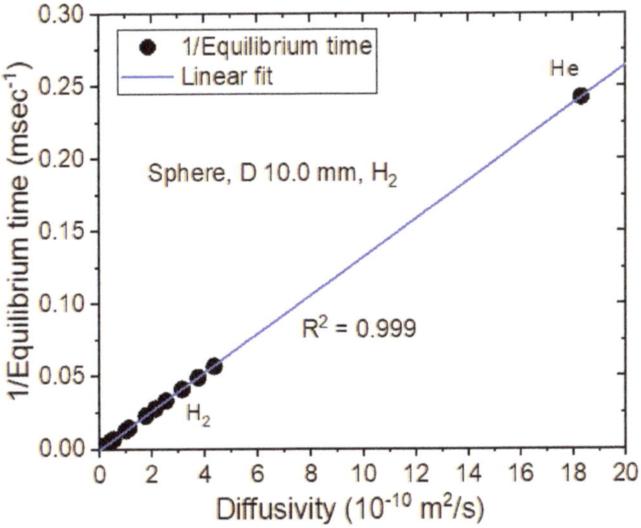

Figure 4. Modeling (**a**–**c**) and experimental (**d**–**f**) results for the sphere-shaped polymer. (**a**) Normalized hydrogen desorption content versus the elapsed time for a fixed diffusivity of 1×10^{-10} m^2/s and five different diameters. (**b**) Linear correlation between the sorption equilibrium time and the squared diameter at five different diffusivities. (**c**) Linear correlation between the reciprocal equilibrium time and the diffusivity with five different diameters. (**d**) Experimental investigations with the desorption equilibrium time versus the squared diameter for three samples. (**e**) Experimental results with linearity between reciprocal equilibrium time versus the diffusivity. (**f**) Experimental results including He gas with linearity between reciprocal equilibrium time versus the diffusivity. D in Figure 4 indicates the diameter of the spherical specimen.

3.3. *Modeling for the Sheet-Shaped Polymer Specimens and Comparison with Experimental Results*

Figure 5 shows the modeling results for the sheet-shaped specimens, together with experimental results. Modelling of the sheet-shaped polymer specimens was also performed using a dedicated diffusion analysis program with five different diffusivities of 1×10^{-8} m^2/s, 1×10^{-9} m^2/s, 1×10^{-10} m^2/s, 1×10^{-11} m^2/s and 1×10^{-12} m^2/s and thicknesses ranging from 1 mm to 15 mm. Figure 5a shows the representative modeling result for normalized hydrogen desorption content versus the elapsed time with a fixed diffusivity of 1×10^{-10} m^2/s and five different thicknesses. The black rectangle marked by blue arrows a, b, c, d and e in Figure 5a indicate the equilibrium times modeled at thicknesses of 1 mm, 2 mm, 4 mm, 7 mm and 15 mm, respectively. Figure 5b shows the desorption equilibrium time; here, a linear relationship was observed with the squared thickness for the five different diffusivities with perfect squared correlation coefficients (R^2 = 1), as predicted in the modeling background section. Figure 5c shows the reciprocal equilibrium time versus diffusivity for the five different thicknesses. The results revealed a linear relationship between reciprocal equilibrium time and diffusivity with perfect squared correlation coefficients (R^2 = 1.00), as predicted in the Section 2. The overall behaviors with different diffusivities were similar to those in Figure 5b,c. Figure 5d shows the experimental investigations with linearity between the desorption equilibrium time versus the squared thickness for fixed diffusivities. The linearity behaviors in experiment were well similar to modeling results in Figure 5b. Figure 5e shows the experimental investigations with linearity between the reciprocal equilibrium time versus the diffusivity. The experimental linear behaviors were good agreement in the modeling results in Figure 5c.

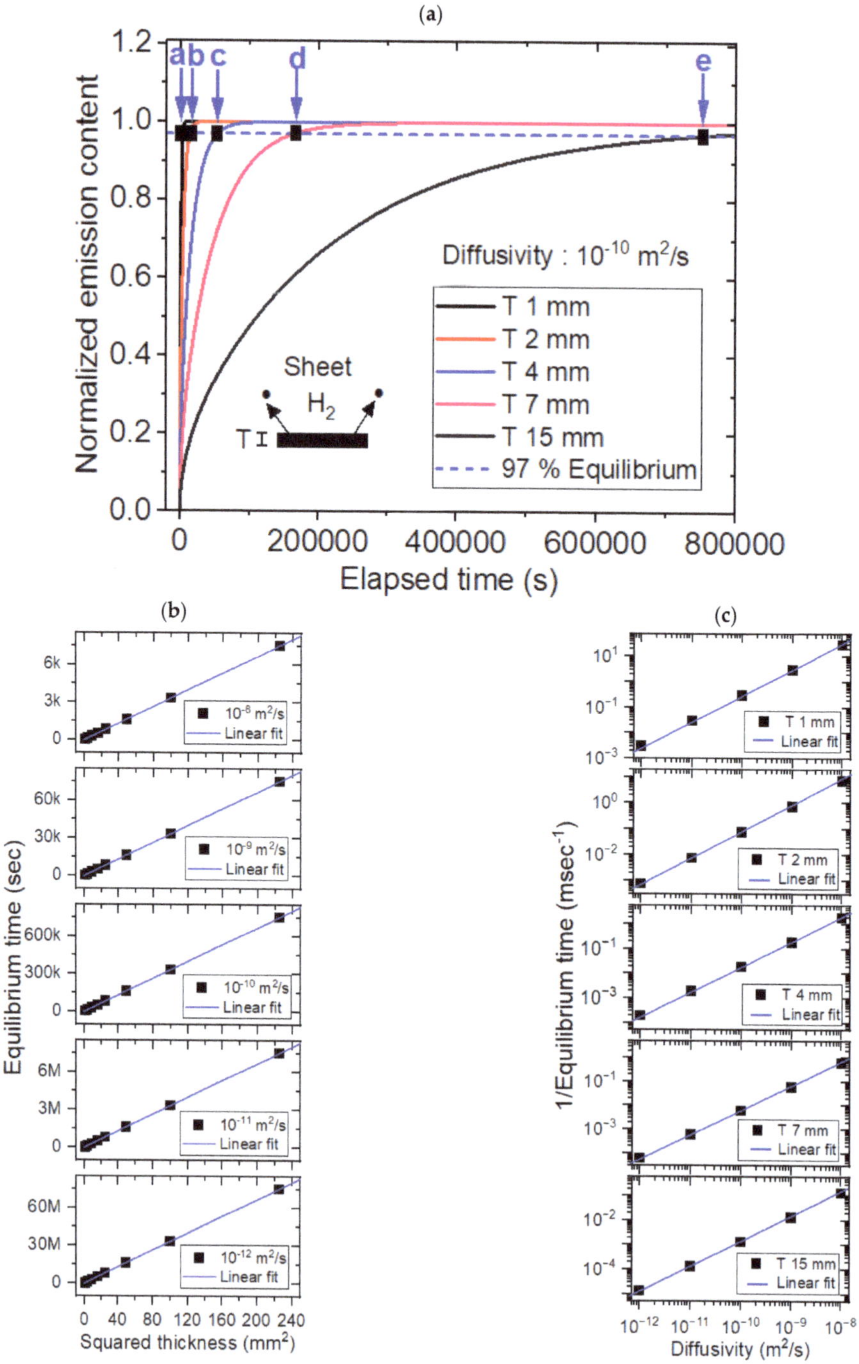

Figure 5. *Cont.*

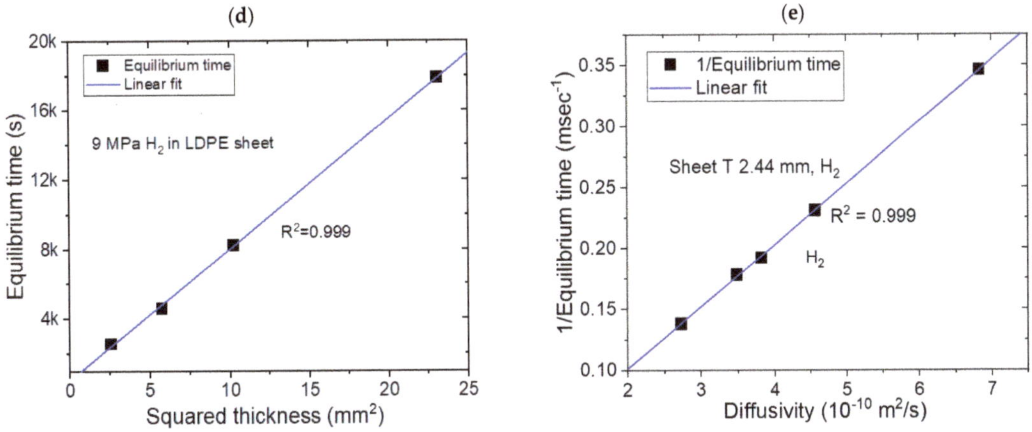

Figure 5. Modeling (**a**–**c**) and experimental (**d**,**e**) results for the sheet-shaped polymer. (**a**) Normalized hydrogen desorption content versus elapsed time for a fixed diffusivity of 1×10^{-10} m^2/s and five different thicknesses. (**b**) Linear correlation between the sorption equilibrium time and the squared thickness for the five different diffusivities. (**c**) Linear correlation between the reciprocal equilibrium time and diffusivity with five different thicknesses. (**d**) Experimental investigations showing linearity between the desorption equilibrium time versus the squared thickness for LDPE. (**e**) Experimental results with the linearity between the reciprocal equilibrium time versus the diffusivity. T in Figure 5 indicates the thickness of the sheet specimen.

As a similar diffusion analysis program, Yang et.al developed the program for calculation of diffusion coefficient based on the unipore model [26]. In order to automatically and time-effectively analyze the sorption–diffusion data, a matlab-based computer program was developed based on a least-squares criterion to regress the experimental gas sorption kinetic data and determine the corresponding diffusion coefficient. The developed program was limitedly applied to the sphere-shaped specimen according to Equation (9). Meanwhile, the developed diffusion analysis program in this work was applied to specimens with cylindrical, spherical and sheet shapes.

4. Conclusions

Modeling of the time-dependent H_2 emission content using a dedicated diffusion analysis program was conducted for cylinder-, sphere- and sheet-shaped polymer specimens. The desorption equilibrium time needed to be determined before the periodic cyclic testing and the high-pressure H_2 exposure. By utilizing the validated analysis program, we also modeled the H_2 desorption equilibrium time with respect to the volume dimension in the cylinder-, sphere- and sheet-shaped polymer specimens with different diffusivities.

The equilibrium time for the desorption processes showed the exponential growth behavior with respect to the squared thickness and the squared diameter of the cylinder-shaped specimen; however, the equilibrium time was found to be linear to the squared diameter for the sphere-shaped specimen and the squared thickness for the sheet-shaped specimen. For all cylinder-, sphere- and sheet-shaped polymers investigated, a linear relationship was observed between the reciprocal equilibrium time and the diffusivity. In summary, the desorption equilibrium times were mainly affected by the following important factors: the diffusion coefficient, sample thickness and diameter.

The modeling results were confirmed by analysis of the solutions with respect to the complicated Fick's second diffusion law. The modeling is in good agreement with experimental investigation results. The coincidence between modeling and experiment in gas diffusion are demonstrated at first time in the work. Thus, the modeling could predict the desorption equilibrium time for any shape of cylinder, sheet and sphere without measurement. Consequently, the time-dependent emitted H_2 concentration behavior of multi-exponential form with a known diffusivity could be predicted for three shaped samples by applying of a diffusion analysis program.

Author Contributions: Conceptualization, J.K.J.; methodology, J.H.L.; validation, J.Y.P.; writing—original draft preparation, J.K.J.; writing—review and editing, S.K.J. All authors have read and agreed to the published version of the manuscript.

Funding: This research was supported by Development of Reliability Measurement Technology for Hydrogen Refueling Station funded by the Korea Research Institute of Standards and Science (KRISS-2024-GP2024-0010).

Institutional Review Board Statement: Not applicable.

Data Availability Statement: The original contributions presented in the study are included in the article, further inquiries can be directed to the corresponding author.

Conflicts of Interest: The authors declare no conflicts of interest.

Appendix A. Flowchart, Application Example and Code for Diffusion Analysis Program

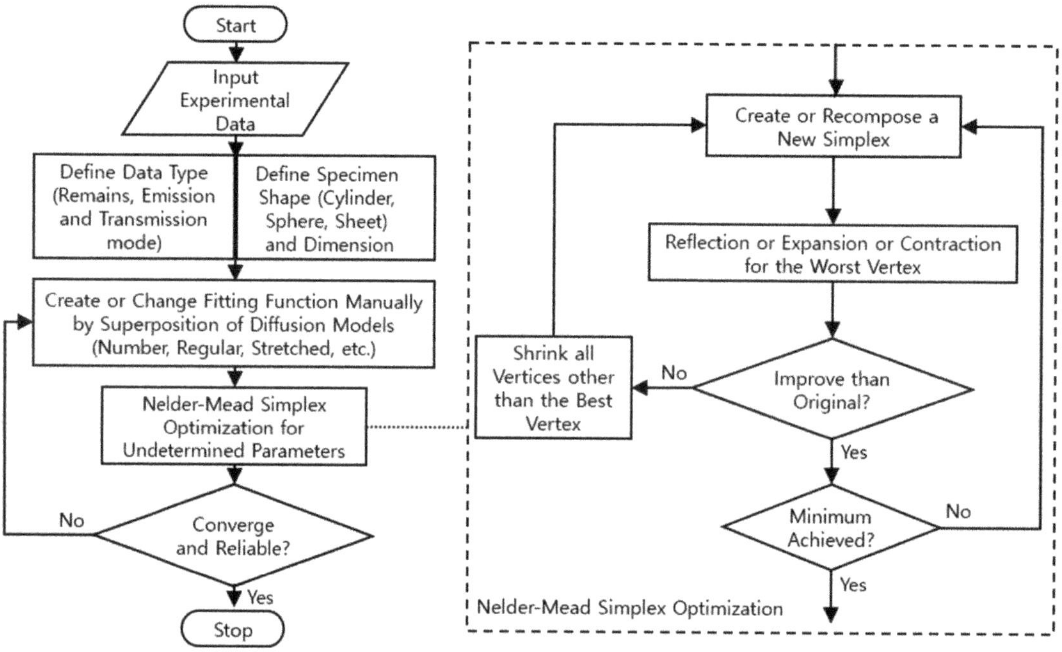

Figure A1. Flowchart for analyzing the gas uptake and diffusivity of specimens with different data types and various shapes with the Nelder–Mead simplex nonlinear optimization algorithm.

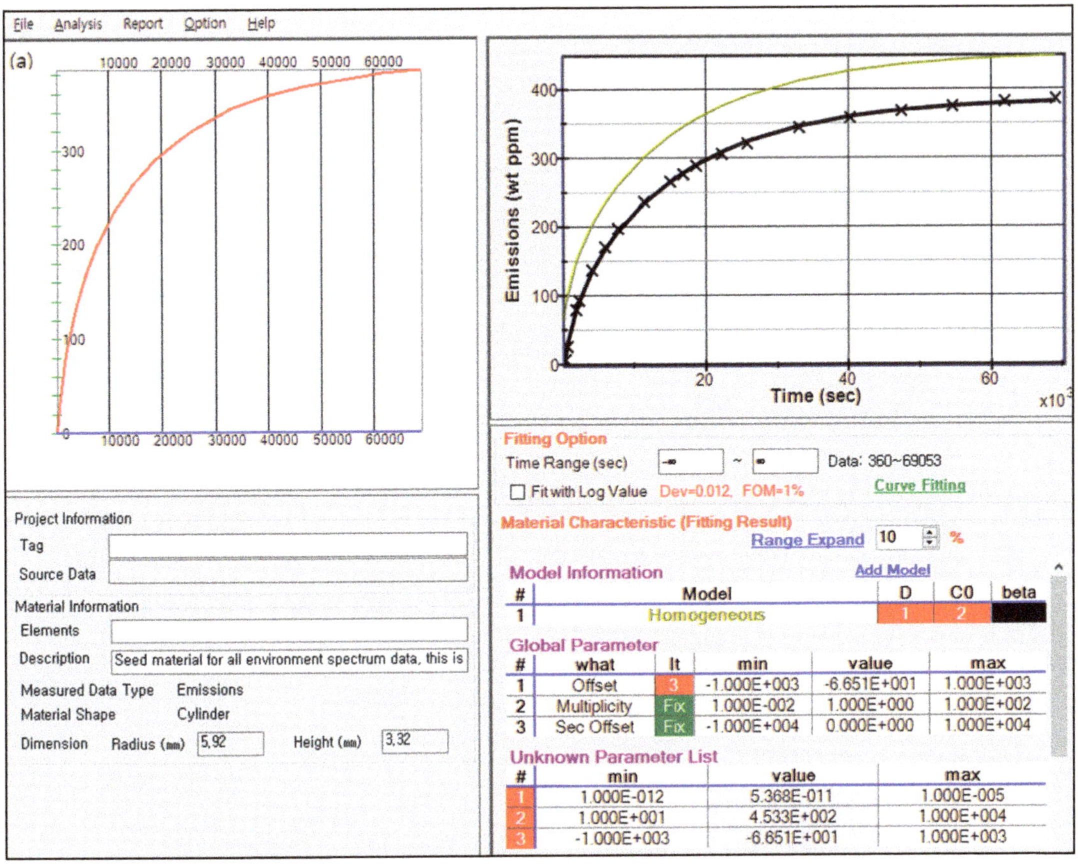

Figure A2. Application example of the diffusion analysis program to obtain the hydrogen uptake and diffusion coefficient for NBR H60 exposed to 8.9 MPa.

<Coding of sheet, sphere- and cylinder-shaped specimen in the emission mode>

```
    // Total amount of diffusing gas which has entered the sheet at time t
    // Ref: Crank et.al. The Mathematics of Diffusion (1975) p. 47, Equation (4.18) [24]
    // D: Diffusion coefficient, L: Sheet thickness
    // M: Emitted total amount by diffusion at infinite time.
    public double Ftn_Sheet(double t) {
        double pi2 = Math.PI × Math.PI;
        double zSum = 0;
        for (int n = 0; n < 50; ++n) {
            int n21 = (2 × n + 1) × (2 × n + 1);
            double add = Math.Exp(−D × n21 × pi2 × t/(L × L))/n21;
            double zSumAdded = zSum + add;
            if (zSum >= zSumAdded) break; // When convergence achieved
            zSum = zSumAdded;
        }
        return M − 8/pi2 × (M × zSum);
}
```

```
    // Total amount of diffusing gas which has entered the sphere at time t.
    // Ref: Crank et.al. The Mathematics of Diffusion (1975) p. 91, Equation (6.20) [24]
    // D: Diffusion Coefficient, a: Radius of sphere.
    // M: Emitted total amount by diffusion at infinite time.
    public double Ftn_Sphere(double t) {
        double pi2 = Math.PI × Math.PI;
        double zSum = 0;
        for (int n = 1; n < 100; ++n) {
            int n2 = n × n;
            double add = Math.Exp(−D × n2 × pi2 × t/(a × a))/n2;
            double zSumAdded = zSum + add;
            if (zSum >= zSumAdded) break; // When convergence achieved
            zSum = zSumAdded;
        }
        return M − 6/pi2 × (M × zSum);
}

    // Total amount of diffusing gas which has entered the cylinder at time t.
    // Ref: Al Demarez, et. al., Acta Metallurgica (1954), Equation (5) [28]
    // D: Diffusion Coefficient, a: Radius of cylinder, L: Length of cylinder
    // M: Emitted total amount by diffusion at infinite time.
    public double Ftn_Cylinder(double t) {
        double pi2 = Math.PI × Math.PI;
        double zSum = 0;
        for (int n = 0; n < 100; ++n) {
            int n2 = (2 × n + 1) × (2 × n + 1);
            double add = Math.Exp(−n2 × pi2 × D × t/(L × L))/n2;
            double zSumAdded = zSum + add;
            if (zSum >= zSumAdded) break; // When convergence achieved
            zSum = zSumAdded;
        }
        double rSum = 0;
        for (int n = 1; n <= 50; ++n) {
            double b = Bessel0root[n − 1];
            double add = Math.Exp(−D × b × b × t/(a × a))/(b × b);
            double rSumAdded = rSum + add;
            if (rSum >= rSumAdded) break; // When convergence achieved
            rSum = rSumAdded;
        }
        return M − 32/pi2 × (M × zSum × rSum);
}

    // Zeros of first kind Bessel function J_0(x)
    public static double[] bessel0root =
        new double[] {2.40482556, 5.52007811, 8.65372791, 11.79153444, 14.93091771,
            18.07106397, 21.21163663, 24.35247153, 27.49347913, 30.63460647,
            33.77582021, 36.91709835, 40.05842576, 43.19979171, 46.34118837,
            49.4826099, 52.62405184, 55.76551076, 58.90698393, 62.04846919,
            65.1899648, 68.33146933, 71.4729816, 74.61450064, 77.75602563,
            80.89755587, 84.03909078, 87.18062984, 90.32217264, 93.46371878,
            96.605267951, 99.74681986, 102.88837425, 106.02993092, 109.17148965,
            112.3130503, 115.45461265, 118.59617663, 121.73774209, 124.87930891,
            128.020877, 131.1624463, 134.3040166, 137.445588, 140.5871604,
            143.7287336, 146.8703076, 150.0118825, 153.153458, 156.2950343};
```

```
// Nelder-Mead Simplex Optimization.
// Return value is fit error
// Optima(class): Manage all optimization procedure and calc. error and FOM etc.
// Simplex(Class): Polytope of N + 1 vertices in N dim. undetermined param. space.
// Vertex(Class): One of N + 1 vertices of Simplex.
// Coord(Class): Having coord. of vertex and methods, reflect(), expand(), ...
// simplexProtocol: ExpFactor = 2.0, ConFactor = 0.5, PrcntInitCoef = 5.0, ...
public double Iterate(double[] unknowns) {
        int b, w;
        int loop = 0; // # of loops
        int dim = unknowns.Length;
        Simplex simplex = new Simplex(unknowns, simplexProtocol.PrcntInitCoef);
        // Calculate errors for each vertexes.
        for (int i = 0; i < dim + 1; ++i)
        simplex.s[i].error = optima.ErrorF(simplex.s[i].coord);
        do { // start iteration
                b = simplex.bestVertex(); // find best and worst vertices
                w = simplex.worstVertex();
                // calculate the midPoint of the non-worst values for later calc's
                double[] mp = simplex.midPoint(w);
                Vertex refl_vrtx = new Vertex(simplex.s[w]); // save the old values
                Coord.reflect(mp, refl_vrtx.coord); // first try a reflection
                refl_vrtx.error = optima.ErrorF(refl_vrtx.coord);

                // is the new vertex better or equal to the old best vertex
                if (refl_vrtx.error < simplex.s[b].error) {
                        // try to expand the reflected vertex next
                        Vertex expd_vrtx = new Vertex(refl_vrtx);
                        Coord.expand(mp, expd_vrtx.coord, simplexProtocol.ExpFactor);
                        expd_vrtx.error = optima.ErrorF(expd_vrtx.coord);
                        // if the expanded vertex is not as good or better than the best
                        // revert to the reflected vertex
                        if (expd_vrtx.error <= refl_vrtx.error) simplex.s[w] = expd_vrtx;
                        else simplex.s[w] = refl_vrtx;
                } else {
                        if (refl_vrtx.error <= simplex.s[w].error) simplex.s[w] = refl_vrtx;
                        else {
                                Vertex cont_vrtx = new Vertex(simplex.s[w]);
                                Coord.contract(mp, cont_vrtx.coord, simplexProtocol.ConFactor);
                                cont_vrtx.error = optima.ErrorF(cont_vrtx.coord);
                                if (cont_vrtx.error <= simplex.s[w].error) simplex.s[w] = cont_vrtx;
                                else {
                                                simplex.contract_all(b, simplexProtocol.ConFactor);
                                                for (int i = 0; i < dim + 1; ++i)
                                                                simplex.s[i].error =
                          otima.ErrorF(simplex.s[i].coord);
                                }; // else
                        }; // else
                };
                ++loop;
        } while (!(simplex.s[b].error <= simplexProtocol.Tolerance ||
        loop >= simplexProtocol.InternalLoop));
        // return best result
        for (int i = 0; i < dim; ++i) unknowns[i] = simplex.s[b].coord[i];
        return simplex.s[b].error;
}
```

References

1. Zheng, Y.; Tan, Y.; Zhou, C.; Chen, G.; Li, J.; Liu, Y.; Liao, B.; Zhang, G. A review on effect of hydrogen on rubber seals used in the high-pressure hydrogen infrastructure. *Int. J. Hydrogen Energy* **2020**, *45*, 23721–23738. [CrossRef]
2. Nishimura, S. Rubbers and elastomers for high-pressure hydrogen seal. *Soc. Polym. Sci.* **2015**, *64*, 356–357.
3. Lipiäinen, S.; Lipiäinen, K.; Ahola, A.; Vakkilainen, E. Use of existing gas infrastructure in European hydrogen economy. *Int. J. Hydrogen Energy* **2023**, *48*, 31317–31329. [CrossRef]
4. Sahin, H.; Gondal, I.; Sahir, M. Prospects of natural gas pipeline infrastructure in hydrogen transportation. *Int. J. Energy Res.* **2012**, *36*, 1338–1345. [CrossRef]
5. Dou, X.; Zhu, T.; Wang, Z.; Sun, W.; Lai, Y.; Sui, K.; Tan, Y.; Zhang, Y.; Yuan, W.Z. Color-tunable, excitation-dependent, and time-dependent afterglows from pure organic amorphous polymers. *Adv. Mater.* **2020**, *32*, 2004768. [CrossRef]
6. Singh, S.B.; De, M. Thermally exfoliated graphene oxide for hydrogen storage. *Mater. Chem. Phys.* **2020**, *239*, 122102. [CrossRef]
7. Pan, M.; Pan, C.; Li, C.; Zhao, J. A review of membranes in proton exchange membrane fuel cells: Transport phenomena, performance and durability. *Renew. Sustain. Energy Rev.* **2021**, *141*, 110771. [CrossRef]
8. Ravera, F.; Dziza, K.; Santini, E.; Cristofolini, L.; Liggieri, L. Emulsification and emulsion stability: The role of the interfacial properties. *Adv. Colloid Interface Sci.* **2021**, *288*, 102344. [CrossRef]
9. Briottet, L.; Moro, I.; Lemoine, P. Quantifying the hydrogen embrittlement of pipeline steels for safety considerations. *Int. J. Hydrogen Energy* **2012**, *37*, 17616–17623. [CrossRef]
10. Huang, Q.; Li, X.; Li, J.; Liu, Y.; Li, X.; Zhu, C. Overview of standards on pressure cycling test for on-board composite hydrogen storage cylinders. In Proceedings of the Volume 1: Codes and Standards; ASME: Nevada, CA, USA, 2022; p. 7.
11. Tanioka, A.; Oobayashi, A.; Kageyama, Y.; Miyasaka, K.; Ishikawa, K. Effects of carbon filler on sorption and diffusion of gases through rubbery materials. *J. Polym. Sci. Polym. Phys. Ed.* **1982**, *20*, 2197–2208. [CrossRef]
12. Sereda, L.; Mar López-González, M.; Yuan Visconte, L.L.; Nunes, R.C.R.; Guimarães Furtado, C.R.; Riande, E. Influence of silica and black rice husk ash fillers on the diffusivity and solubility of gases in silicone rubbers. *Polymer* **2003**, *44*, 3085–3093. [CrossRef]
13. Wang, Z.F.; Wang, B.; Qi, N.; Zhang, H.F.; Zhang, L.Q. Influence of fillers on free volume and gas barrier properties in styrene-butadiene rubber studied by positrons. *Polymer* **2005**, *46*, 719–724. [CrossRef]
14. Barrer, R.M.; Chio, H.T. Solution and diffusion of gases and vapors in silicone rubber membranes. *J. Polym. Sci. C Polym. Symp.* **1965**, *10*, 111–138. [CrossRef]
15. Robeson, L.M.; Liu, Q.; Freeman, B.D.; Paul, D.R. Comparison of transport properties of rubbery and glassy polymers and the relevance to the upper bound relationship. *J. Membr. Sci.* **2015**, *476*, 421–431. [CrossRef]
16. Jung, J.K.; Kim, I.G.; Jeon, S.K.; Kim, K.-T.; Baek, U.B.; Nahm, S.H. Volumetric analysis technique for analyzing the transport properties of hydrogen gas in cylindrical-shaped rubbery polymers. *Polym. Test.* **2021**, *99*, 107147. [CrossRef]
17. Jung, J.K.; Kim, I.G.; Kim, K.T.; Ryu, K.S.; Chung, K.S. Evaluation techniques of hydrogen permeation in sealing rubber materials. *Polym. Test.* **2021**, *93*, 107016. [CrossRef]
18. Jung, J.K. Review of developed methods for measuring gas uptake and diffusivity in polymers enriched by pure gas under high pressure. *Polymers* **2024**, *16*, 723. [CrossRef] [PubMed]
19. Fujiwara, H.; Nishimura, S. Evaluation of hydrogen dissolved in rubber materials under high-pressure exposure using nuclear magnetic resonance. *Polym. J.* **2012**, *44*, 832–837. [CrossRef]
20. Müller, M.; Mishra, R.K.; Šleger, V.; Pexa, M.; Čedík, J. Elastomer-based sealing o-rings and their compatibility with methanol, ethanol, and hydrotreated vegetable oil for fueling internal combustion engines. *Materials* **2024**, *17*, 430. [CrossRef] [PubMed]
21. Philibert, J. One and a half century of diffusion: Fick, Einstein, before and beyond. *Diffus. Fundam.* **2005**, *2*, 1.1–1.10. [CrossRef]
22. Conlisk, A.T. *Essentials of Micro-and Nanofluidics: With Applications to the Biological and Chemical Sciences*; Cambridge University Press: Cambridge, UK, 2012; p. 43.
23. Sander, R. Compilation of Henry's law constants (version 4.0) for water as solvent. *Atmos. Chem. Phys.* **2015**, *15*, 4399–4981. [CrossRef]
24. Crank, J. *The Mathematics of Diffusion*; Oxford University Press: Oxford, UK, 1975.
25. Yamabe, J.; Nishimura, S. Influence of fillers on hydrogen penetration properties and blister fracture of rubber composites for O-ring exposed to high-pressure hydrogen gas. *Int. J. Hydrogen Energy* **2009**, *34*, 1977–1989. [CrossRef]
26. Yang, Y.; Liu, S. Estimation and modeling of pressure-dependent gas diffusion coefficient for coal: A fractal theory-based approach. *Fuel* **2019**, *253*, 588–606. [CrossRef]
27. Nelder, J.A.; Mead, R. A simplex method for function minimization. *Comput. J.* **1965**, *7*, 308–313. [CrossRef]
28. Demarez, A.; Hock, A.G.; Meunier, F.A. Diffusion of hydrogen in mild steel. *Acta Metallurgica* **1954**, *2*, 214–223. [CrossRef]

polymers

MDPI

Article

Study on the Influence of Hygrothermal Aging on the Mechanical Properties of Carbon Fabric/Polyetheretherketone Composites

Xiangyu Xu [1,*], Baoyan Zhang [1], Fenghui Shi [1,*], Kai Liu [2], Gongqiu Peng [1], Liang Gao [1], Junpeng Gao [1] and Yu Du [1]

1 AVIC Manufacturing Technology Institute Composite Technology Center, Beijing 101300, China
2 School of Aerospace Engineering, Beijing Institute of Technology, Beijing 100081, China
* Correspondence: xyxu_buaa@126.com (X.X.); fenghuishi@126.com (F.S.)

Abstract: Owing to its superior mechanical properties and recyclability, the carbon fabric/polyetheretherketone (CFF/PEEK) composite has seen increasing application in engineering domains. However, studies examining the impact of hygrothermal aging on its performance remain relatively limited in the existing literature. To investigate its durability in hygrothermal environments, this study fabricated CFF/PEEK composites with a fiber volume fraction of 55 vol% and subjected them to equilibrium hygroscopic treatment at 70 °C. The hygroscopic behavior of polyetheretherketone (PEEK) resin and CFF/PEEK composites, along with their tensile and compressive properties under dry conditions at room temperature (RTD) and wet conditions at 70 °C (ETW), were systematically evaluated. The results indicated that both PEEK resin and CFF/PEEK composites exhibited Fickian diffusion behavior during the initial stages of aging but diverged in later stages. The equilibrium moisture absorption rates were approximately 0.32% for PEEK resin and 0.19% for CFF/PEEK composites. After aging at 70 °C, the strength of both materials decreased significantly, while the modulus showed only minor changes. Under ETW conditions, the tensile strength retention rate of PEEK resin was 74.92%, and the compressive strength retention rate was 81.85%. For the CFF/PEEK composites, the tensile strength retention rate was approximately 85%, and the compressive strength retention rate was about 95%. The typical failure modes of CFF/PEEK composites did not exhibit notable differences between tensile and compressive specimens after hygrothermal aging. Resin debonding was observed in the moisture-absorbed composite specimens, while no microcracks or delamination were detected. The degradation of mechanical properties is predominantly attributed to the deterioration of the resin matrix and interface characteristics, which are caused by water molecule intrusion and the adverse effects of wet strain mismatch between the resin and fibers.

Keywords: PEEK; thermoplastic composite; hygrothermal aging; mechanical properties; CFRP; carbon fabric

1. Instruction

Carbon fiber reinforced polymer composites (CFRP) have found extensive applications in the aerospace, marine, automotive, and wind power industries due to their exceptional properties such as high strength, low density, superior corrosion resistance, and fatigue endurance [1–3]. Since the initial commercialization of CFRP in the 1960s, its application in aircraft has progressively evolved from non-load-bearing structure components to main load-bearing structural components. The usage of CFRP in aircrafts has become a key indicator of advanced aircraft design [4–6]. Notably, in large commercial aircraft such as the Boeing 787 and Airbus A350, the weight fraction of CFRP exceeds 50% [7–9]. CFRP can be classified into the following two major categories based on their various resin matrices: thermosetting CFRP and thermoplastic CFRP. Currently, thermosetting CFRP dominate the engineering applications in aviation structure components [10,11]. However, with the continuous improvement of requirements for composite materials in the aviation field in terms of performance, low cost, and environmental protection, the drawbacks of thermosetting CFRP have become increasingly prominent. Compared with thermosetting CFRP, thermoplastic CFRP have unique advantages such as recyclability, strong designability, repairability, and easy storage, while also possessing excellent mechanical properties such as high impact resistance, damage tolerance, and high toughness [12,13]. Therefore, thermoplastic CFRP have become a new research hotspot in recent years, and they have gradually been put into practice in aerospace, automotive, marine, bioengineering, and other fields.

Currently, thermoplastic resins utilized in the aviation industry primarily consist of high-temperature resistant and high-performance resin matrices, such as polyetheretherketone (PEEK), polyphenylene sulfide (PPS), polyetherimide (PEI), polyetherketoneketone (PEKK), and polyethersulfone (PES) [14–16]. PEEK is a novel semi-crystalline aromatic thermoplastic engineering polymer that was successfully developed by Imperial Chemical Industries (ICI) in the late 1970s. Owing to its exceptional thermal stability, radiation resistance, corrosion resistance, superior dimensional stability, and electrical properties, as well as its outstanding processability, PEEK is widely acknowledged as one of the best-performing thermoplastic materials globally [17,18]. Currently, PEEK finds extensive application in defense and military industries, aerospace, electronics, energy development and processing, automotive manufacturing, home appliance production, and medical and healthcare sectors.

When utilized in harsh environments, the mechanical properties of resin matrix composites can be substantially diminished, thereby impacting the service life of structural components. Hygrothermal aging is a critical factor influencing the service performance of composite materials and represents one of the most significant failure modes for resin matrix composites [19,20]. In hot and humid conditions, water molecules gradually diffuse into the resin matrix and penetrate along micro-cracks to reach the fiber surface, leading to a reduction in both the mechanical properties of the resin and fibers, as well as the degradation of the interface between them, ultimately resulting in material failure. The diffusion behavior of water molecules into the interior of materials significantly affects the durability of CFRPs in engineering applications [21]. Numerous scholars have investigated the mechanical properties of CFRP in various hygrothermal environments. Ma et al. [22] reported that both the carbon fiber reinforced PEEK and PPS composites exhibit good moisture/temperature resistance and property retention after hygrothermal exposure. Wu et al. [23] found that the long-term mechanical property retention rate of CFRP was superior to that of flax fiber/epoxy resin composites and glass fiber/epoxy resin composites. Li et al. [24] reported that after the dehumidification treatment of the samples that absorbed moisture for 56 days, the retention rate of the interlaminar shear strength recuperated from 77.40% to 91.92%, and the proportion of reversible damage is 63.98%. Feng et al. [25] reported that the tensile strength of T700 carbon fiber/3228 polyurethane resin composites decreased by 14.63%, following aging at 70 °C and 85% relative humidity for one week. Wan et al. [26] found that hygrothermal aging has a significant detrimental effect on the tension–tension fatigue behavior of the 3D5D hybrid braided composites. Micelli et al. [27] noted that each type of FRP has unique components and manufacturing processes. Consequently, conclusions drawn for one material may not be directly applicable to others.

However, due to the nascent stage of thermoplastic composites in engineering applications, there is limited data available on the mechanical properties of CFF/PEEK composites, and a lack of comprehensive durability test databases.

In this study, the CFF/PEEK composite and PEEK resin were subjected to saturated hygroscopic treatment at 70 °C. Subsequently, tensile tests and compressive tests were conducted on the samples both before and after the hygroscopic treatment, yielding mechanical property data under varying environmental conditions. The hygroscopic behavior of PEEK resin and CFF/PEEK composites was analyzed, and a comparative assessment was made of the mechanical properties, failure modes, and fracture morphology of the samples before and after hygroscopic exposure. This research aims to investigate the impact of hygrothermal environments on the mechanical properties of CFF/PEEK composites and provide practical guidance for the engineering applications of thermoplastic composites.

2. Materials and Methods

2.1. Raw Materials and Sample Preparation

Satin carbon fabric and PEEK resin were utilized as raw materials to prepare CFF/PEEK thermoplastic prepreg. This thermoplastic prepreg was manufactured by the AVIC Manufacturing Technology Institute Composite Technology Center. The PEEK resin was supplied by Zhejiang Pfulon Advanced Materials Co., Ltd. (Quzhou, Zhejiang, China). The satin carbon fabric, with a unit area mass of 293 ± 8 g/m^2, was provided by Weihai Expand Fiber Co., Ltd. (Weihai, Shandong, China). The fabric form and the finished prepreg are illustrated in Figure 1.

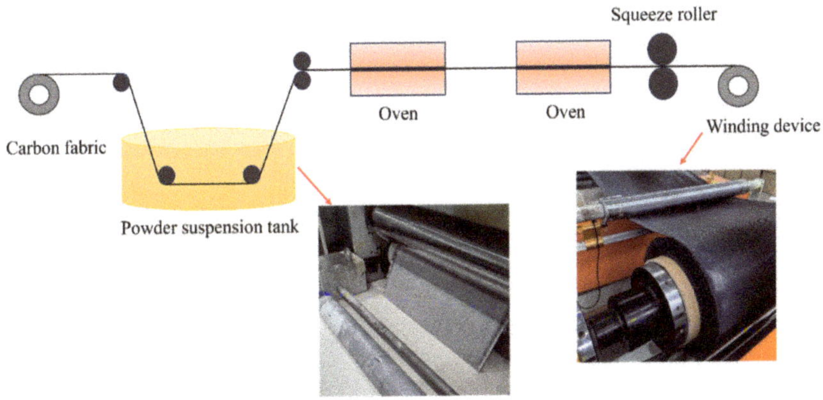

Figure 1. Schematic diagram of CFF/PEEK prepreg preparation process.

The mechanical specimens of PEEK resin were fabricated by injection molding and prepared in accordance with the dimensions specified by the relevant testing standards. The CFF/PEEK composite laminates were manufactured using a hot-pressing molding process. The fiber volume fraction of the composite laminates was approximately 55 vol%, and the layering process was conducted manually. The layering configurations and dimensions for various test specimens are detailed in Table 1. Following the lay-up of the prepregs, they were subjected to heating and pressing in a thermal press, resulting in the formation of composite laminates. The physical properties of CFF/PEEK composite laminates are presented in Table 2. The composite laminates were machined to meet the dimensional requirements stipulated by the testing standards. Prior to being placed in a hygroscopic environment, the edge banding of the composite samples is required.

Table 1. Test specimen layup of CFF/PEEK composite laminates.

Type	Ply Count	Layup	Nominal Lamina Thickness/mm	Size/mm
Warp Tensile Test	8	$[0_f]_8$	2.4	250 × 25
Weft Tensile Test	8	$[90_f]_8$	2.4	250 × 25
Warp Compression Test	8	$[0_f]_8$	2.4	140 × 12
Weft Compression Test	8	$[90_f]_8$	2.4	140 × 12

Notes: Subscript f—fabric prepreg; layup 0 or 90—angle between the fabric warp direction and the x-axis direction.

Table 2. Physical properties of CFF/PEEK composite laminates.

Properties	Resin Content	Density	Fiber Volume Fraction	Glass Transition Temperature
Value	38 wt%	1.57 g/cm^3	55 vol%	146 °C
Test Method	GB/T 3855 [28]	GB/T 1463 [29]	ASTM D3171 [30]	ASTM D 7028 [31]

2.2. Hygrothermal Treatment

In accordance with ASTM D5229, the water absorption behavior of CFF/PEEK composite specimens (50 mm × 50 mm × 2 mm) was evaluated using gravimetric analysis [32]. Initially, the specimens were subjected to drying in an air-drying oven at 40 °C, with periodic weighing conducted every 24 h. The mass loss of the laminate was calculated using Equation (1). Drying was terminated when the daily mass change stabilized below 0.02%, after which the specimens were transferred to a desiccator for further use.

$$M_i = \frac{m_0 - m_i}{m_0} \times 100\% \quad (1)$$

where M_i is the dehumidification amount of the sample; and m_0 and m_i are the measured masses of the sample before and after drying, respectively.

The dried samples were placed in the MHU-408CRSA constant temperature and humidity chamber manufactured by Dongguan Xinbao Instrument Co., Ltd. (Dongguan, Guangdong, China). The samples underwent hygrothermal moisture absorption treatment under conditions of 85% relative humidity and 70 °C ambient temperature. At each aging interval, the samples were removed from the chamber, surface moisture was carefully blotted dry with absorbent paper, and subsequently weighed using an electronic balance with a precision of 0.1 mg. The moisture absorption amount was then calculated according to Formula (2). This procedure was repeated for five samples per group, and the mean value was recorded.

$$W_i = \frac{G_i - G_0}{G_0} \times 100\% \quad (2)$$

where W_i is the moisture absorption capacity of the sample; G_i is the mass of the sample after moisture absorption; and G_0 is the mass of the sample after drying.

2.3. Mechanics Performance Testing

The tensile properties of PEEK resin were evaluated in accordance with the GB/T 1040-2018 standard, while the compressive properties were assessed following the GB/T 1041-2008 standard [33,34]. For CFF/PEEK composites, the tensile properties were tested according to ASTM D3039, and the compressive properties were determined based on ASTM D6641 [35,36]. Prior to the commencement of the mechanical test, strain gauges should be bonded to the working section of the specimens to monitor its strain. All tests were conducted using an Instron-5982 universal testing machine (Instron Instruments Ltd., Boston, MA, USA). The tensile specimens were loaded at a rate of 3 mm/min, and the compressive specimens were subjected to a loading rate of 1 mm/min. The test environment temperature was controlled using a WGDN-7350 high- and low-temperature chamber produced by Lishi (Shanghai) Scientific Instrument Co., Ltd. (Shanghai, China). To ensure data reliability, the mechanical properties of three batches of composites were evaluated, with each batch comprising no fewer than six valid samples. The test results were normalized based on a nominal thickness of 2.4 mm.

2.4. Morphological Analysis

The microscopic morphologies of the fracture surfaces of the mechanical specimens were examined both before and after hygrothermal aging using a Quanta 450 FEG field emission environmental scanning electron microscope (SEM) produced by Thermo Fisher Scientific Ltd. (Waltham, MA, USA). Prior to SEM observation, it was crucial to improve the conductivity of the specimens to ensure optimal imaging quality. This was accomplished by applying a gold coating via sputtering for 60 s using a gold sputtering device.

3. Results and Discussion

3.1. Moisture Absorption Characteristics

The pure PEEK resin and the CFF/PEEK composite laminate specimens were subjected to saturated moisture absorption treatment at 70 °C, respectively. The diffusion behavior of water molecules within the composite material can be characterized by the water transport kinetics equation as follows:

$$M_t/M_m = kt^n \quad (3)$$

where M_t represents the moisture absorption rate at time t, M_m denotes the saturated moisture absorption rate, k signifies the diffusion constant of water molecules, and n is a critical parameter that characterizes the swelling mechanism of the material.

The k and n obtained from fitting the moisture absorption data of PEEK resin and the composite materials using the logarithmic form of the aforementioned equation are summarized in Table 3. Analysis of these data indicates that the values of n for both moisture absorption curves are approximately 0.5. Therefore, it can be reasonably inferred that the initial water absorption behavior of the PEEK resin and CFF/PEEK composite during the early stages of aging is consistent with the Fickian diffusion model [37].

Table 3. Summary of water absorption parameters.

Material	n	k/h^{-n}	M_m/%	k'	D/(10^{-3} mm^2/h^{-1})
PEEK	0.455	0.057	0.32	0.008	2.317
CFF/PEEK composite	0.526	0.063	0.18	0.021	1.353

Notes: n—parameter describing the swelling mechanism; k—diffusion constant; h—specimen thickness, M_m—maximum moisture uptake at equilibrium state; k'—slope of linear portion of the sorption curves; D—diffusion coefficient of the composites.

The solution of the Fickian equation can be described by Equation (4).

$$\frac{M_t}{M_m} = 1 - \frac{8}{\pi^2}\sum_{n=0}^{n=\infty}\frac{1}{(2n+1)^2}exp\left[-\left(\frac{Dt}{h^2}\right)\pi^2(2n+1)^2\right] \quad (4)$$

where D is the diffusion coefficient of the material and h is the thickness of the specimen.

When $M_t/M_m \leq 0.6$, the initial part of the water absorption curve can be described by Equation (5).

$$\frac{M_t}{M_m} = \frac{4}{h}\sqrt{\frac{Dt}{\pi}} \quad (5)$$

When $M_t/M_m > 0.6$, the water absorption curve can be described by Equation (6).

$$\frac{M_t}{M_m} = 1 - exp\left[-7.3\left(\frac{Dt}{h^2}\right)^{0.75}\right] \tag{6}$$

The diffusion coefficient D of water molecules in the material can be given by Equation (7).

$$D = \frac{\pi}{(4Mm)^2}\left(\frac{M_t h}{\sqrt{t}}\right)^2 = \pi\left(\frac{hk'}{4M_m}\right)^2 \tag{7}$$

where k' represents the slope of the linear part of the equation.

Based on the parameters derived from Table 2 of the experiment, the Fickian model fitting the curves for the PEEK resin and CFF/PEEK composite materials were generated and compared with the measured moisture absorption data, as illustrated in Figure 2. The distribution of the fitting curves and data points in Figure 2 indicates that during the initial stage of moisture absorption, the curve gradient is relatively steep, indicating a linear relationship between the moisture absorption rate and $t^{1/2}$. The experimental data exhibit excellent agreement with the Fickian model. In the later stages of moisture absorption, the curve deviates from linearity. This deviation can primarily be attributed to the combined effects of temperature and humidity, which cause both the resin itself and internal defects within the composite material to absorb water, leading to the rapid diffusion of water molecules within the composite material. Following this rapid absorption phase, the moisture absorption rate gradually decreases, and the moisture absorption curve flattens, eventually reaching saturation. The equilibrium moisture absorption rates for the PEEK resin and CFF/PEEK composite material are approximately 0.32% and 0.19%, respectively. Given that carbon fibers mainly consist of carbon elements with a disordered graphite structure [38], they exhibit minimal water absorption. Consequently, in the composite material, only the matrix and interface regions absorb water, resulting in significantly lower saturation moisture absorption for the CFF/PEEK compared to the pure PEEK resin.

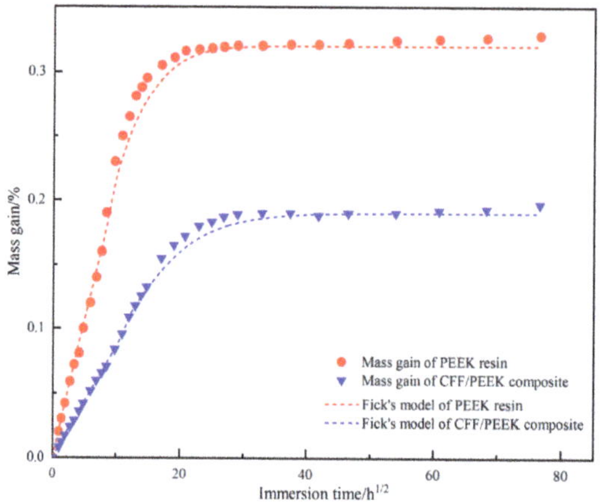

Figure 2. Water absorption curves and Fickian fitting results of PEEK resin and CFF/PEEK composite.

3.2. Influence of the Hygrothermal Environment on the Tensile Properties

The tensile properties of PEEK resin under dry conditions at room temperature (RTD) and wet conditions at 70 °C (ETW) are illustrated in Figure 3. The tensile properties of CFF/PEEK composites under RTD and ETW conditions are presented in Figure 4 (warp direction) and Figure 5 (weft direction). Analysis of the data reveals a notable reduction in tensile properties for both PEEK resin and CFF/PEEK composites under ETW conditions. Table 4 provides the retention rates of the tensile properties of PEEK resin and CFF/PEEK composites. Specifically, the tensile strength of PEEK resin under ETW conditions is approximately 74.92% of that observed under RTD conditions. For CFF/PEEK composites, the warp tensile strength is approximately 84.98%, and the weft tensile strength is approximately 86.08% of their respective RTD values. In contrast, the tensile modulus exhibits less sensitivity to hygrothermal effects. Under ETW conditions, the tensile modulus of PEEK resin is approximately 94.32% of its RTD value, while for CFF/PEEK composites, the warp tensile modulus is approximately 96.94%, and the weft tensile modulus is approximately 95.59% of their RTD values.

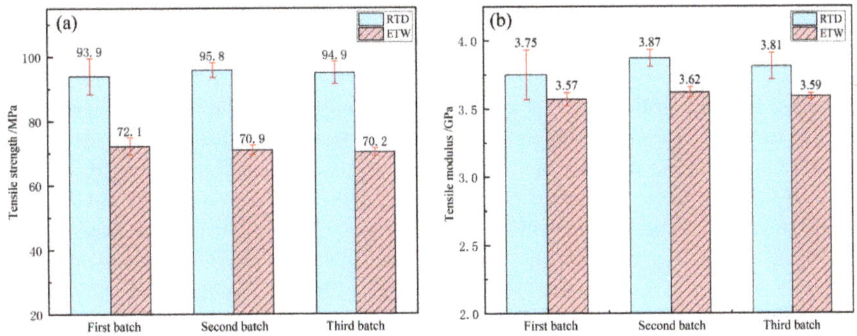

Figure 3. Tensile properties of PEEK resin: (**a**) tensile strength; (**b**) tensile modulus.

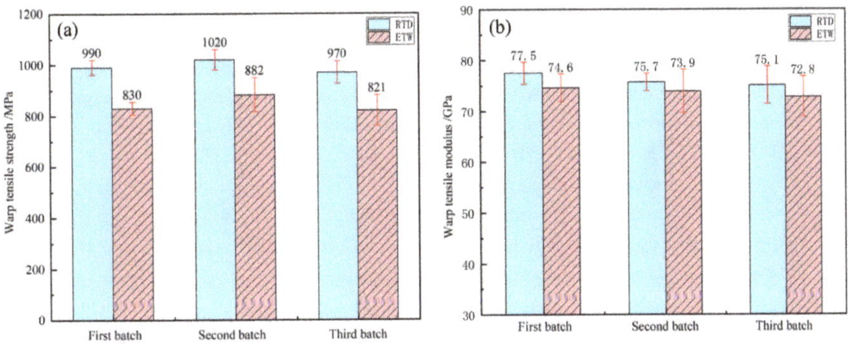

Figure 4. Warp tensile properties of CF/PEEK composite: (**a**) tensile strength; (**b**) tensile modulus.

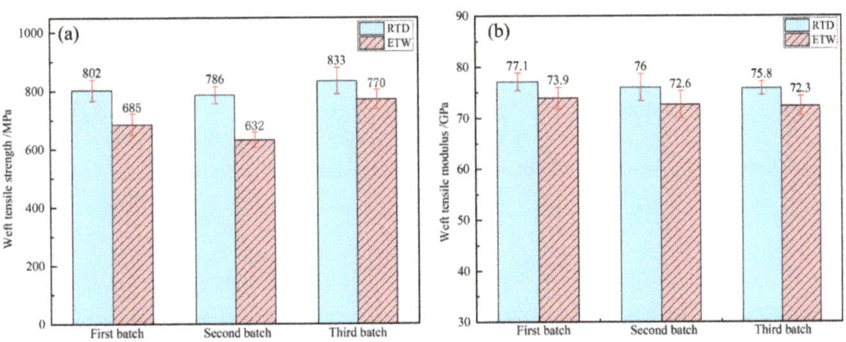

Figure 5. Weft tensile properties of CF/PEEK composite: (**a**) tensile strength; (**b**) tensile modulus.

Table 4. Retention rate of the tensile properties of PEEK resin and CFF/PEEK composites.

Material	Performance	First Batch	Second Batch	Third Batch	Average
PEEK resin	Tensile strength (%)	76.78	74.01	73.97	74.92
	Tensile modulus (%)	95.20	93.54	94.23	94.32
CFF/PEEK composite	Warp tensile strength (%)	83.84	86.47	84.64	84.98
	Warp tensile modulus (%)	96.26	97.62	96.94	96.94
	Weft tensile strength (%)	85.41	80.41	92.44	86.08
	Weft tensile modulus (%)	95.85	95.53	95.38	95.59

The significant influence of the hygrothermal environment on the tensile strength of the composites can be attributed to multiple factors. Firstly, in a wet state, water molecules that penetrate the resin matrix can degrade both the performance of the resin and the interface between the fibers and matrix. Furthermore, water molecules infiltrating through micro-cracks in the resin and reaching the carbon fiber surface can reduce the tensile strength of the fibers to some extent [39]. Additionally, elevated temperatures exacerbate this issue by accelerating water molecule penetration into the composite materials and directly damaging the molecular chains of the resin matrix, thereby weakening the bonding between the matrix and fibers, and reducing the tensile strength of the composites. Moreover, the combined effects of temperature increase and moisture intrusion can cause differential expansion between the fibers and matrix due to differences in their thermal and wet expansion coefficients, leading to internal stresses within the composites, which may contribute significantly to the reduction in tensile strength under ETW conditions. In contrast, hygrothermal environments impact the tensile modulus of composites to a lesser extent. This is because the elastic modulus of carbon fibers remains largely unaffected by temperature and humidity within a certain range, while the modulus of the PEEK resin matrix decreases at high temperatures and under wet conditions. Consequently, the tensile modulus of composites, which is primarily determined by the elastic modulus of carbon fibers, is generally less influenced by hygrothermal environments.

The failure modes of tensile specimens of CFF/PEEK composites are illustrated in Figure 6, while the microscopic morphology of the fracture surfaces is depicted in Figure 7. As observed from Figure 6, tensile failures occur predominantly in the central or terminal regions of the working section, indicating a consistent and effective failure mode. The fracture surfaces exhibit a relatively uniform appearance, with no significant differences noted between the ETW and RTD states. Microscopic examination of the fracture surfaces in Figure 7 reveals that, after moisture absorption, the failure mechanisms involve fiber pull-out and fracture, maintaining a neat fracture surface. Additionally, no evident delamination and cracks were observed after moisture absorption, suggesting minimal chemical alterations during hygrothermal aging. In summary, the hygrothermal aging mechanism, affecting the tensile properties of CFF/PEEK composites following 70 °C saturated moisture absorption treatment, primarily involves the degradation of the matrix and interface due to moisture ingress, as well as property deterioration caused by the differential wet strain between the resin and fibers, leading to wet stress [40].

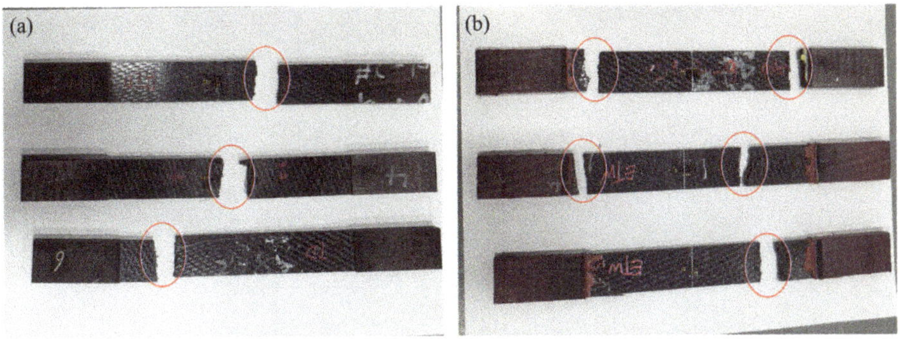

Figure 6. Typical failure modes of tensile test for CFF/PEEK composite: (**a**) CTD; (**b**) ETW.

Figure 7. Morphologies of CFF/PEEK composite tensile fracture under different environmental conditions: (**a**) RTD; (**b**) ETW.

3.3. Influence of Hygrothermal Environment on Compression Properties

The compression properties of PEEK resin under RTD conditions and ETW conditions are illustrated in Figure 8. The compressive properties of CFF/PEEK composites under these same conditions are depicted in Figure 9 (warp direction) and Figure 10 (weft direction). Table 5 presents the compression property retention rates for both PEEK resin and CFF/PEEK composites. Compared to tensile strength, the compression strength of PEEK resin and CFF/PEEK composites exhibits less sensitivity to hygrothermal environments. Meanwhile, the changes in the compression modulus mirror those of the tensile modulus, with both demonstrating relatively high retention rates following hygrothermal aging. Under ETW conditions, the compression strength of PEEK resin is approximately 81.85% of its RTD value, while the compressive modulus retains about 95.49% of its RTD value. For CFF/PEEK composites under ETW conditions, the compressive strength in the warp direction is approximately 95.35% of its RTD value, and in the weft direction, it is about 94.67% of its RTD value. Similarly, the compressive modulus in the warp direction is approximately 98.07% of its RTD value, and in the weft direction, it is about 97.62% of its RTD value.

Under ETW conditions, the compressive performance of PEEK resin exhibits a significant reduction, whereas the compressive strength of CFF/PEEK composite laminates remains relatively stable. This phenomenon can be attributed to water molecules infiltrating the resin matrix in a wet environment, which leads to a marked decline in the compressive properties of the resin matrix and consequently affects the overall compression performance of the composite laminates. Moreover, water molecules also penetrate the interlaminar and carbon fiber/resin interfaces within the composites, causing interface degradation and diminishing their load transfer capability, thereby further compromising their compressive performance [41,42]. The compression strength of composites under hygrothermal conditions serves as a direct indicator of the impact of hygrothermal aging on the compressive strength of the resin matrix.

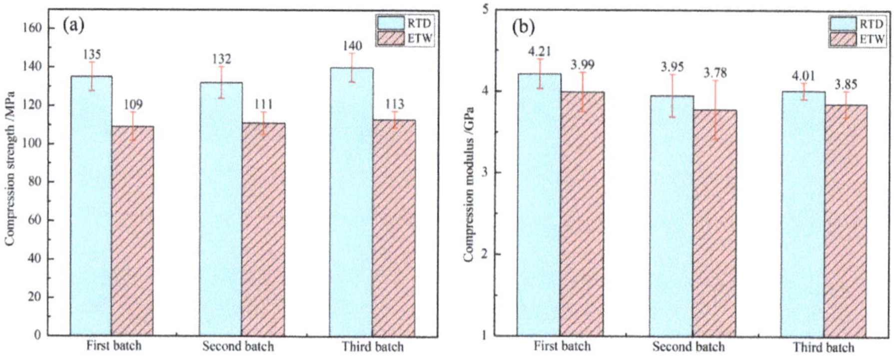

Figure 8. Compression properties of PEEK resin: (**a**) compression strength; (**b**) compression modulus.

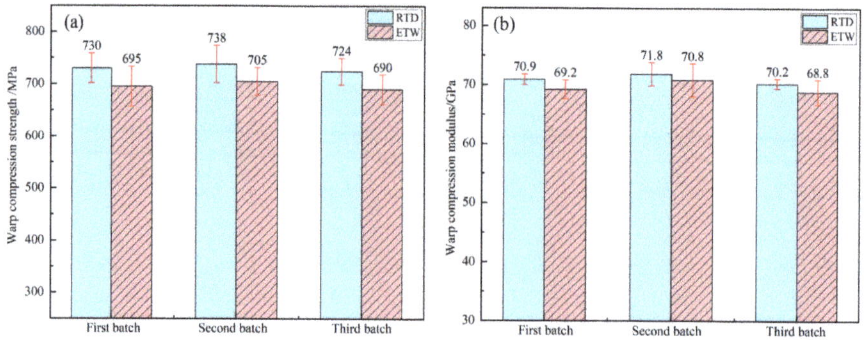

Figure 9. Warp compression properties of CF/PEEK composite: (**a**) compression strength; (**b**) compression modulus.

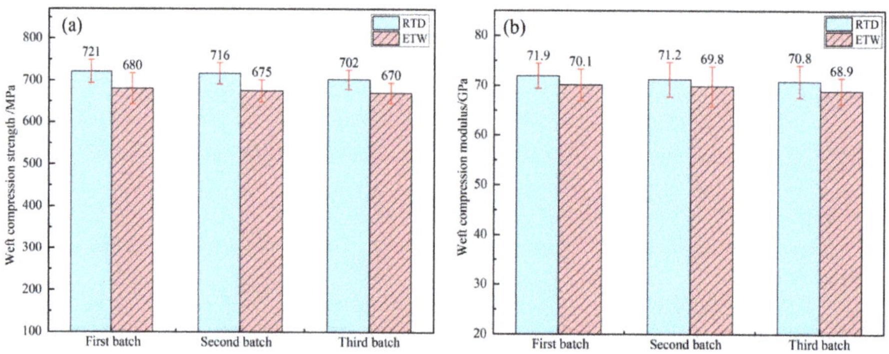

Figure 10. Weft compression properties of CF/PEEK composite: (**a**) compression strength; (**b**) compression modulus.

Table 5. Retention rate of compression properties of PEEK resin and CFF/PEEK composites.

Material	Performance	First Batch	Second Batch	Third Batch	Average
PEEK resin	Compression strength (%)	80.74	84.09	80.71	81.85
	Compression modulus (%)	94.77	95.70	96.01	95.49
CFF/PEEK composite	Warp compression strength (%)	95.21	95.53	95.30	95.35
	Warp compression modulus (%)	97.60	98.61	98.01	98.07
	Weft compression strength (%)	94.31	94.27	95.44	94.67
	Weft compression modulus (%)	97.50	98.03	97.32	97.62

The failure modes of the compression specimens of CFF/PEEK composites are illustrated in Figure 11, while the fracture surface morphologies are depicted in Figure 12. From Figure 11, it is evident that the compression specimens failed within the working section, exhibiting an effective failure mode. Minor delamination was observed at the failure location, and the fracture surfaces displayed wedge-shaped characteristics. No significant differences were noted in the failure modes between the RTD and ETW states. As shown in Figure 12, which presents the fracture surface damage morphology of the specimens, the fracture surfaces before and after moisture absorption remained relatively flat, with no apparent microcracks or delamination. Consequently, it can be inferred that following the 70 °C hygrothermal aging process, no significant chemical reactions occurred in the CFF/PEEK composites. The reduction in compression performance is primarily attributed to the influence of water molecules on the resin matrix properties and the degradation of interfacial performance.

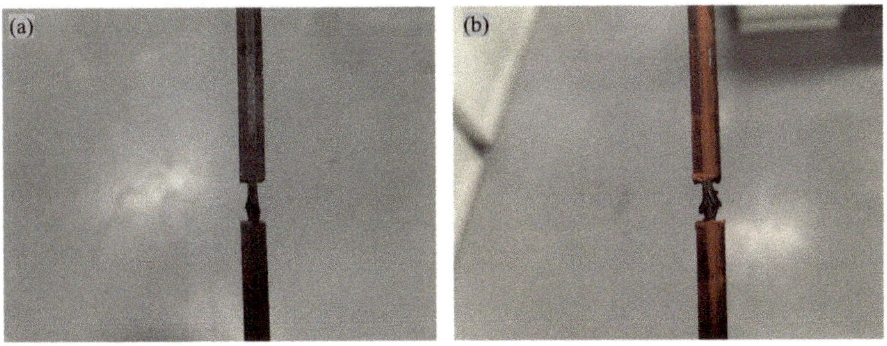

Figure 11. Typical failure modes of compression test for CFF/PEEK composite: (**a**) RTD; (**b**) ETW.

Figure 12. Morphologies of CFF/PEEK composite compression fracture under different environmental conditions: (**a**) RTD; (**b**) ETW.

4. Conclusions

This study investigates the impact of the hygrothermal environment on the mechanical properties of CFF/PEEK composites, examining the changes in these properties before and after moisture absorption. The saturated moisture absorption rate of CFF/PEEK composites at 70 °C is approximately 0.19%, compared to approximately 0.32% for pure PEEK resin. The initial moisture absorption behavior of both materials adheres to Fick's law, although deviations occur in the later stages. After absorbing moisture, the tensile strength retention rate of PEEK resin is approximately 74.92%, while that of CFF/PEEK composites is approximately 84.98% in the warp direction and 86.08% in the weft direction. In contrast, the tensile modulus exhibits less sensitivity to the hygrothermal environment. The compression strength retention rate of PEEK resin is approximately 81.85%, while the compression modulus retention rate stands at approximately 95.49%. For CFF/PEEK composites, compressive performance is relatively less affected by the hygrothermal environment. It was found that the damage mode of CFF/PEEK composites under the ETW state is not significantly different compared to that of the RTD state. Resin debonding was observed in the moisture-absorbed composite specimens, while no microcracks or delamination were detected. The primary reasons for the degradation of the mechanical properties of the CFF/PEEK composites after moisture absorption at 70 °C are attributed to the deterioration of the resin matrix and interface performance, as well as the wet stress induced by the inconsistent wet strain between the resin and fibers.

Author Contributions: Conceptualization, X.X., B.Z. and F.S.; methodology; X.X., B.Z. and F.S.; software, K.L.; validation, B.Z. and Y.D.; formal analysis, X.X.; investigation, X.X.; resources, F.S.; data curation, X.X.; writing—original draft preparation, X.X.; writing—review and editing, L.G. and G.P.; visualization, X.X.; supervision, B.Z. and F.S.; funding acquisition, J.G. and K.L. All authors have read and agreed to the published version of the manuscript.

Funding: This work is supported by National Key R&D Program of China (2022YFC2406103), the Science Foundation of National Key Laboratory of Science and Technology on Advanced Composites in Special Environments (JCKYS2024603C011), and the Key Laboratory Foundation General program (Grant No. 6142906220101).

Institutional Review Board Statement: Not applicable.

Data Availability Statement: All supporting data are contained within the manuscript.

Conflicts of Interest: The authors declare no conflicts of interest.

References

1. Liu, Y.; Lu, F.; Xu, N.; Wang, B.; Yang, L.; Huang, Y.; Hu, Z. Mechanically robust, hydrothermal aging resistant, imine-containing epoxy thermoset for recyclable carbon fiber reinforced composites. *Mater. Des.* **2022**, *224*, 111357. [CrossRef]
2. Liu, Y.; Wang, G.; Shen, Y.; Blackie, E.; He, L. Mode-II fracture toughness of carbon fiber reinforced polymer composites interleaved with polyethersulfone (PES)/carbon nanotubes (CNTs). *Compos. Struct.* **2023**, *320*, 117214. [CrossRef]
3. Zhao, W.; Li, X.; Ye, J.; Liu, K.; Xu, X.; Liu, Z. Semi-destructive methods for evaluating the micro-scale residual stresses of carbon fiber reinforced polymers, Semi-destructive methods for evaluating the micro-scale residual stresses of carbon fiber reinforced polymers. *Compos. Part B Eng* **2024**, *284*, 111722. [CrossRef]
4. Ge, J.; Catalanotti, G.; Falzon, B.; Higgins, C.; McClory, C.; Thiebot, J.-A.; Zhang, L.; He, M.; Jin, Y.; Sun, D. Process characteristics, damage mechanisms and challenges in machining of fibre reinforced thermoplastic polymer (FRTP) composites: A review. *Compos. Part B Eng.* **2024**, *273*, 111247. [CrossRef]
5. Chen, J.; Wang, K.; Zhao, Y. Enhanced interfacial interactions of carbon fiber reinforced PEEK composites by regulating PEI and graphene oxide complex sizing at the interface. *Compos. Sci. Technol.* **2018**, *154*, 175–186. [CrossRef]
6. Sun, G.; Chen, D.; Zhu, G.; Li, Q. Lightweight hybrid materials and structures for energy absorption: A state-of-the-art review and outlook. *Thin-Walled Struct.* **2022**, *172*, 108760. [CrossRef]
7. Avic, R.; Çakıcı, U.G.; Çetinkaya, B.; Öktem, M.F. Effect of atmospheric plasma treatment and wet blast on adhesion characteristics of carbon fiber reinforced LM-PAEK thermoplastic composites. *Compos. Part B Eng.* **2024**, *278*, 111394. [CrossRef]

8. Bilgi, C.; Demir, B.; Aydın, H.; Üstün, B.; Kurtan, Ü. Effect of functionalized graphene nanoplatelet dispersion on thermal and electrical properties of hybrid carbon fiber reinforced aviation epoxy laminated composite. *Mater. Chem. Phys.* **2024**, *325*, 129702. [CrossRef]
9. Harussani, M.M.; Sapuan, S.M.; Nadeem, G.; Rafin, T.; Kirubaanand, W. Recent applications of carbon-based composites in defence industry: A review. *Def. Technol.* **2022**, *18*, 1281–1300. [CrossRef]
10. Toro, S.A.; Gonzálezb, C.; Fernández-Blázquezc, J.P.; Ridruejob, A. Fabrication and mechanical properties of a high-performance PEEK-PEI hybrid multilayered thermoplastic matrix composite reinforced with carbon fiber. *Compos. Part A Appl. Sci. Manuf.* **2024**, *185*, 108308. [CrossRef]
11. Xu, X.; Peng, G.; Zhang, B.; Shi, F.; Gao, L.; Gao, J. Material performance, manufacturing methods, and engineering applications in aviation of carbon fiber reinforced polymers: A comprehensive review. *Thin-Walled Struct.* **2025**, *209*, 112899. [CrossRef]
12. Qiao, Y.; Fring, L.D.; Pallaka, M.R.; Simmons, K.L. A review of the fabrication methods and mechanical behavior of continuous thermoplastic polymer fiber-thermoplastic polymer matrix composites. *Polym. Compos.* **2023**, *44*, 694–733. [CrossRef]
13. Yuan, C.; Ma, Z.; Pu, Y.; Liu, L.; Huang, Y. Soluble semi-aliphatic polyimide sizing in CF/PEI composites for improving mechanical performance and hydrothermal aging resistance. *Compos. Commun.* **2021**, *28*, 100982. [CrossRef]
14. Dabees, S.; Wickramasingha, Y.A.; Akhavan, B.; Hayne, D.J.; Dharmasiri, B.; Henderson, L.C. Tribological behaviour of surface modified carbon-fibre-reinforced polyphenylene sulphide under dry condition. *Tribol. Int.* **2024**, *194*, 109528. [CrossRef]
15. Stoeffler, K.; Andjelic, S.; Legros, N.; Roberge, J.; Schougaard, S.B. Polyphenylene sulfide (PPS) composites reinforced with recycled carbon fiber. *Compos. Sci. Technol.* **2013**, *84*, 65–71. [CrossRef]
16. Matus-Aguirre, M.; Cosson, B.; Garnier, C.; Schmidt, F.; Akué-Asséko, A.C.; Chabert, F. Characterization and modeling of laser transmission welded polyetherketoneketone (PEKK) joints: Influence of process parameters and annealing on weld properties. *J. Adv. Join. Process.* **2024**, *10*, 100252. [CrossRef]
17. Song, Y.-J.; Kim, D.-K.; Han, W.; Choi, S.-H.; Chung, D.-C.; Kim, K.-W.; Kim, B.-J. Facile enhancement of mechanical interfacial strength of recycled carbon fiber web-reinforced polypropylene composites via a single-step silane modification process. *Polymers* **2025**, *17*, 483. [CrossRef]
18. Martin, R.G.; Johansson, C.; Tavares, J.R.; Dubé, M. CF/PEEK skins assembly by induction welding for thermoplastic composite sandwich panels. *Compos. Part B Eng.* **2024**, *284*, 476. [CrossRef]
19. Zhao, Y.; Li, Q.; Zhou, K.; Jing, B.; Zhu, K.; Shi, J.; Li, C. Effects of hygrothermal condition on water diffusion and flexural properties of carbon–glass hybrid fiber-reinforced epoxy polymer winding pipes. *Polymers* **2024**, *16*, 3433. [CrossRef]
20. Li, S.; Chen, Z.; Wang, Y.; Huang, Y.; Yu, T. Quasi-static compression and hygrothermal stability of BMI/CE co-cured composite lattice cylindrical shell. *Compos. Struct.* **2021**, *257*, 113130. [CrossRef]
21. Hassanpour, B.; Karbhari, V.M. Characteristics and models of moisture uptake in fiber-reinforced composites: A topical review. *Polymers* **2024**, *16*, 2265. [CrossRef]
22. Ma, C.C.M.; Lee, C.L.; Chang, M.J.; Tai, N.H. Hygrothermal behavior of carbon fiber-reinforced poly(ether ether ketone) and poly(phenylene sulfide) composites. *Polym. Compos.* **1992**, *13*, 448–453. [CrossRef]
23. Wu, R.; Li, Y.; Yu, T. Comparative study on the hygrothermal durability of different fiber reinforced composites. *Acta Mater.* **2022**, *39*, 4406–4419. [CrossRef]
24. Li, Z.; Liu, Z.; Zhao, Y. Effects of cyclic hygrothermal environments on interfacial behavior and properties of CF/EP composites. *Compos. Commun.* **2025**, *53*, 102229. [CrossRef]
25. Feng, Z.; Mou, H.; Xie, J.; Gong, T. Hygrothermal environment effects on mechanical properties of T700/3228 CFRP laminates. *Mater. Test.* **2019**, *61*, 857–863. [CrossRef]
26. Wan, A.; Shi, J. Hygrothermal aging effect on static and fatigue behavior of three-dimensional five-directional hybrid braided composites under tension. *Mater. Today Commun.* **2024**, *40*, 109843. [CrossRef]
27. Micelli, F.; Nanni, A. Durability of FRP rods for concrete structures. *Constr. Build. Mater.* **2004**, *18*, 491–503. [CrossRef]
28. *GB/T 3855-2005*; Test Method for Resin Content of Carbon Fiber Reinforced Plastics. Standards Press of China: Beijing, China, 2005.
29. *GB/T 1463-2005*; Test Method for Density and relative Density of Carbon Fiber Reinforced Plastics. Standards Press of China: Beijing, China, 2005.
30. *ASTM D3171*; Standard Test Methods for Constituent Content of Composite Materials. American Society of Testing Materials: West Conshohocken, PA, USA, 2022.
31. *ASTM D7028*; Glass Transition Temperature (DMA Tg) of Polymer Matrix Composites by Dynamic Mechanical Analysis (DMA). American Society of Testing Materials: West Conshohocken, PA, USA, 2022.
32. *ASTM D5229*; Standard Test Method for Moisture Absorption Properties and Equilibrium Conditioning of Polymer Matrix Composite Materials. American Society of Testing Materials: West Conshohocken, PA, USA, 2020.
33. *GB/T 1040-2018*; Plastics-Determination of Tensile Properties-Part 1: General Principles. Standards Press of China: Beijing, China, 2018.
34. *GB/T 1040-2008*; Plastics-Determination of Compressive Properties. Standards Press of China: Beijing, China, 2008.

35. *ASTM D3039*; Standard Test Method for Tensile Properties of Polymer Matrix Composite Materials. American Society of Testing Materials: West Conshohocken, PA, USA, 2017.
36. *ASTM D6641*; Standard Test Method for Compressive Properties of Polymer Matrix Composite Materials Using a Combined Loading Compression (GLC) Test Fixture. American Society of Testing Materials: West Conshohocken, PA, USA, 2016.
37. Scida, D.; Assarar, M.; Poilâne, C.; Ayad, R. Influence of hygrothermal ageing on the damage mechanisms of flax-fibre reinforced epoxy composite. *Compos. Part B Eng.* **2013**, *48*, 51–58. [CrossRef]
38. He, F. *Carbon Fibre and Graphite Fibre[M]*; Chemical industry press: Beijing, China, 2010; pp. 364–368.
39. Falkenreck, C.K.; Zarges, J.-C.; Heim, H.-P. Hygrothermal aging behavior of regenerated cellulose fiber-reinforced polyamide 5.10 composites. *Polym. Test.* **2025**, *143*, 108717. [CrossRef]
40. Gupta, A.K.; Kiran, R.; Zafar, S.; Pathak, H. Effect of hygrothermal ageing on the mechanical properties of glass fiber reinforced polymer composite: Experimental and numerical approaches. *Mater. Today Commun.* **2024**, *41*, 111060. [CrossRef]
41. Karbhari, V.M. Influence of Montmorillonite Organoclay Fillers on Hygrothermal Response of Pultruded E-Glass/Vinylester Composites. *Polymers* **2024**, *16*, 2157. [CrossRef]
42. Zhang, Y.; Shi, Y.; Wang, X.; Guan, Y.; Du, C.; Fan, T.; Li, H.; Zhai, J. An Experimental Investigation of the Mechanism of Hygrothermal Aging and Low-Velocity Impact Performance of Resin Matrix Composites. *Polymers* **2024**, *16*, 1477. [CrossRef] [PubMed]

polymers

MDPI

Article

Complex Networks in Phase-Separating Gels: A Computer Simulation Study

Georg Friedrich Beer

Magnetism and Interface Physics & Computational Polymer Physics, Department of Materials, ETH Zurich, 8093 Zurich, Switzerland; 6b33r@pm.me

Abstract: The mesoscopic phase separation in two- and three-dimensional gels has been studied by computer simulation of a bead-spring model of Lennard-Jones particles. The formation of complex networks of high-density phase (HDP) has been investigated and partially explained by competing short- and long-range energies. HDP network formation was found to occur at certain combinations of temperature and spring coefficients, given sufficient particle density. The morphology of the HDP networks changed with these three parameters. HDP networks became more faceted with higher spring coefficients, wider but less dense at higher temperatures, and more voluminous and compact at larger densities. HDP network formation was preceded by a stage of HDP precipitation and followed by a stage of surface minimization.

Keywords: networks; gels; simulation; molecular dynamics; Lennard-Jones particles; bead-spring model; phase separation; high-density phase; morphology; surface optimization

1. Introduction

Polymer networks, so-called gels, can undergo two types of phase transitions: volume transitions and phase separation. This study focuses on phase separation, where the gel separates into regions with high and low polymer concentrations [1]. Up to now, research on phase separation in gels has mainly treated surface effects [2,3]. Thus, the processes occurring within the bulk of the gel are a welcome object of study for simulations, as we still lack proper methods of physical investigation [2,3]. Comprehensive understanding and prediction of microphase separation in gels could be of great use for applications where the gel's microstructure plays a role. Two examples are the development of iongel soft solid electrolytes for solid-state batteries [4–6] and the comprehension of the complex deformation behavior exhibited by gels [7].

Motivated by earlier studies [8], which indicated phase separation with a characteristic length scale significantly larger than a model grid size, Peleg et al. [2] introduced a simplistic model to mimic the collection of long- and short-range forces in a gel. The model consists of particles (representing polymer density) interacting via a Lennard-Jones (LJ) potential (short-range attraction typifying polymer crystallization) connected through harmonic springs (entropy elasticity) in a two-dimensional simple square lattice structure. Their simulations revealed that, under certain circumstances, the gel phase separates into a high-density phase (HDP) and a low-density phase (LDP) with the HDP forming a filamentous, foam-like network.

In [2], Peleg et al. investigated their model for a limited set of system parameters. They quenched the gel from a simple cubic or hexagonal initial state to a few fixed temperatures, while keeping the spring constant $K = 1/10$ ($K = 1/15$ in the hexagonal case) and the density $\rho = 1/3.5$ fixed (both in LJ units [9]). Peleg et al. also found that the system exhibits a hysteresis when varying the temperature. In [3], they present a phase diagram for the same constant density $\rho = 1/3.5$, where they use the fraction of HDP particles as an order parameter. And in [10], Peleg at al. investigated networks on a random lattice.

In this paper, Peleg et al.'s [2] model is generalized to three dimensions, as suggested in [3], where Peleg et al. raised the question of whether a three-dimensional system would show linear HDP filaments connected in a network-like structure, as in two dimensions, or combine into a soap-like structure made of connected surfaces.

We want to establish measures to quantify the morphology of the evolving HDP networks and explore more of the parameter space by varying the density, spring constant, and temperature between simulations to create phase diagrams.

2. Materials and Methods

2.1. Model

In the two-dimensional elastic LJ toy model proposed and investigated by Peleg et al. [2,3,10], the gel is represented by particles initially arranged in a two-dimensional square (or hexagonal) lattice. Using LJ units [9], each particle is attributed a mass of one and is connected permanently to its four initial neighboring particles through weightless harmonic springs. These springs have a spring coefficient, denoted as K, and an equilibrium length of zero. Additionally, pairwise interactions between particles occur via a short-range 12-6-LJ potential, which is truncated beyond a cutoff radius $r_c = 3 \times 2^{1/6}$.

By employing periodic boundary conditions (PBCs), the network of harmonic springs functions to provide long-range repulsive forces, causing the particles to move apart to their simple cubic lattice positions. Conversely, the LJ forces operate at close range, attempting to "glue" the particles together. For three dimensions this model was expanded to a simple cubic lattice (Figure 1a). Here, each particle is connected permanently to its six initial neighboring particles. The dimension-dependent quantities such as distance and temperature were adapted accordingly.

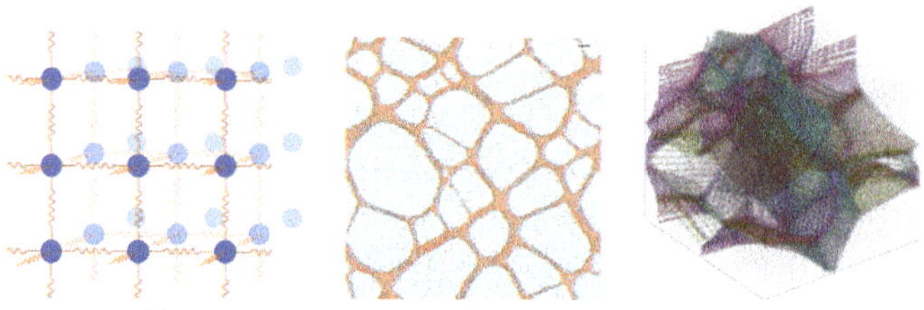

(a) (b) (c)

Figure 1. (**a**) Three-dimensional model of the gel: Simple cubic lattice of LJ particles connected with harmonic springs of equilibrium distance zero, subjected to periodic boundary conditions. (**b**) Two-dimensional gel model simulation after a time of $t = 4000$ with density $\rho = 0.29$, temperature $T = 0.3$, and spring coefficient $K = 0.3$. The springs connecting HDP particles are highlighted in orange for the graphical representation. (**c**) Three-dimensional gel model simulation after a time of $t = 4000$ with simulation parameters $\rho = 0.29$, $T = 0.3$, and $K = 0.1$. In this graphical representation, the springs are colored according to their spatial orientation. All units are dimensionless; reduced LJ units are used [9].

2.2. Molecular Dynamics

The three-dimensional model was studied by a molecular dynamics (MD) simulation in the canonical ensemble. LJ units are used throughout, i.e., physical units were kept dimensionless [9]. The system was initialized as a simple cubic lattice of N particles and the volume of the simulation box $V = L^3$ was set to give the desired number density

$\rho = N/V$. At startup, each particle was attributed a pseudo-random velocity vector $\mathbf{v}$ according to a Maxwell–Boltzmann (MB) distribution $f(\mathbf{v}) \sim e^{-e_{\text{kin}}/T}$ with specific kinetic energy $e_{\text{kin}} = v^2/2$, while the overall momentum was constrained to be zero. Newton's equations of motion were integrated by a Velocity Verlet algorithm with an integration time step $\Delta t = 0.004$. Temperature was kept constant by periodically rescaling the velocities every hundredth time step to the corresponding mean kinetic energy of $\bar{E}_{kin}^{2D} = TN$ in two dimensions and $\bar{E}_{kin}^{3D} = 3TN/2$ in three dimensions.

2.3. Force Calculations

The springs were set up in a way that permanently connected a particle to the nearest periodic-boundary-condition copy of its four/six initial neighbors. In all simulations, spring lengths stayed well below one half of the simulation box length, ensuring consistency. For d dimensions only $d \times N$ springs had to be checked for the spring force calculations. Thus, they used only a small part of the total computing time.

The main bottleneck of these large MD simulations was the calculations of the LJ forces. With more than one hundred thousand particles, and therefore (potentially) several billion LJ force calculations per time step, it became imperative to optimize this process. This was achieved by the implementation of neighbor lists. This effectively reduced the number of LJ force calculations per time step by grouping the particles into neighborhoods. The neighborhood of a given particle was made to consist of all other particles which were likely to interact with it until the next update of the neighbor lists. In the present work, the neighborhood of a particle encompassed all other particles closer than an estimated radius:

$$R_{\text{NL}} = r_c + \Delta R \gtrsim r_c + 12\,\Delta i_{\text{NL}}\,\Delta t\sqrt{\frac{2T}{\pi}}, \tag{1}$$

where r_c denotes the cutoff radius, Δi_{NL} the number of time steps between neighbor list updates, Δt the time step, and T the temperature.

These neighbor lists are most effective at low temperatures facing an even particle distribution at low density. At larger temperatures, the neighborhood radius R_{NL} (1) increases and hence the neighborhoods encompass more particles, which reduces their effectiveness. An uneven particle distribution has a similar effect, increasing the particle count of the neighbor lists of all HDP particles drastically. It was chosen to update the list every hundredth timestep $\Delta i_{\text{NL}} = 100$. We checked that the criterion (1) ensures that all interactions are properly taken into account for all systems studied (see Appendix A).

As programming language and framework, MATLAB (R2017b) was used . All two-dimensional simulations were run and analyzed on a desktop PC (Intel i7-6700 CPU), the three-dimensional simulations were run on the ETH-Zurich Euler cluster. Attempts at using LAMMPS (3 Mar 2020) [11] (Versions) to simulate this model were unsuccessful.

2.4. Methods of Investigation

The evolving HDP networks were of a complex nature. Several methods have been found to describe the gel's structure: (i) graphical representation, (ii) minimal particle distribution (MPD), (iii) spring length distribution (SLD), and (iv) radial shell density distribution (RSD).

(i) Figure 1b,c show a graphical representation of the trajectories of the simulated gel. The connecting springs are drawn with a limited number of points per line. This has the effect of rendering extended springs transparent and contracted springs opaque. In two dimensions, springs smaller than $2 \times 2^{1/6}$ are colored orange, highlighting the HDP. In three dimensions, foam-like structures consisting of faces and filaments evolved. To increase the visibility of these structures, each spring is colored according to its spacial orientation. Overall, this has the effect of highlighting faces and coloring them according to their orientation.

(ii) To determine the MPD of a gel state, for each particle the distance to its closest neighbor was measured. Then, a binning algorithm was applied, with a thousand equally spaced binning intervals ranging from zero to one-fifth of the simulation box length.

(iii) To construct the SLD, this same binning algorithm was used on all current spring lengths instead of the minimal radial particle distances.

(iv) The RSD, which is the radial distribution function not normalized by the bulk density, is usually calculated by measuring all pairwise particle distances, then adding, binning, and normalizing them. With a gel of $N = 60^3$ particles this would have meant calculating and binning 60^6 distances. To keep the calculation effort reasonable, for each particle one percent of the other particles were chosen at random to then measure the distances. The binning was carried out with 2500 binning intervals over half of the simulation box length and therefore with the same binning interval length, or more precisely the same thickness of the spherical binning shell dr, as for the MPD and SLD. The resulting distribution was then normalized by the volume of the binning shell ($4\pi r^2 dr$) and by the number of sample distances per particle ($0.01 \times N$).

To monitor the evolution of the structure over time, several measures were taken to express the gel's structure in one number per time frame. This process is sketched in Figure 2. Any two particles closer to each other than a critical distance of $d_{cd} = 1.1 \times 2^{1/6}$ were declared *HDP particles* (see Figure 2). The simulation box was then regularly separated into *small cubes* of edge length $s_b = 2$. To cancel out random agglomerates, such a small cube needed to encase more than four HDP particles to be declared full. Thereafter, a new grid of *big cubes* with edge length $s_B = 2s_b$ was created–each big cube therefore consisted of eight small cubes in three dimensions or four in two dimensions. A big cube was now declared to be a *surface cube* if at least one, but not all, of its small cubes were full. The above values for agglomerate threshold, the size of a small box, the number of small cubes making up a big cube, and the upper and lower limits for a big cube to be a surface cube were adjusted graphically to fit the HDP structures visible by eye as well as possible.

By this algorithm, the following three numbers were obtained: the number of HDP particles, N_{HDP}; the number of full (small) cubes, b_{full}; and the number of (big) surface cubes, B_{surf}. From these, five quantities of interest could be obtained for two and three dimensions d: the relative HDP surface area,

$$a_{\text{HDP}} = \frac{A_{\text{HDP}}}{L^{d-1}} = \frac{1}{L^{d-1}} \times dB_{\text{surf}} \times s_B^{d-1}, \tag{2}$$

the fraction of HDP particles (only the particles inside of the small cubes were counted),

$$n_{\text{HDP}} = \frac{N_{\text{HDP}}}{N}, \tag{3}$$

the HDP volume fraction,

$$v_{\text{HDP}} = \frac{V_{\text{HDP}}}{L^d} = \frac{1}{L^d} \times b_{\text{full}} \times s_b^d, \tag{4}$$

the density of the HDP,

$$\rho_{\text{HDP}} = \frac{N_{\text{HDP}}}{V_{\text{HDP}}} = \frac{N_{\text{HDP}}}{b_{\text{full}} \times s_b^d}, \tag{5}$$

and lastly, the thickness of the HDP filaments

$$\delta_{\text{HDP}} = \frac{(d-1)V_{\text{HDP}}}{A_{\text{HDP}}}. \tag{6}$$

Equation (6) approximates the filament as an infinitely long cylinder, where the total surface is dominated by the curved surface area. There, the ratio of volume to surface is

$$\frac{V_{\text{cy}}}{A_{\text{cy}}} = \frac{\pi r^2 h}{2\pi r h} = \frac{r}{2} \quad \text{leading to} \quad r = \frac{2V_{\text{cy}}}{A_{\text{cy}}} \tag{7}$$

for the radius r. This is the ratio used in Equation (6) to obtain a rough measure of thickness. The same argument can be made in two dimensions, taking a rectangle instead of a cylinder.

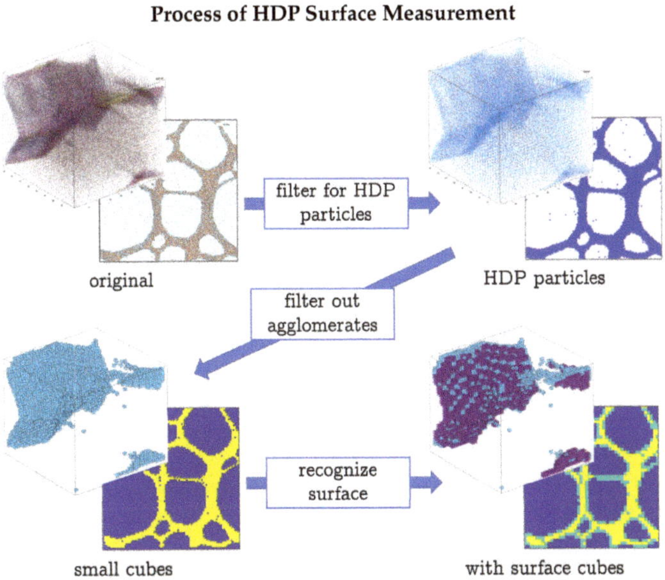

Figure 2. Process of surface measurement of the HDP network of a gel in two and three dimensions. First, HDP particles are determined. Then, all HDP particles which do not fall under the criterion for random agglomerates are put into small blue cubes (yellow squares). From there, the surface is beset with big violet surface cubes (green surface squares), which have the criterion of being partially comprised of small cubes/squares present.

In addition to these structural measures, the total static energy of the system $E_{\text{tot}} = E_{LJ} + E_K$ was calculated as the sum of the total pairwise LJ energy

$$E_{LJ} = 4\sum_{i=1}^{N}\sum_{j>i}^{N}\left(\frac{1}{r_{ij}^{12}} - \frac{1}{r_{ij}^{6}}\right) \tag{8}$$

and the total spring energy

$$E_K = \frac{K}{2}\sum_{i}^{N}\sum_{j:j\smile i}^{d} r_{ij}^2. \tag{9}$$

Here, r_{ij} denotes the scalar distance between particles i and j, and $i\smile j$ symbolizes a spring connection between them.

3. Results

3.1. Evolution of an HDP Network

Under certain conditions, the gel could be observed to phase-separate into a complex network of HDP, enclosing regions of LDP. Throughout this paper, units were kept dimensionless; reduced LJ units were used [9]. Figure 3 shows this for the examples of two- and three-dimensional gels. In both examples, small HDP precipitates could be seen from the start. Over time, they grew and connected to each other to form a complex HDP network.

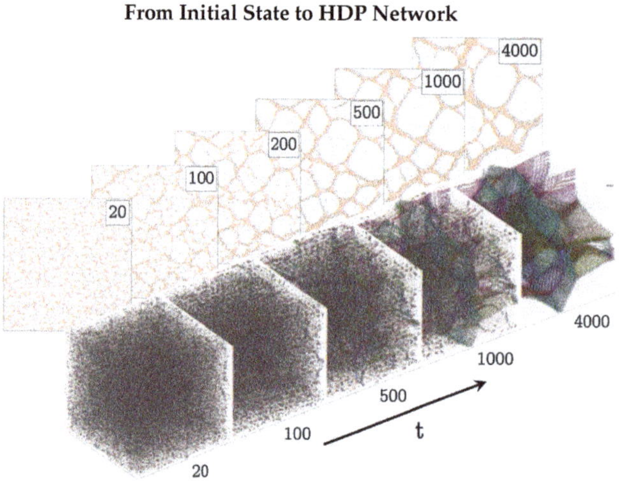

Figure 3. Gel simulations of the two- and three-dimensional models at different points in time t. In both cases, fast phase separation followed by the slower formation of a complex HDP network can be observed. *The two-dimensional* gel model was simulated with $N = 100^2$ particles, at constant density $\rho = 0.3$, temperature $T = 0.3$, and spring coefficient $K = 0.3$. The springs belonging to HDP particles are colored orange. *The three-dimensional* gel simulation was run with $N = 60^3$ particles, at constant density $\rho = 0.1$, temperature $T = 0.3$, and spring coefficient $K = 0.3$. Here, all springs are colored according to their spatial orientation. All units are dimensionless; reduced LJ units are used [9].

To monitor this evolution, the three-dimensional gel's trajectories were analyzed according to the measures established in Section 2.4. Figure 4a shows the particle fraction n_{HDP}, volume fraction v_{HDP}, and density of the HDP. All three measures showed a rapid increase at the beginning which slowed down over time until asymptotic behavior was prevalent. The final volume occupied by the HDP was a little less than 3%, concentrating 73% of all particles to a density of $\rho_{\mathrm{HDP}} = 0.81$ inside the HDP.

The evolution of the relative HDP surface area a_{HDP} is shown in Figure 4b. Two stages can be observed. First, as the HDP precipitated and the precipitates grew larger than the agglomerate threshold, the relative HDP surface area increased. Later, after reaching a maximum of $a_{\mathrm{HDP}} = 1.44$ (HDP surface area was 44% larger than one L^2 side area of the surface cube) between $t = 500$ and $t = 1000$, the relative HDP surface area decreased again, until transitioning to asymptotic behavior towards the end of the simulation. The relative HDP surface area a_{HDP} and HDP volume fraction v_{HDP} are linked by Equation (6) to give the HDP filament thickness δ_{HDP}. This HDP filament thickness decreased to a minimum just before the onset of HDP network formation was visible by eye (see Figure 3) and then increased to an asymptotic value of $\delta_{\mathrm{HDP}} = 5.5$ (simulation box length was about $L = 130$).

The evolution of the gel's static energies showed a decrease in total LJ energy E_{LJ}, with a lesser increase in total spring energy E_{K}, leading to a decrease in total static energy E_{tot} over the course of the simulation (see Figure 4c). After an erratic first phase, the changes in all energies slowed down to become asymptotic towards the end of the simulation.

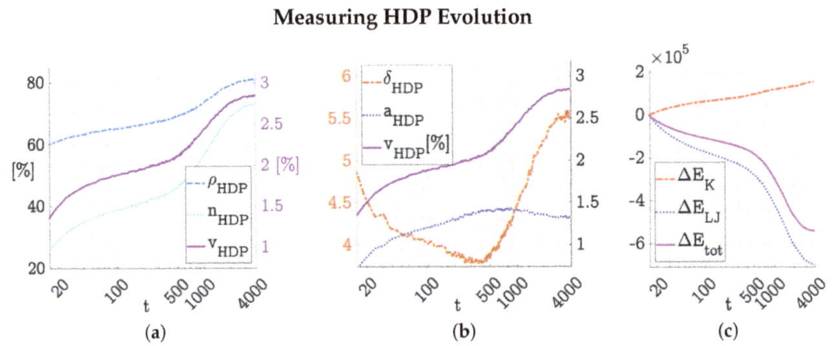

Figure 4. Evolution of the three-dimensional gel depicted in Figure 3 described by the measures established in Section 2.4. In the second half of the simulation, from $t = 2000$ to $t = 4000$, these measures approached an asymptotic state. The gel simulation was run with $N = 60^3$ particles, at constant density $\rho = 0.1$, temperature $T = 0.3$, and spring coefficient $K = 0.3$. (**a**) shows HDP density ρ_{HDP}, HDP particle fraction n_{HDP}, and HDP volume fraction v_{HDP} over time. (**b**) shows the evolution of HDP filament thickness δ_{HDP} together with HDP area fraction a_{HDP} and HDP volume fraction v_{HDP} over time. And (c) shows the change in spring energy ΔE_{K}, LJ energy ΔE_{LJ}, and their sum $\Delta E_{\text{tot}} = \Delta E_{\text{K}} + \Delta E_{\text{LJ}}$. All units are dimensionless; reduced LJ units are used [9].

To further analyze the morphology of the evolving structure, the MPD, SLD, and RSD were examined (see Figure 5). All three distributions showed a peak around the LJ minimum, indicated by a black dashed dotted line at a distance (or length or radius) of $2^{1/6}$. The dashed blue line in the MPD separates HDP from LDP particles. Looking closely at the tip of the MPD at a distance of $d_{\text{MPD}} = 2^{1/6}$, a shift to lower distances can be noticed, indicating an increasing HDP density. After $t = 500$, two peaks at $d_{\text{MPD}} = 1.9$ and around $d_{\text{MPD}} = 4$ became distinct—the right-hand-side peak moving towards larger distances indicates LDP regions of decreasing density. In the SLD, the peaks also grew more distinct over the course of the simulation. At all times, the right-hand side of the SLD showed a steep decrease in the number of large spring lengths $s_{\text{SLD}} > 5$. In this run, the SLD at $s_{\text{SLD}} = 2^{1/6}$ could be observed to decrease over time, from 5×10^4 at the beginning to 2×10^4 after $t = 4000$. As more and more particles joined the HDP over time, the radial shell density grew, and its peaks became more distinct. Drawing the distances of the second to fourth closest neighbors of bcc and fcc structures with minimal distance $2^{1/6}$ into the RSD suggests that a large fraction of the HDP is fcc.

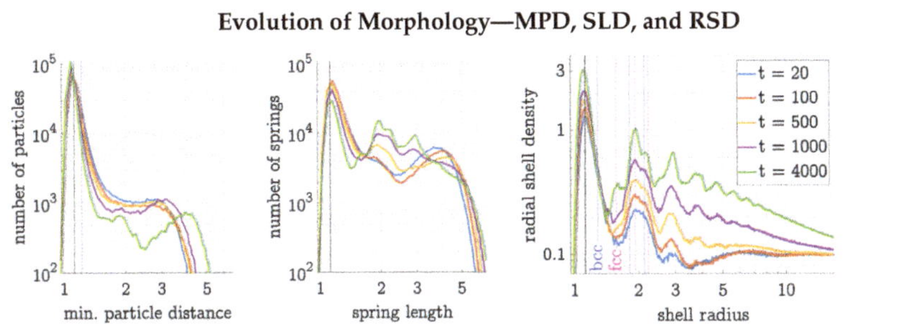

Figure 5. Evolution of the morphology (MPD, SLD, and RSD) of the three-dimensional gel simulation depicted in Figure 3. In all three plots, the dashed dotted black line at $2^{1/6}$ indicates the minimum of the LJ potential. The blue dashed line in the MPD at $1.1 \times 2^{1/6}$ separates HDP from LDP particles. In the RSD, additional lines indicate the first three peaks of bcc and fcc structures with minimal distance of $2^{1/6}$, respectively. The gel simulation was run with $N = 60^3$ particles at a constant density $\rho = 0.1$, temperature $T = 0.3$, and spring coefficient $K = 0.3$. All units are dimensionless; reduced LJ units are used [9].

3.2. Parameter Variations

Originating from the parameter set of the two- and three-dimensional examples treated in Section 3.1, the parameter space spanned by the density ρ, temperature T, and spring coefficient K was further explored. The influence of ρ, T, and K could be investigated separately by choosing one of these parameters to vary while the other two were kept constant.

The two-dimensional (2D) gel model simulations were run with $N = 100^2$ particles, originating from $\rho_{2D} = T_{2D} = K_{2D} = 0.3$; the three-dimensional (3D) simulations with $N = 60^3$ particles, originating from $\rho_{3D} = 0.1$ and $T_{3D} = K_{3D} = 0.3$. Figure 6 shows the graphical representation of these simulations after a time of $t = 4000$.

Influence of Density, Temperature, and Spring Coefficient

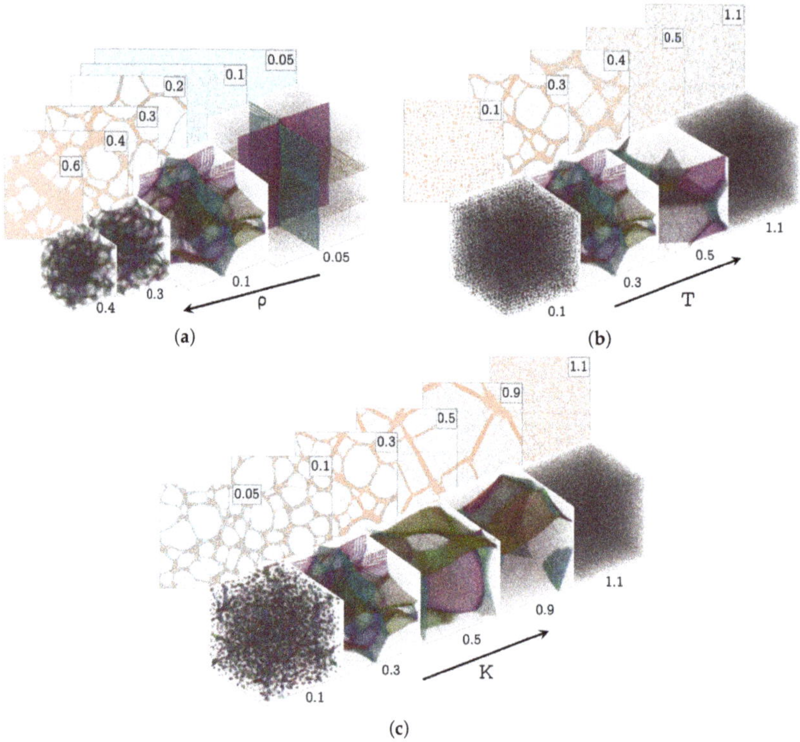

Figure 6. Final state ($t = 4000$) of the gel simulations with one parameter being varied (density ρ (**a**), temperature T (**b**), or spring coefficient K (**c**)) while the other two parameters remained constant. Upper (2D) simulations were conducted with $N = 100^2$ particles and with $\rho_{2D} = T_{2D} = K_{2D} = 0.3$ if not specified. The lower (3D) simulations were conducted with $N = 60^3$ and with constant $\rho_{3D} = 0.1$, $T_{3D} = K_{3D} = 0.3$ if not specified. One simulation was conducted for each set of parameters. All units are dimensionless; reduced LJ units are used [9].

The influence of varying the density on the morphology of the end state of the simulations is shown in Figure 6a. At the lowest densities, no phase separation was observed: HDP did not precipitate. At slightly higher densities, $\rho_{2D} = 0.2$ and $\rho_{3D} = 0.05$, large thin structures could be observed: thin HDP filaments with large LDP holes in between in the 2D simulations; and large thin faces, normal to the initial simple cubic spring lattice, in the 3D simulations. Increasing the density further had the effect of making the HDP network bulkier, especially at the intersections of the filaments (2D simulations) or the faces (3D simulations), and of lowering the observed length scale of the network. At the highest densities, $\rho_{2D} = 0.6$ and $\rho_{3D} = 0.4$, HDP made up the larger part of the simulation box, with only a few LDP holes left.

At temperatures that were too low, $T_{2D} = T_{3D} \leq 0.1$, HDP precipitated but the small precipitates did not connect to a larger network (see Figure 6b). Increasing the temperatures to $T_{2D} = 0.4$ and $T_{3D} = 0.5$, led to an increase in the perceived length scale of the network—fewer but larger faces in the 3D simulation and larger, thicker filaments in the 2D one. Especially in the 2D simulation, the filaments appeared less compact, and fuzzier than in the 'origin' simulations at $T_{2D} = T_{3D} = 0.3$. For larger temperatures, $T_{2D} \geq 0.5$ and $T_{3D} \geq 1$, no phase separation was observed.

If the spring coefficient K is zero, the springs are inactive and the gel becomes an LJ fluid instead; these have already been well researched [12,13]. Therefore, low values of $K < 0.05$ were not of interest. At the lowest investigated spring coefficient, $K_{2D} = 0.05$, of the 2D model, thin filaments could be observed, connecting chunks of HDP (see Figure 6c). The color of the observed HDP was only faintly orange, indicating the presence of many springs longer than $2 \times 2^{1/6}$. The 3D gel at $K_{3D} = 0.1$ showed only HDP precipitates with no apparent connections.

Moving to larger spring coefficients, $K_{2D} = 0.1, 0.3$, the orange coloration of the HDP in the 2D gels became stronger, indicating more short springs inside the HDP. Additionally, the HDP filaments grew thicker, with now only slightly bulkier intersections. A similar morphology could be observed in the 3D gel at $K_{3D} = 0.3$: a network of faces connected at slightly bulkier intersections.

At large spring coefficients, $K_{2D} = 0.5, 0.9$, the filaments of the 2D gel became very straight, with the LDP space in between changing in shape from rounded to having sharp edges . Fewer but larger LDP areas could be seen. The LDP bubbles of the 3D gel behaved in a similar fashion: larger, but still slightly curved faces at $K_{3D} = 0.5$ turning into large planar constricted faces at $K_{3D} = 0.9$. For $K_{2D} \geq 1$ and $K_{3D} \geq 1$, no phase separation was observed.

3.3. Phase Diagrams

To create an overview of which combinations of density ρ, temperature T, and spring coefficient K phase separation and network formation could be found in the three-dimensional model, many small simulations of $N = 30^3$ and $N = 40^3$ particles were started and run until $t = 4000$ (10^6 time steps), or until HDP network growth could be seen. As simulation times in three dimensions were quite high (a few weeks at this size), these simulations were limited to a plane of constant density, $\rho = 0.1$.

Figure 7 shows the phase diagram of the three-dimensional model, indicating where phase separation occurred (pink dots/regions I and II), if the HDP formed a network (blue circles/region II), or if neither thing happened (light blue stars/region III). This phase diagram is to be seen more as a collection of quenched states that include a contribution from the kinetic path than a diagram of "true" thermodynamically equilibrated states. The classification of the gel simulations into these three regions was carried out by the author, judging the graphical representation by eye.

The same approach was used to map the HDP network formation of the two-dimensional model. Simulations of $N = 100^2$ and $N = 150^2$ particles were used to investigate the gel on three planes of constant density: $\rho = 0.1, 0.2,$ and 0.3. Figure 8 shows the three resulting phase diagrams.

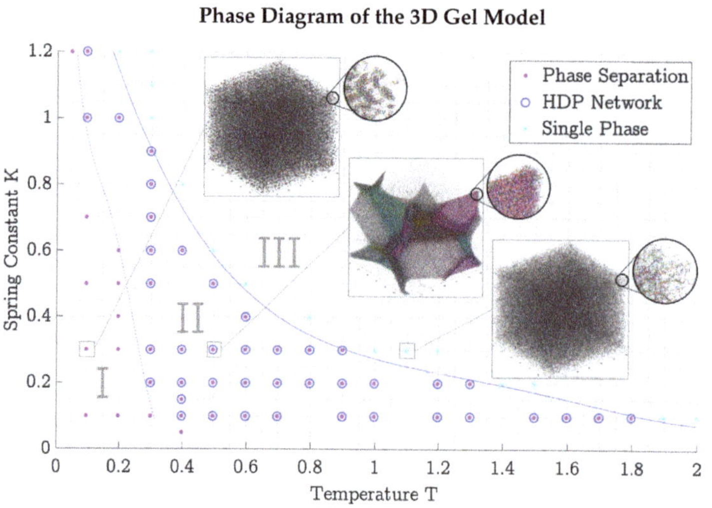

Figure 7. Phase diagram of the three-dimensional gel model at constant density $\rho = 0.1$. The solid blue line separates the upper right single-phase region III from the lower left regions I and II, where phase separation into HDP and LDP was observed. The dashed blue line further divides the simulations into where HDP networks were observed (blue rings/region II) and where only HDP precipitates formed (region I). All units are dimensionless; reduced LJ units are used [9].

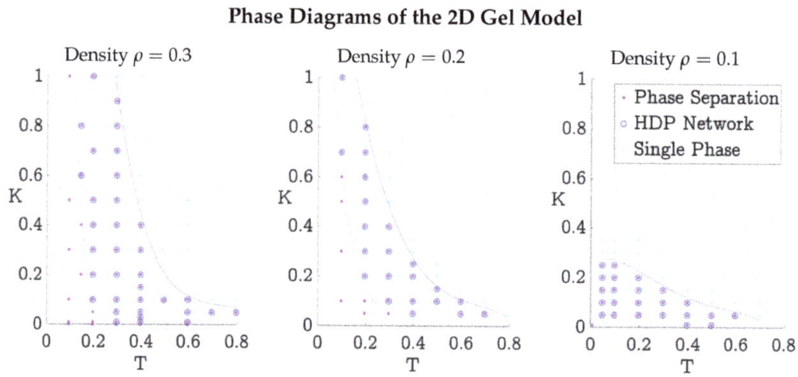

Figure 8. Phase diagrams of the two-dimensional gel model at constant densities: $\rho = 0.3$, 0.2, and 0.1 from left to right. The same three regions as in the phase diagram of the three-dimensional model (Figure 7) could be observed. The solid blue line separates single-phase region III from the lower left regions I and II. The dashed blue line further divides these into region II, where HDP networks could be observed, and region I, where only HDP precipitates formed. All units are dimensionless; reduced LJ units are used [9].

For all three planes of density, a similar shape of the three regions was found, with the HDP network region (II) more askew to the left in the lower-density phase diagrams. Lowering the density had a larger impact on the upper spring coefficient limit than on the temperature limit of region II. The phase separation regions (I and II) of the two-dimensional model phase diagram with $\rho = 0.1$ were smaller than the phase separation regions of the corresponding three-dimensional model phase diagram (Figure 7).

4. Discussion

4.1. Static Energy Balance Between HDP and LDP

The phase separation of the gel can be explained by the competing LJ and harmonic energies. For this, the toy model was simplified to a one-dimensional string of N LJ particles connected through harmonic springs with one group of $n+1$ HDP particles and the rest LDP particles (see Figure 9). In this simplification, the effect of thermal vibrations is neglected—all HDP particles are assumed to be precisely in the LJ minimum at $r_{\min} = 2^{1/6}$ and $E_{\min} = -1$. Therefore, all HDP springs are of equal length $s_1 = s_2 = \cdots = s_n = r_{\min}$. Further, we assume the LDP springs to be spaced equally and to be larger than the LJ cutoff radius. Thus, $s_{n+1} = s_{n+2} = \cdots = (L - ns_1)/(N-n) > r_c$.

The LJ energy of the n HDP particles then is $E_{\mathrm{LJ}} = (-1) \times n$, as only the LJ energy of the HDP particles is nonzero. The energy of the harmonic springs is

$$\begin{aligned} E_{\mathrm{K}} &= \frac{K}{2}\sum_{i=1}^{n} s_i^2 + \frac{K}{2}\sum_{j=n+1}^{N} s_j^2 = \frac{K}{2}\left(ns_1^2 + (N-n)\frac{(L-ns_1)^2}{(N-n)^2}\right) \\ &= \frac{K}{2}\left(ns_1^2 + \frac{(L-ns_1)^2}{(N-n)}\right). \end{aligned} \tag{10}$$

The graph of these two energies and their sum $E_{\mathrm{tot}} = E_{\mathrm{LJ}} + E_{\mathrm{K}}$ is given in Figure 9b. For $N = 100$ particles, a spring coefficient of $K = 1$, a density of $\rho = 0.75$, and a box length of $L = N/\rho = 133$, the total static energy has a minimum at $n = 85$.

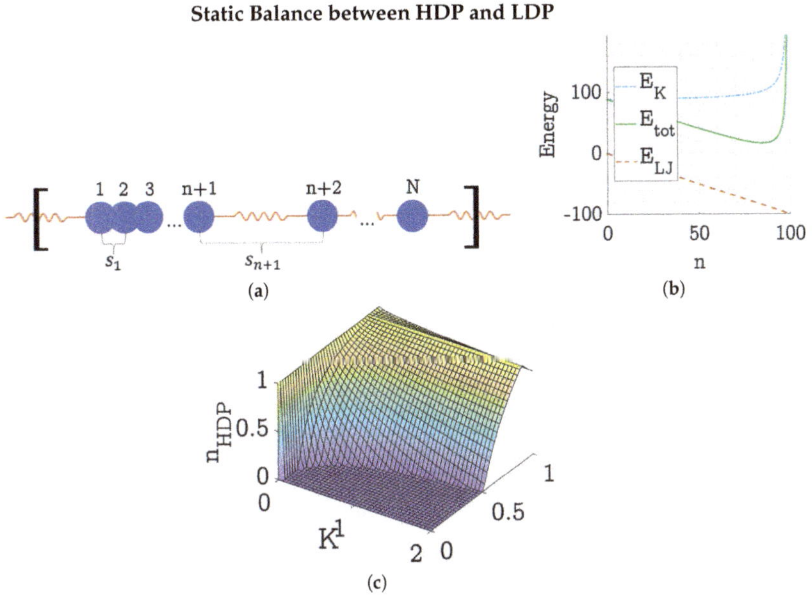

Figure 9. (**a**) Sketch of a one-dimensional simplification of the gel model with particles 1 to $n+1$ in each other's LJ potential grouped together as HDP, and particles $n+2$ to N as LDP. The distances between the particles are denoted with s_n for, e.g., the distance between particle n and particle $n+1$. The brackets stand for the PBC. (**b**) Quantitative plot of the total static energy E_{tot} of the one-dimensional model sketched in (**a**) as the sum of the LJ energy E_{LJ} and the harmonic spring energy E_{K}. Here, with $N = 100$ particles, spring coefficient $K = 1$, density $\rho = 0.75$, and a box length of $L = N/\rho = 133$. The total static energy has a minimum at $n = 85$. (**c**) Surface of minimal total energy in the space spanned by density ρ, spring coefficient K, and HDP particle fraction n_{HDP} for the one-dimensional model sketched in (**a**). All units are dimensionless; reduced LJ units are used [9].

Varying the density and spring coefficient, a surface of minimal total static energy was found—giving for each combination of ρ and K the optimal HDP particle fraction n_{HDP} (see Figure 9c). This shows an energetic motivation for phase separation, driving the system towards an equilibrium state of minimal energy.

Considering one alteration of the model, the observed growth of the network can be explained: If we imagine not one but several groups of HDP particles, for each new group created, one LJ bond is lost and replaced with a stretched spring, raising the total energy. The observed surface optimization is the reverse of this imagined process—decreasing the total energy through replacing long springs with LJ bonds.

4.2. Formation and Dissolution of HDP

To explain the influence of the temperature, and why sometimes phase separation occurred without the formation of an HDP network, let us consider another one-dimensional model (see Figure 10a): One mobile LJ particle is connected with two springs of lengths s_1 and s_2 to its neighbors, which are at a fixed distance of $2s_0 = s_1 + s_2$ from each other. A short calculation reveals the spring energy to be

$$E_K = \frac{K}{2}(s_1^2 + s_2^2) = \frac{K}{2}(r^2 + (2s_0 - r))^2) = K(r^2 - 2rs_0 + 2s_0), \tag{11}$$

where r denotes the distance of the moving particle from its left-hand-side neighbor. Together with the LJ potential, it follows that for $r < s_0$, the total potential of the mobile particle is

$$E_{\text{tot}} = K(r^2 - 2rs_0 + 2s_0) + 4\left(\frac{1}{r^{12}} - \frac{1}{r^6}\right). \tag{12}$$

Both energies E_K and E_{LJ} and the resulting total energy E_{tot} are plotted in Figure 10b.

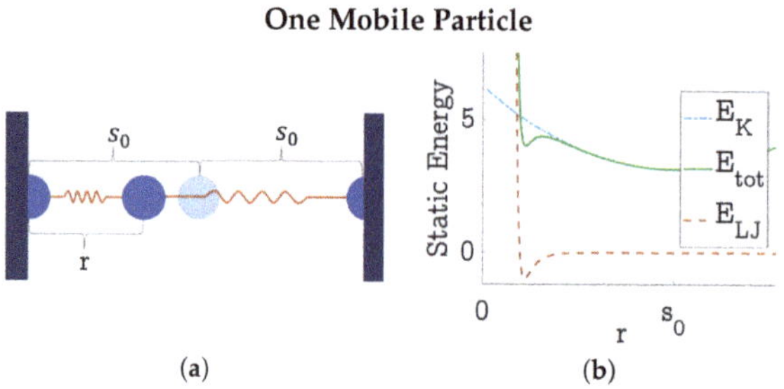

Figure 10. (**a**) Sketch of an LJ particle connected to two other fixed particles over harmonic springs with equilibrium length zero. The mobile particle is at a distance of $r < s_0$ from its left neighbor. (**b**) Harmonic spring energy E_K, LJ energy E_{LJ}, and total energy E_{tot} of the mobile particle for $r < s_0$ with $K = 0.25$ and $s_0 = 5$. All units are dimensionless; reduced LJ units are used [9].

It can be observed, that in order to reach the LJ minimum, which is only a local minimum in the given example, the mobile particle needs to overcome a considerable energy barrier.

If we extended this model to the right-hand side, considering a second, a third, or more mobile particles, those particles could sometimes aid this first one in reaching the left LJ minimum by approaching or by accelerating it to the left, and at other times keep it from reaching the LJ minimum by doing the opposite.

Exact calculations would need to be conducted through a statistical approach, considering the possible vibrations of the gel. For now, the intuition must suffice: It is energetically adverse for one particle to reach the other if it is moving alone. It needs more energy, through either a higher temperature or through neighboring particles vibrating collectively, which can be achieved by slightly increasing the spring coefficient. If there is too much thermal energy or if those neighbors vibrate too much, the particle is in danger of being ripped out of the LJ bond again.

This intuition can be applied to the phase diagrams found for the two- and three-dimensional gel models (see Figures 7 and 8). At low temperatures and low spring coefficients, particles did not have enough vibrational energy to reach each other and form a network. Only small precipitates formed (region I). This could be remedied by reinforcing the connection between the particles by increasing the spring coefficient, to make them more likely to vibrate collectively, thus enabling the precipitates to form a network (region II). Too large spring coefficients then had the reverse effect, forcing the particles apart and dissolving the HDP again (region III). Another possibility of connecting the precipitates from region I was to increase the temperature directly, giving the particles more energy to bridge the gap between precipitates. Too large temperatures then dissolved the HDP again.

4.3. Future Work

Shorter simulation times, allowing more or larger simulations, would be very welcome for future projects. The most promising alternative to the rather slow MATLAB code was found in rewriting the simulation in Julia [14]. In this way, a simulation speed improvement of a factor nine could be achieved. More might be possible by improving the neighbor lists or by parallelizing the calculations.

Future works treating the same model could quantify the influence of the simulation parameters ρ, T, and K on morphology using the final measures from Section 2.4 (HDP surface fraction a_{HDP}, HDP thickness $\delta_{\mathrm{HDP}}, \ldots$), to move away from describing the observed change in morphology from images, as in Section 3.2. This would remove subjectivity and allow possible automatization to, e.g., find the parameters for the largest surface area.

For the same reason, fixed, non-subjective criteria should be developed for associating gels with regions I, II, and III.

One possibility would be to use the small boxes covering the HDP volume (established in Section 2.4). Region II could be defined as a percolating structure of small boxes. This could be implemented with a forest fire algorithm like in [15,16]. Other suitable criteria could be found, taking a certain minimal HDP particle fraction combined with a minimal HDP thickness. Either would enable an automated mapping of phase diagrams like Figures 7 and 8.

Instead of running simulations with constant parameters, outgoing from an initial state, simulations could be started from the final/asymptotic state of a simulation of other parameters. This would serve to further investigate the temperature hysteresis Peleg et al. [2] found to see if a hysteresis could also be observed in the other two parameters (ρ and K), and if the same phase diagram would develop eventually.

The prediction for the optimal HDP particle fraction of a one-dimensional gel model (Section 4.1) could be compared to simulations of a one-dimensional gel model, and generalized to also include the influence of temperature.

To come closer to reality, the model could be altered by, e.g., using a distribution of spring coefficients instead of a uniform K or by randomizing the structure through, e.g., cutting/disabling some of the springs of the initial simple cubic lattice, as in [10]. Another approach would be to insert a second species of solvent particles.

The two-dimensional model of Peleg at al. [2,3] has already been compared to experimental results from investigating biopolymer mixtures in confined geometries [17] and found to be "in good agreement" with their findings.

Another direct link to observable gels might be achieved through changing the three-dimensional model to simulate a surface, by changing one dimension from periodic to non-periodic with a boundary against vacuum or an LJ fluid. The resulting surface could then be compared to scanning electron microscopy (SEM) images or atomic force microscopy (AFM) images of gel surfaces, as obtained in [18–21].

Alternatively, the structure resulting from larger three-dimensional simulations could be compared to gels investigated with micro-computed tomography (micro-CT) [22].

In [3], Peleg et al. investigated the shear modulus of a two-dimensional gel with an HDP network. This work could be continued (also for three dimensions) with an effort to link the system parameters over the gel morphology to its resulting mechanical properties.

5. Conclusions

The generalization of the gel model introduced by Peleg et al. [2] to three dimensions has shown structures which lie somewhere between their two hypotheses. The HDP formed a soap-like structure made of connected surfaces, but depending on the system parameters, the HDP also showed tendencies to form linear filament network-like structures, where the faces connected.

We systematically explored the (ρ, K, T) parameter space for the two- and three-dimensional gel model, sketching the shape of regions with HDP network formation in four phase diagrams (see Figures 7 and 8). Then, in Section 2.4 we established measures allowing quantitative characterization of the gel's structure.

In Sections 4.1 and 4.2, the phase separation occurring in the gel model has been explained and discussed through two one-dimensional simplifications. Section 4.1 explains why there exists an optimal fraction of HDP particles in a static equilibrium and Section 4.2 attempts to justify the shape of the found phase diagrams with a dynamic approach.

We believe that future work on this model should try to connect it to real gels to see if this gel model can be used to predict the morphology of laboratory-produced gels and in this way this help adjust the gel's material properties by tuning its microstructure.

Funding: This research received no external funding.

Institutional Review Board Statement: Not applicable

Data Availability Statement: Trajectory-data available at ETH research collection (https://www.research-collection.ethz.ch/handle/20.500.11850/728415).

Acknowledgments: I am grateful to my supervisor, Martin Kröger for giving me the opportunity to explore this beautiful toy model, and for trusting me, to work in such an independent fashion. Without his insightful feedback this paper would not be of the same quality—without his encouragement it would not exist. Further, I would like to thank Christophe Desarzens and Stefan Schären for their proofreading of this article and their very constructive feedback on my overcomplicated figures.

Conflicts of Interest: The author declares no conflicts of interest.

Appendix A

For a given particle, the neighbor list needs to contain at least all particles in a radius $R_{\mathrm{NL}} = r_c + \Delta R$ that could interact with it in a given time interval $\Delta t \cdot \Delta i_{\mathrm{NL}}$. Where the list is updated every Δi_{NL}-th time step and $r_c = 3 \times 2^{1/6}$ is the cutoff radius above which the LJ potential is set to zero. A reasonable R_{NL} can be estimated as follows. Considering the MB distribution of the velocities, it is very unlikely for particles to have velocities exceeding $4\,\bar{v}$. The probability for that is $p(v \geq 4\,\bar{v}) \leq 10^{-8}$. Accounting for the relative movement of the particles with a factor of two, this gives

$$\Delta R > 2 \times 4\bar{v} \times \Delta i_{NL}\,\Delta t, \tag{A1}$$

where the mean speed is given by $\bar{v} = \sqrt{8T/\pi}$, which yields Equation (1) as an estimate for the radius of the neighbor lists, ensuring that all LJ interactions are properly taken into account.

References

1. Ilyin, S.O. Structural Rheology in the Development and Study of Complex Polymer Materials. *Polymers* **2024**, *16*, 2458. [CrossRef] [PubMed]
2. Peleg, O.; Kröger, M.; Hecht, I.; Rabin, Y. Filamentous networks in phase-separating two-dimensional gels. *Europhys. Lett. (EPL)* **2007**, *77*, 58007. [CrossRef]
3. Peleg, O.; Kröger, M.; Rabin, Y. Model of Microphase Separation in Two-Dimensional Gels. *Macromolecules* **2008**, *41*, 3267–3275. [CrossRef]
4. Alvarez-Tirado, M.; Castro, L.; Guéguen, A.; Mecerreyes, D. Iongel Soft Solid Electrolytes Based on [DEME][TFSI] Ionic Liquid for Low Polarization Lithium-O_2 Batteries. *Batter. Supercaps* **2022**, *5*, e202200049.
5. Chotsuwan, C.; Boonrungsiman, S.; Chokanarojwong, T.; Dongbang, S. Mesoporous soft solid electrolyte-based quaternary ammonium salt. *J. Solid State Electrochem.* **2017**, *21*, 3011–3019. [CrossRef]
6. Mozaffari, K.; Liu, L.; Sharma, P. Theory of soft solid electrolytes: Overall properties of composite electrolytes, effect of deformation and microstructural design for enhanced ionic conductivity. *J. Mech. Phys. Solids* **2022**, *158*, 104621. [CrossRef]
7. Xiao, H.; Li, H.; Wang, Z.L.; Yin, Z.N. Finite Inelastic Deformations of Compressible Soft Solids with the Mullins Effect. In *Advanced Methods of Continuum Mechanics for Materials and Structures*; Naumenko, K., Aßmus, M., Eds.; Springer, Singapore: Singapore, 2016; pp. 223–241. [CrossRef]
8. Sekimoto, K.; Suematsu, N.; Kawasaki, K. Spongelike domain structure in a two-dimensional model gel undergoing volume-phase transition. *Phys. Rev. A* **1989**, *39*, 4912–4914. [CrossRef]
9. Allen, M.P.; Tildesley, D.J. *Computer Simulation of Liquids*; Oxford University Press: Oxford, UK, 2017. [CrossRef]
10. Peleg, O.; Kröger, M.; Rabin, Y. Effect of network topology on phase separation in two-dimensional Lennard-Jones networks. *Phys. Rev. E* **2009**, *79*, 040401. [CrossRef]
11. Thompson, A.P.; Aktulga, H.M.; Berger, R.; Bolintineanu, D.S.; Brown, W.M.; Crozier, P.S.; In 't Veld, P.J.; Kohlmeyer, A.; Moore, S.G.; Nguyen, T.D.; et al. LAMMPS—A flexible simulation tool for particle-based materials modeling at the atomic, meso, and continuum scales. *Comput. Phys. Comm.* **2022**, *271*, 108171. [CrossRef]
12. Schultz, A.J.; Kofke, D.A. Comprehensive high-precision high-accuracy equation of state and coexistence properties for classical Lennard-Jones crystals and low-temperature fluid phases. *J. Chem. Phys.* **2018**, *149*, 204508. [CrossRef] [PubMed]
13. Stephan, S.; Staubach, J.; Hasse, H. Review and comparison of equations of state for the Lennard-Jones fluid. *Fluid Phase Equilibria* **2020**, *523*, 112772. [CrossRef]
14. Bezanson, J.; Edelman, A.; Karpinski, S.; Shah, V.B. Julia: A fresh approach to numerical computing. *SIAM Rev.* **2017**, *59*, 65–98.
15. Albinet, G.; Searby, G.; Stauffer, D. Fire propagation in a 2-D random medium. *J. Phys.* **1986**, *17*, 1–7. [CrossRef]
16. Stauffer, D.; Aharony, A. *Introduction To Percolation Theory: Second Edition*; Taylor & Francis: London, UK, 1992. [CrossRef]
17. Fransson, S.; Peleg, O.; Lorén, N.; Kröger, A.H.M. Modeling and confocal microscopy of biopolymer mixtures in confined geometries. *Soft Matter* **2010**, *6*, 2713–2722. [CrossRef]
18. Kaberova, Z.; Karpushkin, E.; Nevoralová, M.; Vetrík, M.; Šlouf, M.; Dušková-Smrčková, M. Microscopic Structure of Swollen Hydrogels by Scanning Electron and Light Microscopies: Artifacts and Reality. *Polymers* **2020**, *12*, 578. [CrossRef] [PubMed]
19. Hagmann, K.; Bunk, C.; Böhme, F.; von Klitzing, R. Amphiphilic Polymer Conetwork Gel Films Based on Tetra-Poly(ethylene Glycol) and Tetra-Poly(ϵ-Caprolactone). *Polymers* **2022**, *14*, 2555. [CrossRef] [PubMed]
20. Damaschin, R.P.; Lazar, M.M.; Ghiorghita, C.A.; Aprotosoaie, A.C.; Volf, I.; Dinu, M.V. Stabilization of Picea abies Spruce Bark Extracts within Ice-Templated Porous Dextran Hydrogels. *Polymers* **2024**, *16*, 2834. [CrossRef] [PubMed]
21. Dimic-Misic, K.; Imani, M.; Gasik, M. Effects of Hydroxyapatite Additions on Alginate Gelation Kinetics During Cross-Linking. *Polymers* **2025**, *17*, 242. [CrossRef] [PubMed]
22. Cengiz, I.F.; Oliveira, J.M.; Reis, R.L. Micro-CT—A digital 3D microstructural voyage into scaffolds: A systematic review of the reported methods and results. *Biomater. Res.* **2018**, *22*, 26. [CrossRef]

polymers

MDPI

Article

Simulation of the Design Performance of Carbon Fiber/Glass Fiber Hybrid-Reinforced Resin Matrix Composite Rotors

Chong Li [1,*], Jiayou Wang [2], Meng Li [2], Haoyu Wang [2], Yiguo Song [1], Xiangzhe Meng [1] and Ruiliang Liu [2]

[1] College of Mechanical and Electrical Engineering, Harbin Engineering University, Harbin 150001, China; songyiguo@hrbeu.edu.cn (Y.S.); 1423016502@hrbeu.edu.cn (X.M.)
[2] College of Materials Science and Chemical Engineering, Harbin Engineering University, Harbin 150001, China; 17882564891@hrbeu.edu.cn (J.W.); limeng02666@163.com (M.L.); haoyuwang@hrbeu.edu.cn (H.W.); liuruiliang@hrbeu.edu.cn (R.L.)
* Correspondence: lichong@hrbeu.edu.cn; Tel.: +86-181-4534-1314

Abstract: Composite rotors, attributing to their leveraging characteristics of the light weight, high strength, high rigidity, corrosion resistance, and low noise, can significantly reduce the moment of inertia and enhance equipment operational efficiency. Using carbon fiber/glass fiber hybrid-reinforced resin–matrix composites as the rotor base material, the radial stability of a rotor can be effectively increased by regulating the fiber volume content. Meanwhile, the introduction of glass fiber not only enables the transition between the metal hub and composite rim but also optimizes the cost structure of the composite system, overcoming the economic bottleneck of single carbon fiber-reinforced resin–matrix composite rotors. This paper employs the finite element method to analyze a three-dimensional model of a composite rotor, investigating the performance of its metal hub and hybrid-reinforced resin–matrix composite rim. According to the radial stress distribution of the composite rotor during operation, the mixing ratio of carbon fiber/glass fiber is adjusted. The high-speed rotation condition of the composite rotor at 18,000 revolutions per minute is simulated to verify its safety and reliability.

Keywords: composites rotor; hybrid reinforced composite materials; finite element analysis; high-speed rotor

1. Introduction

The flywheel energy storage system (FESS) is a device that utilizes the acceleration and deceleration of rotating rotors to convert electrical energy into mechanical energy and vice versa. It offers several advantages, including a long cycle life, cleanliness, minimal environmental impact, rapid charging and discharging capabilities, high power density, and excellent cycle efficiency. Additionally, the FESS can deliver substantial power output in a short time frame. As a result, it is considered an ideal energy storage solution for the future and has been applied in various fields, such as peak shaving, uninterruptible power supply, power grid stabilization, electric vehicles, and satellite power systems. Furthermore, it is well suited for applications that require frequent charging and discharging over multiple cycles [1–4].

The energy density is a crucial index for evaluating the performance of FESS. The rotating flywheel stores energy in the form of kinetic energy, which depends on the rotational inertia and angular velocity of the flywheel. To maximize the energy storage capacity and energy density of FESSs, it is essential to increase the rotational inertia while limiting the

 https://doi.org/10.3390/polym17121668

angular velocity of the flywheel rotor. The impact of angular velocity is particularly significant, as the energy stored in the flywheel is proportional to the moment of inertia of the rotor and to the square of the maximum rotational speed. The stress in the rotor increases with the rise in rotational speed, leading to significant vibration. This, in turn, compromises the safety of the system. To enhance cost performance, it is essential to reduce the costs associated with the flywheel. The International Renewable Energy Agency estimates that by 2030, the installation cost of FESSs will be 35% lower than the current range of USD 1500 to 6000 per kilowatt-hour [1]. The high cost per kilowatt-hour of the FESS and the strength of the rotor material are the critical factors that are hindering the development of flywheel energy storage. The flywheel rotor is one of the most vital components of FESS. The key to reducing the cost of FESS and improving the material strength of the flywheel rotor lies in selecting the appropriate rotor material and designing an effective rotor structure [5,6].

FESS is generally categorized into low-speed FESS and high-speed FESS. The components of the flywheel rotor are primarily manufactured using metal and composite materials. The metal rotor is typically employed in low-speed energy storage flywheels and benefits from relatively mature design and processing technologies. The structural design of a metal flywheel rotor is often optimized by altering the rotor's shape. Gao [7] utilized metal materials to reduce the manufacturing complexity and cost of FESS. A double-hub combined flywheel was designed to decrease the mass while achieving a greater moment of inertia, thereby enhancing energy storage density. The maximum rotational speed reached up to 13,500 rpm. One of the critical challenges with FESS at high rotational speeds is ensuring the material strength and minimizing the vibration of the flywheel rotor. Higher energy storage density can be attained due to the high specific strength, excellent rigidity, and low specific gravity of composite materials. Consequently, composite materials are the preferred choice for high-speed energy storage flywheels. However, the manufacturing process for composite materials is more complex compared to conventional materials. The composite flywheel rotor has consistently been a focal point of research. In addition to the multi-ring structure, designs incorporating hybrid materials, gradient materials, and pretension fiber winding structures have been developed to enhance energy storage density [7–9].

At present, most flywheel rotors utilize a metal hub combined with a composite rim structure. High-strength aluminum alloys, such as 7075, are commonly employed for the metal hub. The composite rim typically consists of materials like carbon fiber, glass fiber, and epoxy resin. The winding process is applied to form the glass or carbon fiber rim. Most rotors are designed in a disc or cylindrical shape, as fabricating composite rotors with complex geometries using the winding process is challenging. Flywheel rotors are generally constructed as thin discs with larger diameters, as this configuration provides a greater moment of inertia compared to thicker discs [5]. The radial strength of the composite flywheel rotor formed by the winding method is much lower than its circumferential strength due to anisotropy. A large enough value of the initial radial compressive stress of a composite flywheel rotor can be achieved using the press-fit forming process with interference fit. The initial radial compressive stress of the rotor can effectively offset the radial centrifugal tensile stress at high rotational speed. The maximum radial stress of a high-speed rotating composite flywheel rotor is significantly reduced [5,10]. Chen [11] designed a 200 kW FESS for uninterruptible power supply (UPS) and utilized a composite rotor with an interference fit. The maximum rotational speed achieved was 20,000 r/min. The composite rotor demonstrated safe and reliable operation during high-speed testing. Kim [12] designed and manufactured a composite rotor capable of operating at a speed of 15,000 r/min. A new hybrid composite hub, filament-wound in a dome shape, was proposed to meet the strength and deformation requirements of the rotor. A press-fit forming process with an interference fit was employed to minimize the radial tensile stresses at the shaft–hub and hub–rotor interfaces. Variation in hub stiffness from the inner surface to the outer surface was achieved through material hybridization and the careful selection of the winding angle, ensuring proper contact between the shaft–hub and hub–rotor at the maximum rotational speed.

A hybrid composite is a composite material composed of more kinds of fiber reinforcements in the same resin matrix. It utilizes the performance advantages of each material to achieve complementary advantages through their combination. A hybrid composite has special properties that are not easy to obtain with only a single type of fiber reinforcement. Fiber hybrids are becoming more convenient and controllable methods not only to meet the requirements of design performance but also to reduce the cost of products, increase the quality of products, prolong the service life of products, and improve the economic benefits [13–15]. Glass fiber-reinforced polymer (GFRP) composites and carbon fiber-reinforced polymer (CFRP) composites are among the most commonly used composite materials. Carbon fiber is known for its high strength and high modulus; however, it is significantly more expensive than glass fiber. M. Spencer [16] developed glass/carbon fiber hybrid epoxy composite laminates that effectively balanced cost and performance. Tang C.L. [17] designed and manufactured two hybrid composite flywheels consisting of six layers. The radial gradient material was produced by varying the fiber ratio. The hybrid composite flywheel rotor offers advantages such as a smooth radial stress distribution, high structural safety and reliability, and a flexible and feasible hybrid design method.

In summary, this paper establishes a finite element model to verify the feasibility of the rotor design in an energy storage system. The rotor consists of a hybrid-reinforced composite rim made from carbon fiber and glass fiber, along with a metal hub. This study calculates the impact of hybrid-reinforced composite ring layers at various levels on the rotor's stress and determines the maximum rotational speed of the rotor. Furthermore, the safety of the flywheel rotor is confirmed through modal analysis. The findings presented in this paper serve as a valuable reference for the design of high-speed composite rotors.

2. Stress Analysis of a Rotor

A stress analysis to calculate the deformations and strength ratios of a multi-rim rotor can be found in previous publications [18]. However, a summary of the relevant equations is given below. Equation (1) shows the equation governing a rotor rotating at speed ω at radius r in the cylindrical coordinate system:

$$\frac{d\sigma_r}{dr} + \frac{\sigma_r - \sigma_\theta}{r} + \rho r \omega^2 = 0 \tag{1}$$

where σ_r is the radial stress; σ_θ is the circumferential stress; r is the radius of a rotor; ρ is the density; ω is the rotational speed.

The stress–strain relationship can be given as

$$\sigma = Q\varepsilon \tag{2}$$

where σ is the stress vector; ε is the strain vector; Q is the stiffness matrix.

Equation (2) can be rewritten as

$$\begin{pmatrix} \sigma_\theta \\ \sigma_z \\ \sigma_r \end{pmatrix} = \begin{bmatrix} Q_{11} & Q_{12} & Q_{13} \\ Q_{21} & Q_{22} & Q_{23} \\ Q_{31} & Q_{32} & Q_{33} \end{bmatrix} \begin{pmatrix} \varepsilon_\theta \\ \varepsilon_z \\ \varepsilon_r \end{pmatrix} \tag{3}$$

where σ_z is the axial stress; ε_r is the radial strain; ε_θ is the circumferential strain; ε_z is the axial strain. The stiffness matrix Q is given by the material properties as

$$Q = \begin{bmatrix} \frac{1}{E_\theta} & \frac{-v_{\theta r}}{E_r} & \frac{-v_{\theta z}}{E_z} \\ & \frac{1}{E_r} & \frac{-v_{rz}}{E_z} \\ sym. & & \frac{1}{E_z} \end{bmatrix}^{-1} \quad (4)$$

where E_θ is the longitudinal Young's modulus; E_z and E_r are the transverse Young's moduli; $v_{\theta r}$, $v_{\theta z}$, and v_{rz} are the Poisson's ratio.

Assuming the plane stress state, the radial and circumferential strains are expressed as

$$\varepsilon_\theta = \frac{u_r}{r}, \ \varepsilon_r = \frac{\partial u_r}{\partial r} \quad (5)$$

where u_r is the radial displacement.

Between two adjacent rings of the rim (e.g., j and $j + 1$), the compatibility conditions should be satisfied as follows:

$$\sigma_{r_i}^{(j+1)} = \sigma_{r_0}^{(j)}, \ u_{r_i}^{(j+1)} = u_{r_0}^{(j)}, \ (j = 1,2,3, \ldots, N - 1) \quad (6)$$

where r_i and r_0 are the inner and outer radius of the ring, respectively.

3. Finite Element Model

3.1. Structure of Rotor

The structure, materials, and manufacturing processes of energy storage flywheel rotors are typically chosen to maximize energy density and specific energy. Due to the anisotropy of the composite material, the radial strength (perpendicular to the fiber direction) of the composite rim is significantly lower than its circumferential strength (aligned with the fiber direction). This discrepancy makes the rotor susceptible to radial delamination failure during high-speed rotation.

In this paper, the composite rim consists of three rings that are fabricated using a winding forming technique and assembled with an interference fit. The impact of the material and thickness distribution of the three rings on the stress state of the rim is analyzed. Five design schemes, labeled Rotor-A, Rotor-B, Rotor-C, Rotor-D and Rotor-E, are proposed, as illustrated in Table 1. For Rotor-B, Rotor-C, and Rotor E, the material distribution of the composite rim is presented in Figure 1.

Table 1. Material and structure design scheme of the composite rim.

Rim Design Scheme	Material of Inner Ring	Thickness of Inner Ring	Material of Middle Ring	Thickness of Middle Ring	Material of Outer Ring	Thickness of Outer Ring
Rotor-A	Mat.2	50 mm	Mat.5	50 mm	Mat.8	50 mm
Rotor-B	Mat.2 Mat.3	10 mm 40 mm	Mat.3 Mat.4 Mat.5 Mat.6 Mat.7	6 mm 10 mm 18 mm 10 mm 6 mm	Mat.7 Mat.8	36 mm 14 mm
Rotor-C	Mat.2 Mat.3	20 mm 30 mm	Mat.3 Mat.4 Mat.5 Mat.6 Mat.7	6 mm 10 mm 18 mm 10 mm 6 mm	Mat.7 Mat.8	24 mm 26 mm

Table 1. *Cont.*

Rim Design Scheme	Material of Inner Ring	Thickness of Inner Ring	Material of Middle Ring	Thickness of Middle Ring	Material of Outer Ring	Thickness of Outer Ring
Rotor-D	Mat.2	25 mm	Mat.5	50 mm	Mat.8	75 mm
Rotor-E	Mat.2 Mat.3	10 mm 15 mm	Mat.3 Mat.4 Mat.5 Mat.6 Mat.7	6 mm 10 mm 18 mm 10 mm 6 mm	Mat.7 Mat.8	36 mm 39 mm

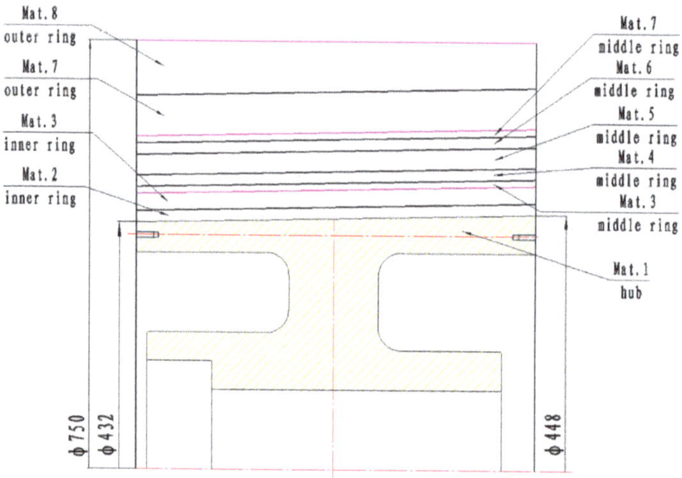

Figure 1. Material distribution of the composite rim of Rotor-B, Rotor-C, and Rotor-E.

3.2. Material Properties of the Metal Hub and the Composite Rim

Aluminum alloy 7075, used for the metal hub, was designated as Mat.1 in ANSYS R16.0, and its properties are presented in Table 2.

Table 2. Material properties of the metal hub [19–21].

Material Number	Material	Density (kg/m^3)	Young's Modulus (GPa)	Poisson's Ratio	Yield Stress (MPa)
Mat.1	Aluminum alloy 7075	2800	72.5	0.3	546

The glass fiber-reinforced composite (GFRC), carbon fiber-reinforced composite (CFRC), and glass/carbon hybrid composite (GCHC) used for the rim of the flywheel rotor were designated as Mat.2–8 in ANSYS. The ring materials' parameters exhibit a gradient variation from the inside to the outside, achieved by adjusting the carbon fiber (CF) and glass fiber (GF) contents during the winding process of the composite rings. The material properties of Mat.2 to Mat.8 are presented in Table 3. The longitudinal Young's modulus, Poisson's ratio, and density of GCHC with varying hybrid ratios increase linearly with the CF ratio. This behavior can be predicted using the linear mixing law. Additionally, the transverse Young's modulus of a GCHC can also be estimated based on the mixing law, which has a minimal impact on the stress and deformation results [17,22]. Tang [17] predicted the material constants of GCHC using Equations (7)–(10), based on the mixing law. The mechanical equations of single-layer and multi-layer composite flywheels were established. A six-layer hybrid flywheel was designed and manufactured, and experimental verification was carried out.

Table 3. Material properties of the composite rim [12,17,23].

Material Number	Material	Longitudinal * Modulus (GPa)	Transversal * Modulus (GPa)	Poisson's Ratio	Density (kg/m^3)
Mat.2	GFRC	38.6	8.27	0.26	1800
Mat.3	83.3% GF + 16.7% CF	58	8.4	0.26	1767
Mat.4	66.7% GF + 33.3% CF	77.4	8.5	0.26	1733
Mat.5	50% GF + 50% CF	96.8	8.6	0.26	1700
Mat.6	33.3% GF + 66.7% CF	116.2	8.7	0.26	1667
Mat.7	16.7% GF + 83.3% CF	135.6	8.9	0.26	1633
Mat.8	CFRC	155	9	0.26	1600

* Longitudinal = fiber direction; transversal = vertical to the fiber direction.

The prediction equations for the longitudinal elastic modulus (E_L), transverse elastic modulus (E_T), Poisson's ratio (v), and density (ρ) are as follows [17]:

$$E_L = V_m E_m + V_g E_g + V_c E_c \quad (7)$$

$$\frac{1}{E_T} = \frac{V_m}{E_m} + \frac{V_g}{E_g} + \frac{V_c}{E_c} \quad (8)$$

$$\mu_L = V_m \mu_m + V_g \mu_g + V_c \mu_c \quad (9)$$

$$\rho = V_m \rho_m + V_g \rho_g + V_c \rho_c \quad (10)$$

where V refers the volume fraction, and the subscripts g, c, and m represent GF, CF, and the matrix, respectively.

3.3. Model and Mesh of Rotor

A quarter model was built because the rotor is an axisymmetric structure. The cyan part illustrates the metal hub, the yellow part illustrates the inner ring, the green part illustrates the middle ring, and the red part illustrates the outer ring, with the material distribution shown in Figure 2. The 8-node hexahedral elements solid 45 and solid 46 were used to mesh the metal hub and the composite rim, respectively, as shown in Figure 3.

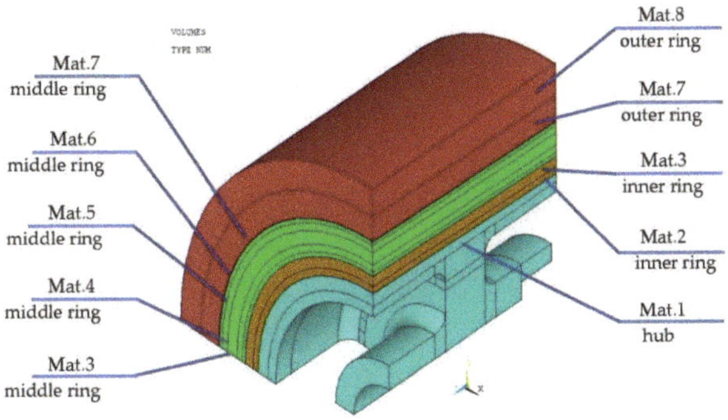

Figure 2. Quarter model of rotor.

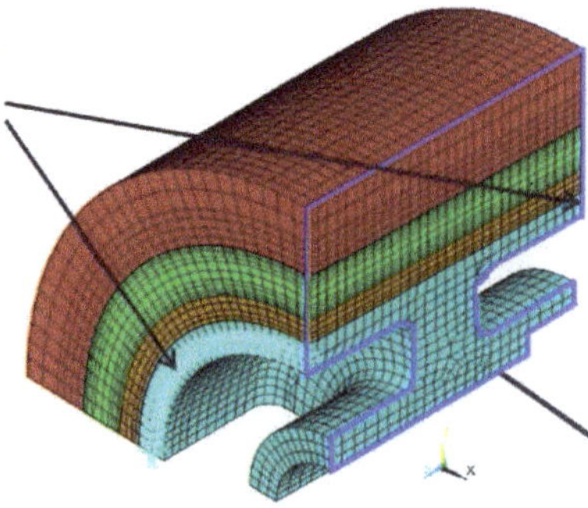

Figure 3. Mesh and boundary conditions of rotor.

3.4. Contact Pair and Boundary Conditions

Contact pairs were established on the contact surfaces between the hub and the inner ring, the inner ring and the middle ring, and the middle ring and the outer ring. Face-to-face contact was utilized to simulate the interference fit of the rotor. The contact surface employed the CONTA174 element, while the target surface used the CONTA170 element. The friction coefficient was set at 0.2, and the effect of temperature on the friction coefficient was not considered.

The level of interference is determined by the offset of the contact surface. If the interference between the rim and the hub, as well as between the rings of the rim, is insufficient, it may lead to the separation or delamination of the rim at high rotational speeds. Conversely, if the interference is excessive, the radial compressive stress on the composite rim's contact surface may surpass the material's radial compressive strength, resulting in damage. In this study, the interference amounts between the hub and the inner ring, the inner ring and the middle ring, and the middle ring and the outer ring were all set at 0.3 mm.

To prevent the complete displacement of the model, constraints were applied to specific nodes in ANSYS. Considering the connection between the hub and the shaft via screws, fixed constraints in the X, Y, and Z directions were applied to the nodes located on the circumference where the screws were positioned. A symmetric constraint was imposed on the circumferential boundary, as illustrated in Figure 3. Centrifugal force was applied to the model by utilizing the rotational angular velocity around the Z-axis with the APDL command, in accordance with the actual operating conditions of the flywheel rotor.

4. Simulation Results and Discussions

The radial strength of the composite ring formed by winding is significantly lower than the circumferential strength, making the rotor with an interference fit susceptible to delamination failure. Consequently, understanding the radial stress distribution of the composite rim is crucial for determining whether a flywheel rotor can operate normally at high rotational speeds. The stress state of the rotor during rotation can be observed in the stress nephograms. Figure 4 illustrates the radial stress nephograms of the inner rings of Rotor-A, Rotor-B, Rotor-C, Rotor-D, and Rotor-E at a speed of 18,000 r/min.

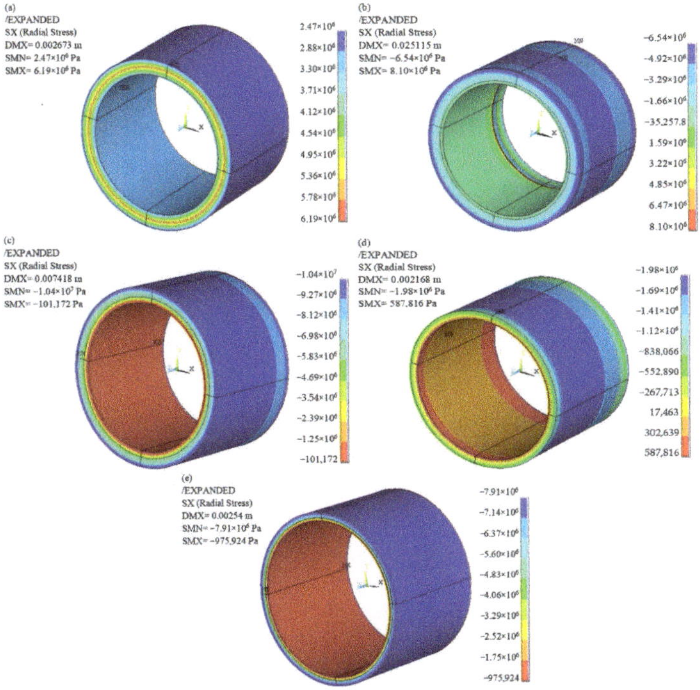

Figure 4. Radial stress distribution of the inner ring: (**a**) Rotor-A; (**b**) Rotor-B; (**c**) Rotor-C; (**d**) Rotor-D; (**e**) Rotor-E. DMX: maximum axial displacement; SMN: minimum radial stress; SMX: maximum radial stress.

4.1. *Influence of Hybrid Composite and Thickness of Rings*

It is evident that the radial stress values on the inner and outer contact surfaces of the inner ring of Rotor-A are all positive, indicating tensile stress with no compressive stress present, as illustrated in Figure 4a. The rotor may fail due to the separation of the hub and the rim at a speed of 18,000 r/min.

Compared to Rotor-A, both tensile and compressive radial stress zones coexist simultaneously on the contact surface between the hub and the inner ring of Rotor-B, as shown in Figure 4b. The presence of tensile stress on the contact surface indicates that the material or structural design of the rotor should be further optimized to ensure that the radial stress on all contact surfaces is compressive, allowing the rotor to operate normally.

In contrast, both the inner and outer contact surfaces of the inner ring of Rotor-C exhibit only compressive stress, as illustrated in Figure 4c. This design helps prevent the separation of the hub, inner ring, and middle ring, thereby avoiding rotor failure at a speed of 18,000 r/min. The gradient change in the material properties of the hybrid composite improves the radial stress distribution of the rotor. The low specific modulus within the rotor leads to significant deformation, while the high specific modulus on the outer layer results in minimal deformation. Consequently, the deformation of the inner layer is greater than that of the outer layer during rotation. The interaction between the inner and outer layers generates compressive stress, which further mitigates centrifugal tensile stress [17]. To improve the stress distribution of the rotor, the gradient change in the specific modulus of the hybrid composite should be made smoother by adjusting the ratio of the low-modulus GF to the high-modulus CF.

The thickness ratio of the rings from the inside to the outside of Rotor-D is 1:2:3. In comparison to Rotor-A, which has three rings of uniform thickness, the contact surface between the hub and the inner ring of Rotor-D remains within the tensile stress zone. The contact surface between the inner ring and the middle ring of Rotor-D experiences compressive stress, as illustrated in Figure 4d. The increasing thickness of the rings from the inside to the outside improves the radial stress distribution of the rotor to a certain extent.

Analyzing Rotor-C, it can be observed that although both the inner and outer surfaces of the inner ring of the rotor experience compressive stress, the displacement reaches a maximum of 7.418 mm. Consequently, the stress state of Rotor-E, which features a smooth gradient change in the specific modulus and a thickness ratio of rings of 1:2:3 from the inside to the outside, was examined further at a rotational speed of 18,000 r/min. Both the inner and outer contact surfaces of the inner ring of Rotor-E also exhibit compressive stress, as illustrated in Figure 4e. It is evident that the displacement of the inner ring of Rotor-E is 2.54 mm, which is less than that of Rotor-C. Compared to Rotor-C, Rotor-E represents a superior design scheme.

4.2. Maximum Rotational Speed Analysis

The maximum rotational speed of the flywheel rotor was analyzed when further reaching 18,400 r/min. The radial stress distribution of the inner and outer rings of the rim as well as the von Mises equivalent stress distribution of the hub are illustrated in Figure 5 at a speed of 18,400 r/min. It is evident that the contact surfaces between the hub, the inner ring, the middle ring, and the outer ring of the rotor all experience radial compressive stress, as shown in Figure 5a–c.

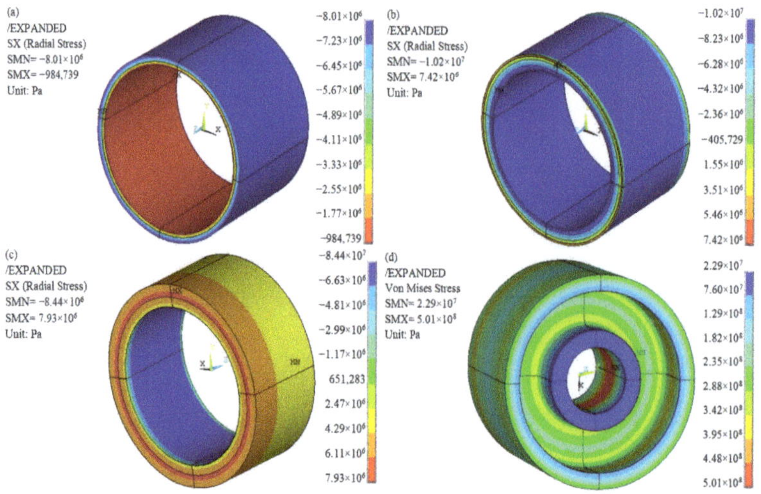

Figure 5. Stress distribution of Rotor-E: (**a**) radial stress distribution of the inner ring; (**b**) radial stress distribution of the middle ring; (**c**) radial stress distribution of the outer ring; (**d**) von Mises equivalent stress distribution of the hub.

The tensile strength of the hybrid fiber composite is greater than that of two single fiber composites [22]. The ultimate tensile and compressive strengths in the transverse direction of the glass fiber-reinforced composite (GFRC) are 31 MPa and 118 MPa, respectively [12,17,23]. The radial tensile and compressive stresses of the rim are 7.93 MPa and 10.2 MPa, respectively, which are significantly lower than the ultimate tensile and compressive strengths in the transverse direction of the GFRC. The von Mises equivalent stress distribution is utilized to analyze the operational state of the metal hub. As shown in Figure 5d, the maximum equivalent stress of the hub is 501 MPa, which is lower than the yield stress of 7075 aluminum alloy (546 MPa). This maximum stress occurs at the internal surface of the central hole of the hub. To ensure the safe operation of the rotor, it can be concluded that the rotational speed of Rotor-E should not exceed 18,000 r/min.

An important index to evaluate the performance of FESS, the energy storage density e, can be expressed as [24]

$$e = \frac{E_{max}}{m} = \frac{J\omega_{max}^2}{2m} \tag{11}$$

where E_{max} is the maximum energy stored by the flywheel; m is the mass of the flywheel; J is the moment of inertia of the flywheel; ω_{max} is the maximum angular velocity of the flywheel rim.

$$J = \frac{1}{2}m\left(r_i^2 + r_e^2\right) = \frac{1}{2}mr_e^2\left(\alpha^2 + 1\right) \tag{12}$$

$$e = \frac{(1+\alpha^2)v_{max}^2}{4} \tag{13}$$

where r_i is the internal diameters; r_e is the external diameters; α is the ratio of internal to external diameters; v_{max} the maximum linear velocity of the flywheel rim.

According to Equation (13), the energy storage density of the flywheel rotor is proportional to the square of the rim's linear velocity. The linear speed of the rotor is constrained by the material's strength. The maximum linear velocity of a metal flywheel rotor typically ranges from 300 to 500 m/s, as indicated in Table 4 [18,24]. N. Hiroshima [25] employed a three-dimensional CF-reinforced composite material as the rotor material, and the finite element analysis method was utilized to assess the stress distribution of the rotor during high-speed rotation. High-speed spin tests confirmed that the tensile stress generated by the rotor at a speed of 35,900 rpm (526 m/s) led to the rotor's failure. In contrast, the linear speed of the flywheel rotor analyzed in this paper reaches 706.5 m/s, resulting in a higher energy storage density.

Table 4. Comparison of metal and hybrid composite flywheel rotors [18,24,25].

Material	Outer Radius (mm)	Rotation Speed (r/min)	Linear Velocity (m/s)
Al 7075 T6	200	20,156	421.9
Steel 18Ni 300	200	15,460	323.6
Stainless steel 455	200	22,087	462.3
Al 6061 T6 kevlar49 epoxy	200	17,006	356
3D CF-reinforced composite	140	35,900	526
Hybrid composite (Rotor-E)	375	18,000	706.5

4.3. Modal Analysis

A flywheel rotor having an operating frequency range that is close to its natural frequency may lead to structural failure. Therefore, it is essential to conduct a modal analysis during the design of the flywheel rotor to ensure that the natural frequency of the rotor is higher than the operating frequency, thereby preventing mechanical resonance. The modal analysis of the rotor is performed after the static analysis to account for the effects of centrifugal force on the modal characteristics. Once the static analysis is complete, the modal analysis can be initiated by selecting the type, which allows for direct solving without the need to redefine the boundary conditions. The displacement constraint conditions remain the same as those used in the static analysis, and the rotational speed is disregarded in the modal analysis.

The relationship between the rotational speed and the natural frequency can be given as

$$n = 60 \times f \tag{14}$$

where n is the critical speed; f is the natural frequency.

The natural frequencies and critical speeds of Rotor-E are presented in Table 5. At the critical speed, the system may experience resonance or instability, potentially leading to flywheel failure. It is evident that the minimum critical speed of the flywheel is 102,756 r/min, which significantly exceeds the designed speed of 18,000 r/min. Consequently, the flywheel operates outside the resonance region during its functioning.

Table 5. First 10th-order natural frequencies and critical speeds of the composite rotor.

Set	Natural Frequency, f (Hz)	Critical Speed, n (r/min)
1	1712.6	102,756
2	1906.4	114,384
3	2294.0	137,640
4	2664.1	159,846
5	3255.6	375,336
6	3401.2	204,072
7	3483.9	209,034
8	3751.2	225,072
9	3979.6	238,776
10	4030.0	241,800

The vibration mode diagram of the flywheel rotor illustrates the deformation of the rotor when its rotational speed resonates with the rotor's natural frequency. These vibration modes depend on the specific design and properties of the rotor. The vibration mode diagram of Rotor-E is shown in Figure 6. This information is valuable for designing and optimizing the flywheel rotor to prevent excessive vibrations.

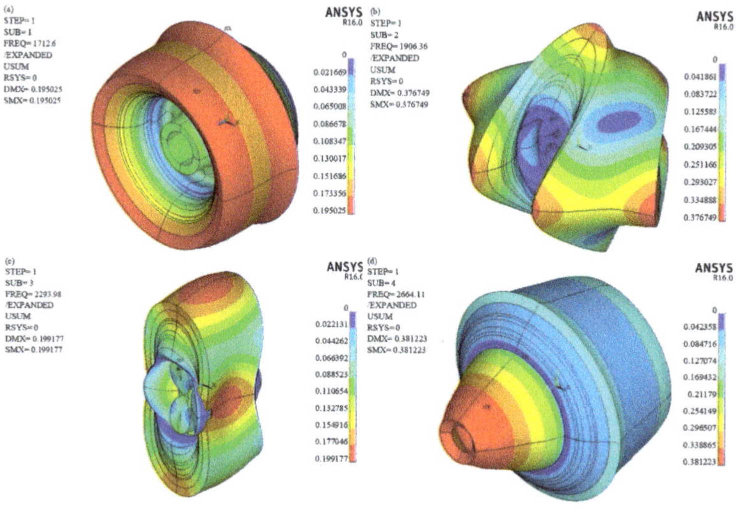

Figure 6. Vibration mode diagrams of Rotor-E: (**a**) First-order vibration mode; (**b**) Second-order vibration mode; (**c**) Third-order vibration mode; (**d**) Fourth-order vibration mode.

5. Conclusions

In this paper, we discuss the structure and material design of a flywheel rotor featuring a metal hub and a composite rim, utilizing the 3D finite element method. The composite rim consists of three wound rings. The material properties of the hybrid composite gradually vary from the inner to the outer layers by adjusting the GF and CF contents during the winding process of the composite rings. Aluminum alloy 7075 is employed for the metal hub.

The gradient variation in the specific modulus in the hybrid composite can enhance the radial stress distribution of the rotor. A smooth gradient change in the specific modulus of the hybrid composite rim, achieved by adjusting the ratio of GF to CF, represents a more effective approach. Furthermore, increasing the thickness of the rings from the inside to the outside can further optimize the radial stress distribution of the rotor to a certain extent.

Based on the analysis results, the material and structural design of the rotor are finalized. The optimal design of the rim consists of hybrid composite rings with the following thicknesses: the inner ring is 25 mm, the middle ring is 50 mm, and the outer ring is 75 mm, that is, a thickness ratio of 1:2:3 for the inner, middle, and outer rings. The inner ring transitions from 100% GF to 83.3% GF and 16.7% CF, the middle ring transitions from 83.3% GF and 16.7% CF to 16.7% GF and 83.3% CF, and the outer ring transitions from 16.7% GF and 83.3% CF to 100% CF. The hub, inner ring, middle ring, and outer ring are assembled using an interference fit, with an interference amount of 0.3 mm for all components. Through modal analysis, the natural frequencies and mode shapes of the flywheel rotor are established. The rotor avoids the resonance region during operation, indicating that the flywheel is safe to operate at a rotational speed of 18,000 r/min.

Author Contributions: Conceptualization, C.L., J.W. and H.W.; methodology, C.L., M.L. and H.W.; software, H.W., R.L. and Y.S.; validation, C.L., X.M. and H.W.; formal analysis, C.L. and R.L.; investigation, C.L., H.W., Y.S. and X.M.; resources, C.L.; data curation, Y.S. and X.M.; writing—original draft preparation, H.W.; writing—review and editing, C.L., R.L. and H.W.; visualization, R.L., H.W. and Y.S.; supervision, C.L.; project administration, C.L.; funding acquisition, C.L. All authors have read and agreed to the published version of the manuscript.

Funding: This research was funded by the Special Project for Basic Product Innovation (QT1451-0415045), The Key Research and Development Program of Heilongjiang Province (grant No. GA23A904), and the Civilian Ship Special Project (CBG5N21-3-3).

Institutional Review Board Statement: Not applicable.

Data Availability Statement: The data presented in this study are available upon request from the corresponding author. The data are not publicly available as the data are part of an ongoing study.

Conflicts of Interest: The authors declare no conflicts of interest.

Abbreviations

The following abbreviations are used in this manuscript:

FESS	Flywheel Energy Storage System
GFRP	Glass Fiber-Reinforced Polymer
CFRP	Carbon Fiber-Reinforced Polymer
GFRC	Glass Fiber-Reinforced Composite
GCHC	Glass/Carbon Hybrid Composite
GF	Glass Fiber
CF	Carbon Fiber

References

1. Li, X.J.; Palazzolo, A. A review of flywheel energy storage systems: State of the art and opportunities. *J. Energy Storage* **2022**, *46*, 103576. [CrossRef]
2. Aydogmus, O.; Boztas, G.; Celikel, R. Design and analysis of a flywheel energy storage system fed by matrix converter as a dynamic voltage restorer. *Energy* **2022**, *238*, 121687. [CrossRef]
3. Karrari, S.; De Carne, G.; Noe, M. Model validation of a high-speed flywheel energy storage system using power hardware-in-the-loop testing. *J. Energy Storage* **2021**, *43*, 103177. [CrossRef]

4. Jo, J.H.; Ryu, Y.G.; Choe, Y. Simulation on modified multi-surface levitation structure of superconducting magnetic bearing for flywheel energy storage system by H-formulation and Taguchi method. *Phys. C Supercond. Its Appl.* **2023**, *613*, 1354305. [CrossRef]
5. Li, X.H.; Zhu, Z.W. Nonlinear dynamic characteristics and stability analysis of energy storage flywheel rotor with shape memory alloy damper. *J. Energy Storage* **2022**, *45*, 103392. [CrossRef]
6. Bamisile, O.; Zheng, Z.; Adun, H.; Cai, D.S.; Ting, N.; Huang, Q. Development and prospect of flywheel energy storage technology: A citespace-based visual analysis. *Energy Rep.* **2023**, *9*, 494–505. [CrossRef]
7. Gao, J.Z.; Zhao, S.D.; Liu, J.J.; Du, W.; Zheng, Z.H.; Jiang, F. A novel flywheel energy storage system: Based on the barrel type with dual hubs combined flywheel driven by switched flux permanent magnet motor. *J. Energy Storage* **2022**, *47*, 103604. [CrossRef]
8. Kale, V.; Secanell, M. Rotor Design and Optimization of Metal Flywheels. In *Encyclopedia of Energy Storage*; Cabeza, L.F., Ed.; Elsevier: Oxford, UK, 2022; pp. 41–56.
9. Arani, A.A.K.; Karami, H.; Gharehpetian, G.B.; Hejazi, M.S.A. Review of Flywheel Energy Storage Systems structures and applications in power systems and microgrids. *Renew. Sustain. Energy Rev.* **2017**, *69*, 9–18. [CrossRef]
10. Tzeng, J.; Emerson, R.; Moy, P. Composite flywheels for energy storage. *Compos. Sci. Technol.* **2006**, *66*, 2520–2527. [CrossRef]
11. Chen, Y.G.; Zang, B.Q.; Wang, H.T.; Liu, H.X.; Li, H.R. Research on Composite Rotor of 200kW Flywheel Energy Storage System High Speed Permanent Magnet Synchronous Motor for UPS. In Proceedings of the 24th International Conference on Electrical Machines and Systems (ICEMS), Gyeongju, Republic of Korea, 31 October–3 November 2021; pp. 398–403.
12. Kim, S.J.; Hayat, K.; Nasir, S.U.; Ha, S.K. Design and fabrication of hybrid composite hubs for a multi-rim flywheel energy storage system. *Compos. Struct.* **2014**, *107*, 19–29. [CrossRef]
13. Kudva, A.; Kotian, A.; Mahesha, G.T.; Pai, D. Experimental investigation of mechanical properties of bamboo/carbon fiber reinforced hybrid polymer matrix composites. *Mater. Today Proc.* **2023**. [CrossRef]
14. Andoko, A.; Gapsari, F.; Wijatmiko, I.; Diharjo, K.; Rangappa, S.M.; Siengchin, S. Performance of carbon fiber (CF)/Ceiba petandra fiber (CPF) reinforced hybrid polymer composites for lightweight high-performance applications. *J. Mater. Res. Technol.-JmrT* **2023**, *27*, 7636–7644. [CrossRef]
15. Dong, T.T.; Zhao, Z.Y.; Zheng, B.P.; Chen, C.Y.; Ma, P.B. Response and damage mechanism of carbon/aramid intra-ply hybrid weft-knitted reinforced composites under low-velocity impact. *Thin-Walled Struct.* **2024**, *196*, 111495. [CrossRef]
16. Spencer, M.; Chen, X. Static and fatigue cracking of thick carbon/glass hybrid composite laminates with complex wrinkle defects. *Int. J. Fatigue* **2023**, *177*, 107963. [CrossRef]
17. Changliang, T.; Xingjian, D.; Yong, W. Mechanical design and spin test of a multilayer commingled composite flywheel. *J. Tsinghua Univ.* **2015**, *55*, 361–367.
18. Kale, V.; Secanell, M. A comparative study between optimal metal and composite rotors for flywheel energy storage systems. *Energy Rep.* **2018**, *4*, 576–585. [CrossRef]
19. Rastegarzadeh, S.; Mahzoon, M.; Mohammadi, H. A novel modular designing for multi-ring flywheel rotor to optimize energy consumption in light metro trains. *Energy* **2020**, *206*, 118092. [CrossRef]
20. Singh, A.K.; Datta, S.; Chattopadhyay, A.; Phan, N. Effects of overload mode-mixity on fatigue damage behavior and governing micromechanisms in AA7075 under biaxial fatigue loading. *Int. J. Fatigue* **2021**, *145*, 106141. [CrossRef]
21. Sundaram, S.K.; Bharath, A.G.; Aravind, B. Influence of target dynamics and number of impacts on ballistic performance of 6061-T6 and 7075-T6 aluminum alloy targets. *Mech. Based Des. Struct. Mach.* **2022**, *50*, 993–1011. [CrossRef]
22. Sharma, K.K.; Shrivastava, Y.; Neha, E.; Jain, A.; Singh, B. Evaluation of flexural strength of hybrid FRP composites having three distinct laminates. In Proceedings of the 2nd International Conference on Future Learning Aspects of Mechanical Engineering (FLAME), Amity University, Noida, India, 5–7 August 2020; pp. 418–422.
23. Arjun, K.P.; de Barros, S.; Budhe, S. Theoretical determination of elastic and flexural modulus for inter-ply and intra-ply hybrid composite material. *Compos. Struct.* **2022**, *281*, 114971. [CrossRef]
24. Wang, P.; Gu, T.; Sun, B.; Liu, R.; Zhang, T.; Yang, J. Design and Performance Analysis of Super Highspeed Flywheel Rotor for Electric Vehicle. *World Electr. Veh. J.* **2022**, *13*, 147. [CrossRef]
25. Hiroshima, N.; Hatta, H.; Koyama, M.; Yoshimura, J.; Nagura, Y.; Goto, K.; Kogo, Y. Spin test of three-dimensional composite rotor for flywheel energy storage system. *Compos. Struct.* **2016**, *136*, 626–634. [CrossRef]

polymers

Article

Systematic Investigation on the Swelling Response and Oil Resistance of NBR Using the Prediction Models Determined by the Modified Flory–Huggins Interaction Parameter

Yiran Jing and Guangyong Liu *

Key Laboratory of Rubber-Plastics, Ministry of Education, Qingdao University of Science & Technology, Qingdao 266042, China; 17863957197@163.com
* Correspondence: liuguangyong@qust.edu.cn

Abstract: The equilibrium swelling test was employed to determine the swelling response of Nitrile Butadiene Rubber (NBR) with various acrylonitrile (ACN) contents, and the three-dimensional solubility parameter (HSP) and modified Flory–Huggins interaction parameter (χ_{HSP}) were used to establish the prediction model of the oil-resistant property. The results indicate that the energy difference (Ra) between NBR and solvents calculated by HSP values can be correlated with the swelling response qualitatively with an inversed "S-shape", and high swelling response occurs at Ra < 8 $MPa^{1/2}$ for NBR. For the purpose of establishing the prediction model, the new modified χ_{HSP} value has been calculated and fitted with the swelling response using exponential and logarithmic fittings, respectively. Two prediction models considering all the possible influencing factors have been obtained to determine the swelling response and oil resistance of NBR-based rubber products in bio-fuels, represented by the bio-diesel and IRM 903 test oil in this work. The swelling response of NBR can be evaluated precisely, and high swelling regions can be predicted and avoided in the new emerging fuels through the prediction models. Thus, the oil resistance of NBR-based rubber products, such as seals, holes and gaskets can be well predicted now.

Keywords: three-dimensional solubility parameter; Flory–Huggins interaction parameter; prediction model; swelling response; NBR

1. Introduction

Nitrile butadiene rubber (NBR) is an unsaturated polar polymer which is copolymerized with butadiene and acrylonitrile monomers [1,2]. The polarity and unsaturation of NBR are related to the acrylonitrile content. With the increase in acrylonitrile content, the molecular polarity, intermolecular force and cohesion energy density increase, leading to the improvement of the anti-static property and non-polar oil resistance [3–5]. Due to the good resistance to non-polar media (gasoline, kerosene, etc.), NBR is mainly used in a variety of oil resistant and antistatic products, such as oil pipes, sealing strips, textile rollers, and so on [6–8]. Therefore, the study of the diffusion and swelling behaviors of NBR in different media becomes necessary.

Hildebrand and Scott [9] first introduced the concept of a solubility parameter (δ) to characterize the thermodynamic behavior of non-electrolytes in binary systems, which can be used to predict the polarity, solvent resistance, and polymer–polymer compatibility of substances.

$$\delta = \left[\frac{E_{coh}}{V_{mol}}\right]^{1/2} \tag{1}$$

where δ is the solubility parameter with the unit of $MPa^{1/2}$, E_{coh} is the cohesive energy of a substance, representing the amount of energy required to hold one mole of molecules together, and V_{mol} is the molar volume.

This formula has limitations in applications when there are polar interactions because the dissolution becomes more complex (even the solvation process occurs when the solubility parameters of the solvent and polymer are quite different) [10–12]. Based on Hildebrand's regular solution theory, Hansen extended the solubility parameter to polar and conjugated systems, and established a three-dimensional solubility parameter system, abbreviated as HSP [13]. The cohesion energy of a substance is considered to be the sum of various interaction forces between molecules, including the van der Waals force, covalent bond, hydrogen bond, ionic bond, electrostatic action, the permanent dipole moment and the induced dipole moment. Hansen divided the cohesion energy of a substance into the sum of the contribution values of three components: the dispersion force (E_d), the polarity force (Ep) and the hydrogen bond force (E_h) [14,15]. By extending the solubility parameter to a polar system and an association system, a three-dimensional solubility parameter system is established as follows:

$$\delta_d = \left(\frac{E_d}{V}\right)^{\frac{1}{2}},\ \delta_p = \left(\frac{E_p}{V}\right)^{\frac{1}{2}},\ \delta_h = \left(\frac{E_h}{V}\right)^{\frac{1}{2}} \tag{2}$$

$$\delta_t = \delta_d^2 + \delta_p^2 + \delta_h^2 \tag{3}$$

where δ_t is the Hildebrand solubility parameter, δ_d, δ_p, and δ_h correspond to the three partial solubility parameters.

It provides an easy method for predicting the compatibility of polymer blends, selecting solvents, and predicting polymer–solvent interactions, which has been widely used in polymer solutions [16–18].

The concept of HSP can be more intuitively reflected in a three-dimensional spatial graph: δd, δp, and δh are the three coordinate axes with the polymer as the spheric center. The interaction strength between polymer and solvent is taken as the radius (R_0) to obtain a sphere named as the solubility sphere of polymer. The interaction between polymer and solvent can be expressed in terms of their spatial distance Ra (energy difference). The solvent can dissolve or swell the polymer quickly when it is inside of the solubility sphere ($Ra < R_0$) [19].

$$Ra = \left[a\left(\delta_d^P - \delta_d^S\right)^2 + \left(\delta_p^P - \delta_p^S\right)^2 + \left(\delta_h^P - \delta_h^S\right)^2\right]^{1/2} \tag{4}$$

where δi (i = d, p, h) for each of the three components of the HSP values, and the superscripts P and S represent the polymer and solvent, respectively.

The solubility parameter is an important thermodynamic parameter of polymers, which can predict the polarity of a substance, the solvent resistance, and the polymer–polymer compatibility [20–22]. The solubility parameter of a low molecular solvent can be obtained directly from the enthalpy of evaporation [23,24]. Due to the high molecular weight of polymers, as well as the possible existence of the macromolecular crystallization and degradation and other factors, the solubility parameter of polymers cannot be directly obtained by evaporation. Therefore, indirect methods have to be used to determine the solubility parameter, including the optical refractive index method [25], the turbidimetric titration method [26], gas chromatography [27], the intrinsic viscosity method [28] and the equilibrium swelling method [29–31].

The ACN content determines the key properties of NBR vulcanizates, for example, the oil resistance and low-temperature properties. The ACN content of NBR determines its polarity, which in turn determines the resistance to apolar oils and greases. Thus, due to the large range of different compositions available, adequate NBR grades can be selected for many applications. In this work, three different NBR vulcanizates with incremental ACN contents are selected for determining the swelling responses in various solvents through the equilibrium swelling tests. The HSP values and the modified χ_{HSP} parameters are calculated to establish the prediction models, which are expected to predict the swelling responses and oil resistance of NBR-based rubber products in newly emerged bio-fuels.

2. Experimental Section

2.1. Materials

The Nitrile butadiene rubber with the ACN contents of 34 wt% (NBR34), 39 wt% (NBR39) and 44 wt% (NBR44), respectively, are supplied by ARLANXEO (The Hague, The Netherlands). The crosslinking system composing of 2, 4-di-tert-butylisopropylbenzene peroxide (Luperox F40) and triallyl isocyanurate (TAIC) are purchased from Arkema (Paris, France). Organic solvents (analytical reagent) used for the swelling tests, including alkanes, aromatics, ethers, esters, ketones, nitriles and amides are commercially available industrial products.

2.2. Equilibrium Swelling Test

Only the crosslinking system (Luperox F40 6 phr, TAIC 2 phr) was used to reduce the influence of other factors. NBR was mixed with the crosslinking system directly using a two-roll mill (BOLON Precision Testing Machines Co. Ltd., BL-6175, Dongguan, China) at 50 °C with a roller speed ratio of 1:1.2. Then, uniformly dispersed NBR compounds were obtained. The vulcanized rubber sheets of a 2 mm thickness were obtained by a heated curing press at 180 °C and t90 + 5 min as the curing time determined by the Moving Die Rheometer (MDR2000, Alpha Technologies, Bellingham, WA, USA).

Circular samples with the diameter of 12 mm were cut from the vulcanized sheets for the swelling tests. Five samples were selected for each test to reduce the weighing error. These samples were first weighed for the initial weight, then immersed in the selected solvents for the swelling tests. At specifical intervals, the samples were taken out, and extra solvent on the surface was removed quickly with filter paper. Then, the samples were weighed immediately. After weighing, the samples were placed back into the original test bottles. Such operations were conducted several times until reaching equilibrium swelling and then the samples were weighed again to obtain the final weight. The equilibrium swelling tests were carried out to calculate the swelling ratio (q) of the NBR samples in the well-selected solvents, as shown in Table 1.

Table 1. The HSP values of the solvents and the swelling ratios of the NBR vulcanizates.

Solvents		δ_d, MPa$^{1/2}$	δ_p, MPa$^{1/2}$	δ_h, MPa$^{1/2}$	q, NBR34	q, NBR39	q, NBR44
Alkanes	Isooctane	14.10	0.00	0.00	0.03	0.01	0.01
	Heptane	15.30	0.00	0.00	0.09	0.03	0.02
	Cyclohexane	16.80	0.00	0.20	0.16	0.05	0.02
Aromatics	Ethylbenzene	17.80	0.60	1.40	1.81	1.24	1.03
	Toluene	18.00	1.40	2.00	2.33	1.68	1.42
Ethers	Diethyl ether	14.50	2.90	4.60	0.51	0.36	0.29
	THF	16.80	5.70	8.00	4.59	4.12	3.86
Esters	Butyl acetate	15.60	6.20	4.90	3.35	3.19	3.14
	Methyl acetate	15.50	7.20	7.60	2.63	2.71	2.79
Ketones	2-butanone	16.00	9.00	5.10	4.58	4.33	4.27
	Acetone	15.50	10.40	7.00	1.94	1.41	1.17
	Cyclohexanone	17.80	8.40	5.10	2.19	2.01	1.95
Nitriles	Butyronitrile	15.30	12.40	5.10	2.96	2.98	3
	Benzonitrile	18.80	12.00	3.30	4.39	4.46	4.45
	Acetonitrile	15.30	18.00	6.10	1.07	1.4	1.68
Amides	DMF	17.40	13.70	11.30	3.59	3.82	4.04

The swelling process in detail can be found in our previous study [32], and q is determined as follows:

$$q = \frac{W_S/\rho_S}{W_R/\rho_R} \tag{5}$$

where W_S is the mass uptake of the solvent at an equilibrium state, W_R is the initial mass of the rubber sample before swelling, ρ_S and ρ_R are the densities of the solvent and rubber sample, respectively. Herein, q can be defined as the volume of the absorbed solvent per unit volume of the rubber sample in units of mL/cm^3.

3. Results and Discussion

3.1. Correlation of Swelling Ratio with Solubility Parameter

The swelling ratio (q) of NBR varies greatly in different solvents, which may be related to the one-dimensional solubility parameter (δ_t) of the solvent. In order to better compare this relationship, q values of three NBR vulcanizates (NBR34, NBR 39, and NBR44) were preliminarily correlated with the δ_t values of solvents, as shown in Figure 1.

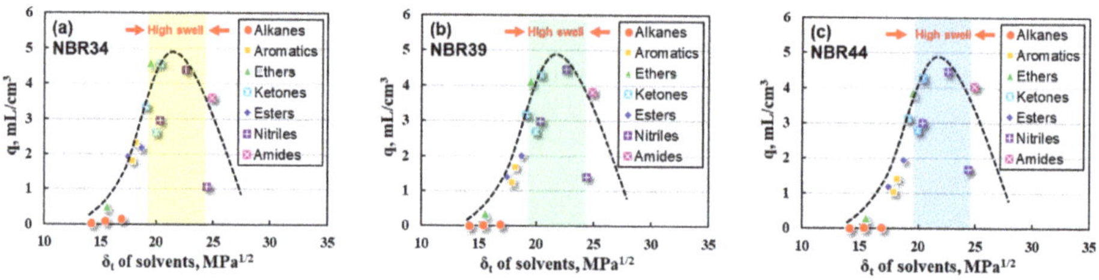

Figure 1. Plots of q values for NBR with δ_t values for solvent (**a**) NBR34, (**b**) NBR39, and (**c**) NBR44.

It can be clearly seen that the q values of NBR vulcanizates increase first and then decrease with the increase in the δ_t values of the solvents, passing through high swell regions corresponding to the range of δ_t values from 20 to 25 MPa$^{1/2}$ (dashed line in Figure 1). Therefore, the δ_t value corresponding to the maximum q value can be regarded as the one-dimensional solubility parameter (δ_t) of NBR, which is estimated within the range of 21~24 MPa$^{1/2}$. However, it is impossible to distinguish the specific δ_t values for the three different NBR samples, i.e., NBR34, NBR39, and NBR44.

For this reason, a computer program was employed to calculate the HSP values of NBR by inputting the swelling data. The swelling ratios (q) of NBR in these solvents were used directly as input data into the computer program to determine the specific three-dimensional solubility parameters. The range of the swelling ratio was from 0 to 5 mL/cm^3 and, therefore, the calculation routine of the program assumed the intermediate value of ca 2.5 mL/cm^3 as the starting point and the boundary condition between a high and low swelling response. The program analyses and analog computes the swelling ratio, and outputs HSP values along with HSP graphs of NBR34, NBR39, and NBR44, as shown in Table 2. The computational procedure of the program has been described in detail in our previous study [28].

Table 2. HSP values of NBR vulcanizates.

NBR	δ_d, MPa$^{1/2}$	δ_p, MPa$^{1/2}$	δ_h, MPa$^{1/2}$	δ_t, MPa$^{1/2}$
NBR34	19.6	8.5	6.7	22.4
NBR39	19.6	9.6	6.7	22.8
NBR44	19.6	10.5	7.2	23.4

The dispersion force component (δ_d) of NBR34, NBR39, and NBR44 rubber is basically the same, the difference in the hydrogen bond force component (δ_h) is small, and the polar force component (δ_p) is quite different, which is mainly caused by the incremental ACN content.

In high swelling regions, however, there are still some solvents, such as acetone and DMF, showing relatively low swelling responses even if their δ_t values are within the same range of 21~24 MPa$^{1/2}$. This may be because the δ_p and δ_h values of these two solvents are quite different from those of NBR.

Further comparison of the q value of NBR with the $\Delta\delta_t$ between NBR and solvent was studied as shown in Figure 2.

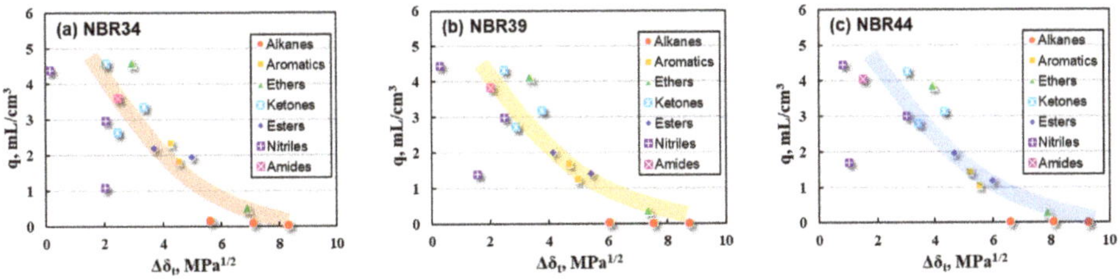

Figure 2. Correlation of q with $\Delta\delta_t$ for NBR vulcanizates (**a**) NBR34, (**b**) NBR39, and (**c**) NBR44.

The color bands are added into Figure 2 to illustrate the possible trends in q with $\Delta\delta_t$ values. There is a rough relationship between q and $\Delta\delta_t$ for NBR vulcanizates, indicating that a higher $\Delta\delta_t$ value relates to a lower q value, and vice versa. However, there are still many discrete points in this correlation, especially for nitrile solvents exhibiting a small q value even if the $\Delta\delta_t$ is low. The $\Delta\delta_t$ cannot explain such a phenomenon, and hence, it is necessary to further use the HSP values to solve this problem.

Ra values calculated by HSP values were used to correlate the swelling ratios of NBR vulcanizates and the results were shown in Figure 3.

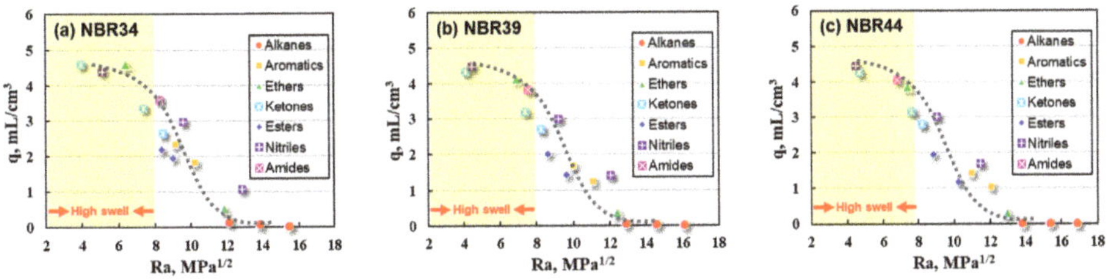

Figure 3. Correlation of q values with Ra for NBR vulcanizates (**a**) NBR34 (**b**) NBR39 (**c**) NBR44.

The q values of NBR34, NBR39, and NBR44 vulcanizates increase with the decrease in Ra values, showing inversed S-shapes (dashed lines in Figure 3). The swell response is in the high swell region (yellow band in Figure 3) when the Ra is less than 8 $MPa^{1/2}$; on the contrary, the swelling ratio becomes lower at a higher Ra > 8 $MPa^{1/2}$, and it becomes quite small especially when the Ra is greater than 12 $MPa^{1/2}$. Better compatibility between the NBR and solvent contributes to a higher swelling response which can be reflected in a smaller Ra value as well. Further observation of Figure 3 shows that there is no obvious discrete point on the q~Ra curve, that is, the correlation of q with Ra is more acceptable compared to the q~$\Delta\delta_t$ curve. NBR is usually used to make protective gloves, seals, holes and gaskets, and thus, the liquid medium resistance is very important for NBR-based products. The Ra value shows better advantages in predicting the oil resistance of NBR vulcanizates and avoiding the potential risks of liquid leakage.

However, the inversed "S-shape" curve is still difficult to express precisely or be quantitatively fitted by a mathematical equation, which makes it impossible for us to establish the prediction model of the swelling response for an NBR vulcanizate.

3.2. *Flory–Huggins Interaction Parameters of NBR/Solvent*

The increasing use of new fuels and oils such as bio-diesel and bio-gasoline and their effects on rubber materials have aroused widespread concern in the industry. NBR-based seals and gloves are often in contact with organic liquids with great differences in polarity, and the interaction between NBR and different liquids should be fully considered. The three-dimensional solubility parameters and Flory–Huggins interaction parameters (χ) provide a concise method for studying the interactions existing in NBR swelling systems.

The χ parameter was determined originally based on the corresponding state theory (CST), which was used to calculate the χ values of non-polar systems (non-polar rubber and non-polar solvents) and to characterize the thermodynamic state of rubber macromolecular solutions [33,34]. The equation is shown as follows:

$$\chi = \chi_H + \chi_S \quad (6)$$

where χ_S is the entropy correction term with a constant of 0.34 for a strictly nonpolar system. χ_H refers to the contribution of the enthalpy change which can be calculated by the δ_t values of the rubber and solvent:

$$\chi_H = \frac{V}{RT}(\delta_p - \delta_s)^2 \quad (7)$$

Therefore, for the non-polar rubber solution system, the following calculation formula can be obtained:

$$\chi = 0.34 + \frac{V_m}{RT}(\delta_p - \delta_s)^2 \quad (8)$$

where V_m is the molar volume of a small molecule solvent, R is the ideal gas constant, T is the absolute temperature, and δp and δs are one-dimensional solubility parameters of the rubber and solvent, respectively.

According to Flory [35–37], a provided condition under which a rubber and a solvent are expected to be completely miscible through the entire composition range is the following,

$$\chi < \frac{1}{2}\left[1 + \left(\frac{V_s}{V_p}\right)^{1/2}\right]^2 \quad (9)$$

Therefore, it can be inferred that a critical condition or a critical interaction parameter (χ_C) for rubber–solvent miscibility is,

$$\chi_c = \frac{1}{2}\left[1 + \left(\frac{V_s}{V_p}\right)^{1/2}\right]^2 \quad (10)$$

where V_P and V_S are the molar volumes of rubber macromolecules and a small molecule solvent, respectively.

Comparing V_P and V_S values, it can be considered that V_S/V_P tends to be zero. Therefore, the condition of a rubber dissolving in a solvent is $\chi < 0.5$ for the non-polar system. The δp and δs values in Equation (8) need to be very close to each other to meet such a condition that the rubber is sufficiently dissolved in the solvent.

However, Equation (8) shows obvious limitations in describing the polar rubber solution system because it does not consider the contributions of the polar and the hydrogen bond forces to the rubber dissolution process. It is attempted to solve this problem by using a modified Flory–Huggins interaction parameter (χ_{HSP}) that can be calculated by the HSP values for the polar rubber solution system, as follows.

$$\chi_{HSP} = \frac{V_S}{4RT}\left[4\left(\delta_d^P - \delta_d^S\right)^2 + \left(\delta_p^P - \delta_p^S\right)^2 + \left(\delta_h^P - \delta_h^S\right)^2\right] \quad (11)$$

$$\chi_{HSP} = \frac{V_S}{4RT}Ra^2 \quad (12)$$

Equation (12) for calculating the χ_{HSP} value includes the contributions from the dispersion force, the polar force and the hydrogen bond force and takes into account the influences of molar volume and temperature. For real rubber solutions or swelling systems, special interactions such as polarity and hydrogen bonding contribute more to the χ value, resulting in some unpredictable miscibility or swelling properties. Therefore, in most cases, Equation (12) is more suitable for the evaluation of the compatibility between rubber and solvent when handling the swelling and dissolving processes [38,39].

3.3. Prediction Model for the Swelling Response of NBR

In many practical applications, NBR-based rubber products are often exposed to a variety of polar fluids, and the prediction of the swelling properties in various solvents or solvent mixtures becomes particularly important. There are many factors affecting the swelling behaviors of NBR, including the structures of the NBR and the liquid, the size of the liquid, and the temperature. The new χ_{HSP} value provides an effective way to construct the prediction model of the swelling properties by fully considering all the possible influencing factors. Therefore, it is attempted to correlate q with the χ_{HSP} value for NBR vulcanizate, and the results are shown in Figure 4.

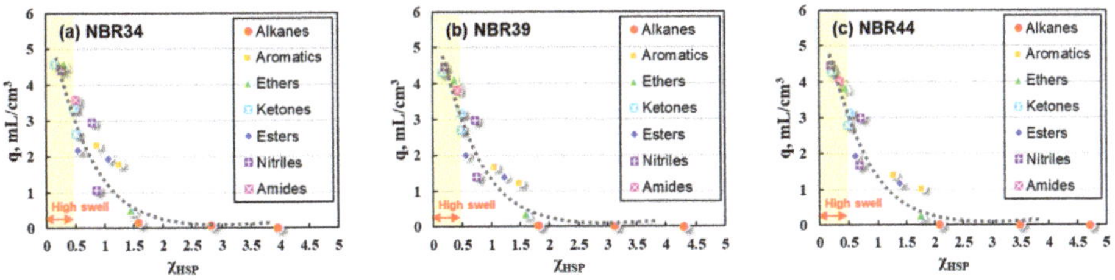

Figure 4. Plots of q with χ_{HSP} values for (**a**) NBR34, (**b**) NBR39, and (**c**) NBR44.

It can be seen from Figure 4 that q increases with the decrease of the χ_{HSP} value (dashed line in Figure 4). Assuming q = 2.5 mL/cm^3 as the boundary condition, the high swelling region (yellow band in Figure 4) corresponds to the χ_{HSP} value being less than 0.5. This result agrees well with the theory based on the Hildebrand regular solution model, which holds that for the actual rubber swelling system, the swelling becomes more complex, but the χ value must be less than 0.5 for the rubber–solvent system with a completely miscible or high swelling state.

Furthermore, mathematical fitting between q and the χ_{HSP} value was made for the purpose of predicting the swelling response of NBR quantitatively by theoretical calculation. It can be preliminarily judged from Figure 4 that there might be an exponential or logarithmic relationship between q and χ_{HSP}, as shown in Figure 5. Similar mathematical fittings have been adopted to Fluororubber (FKM) and Hydrogenated Nitrile rubber (HNBR) due to the relatively high degree of fittings in our previous study [28].

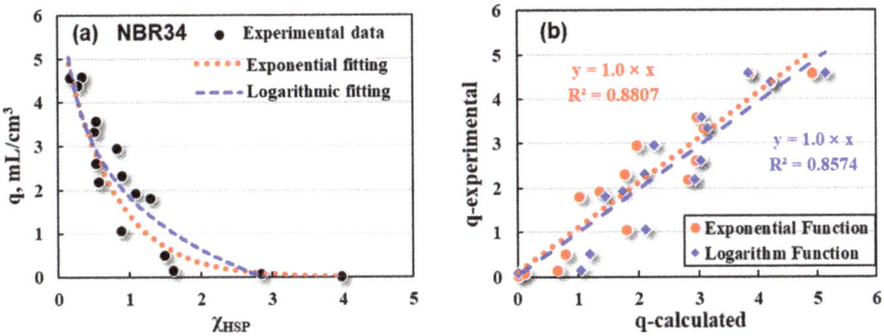

Figure 5. Mathematical fittings for NBR34's (**a**) q and χ_{HSP}, and its (**b**) q values between experimental and calculation data.

As shown in Figure 5a, exponential (red dotted line) and logarithmic (blue dashed line) fittings are performed on the relationships of q (black dots) with the χ_{HSP} as follows:

$$\text{Exponential: } q = 6.2 \times e^{(-1.4 \times \chi_{HSP})}\ R^2 = 0.8789 \quad (13)$$

$$\text{Logarithm: } q = 1.9 - 1.8 \times \ln(\chi_{HSP})\ R^2 = 0.8574 \quad (14)$$

Both exponential fitting and logarithmic fitting show relative high degree of fittings, which are quite different from the linear relationship of non-polar styrene-butadiene rubber (SSBR) [40,41]. One of the main purposes of the theoretical calculation of χ_{HSP} is to predict the swelling response and oil resistance of NBR parts in different fluids.

Therefore, the swelling ratios (q-calculated) of NBR34 in each solvent were calculated quantitatively by using exponential Equation (13) and logarithmic Equation (14), respectively, and compared with the experimental ones (q-experimental), as shown in Figure 5b. q-calculated and q-experimental values have some discreteness, but basically show a good linear relationship. The fitting degree of the exponential relationship is slightly higher than that of logarithmic fitting. It can be deduced that the logarithmic fitting may have negative q values when χ_{HSP} values are large (poor compatibility). Such a phenomenon is called swelling shrinkage, which is the case of the oil-resistant rubber products in some extreme fluids.

Then, exponential and logarithmic fittings of q~χ_{HSP} were made for NBR39 and NBR44, respectively, and the swelling ratios were calculated theoretically. The results are shown in Figures 6 and 7.

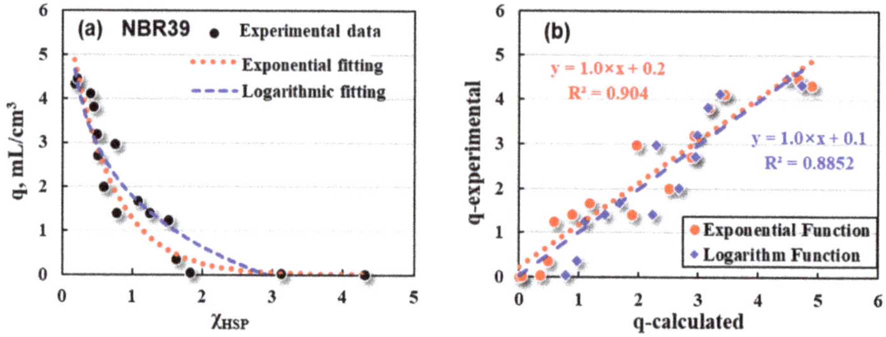

Figure 6. Mathematical fittings for NBR39's (**a**) q and χ_{HSP}, and its (**b**) q values between experimental and calculation data.

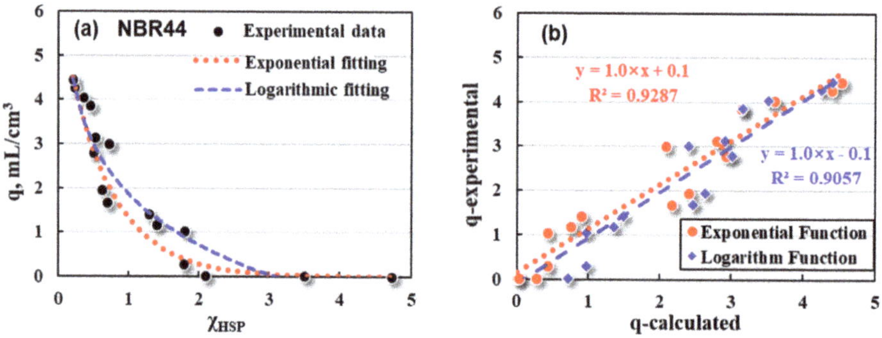

Figure 7. Mathematical fittings for NBR44's (**a**) q and χ_{HSP}, and its (**b**) q values between experimental and calculation data.

Exponential (red dotted line) and logarithmic (blue dashed line) fittings between the q (black dots) and χ_{HSP} for NBR39 are described as follows:

$$\text{Exponential: } q = 6.5 \times e^{(-1.6\times\chi_{HSP})}\ R^2 = 0.9022 \tag{15}$$

$$\text{Logarithm: } q = 1.8 - 1.7 \times \ln(\chi_{HSP})\ R^2 = 0.8852 \tag{16}$$

Exponential (red dotted line) and logarithmic (blue dashed line) fittings between the q (black dots) and χ_{HSP} for NBR44 are described as follows:

$$\text{Exponential: } q = 6.2 \times e^{(-1.5\times\chi_{HSP})}\ R^2 = 0.9276 \tag{17}$$

$$\text{Logarithm: } q = 1.9 - 1.6 \times \ln(\chi_{HSP})\ R^2 = 0.9057 \tag{18}$$

It can be seen that for NBR39 and NBR44, both exponential and logarithmic fittings show a high degree of fittings. Similarly, the swelling ratios (q-calculated) of NBR39 and NBR44 in each solvent were calculated mathematically and compared with the experimental ones (q-experimental). The results are shown in Figures 6b and 7b. Both the calculated and the experimental q values are discrete to a certain extent, but basically show a good linear relationship. In short, exponential and logarithmic fittings are expected to be suitable for the basic prediction models in evaluating the swelling responses of NBR-based rubber products.

3.4. Basic Prediction Models of Swelling Responses

As discussed above, there are exponential and logarithmic relations between q~χ_{HSP}. For the three exponential relations, the pre-exponential factors are in the range of 6.2 to 6.5 and the exponential factors are 1.4~1.6. For the three logarithmic relations, the addition factors are in the range of 1.6 to 1.8 and the logarithmic factors are 1.8~1.9. Herein, two general prediction models can be established preliminarily as follows:

$$\text{Exponential: } q = 6.3a \times e^{(-1.5m\times\chi_{HSP})} \tag{19}$$

$$\text{Logarithm: } q = 1.8b - 1.8n \times \ln(\chi_{HSP}) \tag{20}$$

where a and m represent the correction coefficients of the pre-factors for an exponential model, and n and b are the correction coefficients of pre-factors for a logarithmic model, respectively.

It has been discussed above that the swelling ratio of rubber is not only related to the Flory–Huggins interaction parameter (χ_{HSP}), but also to the crosslinking density (ν), swelling temperature, filler dosage, and other factors. The temperature and molar volume (Vmol) of the solvent have been considered in the calculation of the χ_{HSP} value, and the crosslink density and filler content need to be considered in the prediction model as well.

The chemical crosslink density (ν_c) can be determined by the equilibrium swelling test with the aid of the Flory–Huggins theory. Fillers (such as carbon black and silica) have a positive correlation effect on the total crosslink density, which in fact mainly increases the physical crosslink density of rubber (ν_p). Therefore, the crosslink density needs to be taken into account when establishing a prediction model of rubber swelling. The crosslinking network of rubber is extremely complicated, especially in that the number of the crosslinking points is difficult to be quantified. According to Flory–Rehhner theory, the crosslink density (ν) is related to the volume fraction (v_r) of rubber, which directly affects the maximum swelling ratio (q) instead of the variation trend between q~χ_{HSP}.

Therefore, the modified prediction models of NBR can be obtained as follows:

$$\text{Exponential: } q = 6.3a \times e^{(-1.5m\times\chi_{HSP})} + \alpha\nu \tag{21}$$

$$\text{Logarithm: } q = 1.8b - 1.8n \times \ln(\chi_{HSP}) + \beta\nu \tag{22}$$

where ν is the crosslink density, α and β are correction coefficients based on the crosslink density (chemical and physical crosslink densities).

Actually, these correction coefficients, including a, m, b, and n are related to the swelling system with certain relationships.

3.5. Application of Prediction Model in Oil Resistance of NBR

The swelling behavior and oil resistance of NBR-based products in new emerging bio-fuels, bio-gasolines and bio-diesels for example, have become important research topics in both academic and industrial fields. Some fuel mixtures containing small amounts of alcohol, such as ethanol gasoline, have a serious effect on the swelling responses of rubber parts. NBR products are widely used in such aspects, hence, it is very important to determine the swelling responses of NBR in these new liquids.

Bio-diesel is a kind of diesel with different components based on the plant raw materials used, which is considered to be low viscosity, low toxicity and non-flammable environmental diesel. Therefore, bio-diesel is an ideal alternative to petroleum diesel as a fuel source. IRM 903 (ASTM No. 3) is a test oil commonly used in laboratories to evaluate the oil resistance of polar rubber products.

New bio-fuels are configured, respectively, by the bio-diesel and IRM 903 test oil supplemented with anhydrous ethanol, and the oil resistance of NBR in such fuels is explored. One of the applications of χ_{HSP} is to predict the swelling behavior or oil resistance of rubber products in new fuels, especially in extreme environments, for example, at very high and low temperatures. IRM 903 and bio-diesel are mixtures with a carbon atom composition similar to that of the "naphtha high-flash" substance. The HSP values of the two fuels are determined according to the previous work as shown in Table 3.

Table 3. The HSP values of IRM 903 and bio-diesel.

Fluids	δ_d, MPa$^{1/2}$	δ_p, MPa$^{1/2}$	δ_h, MPa$^{1/2}$	δ_t, MPa$^{1/2}$	Vmol, mL/mol
IRM 903	17.9	0.7	1.8	18.0	350
Bio-diesel	16.5	0.3	0.9	16.5	152
Ethanol	15.8	8.8	19.4	26.5	59

The HSP values follow the linear additive property, and thus the HSP values of the fluid mixtures, IRM 903/Ethanol and bio-diesel/Ethanol, can be calculated as follows:

$$\delta_{i,mix} = \varphi_1\delta_{i,1} + \varphi_2\delta_{i,2} \quad (23)$$

where $\delta_{i,mix}$ represents the $\delta_{d,mix}$, $\delta_{p,mix}$, and $\delta_{h,mix}$ values of each component, respectively. φ_1 and φ_2 are the volume fractions of components 1 and 2, respectively.

From the point of view of reducing costs and improving exhaust emissions, organic solvents such as ethanol are often added into bio-diesel or bio-gasoline fuels. For these bio-fuel blends, however, their composition and HSP values will be changed accordingly, which has unforeseen effects on the swelling properties of rubber, and this unforeseeable effect may become more serious especially under high-temperature conditions. For this reason, the swelling responses of NBR in these bio-fuel blends have been calculated by using the prediction models obtained from Equations (21) and (22). Herein, the three NBR vulcanizates, NBR34, NBR39, and NBR44, are fixed with the same crosslink density and other possible factors so as to make the four correction coefficients (a, b, m, and n) equal 1 and the other two crosslinking-related correction coefficients (α, β) equal 0. The swelling ratios (q-cal.) predicted by the prediction models are firstly correlated with the ethanol contents for NBR vulcanizates, as shown in Figure 8.

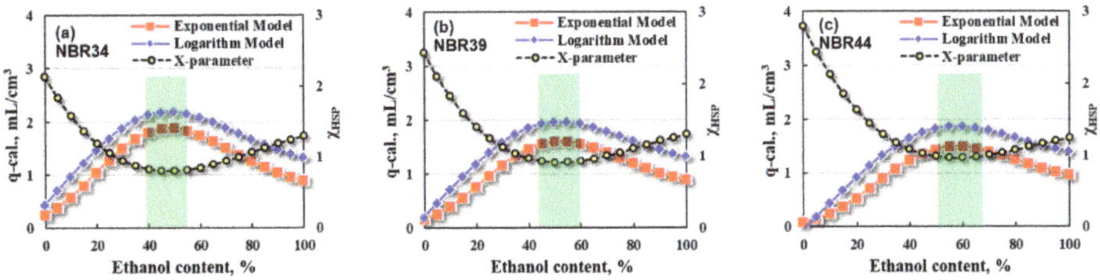

Figure 8. Correlations of q-cal. with ethanol contents in bio-diesel/ethanol blends of (**a**) NBR34, (**b**) NBR39, and (**c**) NBR44.

It can be seen from the bio-diesel/ethanol blends in Figure 8 that the q-cal values obtained by the exponential model (red square) are slightly lower than those calculated by the logarithm model (blue diamond). In addition, with the increase in ethanol content, the q-cal value increases first and then decreases, passing through a maximum swelling region. The corresponding χ_{HSP} value shows the opposite trend, that is, the χ_{HSP} value decreases first and then increases with the increasing ethanol content. The phenomenon is similar but the composition of the bio-diesel/ethanol blends is different when the maximum swelling region appears for the three NBR vulcanizates. The maximum swelling regions correspond to the ethanol contents of 40~55 vol %, 45~60 vol %, and 50~70 vol% for NBR34, NBR39, and NBR44, respectively.

Further calculations are made using the prediction models for the IRM903/ethanol blends to determine the relationship between the q-cal and ethanol content, as shown in Figure 9.

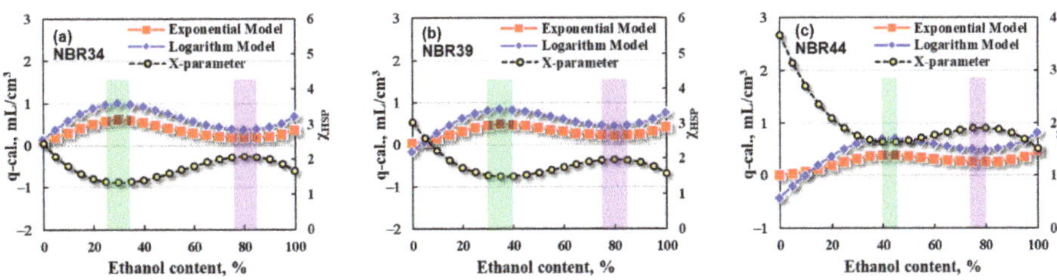

Figure 9. Correlations of q-cal. with ethanol content in IRM903/ethanol blends of (**a**) NBR34, (**b**) NBR39, and (**c**) NBR44.

For the IRM903/ethanol blends, the q-cal values determined by the exponential model (red square) are slightly lower than those obtained by the logarithm model (blue diamond) when the ethanol contents are higher than the three critical values, 10 vol%, 15 vol%, and 20 vol%, respectively, for NBR34, NBR39, and NBR44. However, there exist negative q-cal. values determined from the logarithm model when the ethanol content is less than 10 vol% for NBR34, 15 vol% for NBR39, and 25 vol% for NBR44, respectively. Such a phenomenon is known as shrinkage, that is, the swelling ratio is less than zero, which is often encountered in oil-resistant rubber hose products. The shrinkage becomes more evident while the compatibility between rubber and liquid is worse. In addition, there exists a high swelling region (max q-cal.) and a low swelling region (min q-cal.) with the increase in ethanol content, and the corresponding χ_{HSP} value shows an opposite trend. It can be further found that the low swelling region becomes less noticeable for the NBR with a higher ACN content, and even a swelling platform comes out for the NBR44 which has the highest ACN content.

Additionally, the high swelling and low swelling regions of NBR vulcanizates correspond to different ethanol contents of IRM903/ethanol blends, which have been depicted in Figure 9 (green and purple shadows). Taking NBR34 as an example, a high swelling region appears at the ethanol content of 40~50 vol% and a low swelling region appears at 80~90 vol%. The high swelling region shifts to the direction of a higher ethanol content when the ACN content of NBR increases, or the polarity of NBR is higher.

The q-cal. values determined by the exponential and logarithmic models are further correlated with χ_{HSP} values for bio-diesel/ethanol and IRM903/ethanol blends, respectively. The results are shown in Figure 10.

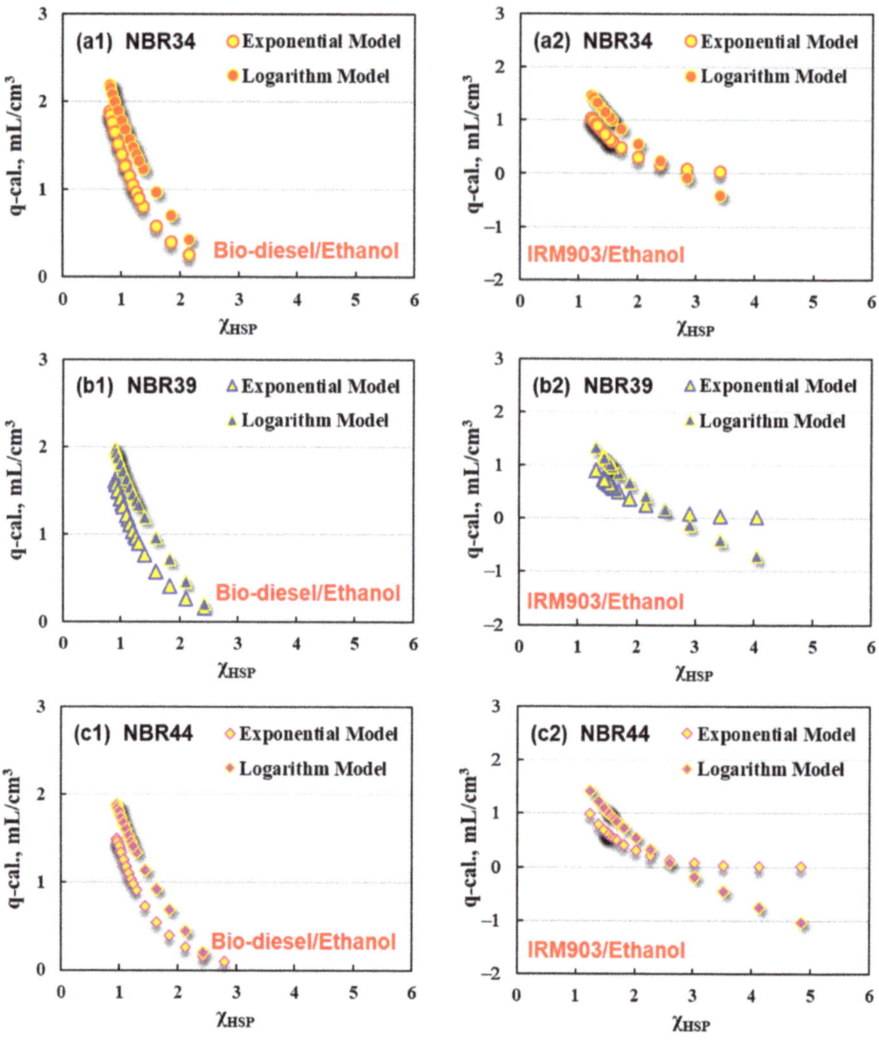

Figure 10. Correlation of q-cal. with χ_{HSP} for (**a1**) NBR34 in bio-diesel/ethanol, (**a2**) NBR34 in IRM903/ethanol, (**b1**) NBR39 in bio-diesel/ethanol, (**b2**) NBR39 in IRM903/ethanol, (**c1**) NBR44 in bio-diesel/ethanol, and (**c2**) NBR44 in IRM903/ethanol.

It can be observed for the bio-diesel/ethanol blends (Figure 10a1,b1,c1) that q-cal. values determined by the exponential model present a being slightly lower than the q-cal. values obtained by the logarithm model. Additionally, the NBR with the lowest ACN content displays the highest swelling response (highest q-cal. value).

It becomes more complicated for the IRM903/ethanol blends in regard to the relationship between the q-cal. and χ_{HSP} value, as shown in Figure 10a2,b2,c2. The q-cal. value decreases with the increasing χ_{HSP} value, but the q-cal. ~χ_{HSP} curves obtained from exponential and logarithm models show an intersection point (χ_{IN}) at a certain χ_{HSP} value. q-cal. values calculated by the logarithm model are higher than those of the exponential model when χ_{HSP} is less than χ_{IN}, and vice versa. What is interesting is that χ_{IN} is 2.5, the same value for all the NBR vulcanizates, which may provide important theoretical guidance for predicting the oil resistance of NBR-based rubber products in practical applications.

4. Conclusions

The increasing use of new bio-fuels, like bio-gasoline and bio-diesel have imposed more challenges on rubber materials. Therefore, the prediction of the swelling response and oil resistance of rubber products in such bio-fuels has become very desirable. Based on this, prediction models of the swelling responses for NBR vulcanizates have been established using modified χ_{HSP} parameters. The swelling ratios increase with the decrease in χ_{HSP} values following both exponential and logarithmic relationships. The prediction models are developed for the purpose of predicting and quantitative evaluating the swelling responses of NBR vulcanizates in bio-fuels which, in this work, are represented by bio-diesel/ethanol and IRM903/ethanol, respectively. Different phenomena, including dilation and shrinkage after swelling, have been predicted for NBR-based rubber products, which are frequently used in industrial and automotive applications, such as in bio-fuels, hydraulic fluids, and lubricants. Potentially, the prediction models obtained in this work can provide a concise way to pre-screen possible elastomers to meet the oil-resistant performance criterion, especially in some extreme application environments.

Author Contributions: Methodology, G.L.; Software, Y.J.; Validation, G.L.; Data curation, Y.J.; Writing—original draft, Y.J.; Writing—review & editing, G.L.; Supervision, G.L. All authors have read and agreed to the published version of the manuscript.

Funding: This research was funded by the Shandong Provincial Natural Science Foundation (ZR2020QE079).

Institutional Review Board Statement: Not applicable.

Data Availability Statement: The original contributions presented in the study are included in the article, further inquiries can be directed to the corresponding author.

Conflicts of Interest: The authors declare no conflict of interest.

References

1. Yang, S.; Che, R.; Chen, J.; He, Z.; Lai, T.; Liang, Y.; Bian, X. A Feasible Way to Modify Microcrystalline Cellulose Powder and Its Reinforcing Effect for NBR Composites. *Polym. Compos.* **2024**, *45*, 4709–4724. [CrossRef]
2. Yang, T.; Han, C.; Deng, Q.; Wen, F. Optimized Tribological Performance of Nitrogen-Doped Diamond-like Carbon Films on NBR: Influence of Bias Voltage of DC Magnetron Sputtering. *Nanomaterials* **2024**, *14*, 564. [CrossRef] [PubMed]
3. Wang, M.; Li, Y.; Qian, C.; Wang, S.; Liu, D. Relationship between the Aging Thermal Oxygen and Mechanical Properties of Nitrile Butadiene Rubber Reinforced by RD-Loaded Carboxylated Carbon Nanotubes. *Appl. Surf. Sci.* **2024**, *655*, 159569. [CrossRef]
4. Xie, X.; Yang, D. Simultaneously Improved Thermal Conductivity and Dielectric Properties of NBR Composites by Constructing 3D Hybrid Filler Networks. *Macromol. Mater. Eng.* **2023**, *308*, 2300016. [CrossRef]
5. Hao, X.; Wang, G.; Wang, X.; Li, K.; He, Q. The Thermal Stability and Flame Retardant Effect of $Al(OH)_3$ on NBR Containing Intumescent Flame Retardant. *J. Polym. Res.* **2023**, *30*, 410. [CrossRef]
6. Ammineni, S.P.; Nagaraju, C.; Lingaraju, D. Thermal Degradation of Naturally Aged NBR with Time and Temperature. *Mater. Res. Express* **2022**, *9*, 065305. [CrossRef]
7. Ghowsi, M.A.; Jamshidi, M. Using a Facile Method to Predict Properties of Recycled Waste Nitrile Rubber (NBR) through Devulcanization. *Sci. Rep.* **2023**, *13*, 15361. [CrossRef]
8. Da Silva Ruiz, N.M.; de Sousa, A.M.F.; Da Costa Marques Calderari, M.R.; de Figueiredo, M.A.G.; Lima, A.P.; de Azevedo, E.R. Time Domain NMR Evaluation of Thermal and Thermochemical Aging of Nitrile Rubber on Crosslinking and Mechanical Properties. *Polym. Degrad. Stab.* **2024**, *224*, 110727. [CrossRef]
9. Schmidt, F.C. The Solubility of Nonelectrolytes (Hildebrand, Joel H.). *J. Chem. Educ.* **1952**, *29*, 51. [CrossRef]

10. Jiang, X.; Yuan, X.; Guo, X.; Zeng, F.; Liu, G. Determination of three-dimensional solubility parameters of HNBR/EPDM blends and the transport behaviors in ester solvents. *J. Appl. Polym. Sci.* **2022**, *139*, e52881. [CrossRef]
11. Jing, Y.; Cui, Z.; Zou, H.; Tu, J.; Jiang, X.; Shi, X.; Yong, Z.; Liu, S.; Liu, G. Three-dimensional solubility parameters of natural rubber and its predictive power in diffusion coefficients. *J. Appl. Polym. Sci.* **2022**, *139*, 51473. [CrossRef]
12. Seymour, R.B.; Sosa, J.M. Estimation of Solubility Parameters of Nonelectrolytes. *Nature* **1974**, *248*, 759–761. [CrossRef]
13. Nielsen, T.B.; Hansen, C.M. Elastomer Swelling and Hansen Solubility Parameters. *Polym. Test.* **2005**, *24*, 1054–1061. [CrossRef]
14. Murase, M.; Ohta, R.; Iseki, T. Liquid-Repellent Films Comprising Octamethylsilsesquioxane Selected Based on Three-Dimensional Solubility Parameters. *Langmuir* **2020**, *36*, 2247–2254. [CrossRef]
15. Wang, Y.Z.; Bi, L.Y.; Zhang, H.J.; Zhu, X.T.; Liu, G.Y.; Qiu, G.X.; Liu, S.S. Predictive power in oil resistance of fluororubber and fluorosilicone rubbers based on three-dimensional solubility parameter theory. *Polym. Test.* **2019**, *75*, 380–386. [CrossRef]
16. Liu, G.; Hoch, M.; Liu, S.; Kulbaba, K.; Qiu, G. Quantitative exploration of the swelling response for carbon black filled hydrogenated nitrile rubber with three-dimensional solubility parameters. *Polym. Bull.* **2015**, *72*, 1961–1974. [CrossRef]
17. Wiśniewski, R.; Śmieszek, E.; Kamińska, E. Three-Dimensional Solubility Parameters: Simple and Effective Determination of Compatibility Regions. *Prog. Org. Coat.* **1995**, *26*, 265–274. [CrossRef]
18. Huang, J.-C.; Lin, K.-T.; Deanin, R.D. Three-Dimensional Solubility Parameters of Poly(ε-Caprolactone). *J. Appl. Polym. Sci.* **2006**, *100*, 2002–2009. [CrossRef]
19. Larson, B.K.; Hess, J.M.; Williams, J.M. Procedure for Estimating Oil Three-Dimensional Solubility Parameters. *Rubber Chem. Technol.* **2017**, *90*, 621–632. [CrossRef]
20. Su, R.; Liu, G.; Sun, H.; Yong, Z. A new method to measure the three-dimensional solubility parameters of acrylate rubber and predict its oil resistance. *Polym. Bull.* **2022**, *79*, 971–984. [CrossRef]
21. Jiang, X.; Hao, Y.; Wang, H.; Tu, J.; Liu, G. Application of Three-Dimensional Solubility Parameter in Diffusion Behavior of Rubber-Solvent System and Its Predictive Power in Calculating the Key Parameters. *Macromol. Res.* **2022**, *30*, 271–278. [CrossRef]
22. Wang, H.; Liu, S.; Liu, G. Investigation on the Thermo-Oxidative Aging Resistance of Nitrile-Butadiene Rubber/Polyamide Elastomer Blend and the Swelling Behaviors in Fuels Predicted via Hansen Solubility Parameter Method. *Polym. Degrad. Stab.* **2023**, *217*, 110512. [CrossRef]
23. Mieczkowski, R. The Determination of the Solubility Parameter Components of Polystyrene. *Eur. Polym. J.* **1989**, *25*, 1055–1057. [CrossRef]
24. Chen, J.; Farajtabar, A.; Jouyban, A.; Acree, W.E., Jr.; Zhao, H. Solubility, Three-Dimensional Hansen Solubility Parameters, and Solution Thermodynamics of 3,3′-Diaminodiphenyl Sulfone in 14 Neat Solvents from 283.15 to 328.15 K. *J. Chem. Eng. Data* **2021**, *66*, 2167–2176. [CrossRef]
25. Koenhen, D.M.; Smolders, C.A. The Determination of Solubility Parameters of Solvents and Polymers by Means of Correlations with Other Physical Quantities. *J. Appl. Polym. Sci.* **1975**, *19*, 1163–1179. [CrossRef]
26. Marco, C.; Bello, A.; Fatou, J.G.; Garza, J. Solubility Parameter of Poly(1,3-Dioxepane). *Die Makromol. Chem.* **1986**, *187*, 177–185. [CrossRef]
27. DiPaola-Baranyi, G. Estimation of Polymer Solubility Parameters by Inverse Gas Chromatography. *Macromolecules* **1982**, *15*, 622–624. [CrossRef]
28. Siemann, U. The Solubility Parameter of Poly (Dl-Lactic Acid). *Eur. Polym. J.* **1992**, *28*, 293–297. [CrossRef]
29. Flory, P.J.; Rehner, J., Jr. Statistical Mechanics of Cross-Linked Polymer Networks II. Swelling. *J. Chem. Phys.* **1943**, *11*, 521–526. [CrossRef]
30. Liu, G.; Hoch, M.; Wrana, C.; Kulbaba, K.; Liu, S.; Bi, W.; Zhao, S. Investigation of the swelling response and quantitative prediction for hydrogenated nitrile rubber. *Polym. Test.* **2014**, *34*, 72–77. [CrossRef]
31. Hashimoto, I. Relation Between the Swelling and Solution Behaviors and the Three Dimensional Solubility Parameters for Various Polymers in Organic Solvents. *Sen'i Gakkaishi* **1978**, *34*, T469–T473. [CrossRef]
32. Liu, S.; Jing, Y.; Tu, J.; Zou, H.; Yong, Z.; Liu, G. Systematic Investigation on the Swelling Behaviors of Acrylonitrile-butadiene Rubber via Solubility Parameter and FLORY-HUGGINS Interaction Parameter. *J. App. Polym. Sci.* **2022**, *139*, 52172. [CrossRef]
33. Tang, S.; Zhang, R.; Liu, F.; Liu, X. Hansen Solubility Parameters of Polyglycolic Acid and Interaction Parameters between Polyglycolic Acid and Solvents. *Eur. Polym. J.* **2015**, *72*, 83–88. [CrossRef]
34. Pajula, K.; Taskinen, M.; Lehto, V.-P.; Ketolainen, J.; Korhonen, O. Predicting the Formation and Stability of Amorphous Small Molecule Binary Mixtures from Computationally Determined Flory−Huggins Interaction Parameter and Phase Diagram. *Mol. Pharm.* **2010**, *7*, 795–804. [CrossRef] [PubMed]
35. Lindvig, T.; Michelsen, M.L.; Kontogeorgis, G.M.A. Flory–Huggins Model Based on the Hansen Solubility Parameters. *Fluid Phase Equilibria* **2002**, *203*, 247–260. [CrossRef]
36. Clarke, C.J.; Eisenberg, A.; La Scala, J.; Rafailovich, M.H.; Sokolov, J.; Li, Z.; Qu, S.; Nguyen, D.; Schwarz, S.A.; Strzhemechny, Y.; et al. Measurements of the Flory−Huggins Interaction Parameter for Polystyrene−Poly(4-Vinylpyridine) Blends. *Macromolecules* **1997**, *30*, 4184–4188. [CrossRef]
37. Huang, J. Probe Dependency of Flory–Huggins Interaction Parameters between Solvents: Cases of Hydrocarbons and Isosteric Derivatives. *J. Appl. Polym. Sci.* **2013**, *127*, 5000–5006. [CrossRef]

38. Liu, S.; Jing, Y.; Tu, J.; Zou, H.; Yong, Z.; Gao, Y.; Tang, Z.; Liu, G. New FLORY-HUGGINS Interaction Parameter between SBR and P-phenylenediamine Antioxidants and the Predictive Power in Molecular Structure and Rubber Formula Design. *J. Appl. Polym. Sci.* **2022**, *139*, 51840. [CrossRef]
39. Jiang, X.; Yuan, X.; Guo, X.; Zeng, F.; Wang, H.; Liu, G. Study on the Application of Flory–Huggins Interaction Parameters in Swelling Behavior and Crosslink Density of HNBR/EPDM Blend. *Fluid Phase Equilibria* **2023**, *563*, 113589. [CrossRef]
40. Chu, H.Z.; Liu, D.; Cui, Z.W.; Wang, K.; Qiu, G.X.; Liu, G.Y. Effect of Crosslink Density on Solubility Parameters of Styrene Butadiene Rubber and the Application in Pre-Screening of New Potential Additives. *Polym. Test.* **2020**, *81*, 106253. [CrossRef]
41. Liu, S.S.; Li, X.P.; Qi, P.J.; Song, Z.J.; Zhang, Z.; Wang, K.; Qiu, G.X.; Liu, G.Y. Determination of Three-Dimensional Solubility Parameters of Styrene Butadiene Rubber and the Potential Application in Tire Tread Formula Design. *Polym. Test.* **2020**, *81*, 106170. [CrossRef]

polymers

MDPI

Article

The Network Construction of a New Byproduct-Free XLPE-Based Insulation Using a Click Chemistry-Type Reaction and a Theoretical Study of the Reaction Mechanism

Yang Du [1], Hui Zhang [1,*], Wei Han [1], Xia Du [1], Yan Shang [1], Hongda Yang [2], Xuan Wang [1,*], Qingguo Chen [1] and Zesheng Li [3]

[1] Key Laboratory of Engineering Dielectrics and Its Application of Ministry of Education & School of Material Science and Chemical Engineering, Harbin University of Science and Technology, Harbin 150080, China; duyang950711@163.com (Y.D.); charles_han@fjnu.edu.cn (W.H.); duxia62@126.com (X.D.); shangyan1972@126.com (Y.S.); qgchen@hrbust.edu.cn (Q.C.)

[2] Electric Power Research Institute, State Grid Heilongjiang Electric Power Co., Ltd., Harbin 150030, China; yanghongda_phd16@hrbust.edu.cn

[3] Key Laboratory of Cluster Science of Ministry of Education & School of Chemistry, Beijing Institute of Technology, Beijing 100081, China; zeshengli@bit.edu.cn

* Correspondence: hust_zhanghui11@hotmail.com (H.Z.); wangxuan@hrbust.edu.cn (X.W.)

Abstract: Cross-linked polyethylene (XLPE) is applied in most advanced high-voltage direct-current (HVDC) power cable insulations, which are produced via dicumyl peroxide (DCP) technology. The electrical conductivity of insulation material can be increased by cross-linking byproducts from the DCP process. Hence, currently much attention is being paid to a new process to produce cross-linking byproduct-free XLPE. The cross-linking in situ between ethylene–glycidyl methacrylate copolymer and 1,5-disubtituted pentane via reactive compounding is a substitute for DCP. The reaction potential energy information of the eighteen reaction channels was obtained at the B3LYP/6-311+G(*d*,*p*) level. Results demonstrated that epoxy groups and 1,5-disubtituted reactive groups can react in situ to realize the XLPE-based network structure via covalent linking, and epoxy ring openings yielded ester. 1,5-disubtituted pentane played a cross-linker role. The reactivity of the carboxyl group was stronger than that of the sulfydryl or hydroxyl group. The reaction channel RTS1 was more kinetically favorable due to the lower reaction Gibbs energy barrier height of 1.95 eV. The cross-linking network construction of the new XLPE insulation without byproducts opens up the possibility of DCP substitution, which is beneficial to furthering the design of thermoplastic insulation materials for power cables in the future.

Keywords: power cable; polyethylene; byproduct-free; cross-linking reaction mechanism; 1,5-disubtituted pentane; 2-ethylbutyric acid oxiranylmethyl ester

1. Introduction

The transition from using fossil energy sources to renewables is urgently required for carbon emission reduction. Usually, wind and solar energy are most abundant in regions far away from end-users, which requires long-distance and efficient power transportation. Transmission lines need reverse energy flow direction because renewable energy sources are intermittent. High-voltage direct-current (HVDC) transmission lines meet these requirements, which will be a vital part of power grids in the future to integrate renewables seamlessly [1]. When transmission lines pass through densely populated areas or large bodies of water, it is necessary to submerge or bury power cables underground, and it is essential for transmission lines to be surrounded by insulation layers. A large amount of research is focused on upgrading the quality of insulated cables so as to increase transmission voltage outside the present ultimate limit (640 kV) [2]. Because at low temperatures polyethylene becomes too brittle and at 70 °C becomes too stiff, polyethylene by itself fails to be an appropriate insulation material. Therefore, developing the above all insulation for advanced high-voltage power cable involves obtaining cross-linked polyethylene (XLPE) with outstanding electrical and mechanical behaviors.

Recently, several strategies, such as grafting a voltage stabilizer or antioxidant onto polyethylene, introducing deep traps, and modifying montmorillonite, have been proposed as means to improve electrical tree resistance [3], thermal aging properties [4], and space charge behavior [5]. In order to improve material properties [6] and enhance material functionality [7], nanotechnology and organic–inorganic hybrids are often used as effective methods.

However, XLPE production often uses dicumyl peroxide (DCP) [8] as a cross-linker, which does lead to the emission of some volatile or hazardous byproducts [9] (methane, acetophenone, cumyl alcohol, *α*-methyl styrene, etc.,) and usually needs a time- and energy-consuming degassing procedure to remove these unwanted byproducts [10]. Even trace amounts of residual impurities can raise the electrical conductivity of polyethylene [11]. The overall quality of XLPE insulation improves with any reduction in the amount of additives used. Obviously, it would be desirable to find preparation strategies for polyethylene copolymer blends that can replace DCP cross-linking and entirely prevent the emission of harmful byproducts, and thereby improve the total quality of the insulation material.

Müller and colleagues prepared XLPE insulation via blending two polyethylene copolymers, p(E-stat-GMA) (ethylene–glycidyl methacrylate copolymer) and p(E-stat-AA) (ethylene–acrylic acid copolymer) [12]. The epoxy and carboxyl groups, respectively, from these two polyethylene copolymers react to form covalent bonds at 160 °C. The resulting cross-linking product meets the demands of polyethylene resin, and at the same time, its electrical conductivity is equivalent to the measured value of commercial XLPE. Müller's research group explored the preparation process of a polyethylene–polypropylene-type (PE-PP-type) copolymer in situ blended at 170 °C [13], and confirmed their recyclability via a reprocessing process at the same temperature. Covalent bonds form between two polyethylene copolymers: one is p(E-stat-GMA) and the other is maleic anhydride-grafted polypropylene (PP-graft-MA). They also proved that polyethylene copolymers containing epoxy resin and glycidyl methacrylate can be efficiently cross-linked with three bifunctional curing agents (amines, carboxylic acids, and hydrazides), without byproduct formation, at 160 to 200 °C [14]. This promising approach inspired us to explore new cross-linking network construction of thermoplastic insulation materials for power cables.

In this study, the design of a new reaction system to produce XLPE in situ without additives or the release of any byproducts was established. The cross-linking reaction in situ between epoxy groups with 1,5-disubtituted reactive groups replaced the DCP cross-linking process, and electronic effects were estimated. Molecular formulas, molecular names, and corresponding abbreviations of 1,5-disubtituted pentane molecules are listed in Table 1. 1,5-disubtituted pentane can be used as a cross-linker.

Table 1. Molecular names, molecular formulas, and corresponding abbreviations (ab.) of 1,5-disubtituted pentane molecules.

Molecular Formula	Molecular Name	ab.
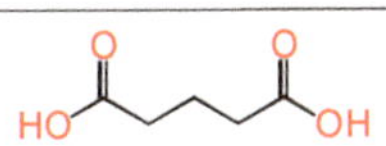	glutaric acid	PDA
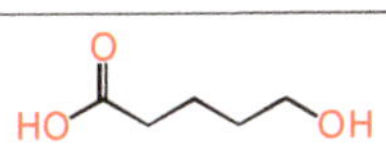	5-hydroxypentanoic acid	HPA
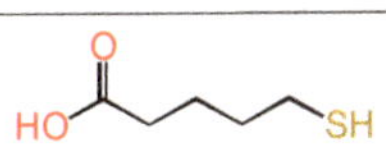	5-mercaptopentanoic acid	MPA
	1,5-pentanediol	PDOL

Table 1. *Cont.*

Molecular Formula	Molecular Name	ab.
	5-mercaptopentanol	MPOL
	1,5-pentanedithiol	PDM

As far as we know, theoretical research on the reaction mechanisms of the synergistic cross-linking reaction between epoxy groups and 1,5-disubtituted reactive groups at molecular and atomic levels has been rarely reported. These reaction processes are described in detail. The novel proposed reaction mechanism would be beneficial for opening up the possibility of cross-linking network construction and expanding the applicative range of polyethylene-based insulation materials for advanced power cables in the future.

2. Computational Methods

The density functional theory (DFT) [14] was used to study the reaction mechanism of the synergistic cross-linking reaction. At the B3LYP/6-311+G(*d,p*) level [15], geometry structures were optimized, and along the minimum Gibbs free energy path (MEP) [16], energies, gradients and force constants of reactants, transition states, and products were calculated. According to the results of these stationary electronic structure calculations, potential energy surface information was obtained. This research method has been proven reliable and suitable for current research [17]. Therefore, our objective was to study the reaction mechanism of the synergistic cross-linking reaction and estimate the electronic effects of substituted groups. When a gradient step-size was 0.05 $(amu)^{1/2}$ bohr, the MEP was achieved by intrinsic reaction coordinate (IRC) theory [18]. The transition state and reaction channel were abbreviated to the corresponding TS and R, separately. These calculations of electronic structures were completed by the GAUSSIAN09 program package [19].

3. Results and Discussion

3.1. Stationary Point Geometries

In this work, theoretical study of the reaction behavior of a synergistic cross-linking reaction was completed at the B3LYP/6-311+G(*d,p*) level. In Table 2, reaction equations for the eighteen studied reaction channels are plotted; 2-ethylbutyric acid oxiranylmethyl ester was selected as a model molecule of an ethylene–glycidyl methacrylate copolymer. The labels of transition states and reaction channels correspond one-to-one. In Figure 1, geometry structures of the eighteen transition states after optimization are provided. In Electronic Supplementary Materials (ESM), the corresponding standard orientations are offered. Table 2 also lists the bond lengths (b/f) for transition state cleavage and bond formation after optimization, the corresponding bond lengths in the equilibrium geometries of reactants and products, and the imaginary frequency values (freq.) for transition states. For the transition state TS1, the length of the breaking bond (b) O-H on the carboxyl group was 1.233 Å, and the lengths of forming bonds (f) O-H and O-C were 1.169 and 2.299 Å. As seen in Table 2, for synergistic reaction channels TS1, TS2, and TS3, the structures of transition states shared common characteristics. In the equilibrium molecules, compared with the elongation of the forming bonds, the elongation of the breaking bonds was relatively smaller, separately. That is to say, the hydrogen abstraction reactions were all reactant-like. These reaction channels proceeded via "early" transition states and applied to an exothermic reaction, which is in accordance with Hammond's postulate [20].

Table 2. Optimized bond lengths of breaking/forming bonds (b/f) for transition state, reactants (r) and products (p) (in Å), together with calculated breaking/forming bond frequencies (in cm^{-1}), calculated reaction Gibbs free energies (ΔG), and Gibbs potential barrier heights ($\Delta G^{\neq}$) (in eV).

Number	Reaction Equation	r	b/f	p	freq.	$\Delta G^{\neq}$	ΔG
1	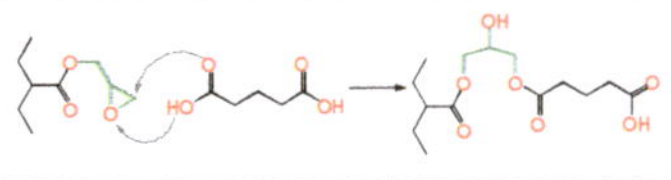	0.969	1.233/OH1.169 1.233/OC2.299	OH0.963 OC1.445	874 *i*	1.95	−0.36
1-a	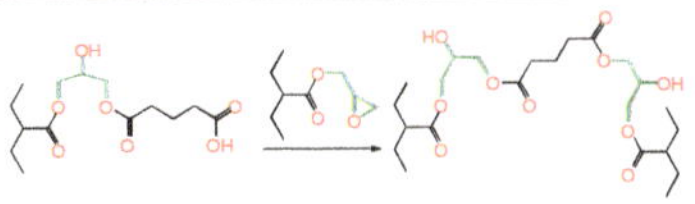	0.969	1.184/OH1.214 1.184/OC2.278	OH0.963 OC1.440	880 *i*	2.67	0.27
2	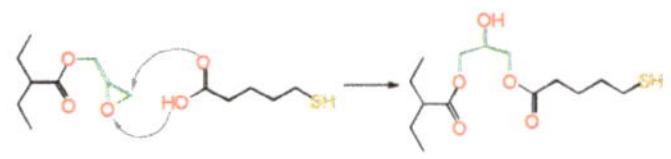	0.969	1.244/OH1.160 1.244/OC2.306	OH0.964 OC1.441	874 *i*	2.05	−0.42
2-a	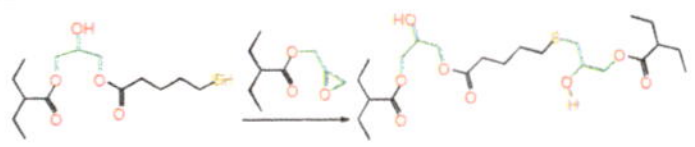	1.348	1.454/OH1.499 1.454/SC2.766	OH0.970 SC1.832	815 *i*	2.59	−0.64
3	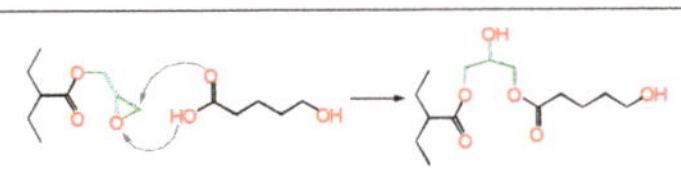	0.969	1.219/OH1.181 1.219/OC2.296	OH0.963 OC1.445	898 *i*	2.13	−0.41
3-a	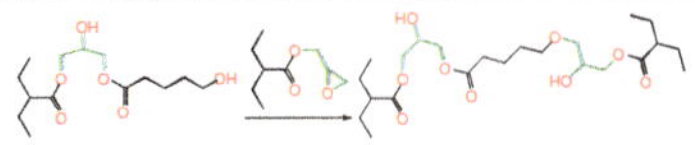	0.962	1.017/OH1.553 1.017/OC2.184	OH0.964 OC1.414	407 *i*	2.53	−0.37
4	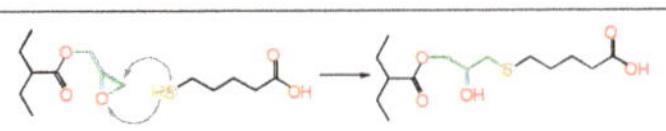	1.349	1.454/OH1.503 1.454/SC2.763	OH0.970 SC1.832	797 *i*	2.54	−0.64
4-a	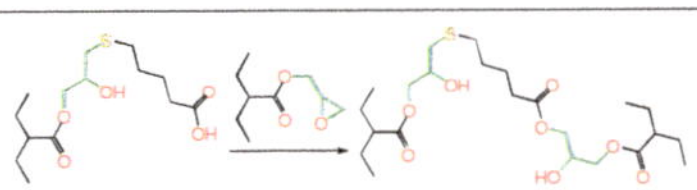	0.969	1.224/OH1.176 1.224/OC2.299	OH0.963 OC1.441	892 *i*	2.61	0.12
5	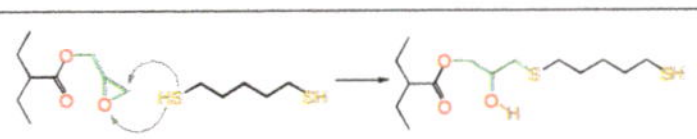	1.349	1.442/OH1.533 1.442/SC2.748	OH0.970 SC1.831	728 *i*	2.63	−0.62
5-a	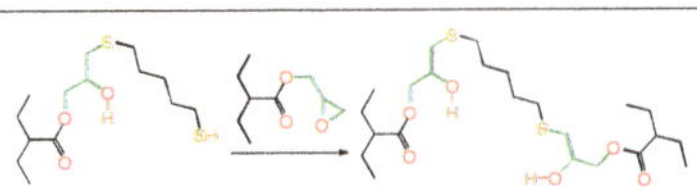	1.349	1.434/OH1.557 1.434/SC2.737	OH0.970 SC1.832	674 *i*	2.81	−0.51
6	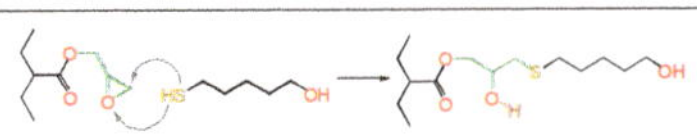	1.349	1.435/OH1.554 1.435/SC2.742	OH0.967 SC1.842	670 *i*	2.62	−0.66
6-a	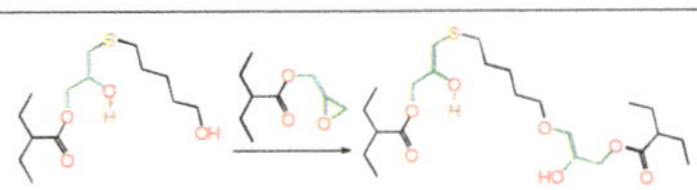	0.962	1.017/OH1.553 1.017/OC2.183	OH0.964 OC1.414	406 *i*	2.55	−0.37
7	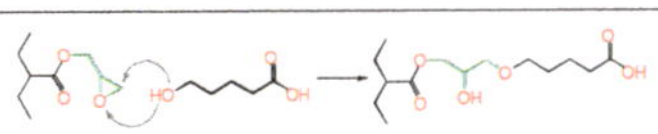	0.962	1.018/OH1.547 1.018/OC2.183	OH0.967 OC1.412	406 *i*	2.45	−0.40
7-a	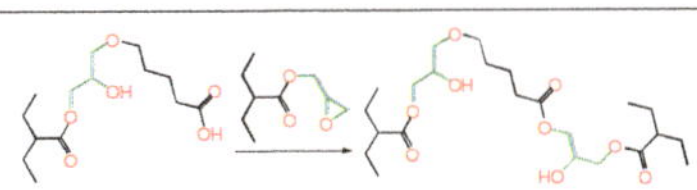	0.969	1.233/OH1.169 1.233/OC2.302	OH0.963 OC1.440	877 *i*	2.25	−0.12
8	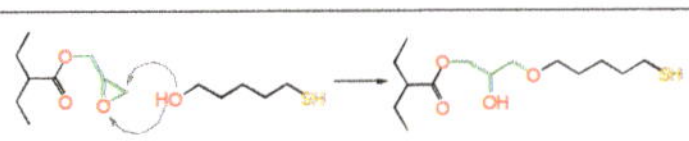	0.962	1.017/OH1.550 1.017/OC2.184	OH0.963 OC1.413	405 *i*	2.43	−0.47

Table 2. *Cont.*

Number	Reaction Equation	r	b/f	p	freq.	$\Delta G^{\neq}$	ΔG
8-a	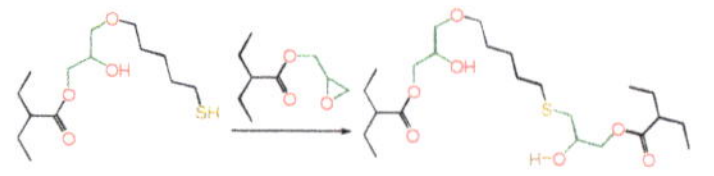	1.349	1.429/OH1.572 1.429/SC2.729	OH0.964 SC1.832	645 *i*	2.66	−0.63
9	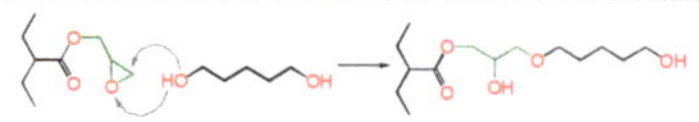	0.962	1.016/OH1.555 1.016/OC2.184	OH0.967 OC1.442	403 *i*	2.42	−0.39
9-a	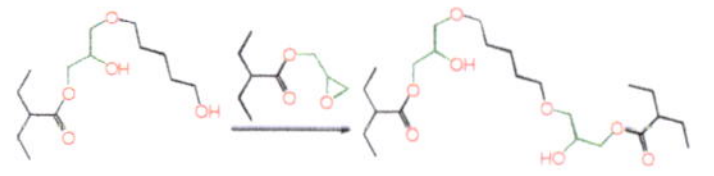	0.962	1.017/OH1.551 1.017/OC2.181	OH0.970 OC1.413	410 *i*	2.49	−0.44

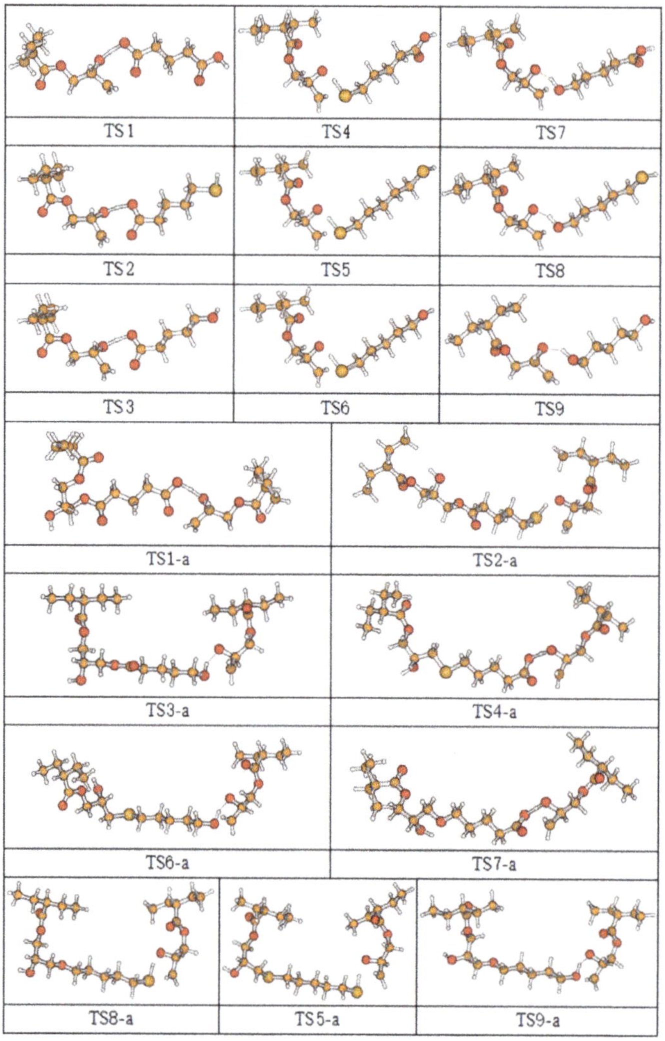

Figure 1. Optimized geometric structures of the eighteen transition states.

3.2. The Minimum Gibbs Free Energy Path

In Figure 2, three key parameters along the MEP, including bond lengths, bond angles, and dihedral angles of reaction channel R1 at the B3LYP/6-311+G(*d,p*) level, are depicted. The synergistic reaction channel (R1) involves the H-transfer of a carboxyl group to O on epoxy and the O-attack of a carboxylic group to a C atom of the -CH- site on epoxy via transition state TS1, at time step 0. From reactant (step −100) to product (step 100), five key steps are included here.

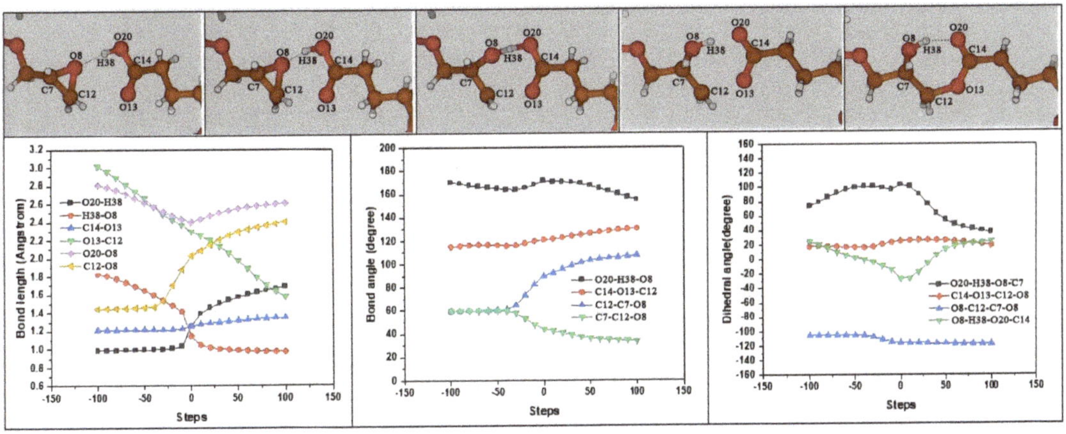

Figure 2. Changes in bond lengths, bond angles, and dihedral angles of reaction channel R1 along minimum Gibbs free energy path (MEP).

First, at IRC step −100, the H of the carboxyl group transferred to an O atom on epoxy; the bond length of O20-H38 on the carboxyl group was 0.987 Å; it was 3.012 Å between O13 on the carbonyl group and C12 in the -CH_2- site on epoxy; the distance between H38 and O8 on epoxy was 1.828 Å; and the bond length of C12-O8 on epoxy was 1.444 Å. Second, at IRC step −30, the H38 gradually transferred to the O8 atom; the distance between H38 and O8 decreased to 1.539 Å; and at the same time the epoxy ring opened, the distance between O13 and C12 was 2.483 Å and the distance between C12 and O8 was 1.528 Å. Third, at IRC step 0, the transition state TS1 presented stretch oscillating of H38 between O8 and O20; the O13 atom continued to approach C12; and O8, H38, and O20 nearly became a straight line, ∠O20-H38-O8, which peaked at 171.42°. Meanwhile H38, O20, C14, O13, O8, and C12 participated in the reaction and formed a hexatomic ring structure, which was almost co-planed. Fourth, at IRC step 25, the H38 transferred to the O8 atom, the bond of O20 and H38 broke, the bond length of H38-O8 continued to decrease to 1.011 Å, and it was 2.178 Å between O13 and C12. Last, at IRC step 100, O13 attacked the C12 atom to form an O-C bond. Now, with a bond length of 1.576 Å, the distance between the ring-opened C12 and O8 was 2.400 Å, while the bond length of the carbonyl group O13-C14 almost remained unchanged throughout the entire reaction process.

3.3. Energetics

In Table 1, at the B3LYP/6-311+G(*d,p*) level, the reaction Gibbs free potential barrier heights ($\Delta G^{\neq}$) and the reaction Gibbs free energies (ΔG) at 298 K of the eighteen reaction channels calculated are presented. For the reaction of epoxy alcohol with carboxylic acid, the reaction of ring-opening and cyclization of epoxy alcohol can be catalyzed by the carboxylic acid [21]. For the reaction of epoxy groups in epoxy resin containing amino groups, it is generally agreed to generate new C-N and O-H groups throughout the curing reaction process [22,23]. The synergistic cross-linking reaction refers to this process, in which the hydrogen in active functional groups (a hydroxyl or sulfydryl group) transfers to the oxygen on epoxy, and meanwhile the oxygen of the carbonyl group (or the oxygen of the hydroxyl group or the sulfur of the sulfydryl group) attacks the carbon of -CH_2- on epoxy.

As seen in Table 1, the synergistic cross-linking reaction between epoxy groups and 1,5-disubtituted reactive groups was a nucleophilic addition reaction. Studied 1,5-disubtituted (bifunctional groups) pentanes (PDA, HPA, MPA, PDOL, MPOL, and PDM) can be used as the attacking reagent. One side functional group reacts with the epoxy compound, and then the other side functional group also undergoes nucleophilic addition, with another molecule epoxy compound being considered.

First, for reactions 1, 4 and 7, one side functional group (-COOH, -SH, or OH) reacted with epoxy compound, and the other side functional group was a carboxyl group. The reaction Gibbs free energy barriers of reaction channel R1 (the one side functional group was -COOH) was the lowest ($\Delta G^{\neq}_{R1}$ = 1.95 eV), followed by reaction channels R4 (the one side functional group was -OH) ($\Delta G^{\neq}_{R4}$ = 2.45 eV) and R7 (the one side functional group was -SH) ($\Delta G^{\neq}_{R7}$ = 2.54 eV). According to the report, it was easier for the carboxyl group to provide hydrogen for the protonation of the epoxy group [21]. Because the stronger the acidity, the higher the reactivity, the lower the value of $\Delta G^{\neq}$. The acidity of the -COOH group was the strongest among the three kinds of substituted groups (-COOH, -SH, and OH); thus, $\Delta G^{\neq}$ (R1) was the lowest, and the reaction was the easiest. Similar reactivity rules were also found when the other side functional groups were hydroxyl and sulfydryl groups.

Meanwhile, for reactions 4 and 7, the active functional groups (-SH and -OH) attacked the O atom on the epoxy group (see Figure 3), and the O-H bond of the -OH or the S-H bond of -SH broke. The active H of active functional groups transferred to O on the epoxy group. The oxygen atom of -OH or the sulfur atom of -SH attacked the carbon atom in -CH_2- sites on the epoxy group. Tetra-atomic ring structures were exhibited on transition states TS4 and TS7, and these structures were unstable; because of their large amounts of tension, they possessed higher energy. In reaction 1, the active H and acyl O in carboxyl groups attacked the oxygen and carbon atoms in the epoxy group; separately, a hex-atomic ring structure was exhibited on the transition state TS1, the structure was almost co-plane and nearly non tension, and the structure of the transition state was more stable. Thus, the reaction Gibbs free potential barriers of TS1 were lower than those of TS4 and TS7.

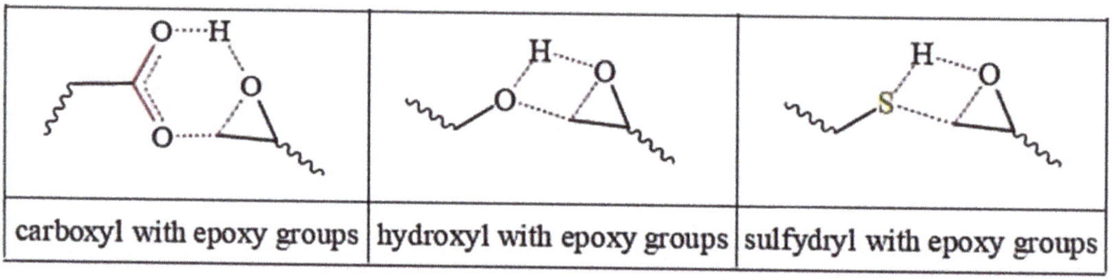

Figure 3. Schematic diagrams of synergistic cross-linking reaction of functional groups with epoxy groups.

Second, for reactions 7-a, 8-a, and 9-a, after one side functional group (OH) reacted with the epoxy group compound, the other side functional groups (-COOH, -SH, or OH) underwent nucleophilic addition with another molecule epoxy compound. The reaction Gibbs free energy barriers of reaction channel R7-a was the lowest ($\Delta G^{\neq}_{R7\text{-}a}$ = 2.25 eV), followed by reaction channels R9-a ($\Delta G^{\neq}_{R9\text{-}a}$ = 2.49 eV) and R8-a ($\Delta G^{\neq}_{R8\text{-}a}$ = 2.66 eV). Because the volume of the hydroxyl group was relatively small, the reactivity mainly depended on the acidity of functional groups. The cross-linking reaction was more likely to occur due to the lowest $\Delta G^{\neq}$ (R7-a).

Third, for reactions 4-a, 5-a, and 6-a, after one side functional group (SH) reacted with the epoxy group compound, the other side functional groups (-COOH, -SH, or OH) reacted with another molecule epoxy compound. The reaction Gibbs free energy barrier was affected by two main factors. On the one hand, the larger the volume of the attacking reagent, the greater the steric hindrance, and the reaction proceeded with difficulty. On the other hand, it was determined by the acidity of the reagent, which was strong, and the reaction was prone to occur. The hydroxyl group possessed the smallest size, the carboxyl group had the strongest acidity, and comprehensively considering these two factors, as a result, the reaction Gibbs free energy barriers followed the trend $\Delta G^{\neq}_{R6\text{-}a}$ (2.55 eV) < $\Delta G^{\neq}_{R4\text{-}a}$ (2.61 eV) < $\Delta G^{\neq}_{R5\text{-}a}$ (2.81 eV). Eslami and Müller-Plathe [24] reported that free-energy barriers for the nucleation of different crystals are dramatically different, and a high energy barrier may be due to the rearrangement of particles during the crystalline process. In this study, the network construction of insulation materials was carried out through melt blending. However, this method involves the process of cooling and extrusion molding, which can form an amorphous structure. This amorphous structure can lead to uneven crystallization, with no lattice forming. In the next step, we will try to further study two molecules of 1,5-disubstituted pentane participated reactions.

Müller and colleagues established a new preparation process using the blending reaction of two polyethylene copolymers at 160 °C [12]. They explored a recycle approach for the in situ formation of a polyethylene–polypropylene-type copolymer via ternary blending at 170 °C [13]. They also demonstrated the efficient cross-linking of an ethylene–glycidyl methacrylate copolymer with bifunctional curing agents (amine, carboxylic acids, or hydrazides) at 160 to 200 °C [14]. Our theoretical calculation results demonstrate that the nucleophilic addition reaction channel RTS1 epoxy-bearing polyethylene copolymer and -COOH group of 1,5-disubtituted pentane (PDA) is more favorable with lower reaction Gibbs free energy barriers. Synergistic cross-linking forms covalent bonds with a click chemistry-type reaction.

4. Conclusions

In this study, a new cross-linking network of XLPE-based insulation was constructed with a click chemistry-type reaction based on theoretical calculation results of density functional theory, and a theoretical study of the cross-linking reaction mechanism between epoxy-bearing polyethylene copolymer and 1,5-disubtituted pentane was proposed for the first time. The reactivity of the carboxyl group was the highest, and stronger than those of hydroxyl and sulfydryl groups. The reaction channel R1 of epoxy-bearing polyethylene copolymer and carboxyl group substitution diacid PDA was conducive to the progress of the crosslinking reaction with the reaction Gibbs energy barrier as low as about 1.95 eV at the B3LYP/6-311+G(*d*,*p*) level. Cross-linking reactions processed in situ underwent epoxy ring-opening and nucleophilic addition reaction. Evidently, the designed cross-linking enabled the realization of a new type of network XLPE-based insulation material, with an atomic economic benefit of 100%. The use of reactive compounding with click chemistry is a promising approach, which sets up a byproduct-free alternative to conventional peroxides cross-linking. Further research is underway on electron injection and storage behavior in novel network XLPE-based insulating materials.

Supplementary Materials: The following supporting information can be downloaded at: https://www.mdpi.com/article/10.3390/polym16243536/s1, Text S1: The optimized standard orientations of transition states of the eighteen reaction channels at the B3LYP/6-311+G(*d*,*p*) level.

Author Contributions: All authors contributed to this study's conceptualization. Y.D. and W.H. conducted formal analysis and visualization. H.Z. was responsible for funding acquisition, project administration, and writing—review and editing. X.D. and Y.S. were responsible for investigation and methodology. X.W. and Q.C. conducted supervision and validation. H.Y. and Z.L. were responsible for data curation, resources, and software. The original draft of this manuscript was written by Y.D., and all authors commented on previous versions of this manuscript. All authors have read and agreed to the published version of the manuscript.

Funding: This work was supported by the Key Project of the Joint Fund for Regional Innovation and Development of the National Natural Science Foundation of China (Grant number U20A20307).

Institutional Review Board Statement: Not applicable.

Data Availability Statement: Original contributions presented in this study are included in this article and Supplementary Materials. Further inquiries can be directed to the corresponding authors.

Conflicts of Interest: Hongda Yang was employed by the company Electric Power Research Institute, State Grid Heilongjiang Electric Power Co., Ltd. The remaining authors declare that the research was conducted in the absence of any commercial or financial relationships that could be construed as a potential conflict of interest.

References

1. Mazzanti, G.; Marzinotto, M. *Extruded Cables for High-Voltage Direct-Current Transmission*; Wiley: Hoboken, NY, USA, 2013.
2. Montanari, G.C.; Morshuis, P.H.F.; Zhou, M.; Stevens, G.C.; Vaughan, A.S.; Han, Z.; Li, D. Criteria influencing the selection and design of HV and UHV DC cables in new network applications. *IET High Volt.* **2018**, *3*, 90–95. [CrossRef]
3. Li, C.Y.; Zhang, C.C.; Zhao, H.; Zhang, H.; Wang, X.; Han, B.Z. Grafted UV absorber as voltage stabilizer against electrical degradation and breakdown in cross-linked polyethylene for high voltage cable insulation. *Polym. Degrad. Stabil.* **2021**, *185*, 109498. [CrossRef]
4. Zhang, C.C.; Wang, T.T.; Sun, W.F.; Li, C.Y.; Zhao, H. Grafting of antioxidant onto polyethylene to improve DC dielectric and thermal aging properties. *IEEE Trans. Dielectr. Electr. Insul.* **2020**, *28*, 541–549. [CrossRef]
5. Gao, J.G.; Yao, Z.H.; Liu, L.W.; Ju, H.C.; Li, L.L.; Guo, N.; Niu, J.C. The effect of polyethylene-based nano-montmorillonite composite interfaces on charge transport. *J. Mater. Sci. Mater. Electron* **2023**, *34*, 860–873. [CrossRef]
6. Drobota, M.; Ursache, S.; Aflori, M. Surface functionalities of polymers for biomaterial applications. *Polymers* **2022**, *14*, 2307. [CrossRef]
7. Shahani, S.; Gao, Z.; Qaisrani, M.A.; Ahmed, N.; Yaqoob, H.; Khoshnaw, F.; Sher, F. Preparation and characterisation of sustainable wood plastic composites extracted from municipal solid waste. *Polymers* **2021**, *13*, 3670. [CrossRef]
8. Smedberg, A.; Hjertberg, T.; Gustafsson, B. Crosslinking reactions in an unsaturated low density polyethylene. *Polymer* **1997**, *38*, 4127–4138. [CrossRef]
9. Sahyoun, J.; Crepet, A.; Gouanve, F.; Keromnes, L.; Espuche, E. Diffusion mechanism of byproducts resulting from the peroxide crosslinking of polyethylene. *J. Appl. Polym. Sci.* **2017**, *134*, 44525–44536. [CrossRef]
10. Andrews, T.; Hampton, R.N.; Smedberg, A.; Wald, D.; Waschk, V.; Weissenberg, W. The role of degassing in XLPE power cable manufacture. *IEEE Electr. Insul. Mag.* **2006**, *22*, 5–16. [CrossRef]
11. Nilsson, U.H.; Andersson, J.; Englund, V.; Eriksson, V.; Hagstrand, P.O.; Smedberg, A. The role and measurement of DC conductivity for HVDC cable insulation materials. *IEEE Ann. Rep. Conf. Electr. Insul. Dielectr. Phen.* **2015**, *2015*, 31–34. [CrossRef]
12. Mauri, M.; Peterson, A.; Senol, A.; Elamin, K.; Gitsas, A.; Hjertberg, T.; Matic, A.; Gkourmpis, T.; Prietod, O.; Müller, C. Byproduct-free curing of a highly insulating polyethylene copolymer blend: An alternative to peroxide crosslinking. *J. Mater. Chem. C* **2018**, *6*, 11292–11302. [CrossRef]
13. Ouyang, Y.W.; Mauri, M.; Pourrahimi, A.M.; Östergren, I.; Lund, A.; Gkourmpis, T.; Prieto, O.; Xu, X.D.; Hagstrand, P.O.; Müller, C. Recyclable polyethylene insulation via reactive compounding with a maleic anhydride-grafted polypropylene. *ACS Appl. Polym. Mater.* **2020**, *2*, 2389–2396. [CrossRef]
14. Mauri, M.; Tran, N.; Prieto, O.; Hjertberg, T.; Müller, C. Crosslinking of an Ethylene-Glycidyl methacrylate copolymer with amine click chemistry. *Polymer* **2017**, *111*, 27–35. [CrossRef]
15. Truong, T.N.; Duncan, W.T.; Bell, R.L. Direct ab, initio dynamics methods for calculating thermal rates of polyatomic reactions. *J. Am. Chem. Soc.* **1996**, *85*, 629. [CrossRef]
16. Lee, C.; Yang, W.; Parr, R.G. Development of the Colle-Salvetti correlation-energy formula into a functional of the electron density. *Phys. Rev. B* **1988**, *37*, 785–789. [CrossRef]
17. Miehlich, B.; Savin, A.; Stoll, H.; Preuss, H. Results obtained with the correlation energy density functionals of Becke and Lee, Yang and Parr. *Chem. Phys. Lett.* **1989**, *157*, 200–206. [CrossRef]
18. Tirado-Rives, J.; Jorgensen, W.L. Performance of B3LYP density functional methods for a large set of organic molecules. *J. Chem. Theory Comput.* **2008**, *4*, 297–306. [CrossRef]

19. Hratchian, H.P.; Schlegel, H.B. Accurate reaction paths using a Hessian based predictor-corrector integrator. *J. Chem. Phys.* **2004**, *120*, 9918–9924. [CrossRef]
20. Frisch, M.J.; Trucks, G.W.; Schlegel, H.B.; Scuseria, G.E.; Robb, M.A.; Cheeseman, J.R.; Scalmani, G.; Barone, V.; Mennucci, B.; Petersson, G.A.; et al. *Revision A.02*; Gaussian, Inc.: Wallingford, CT, USA, 2009.
21. Hammond, G.S. A correlation of reaction rates. *J. Am. Chem. Soc.* **1955**, *77*, 334–338. [CrossRef]
22. Xu, L.; Fang, G.Y.; Tao, J.B.; Ye, Z.H.; Xu, S.N.; Li, Z.Y. Molecular mechanism and solvation effect of supramolecular catalysis in a synthetic cavitand receptor with an inwardly directed carboxylic acid for ring-opening cyclization of epoxy alcohols. *ACS Catalysis* **2018**, *8*, 11910–11925. [CrossRef]
23. He, Z.P.; Lv, W.; Gao, G.L.; Yin, Q.W. Investigation of the chemical changes and mechanism of the epoxy-amine system by in situ infrared spectroscopy and two-dimensional correlation analysis. *Polym. J.* **2022**, *54*, 1445–1452. [CrossRef]
24. Eslami, H.; Müller-Plathe, F. Self-Assembly Pathways of Triblock Janus Particles into 3D Open Lattices. *Small* **2024**, *20*, 2306337. [CrossRef]

polymers

MDPI

Article

Dynamic Mechanical Properties and Energy Absorption Capabilities of Polyureas Through Experiments and Molecular Dynamic Simulation

Ke Yang [1,2,*], Shanda Wang [1,2], Yanru Chen [1,2], Hanhai Dong [1,2], Quanguo Wang [1,2] and Qingli Cheng [1,2]

1 State Key Laboratory of Chemical Safety, Qingdao 266000, China; donghh.qday@sinopec.com (H.D.)
2 Sinopec Research Institute of Safety and Engineering Co., Ltd., Qingdao 266000, China
* Correspondence: yangk.qday@sinopec.com

Abstract: Polyurea (PUR) has been widely used as a protective coating in recent years. In order to complete the understanding of the relationship between PUR microstructure and its energy absorption capabilities, the mechanical and dynamic performance of PURs containing various macrodiol structural units were compared using material characterization techniques and molecular dynamic simulation. The results showed that the PUR polycarbonate diols formed as energy absorbing materials showed high tensile strength, high toughness, and excellent loss factor distribution based on the comparison of stress–strain tensile curves, glass transition temperatures, phase images, and dynamic storage loss modulus. External energy from simple shear deformation was absorbed to convert non-bond energy, in particular, based on fractional free volume, interaction energy, and total energy and hydrogen bond number change from the molecular dynamic simulation. Hydrogen bonds formed between soft segments and hard segments in the PURs have been proven to play a significant role in determining their mechanical and dynamic performance. The mechanical and dynamic properties of PURs characterized and tested using experimental techniques were quantified effectively using molecular dynamic simulation. This is believed to be an innovative theoretical guidance for the structural design of PURs at the molecular level for the optimization of energy absorption capabilities.

Keywords: polyurea; mechanical property; molecular dynamic simulation; energy absorption capabilities

1. Introduction

The number of the types and outputs of dangerous chemicals are progressively increasing with the rapid growth of the chemical industry all over the world, which leads to a dramatic increase in fire and explosion accidents [1–4]. The United Nations International Labor Organization stated that 2.78 million people die every year from chemical-related occupational diseases and accidents. The cost of these diseases and accidents is nearly USD 2.9 trillion [5]. In order to reduce the damage caused by chemical explosions, the protective materials with high strength and high toughness are extensively used in buildings, vehicles, and storage tanks [6–9].

In recent years, polyurea (PUR) has been widely used as a protective coating for bulletproofing and reducing shock waves because of its excellent dynamic and high-mechanical properties [10–13]. The dynamic mechanical properties and energy absorption mechanism of PUR have attracted researchers' attention [14–16]. This is because PUR shows high-strain-hardening phenomenon under the high-speed load condition, and the dynamic mechanical response of the chain section to the glass state is distinct from that of rubber accompanied by a large amount of energy dissipation [17,18]. In addition, PUR behaves like block copolymers: the soft phase of PUR leads to rapid deformation to form a smaller and obscure microphase structure with a tensile deformation increased, while the hard phase of PUR rotated to form new microphase structures along the stretching direction [19–21]. Moreover, the tensile deformation of PUR could cause hydrogen bond cleavage and rearrangement in its hard phase, leading to energy absorption and dispersion [22–25].

Microstructures of polymers, such as chemical composition [26–30], molecular chain length [31,32], crystallization [33,34], and microphase separation [35–38], can affect the mechanical and dynamic properties of PUR. Although there are many studies on the dynamic properties and energy absorption of PUR, there are few discussions on the quantitative relationship of the microstructure of PUR and the dynamic mechanical properties of energy absorption [11,39]. Molecular dynamic simulation is a powerful tool to establish a direct quantitative relationship between microstructure and macroscopic properties, and it could also allow for the observation of microstructure changes in materials and the prediction of material properties [40–45].

Our team has also conducted detailed research on the resistance of polyurea to hydrocarbon explosions [46,47]. To further enhance the explosion resistance of polyurea materials, it is imperative to conduct an in-depth investigation into the energy absorption mechanisms of polyurea materials. In this article, the mechanical and dynamic properties of various PURs with four typical macrodiol structural units were investigated using characterizations and tests, as well as molecular dynamic simulation (MD). The performances of PURs with various macrodiol structural units were quantified effectively at the molecular level to reveal an energy absorption mechanism with the combination of experimental results and simulation data.

2. Materials and Methods

2.1. Materials

Polycarbonate diol (PC1000, M_n = 1000) from Covestro, Polycaprolactone diol (PCL1000, M_n= 1000), Poly(propylene glycol) (PPG1000, M_n = 1000), and Polytetramethylene ether glycol (PTMG1000, M_n=1000) from Aladdin Bio-Chem Technology were all dried for 2 h under vacuum at 100 °C and then stored in N_2 atmosphere. Poly(propylene glycol) bis(2-aminopropyl ether) (D2000, M_n = 2000), Poly(propylene glycol) bis(2-aminopropyl ether) (D230, M_n = 230), and Isophorone diisocyanate (IPDI) were purchased from Aladdin Bio-Chem Technology, used as received. N, N-Dimethylformamide (DMF) was dried using 3-Åmolecular sieves.

2.2. PUR Synthesis and Film Preparation

The four kinds of polymerization reactions to form PURs between four kinds of diols (PC1000, PCL1000, PPG1000, or PTMG1000) and IPDI were conducted in a glove box under an argon atmosphere. These four PURs were named using the abbreviation of the corresponding diol. A typical polymerization process of PPG1000-PUR is shown as follows. IPDI (5.55 g, 25 mmol) was first dissolved in DMF (5 mL) in a 100 mL glass flask, and then PPG1000 (5.00 g, 5 mmol) was added into the flask. The whole solution was stirred for 4 h at 80 °C to form a pre-polymer, and then the solution was cooled down to room temperature. Another 10 g of DMF was then added to the solution to adjust its viscosity. Finally, the pre-polymer was added to the mixture of D2000 (9.00 g, 4.5 mmol) and D230 (3.56 g, 15.5 mmol) at room temperature for a further stirring of 2 h to form PPG1000-PUR.

To prepare samples for characterization and tests, the PPG1000-PUR solution was poured into a Teflon square mold (150 mm × 150 mm × 1 mm), and then gradually heated from 25 °C to 100 °C for 48 h. Residual solvent was removed under vacuum at 80 °C for 24 h. The thickness of the four PURs were within the range of 0.6 to 0.8 mm. Based on the SEM of cross sections, no bubble was observed inside the samples (Figure S1).

2.3. Characterization and Measurements

Molecular weights were measured using a Gel Permeation Chromatography (GPC, RI 2414, Waters, Milford, MA, USA) at a flow rate of 1.0 mL/min in THF at 35 °C with polystyrene calibration and a Multi-Angle Laser Light Scattering (SLS, Wyatt, Santa Barbara, CA, USA) at a flow rate of 1.0 mL/min in THF at 35 °C. A Fourier-transform Infrared Spectroscopy (FTIR, Nicolet 6700, Thermo Fisher Scientific, Waltham, MA, USA) fitted with a diamond ATR cell was used to obtain infrared spectra of the absorption of the materials. Optical photos were collected with polarized light using a Polarizing Microscope (POM, S60, Hui Tong Optical Instrument Co. Ltd., Shanghai, China) at room temperature. A X-ray Powder Diffraction (XRD, D8, Bruker, Karlsruhe, Germany) with a copper-point-focused source (λ = 0.154 nm) at 50 kV was used for phase identification of the crystalline material and provided information on interdomain spacings. Micron morphology was characterized using a Scanning Electron Microscope (SEM, S3400, Hitachi, Japan). A Atomic Force Microscope (AFM, SPM9700, SHIMADZU, Kyoto, Japan) in tapping mode was used to collect phase images. Thermal properties were characterized using a Differential Scanning Calorimeter (DSC, Q2000, TA, Santa Fe Springs, CA, USA) and a Thermal Gravimetric Analyzer (TGA, Q500, TA, Santa Fe Springs, CA, USA) under nitrogen with the heating rate of 10 °C min^{-1}. Mechanical performance was tested using an Universal Testing Machine (Labthink 1510, Jinan, China) in tensile test mode at 100 mm/min. Dynamics were tested using a Dynamic Mechanical Analyzer (DMA, Q800, TA, Santa Fe Springs, CA, USA), including loss factor (tan δ), storage modulus, and loss modulus, in tension mode with a temperature range of −100–80 °C and a frequency range of 0–100 Hz.

2.4. Computational Section

All data obtained in this paper are from Materials studio (2020, BIOVIA, San Diego, CA, USA). The COMPASS force field was chosen for the analog modules Forcite and Amorohous Cell and the Condensed-Phase Optimized Molecular Potential for atomistic simulation studies [44]. PUR chains were created from PPG1000, PTMG1000, PCL1000, PC1000, IPDI, D2000, and D230 molecules with the ratios provided by Build module. The potential energy minimization of the PUR chains within the Forcite module was carried out at 298 K using Smart algorithms (combining the Steepest Decent, Conjugate Gradient, and the Newton's minimization algorithms). PUR simulation boxes were created using Amorphous Cell module and were annealed from 298 K to 598 K for 5 cycles to relax polymer chains. MD simulation was conducted for 1000 ps at NPT condition (0.1 Mpa and 298 K) until the box size became constant, and then NVT condition (constant size and 298 K) was used for 1000 ps and equilibrium was reached. Finally, the computation cell of size was 3.4 nm × 3.4 nm × 3.4 nm, and the cutoff radius is 1.55 nm [48]. All of the computation results were the average value from 100 MD simulation frames.

These model errors were revealed to be lower than 2% when compared to $Density_{MD}$ from the MD simulation and $Density_{Exp}$ from the experiments (Table 1).

Table 1. Density of PURs determined through experiment and MD simulation.

Sample Name	PPG1000-PUR	PTMGG1000-PUR	PCL1000-PUR	PC1000-PUR
$Density_{Exp}$ (g/cm^3)	1.021	1.007	1.051	1.053
$Density_{MD}$ (g/cm^3)	1.041	1.019	1.045	1.055
Error (%)	1.9	1.2	0.6	0.2

3. Results

3.1. Polymerization Result Analysis

The synthetic routes of the PURs are shown in Figure 1. A series of macrodiols including PPG1000, PTMG1000, PCL1000, and PC1000 were reacted with the given IPDI-DMF solution to synthesize the pre-polymer solution. Then, D2000 and D230 were added into the pre-polymer solution to form the four PURs as chain extenders. PPG1000-PUR and PTMG1000-PUR are polyether-type PURs, while PCL1000-PUR and PC1000-PUR are polyester-type PURs. Because the high reactivity of the amine group and isocyanate group bring uncontrollable polymerization, the pre-polymerization of diisocyanate and macrodiols as capping agents is necessary to control the synthesis of PUR. In an effort to gain further insight into the effect of the macrodiol structural units on material properties, the macrodiol structural units in the PURs were considered as the macrodiol segments, and the other structural units including IPDI, D2000, and D230 were considered as the resin segments. The macrodiol segments of the four PURs still remained at 24 wt% by controlling feed ratio. Detailed results after polymerization are shown in Table 2. The molecular weights (M_n) of each of the four PURs were quite similar, and they all had a unimodal and relatively narrow molecular weight distribution (MWD) in the GPC curves (Figure S2). This indicates that no gelation and other side reactions were carried out during polymerization; hence, the four PURs all had comparable molecular weight and topological structure.

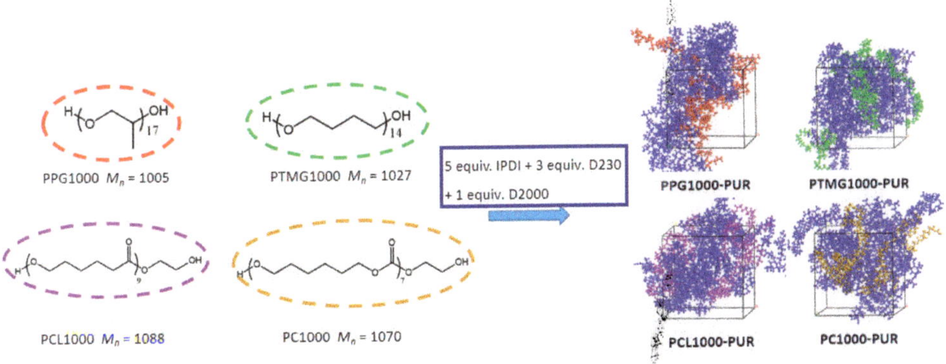

Figure 1. Schematic photo of synthetic routes and molecular models of the four types of PURs.

Table 2. Polymerization results and polymer information.

Sample Name	M_n [a] ($\times 10^3$ g/mol)	MWD [a]	M_n [b] ($\times 10^3$ g/mol)	MWD [b]	$T_{g\,DSC}$ [c] (°C)	$T_{g\,DMA}$ [d] (°C)	T_d [e] (°C)	d [f] (nm)
PPG1000-PUR	10.87	1.62	10.45	1.33	−60	−45	243	14.9
PTMG1000-PUR	10.19	1.73	9.51	1.42	−65	−48	292	14.8
PCL1000-PUR	9.33	1.77	7.97	1.48	−57	−35	282	14.5
PC1000-PUR	11.72	1.84	10.81	1.55	−56	−22	270	14.8

[a] determined by GPC. [b] determined by SLS. [c] determined by DSC. [d] determined by DMA. [e] determined by TGA. [f] calculated from the peak maxima (q_{max}) of SAXS curves, $d = 2\pi/q_{max}$.

3.2. Thermal Property Analysis

The thermal properties of all the resulting PURs were characterized using DSC. In the first DSC heating scan of heat history (Figure S3), no clear thermal transition behavior of PURs was observed. In the second heating scan (Figure S4), the four PURs had similar macrodiol segments $T_{g\,DSC}$ about −60 °C, and no endothermic or exothermic transitions aside from $T_{g\,DSC}$ are evident in these PURs, which suggests there is no typical thermal transition and T_ms in these PURs. Based on TGA curves (Figure S5), PPG1000-PUR had a relatively low thermal decomposition temperature (T_d = 243 °C), and the T_ds of other PURs was approximately 270–290 °C.

3.3. Microscopic Morphology Analysis

The POM photos of the four PURs are presented in Figure 2. On the one hand, no clear crystallizing phase was supposed to be found in the POM photos of PPG1000-PUR or PC1000-PUR. On the other hand, the crystallizing phase were observed in the POM photos of PTMG1000-PUR and PCL1000-PUR, which synthesized because of semi-crystalline PTMG1000 and PCL1000. Especially, PCL1000-PUR showed the typical extinction of crystalline. This indicates that the semi-crystalline macrodiol soft segment could form a crystallizing phase in the PURs. Moreover, in the XRD profile of PCL1000-PUR, there was a sharp peak at 21.68°, which proved that the crystallizing phase existed in the PCL1000-PUR (Figure S6). However, the T_ms of PTMG1000-PUR and PCL1000-PUR were not detected in the DSC curves (Figure S3), probably because of low crystallizing degree.

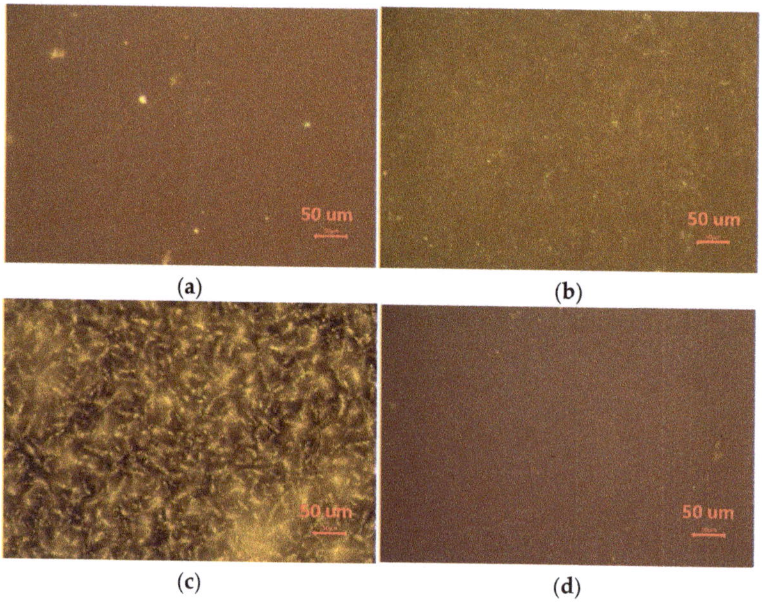

Figure 2. Optical photos of the four PURs: (**a**) PPG1000-PUR, (**b**) PTMG1000-PUR, (**c**) PCL1000-PUR, (**d**) PC1000-PUR.

ATR-FTIR spectra of all PURs showed absorption peaks at 1630 cm^{-1} that corresponded to the C=O bond of isocyanate groups, and PCL1000-PUR and PC1000-PUR also featured the two special peaks at 1724 cm^{-1} and 1738 cm^{-1}, which corresponded to the C=O bond of ester groups of caprolactone and carbonate, respectively (Figures 3a and S7). Moreover, N−H stretching regions in the ATR-FTIR of PUR could be used to infer H-bond association in PUR [32]. The N−H bond assignments are 3310 cm^{-1} and 3368 cm^{-1} for ordered and disordered N−H groups, respectively. The wide N−H stretching peaks in the ATR-FTIR of PUR were separated into two special peaks for ordered and disordered N−H groups by Gauss fitting (Figure 3b). Based on the two peaks' area ratio (ordered/disordered), the H-bond association values of PUR increased in the order PCL1000-PUR (1.06) < PTMG1000-PUR (1.57) < PPG1000-PUR (1.69) < PC1000-PUR (1.71). Perhaps because the soft segment of PCL1000-PUR was crystallized to restrict the H-bond formation, the H-bond association values of PCL1000-PUR were relatively low.

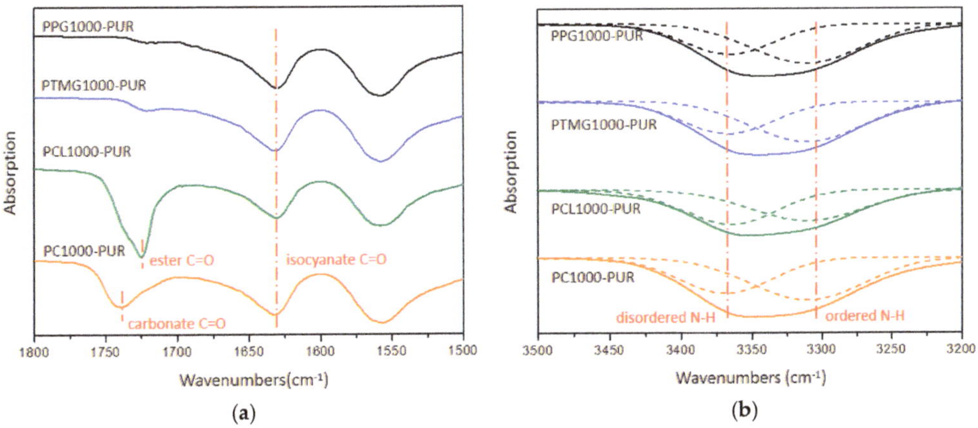

Figure 3. ATR-FTIR spectra of the PURs: (**a**) attribution of C=O peaks, (**b**) state of N-H association.

Tapping-mode AFM-phase images showed the microphase distribution of these PURs (Figure S8). In the AFM-phase images of PTMG1000-PUR and PCL1000-PUR, there were several bright and dark regions considered as hard and soft phase, respectively. But, in the AFM-phase images of PPG1000-PUR and PC1000-PUR, the bright and dark regions were hardly recognized. Moreover, unlike aromatic PUR with clear nanoscale ribbon morphology [19,49], these PURs had relatively indistinct microphase morphology. For quantitative insight into the degree of hard and soft segment microphase morphology, the mean interdomain spacings (d) of all these four PURs are calculated from approximately 14 nm through to absolute intensity SAXS in Table 2 and Figure S9. These results indicated that the macrodiol segments could not play an important role in phase separation in the PURs.

3.4. Mechanical Property Analysis

Surface hardness and tensile tests were performed to evaluate the mechanical properties of these PURs. Based on Shore A values of material surface, the hardness of the PURs from low to high is as follows: PTMG1000-PUR < PCL1000-PUR < PC1000-PUR < PPG1000-PUR. The tensile test was conducted using a universal tensile machine at a draw rate of 100 mm/min. Young's modulus, ultimate tensile strength, elongation at break, and toughness are summarized in Table 3 and Figure S10. The ultimate tensile strength of these PURs from low to high is as follows: PTMG1000-PUR < PCL1000-PUR < PPG1000-PUR < PC1000-PUR, and the ultimate tensile strength and toughness of PC1000-PUR achieved 12.01 MPa and 44.27 MJ/m^3, respectively, which were both much higher than the other three PURs. As a result, it implied that material mechanical properties were dependent on the soft segment.

In order to further investigate the effect of the soft phase in the PUR on its mechanical properties, the scanning electron microscope (SEM) images of these PURs before and after the tensile test are compared (Figure 4). Before the tensile test, the surface morphology of the four PURs were flat and smooth and no obvious crack was observed, as expected. However, the sample surface morphology of the four PURs became different after the same tensile test.

Table 3. Mechanical properties of the PURs.

Sample Name	Surface Hardness	Young's Modulus (MPa)	Tensile Strength (MPa)	Elongation at Break (%)	Toughness (MJ/m^3)
PPG1000-PUR	88	80	5.19	182	8.39
PTMG1000-PUR	78	20	2.71	262	5.91
PCL1000-PUR	82	21	2.85	248	6.31
PC1000-PUR	87	29	12.01	688	44.27

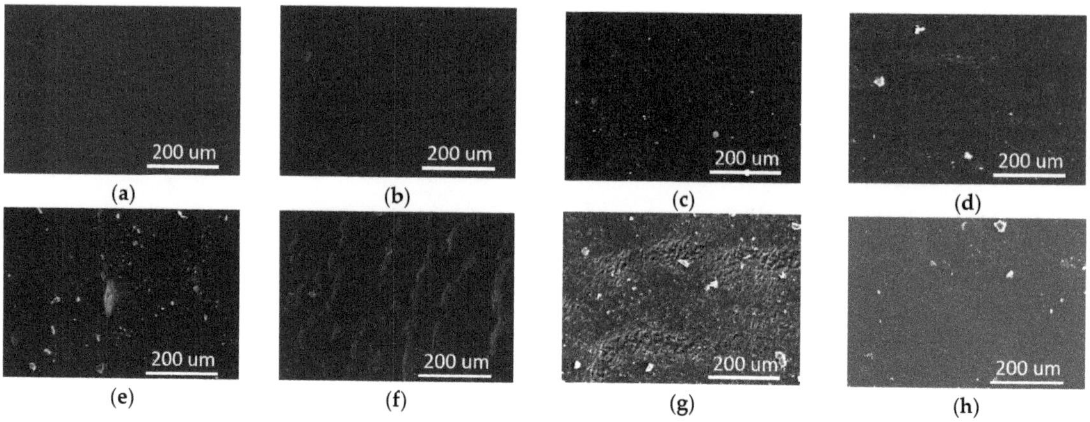

Figure 4. SEM images of the PURs: (**a**) PPG1000-PUR, (**b**) PTMG1000-PUR, (**c**) PCL1000-PUR, (**d**) PC1000-PUR before tensile testing, (**e**) PPG1000-PUR, (**f**) PTMG1000-PUR, (**g**) PCL1000-PUR, (**h**) PC1000-PUR after tensile testing.

Although PPG1000-PUR had a few cracks about 100 μm long on the surface, the entire surface was still kept flat and smooth. There were lots of similar cracks on the surface of PTMG1000-PUR, and many cracks were attached together to form longer cracks on the surface. Compared to PTMG1000-PUR, PCL1000-PUR showed more and smaller cracks to form a crazing zone on the surface. Interestingly, no crack was observed on the surface of PC1000-PUR. In other words, the rest parts of PC1000-PUR still kept good integrality, and the material had even been broken under tensile force. As above, the results suggested that relatively homogeneous microphase morphology was profitable to improve the mechanical properties of the PURs. The tensile fracture surfaces of four types of PUR materials were precisely analyzed (Figure 5). The tensile fracture surface of PC1000-PUR exhibited a significant amount of feather structures, indicating a relatively typical ductile fracture, which can absorb a large amount of energy. Similarly, the tensile fracture surfaces of PPG1000-PUR and PTMG1000-PUR materials also showed clear feather structures, while the feather structures on the tensile fracture surface of PTMG1000-PUR were smaller in size. In contrast, no feather structures were observed on the fracture surface of the PCL1000-PUR material. Instead, there were numerous clusters of particles approximately 5 um in size. These particles are likely the phase-separated products resulting from the crystallization of PCL1000 macrodiol, and these distinct particle clusters reduced the tensile performance of the material.

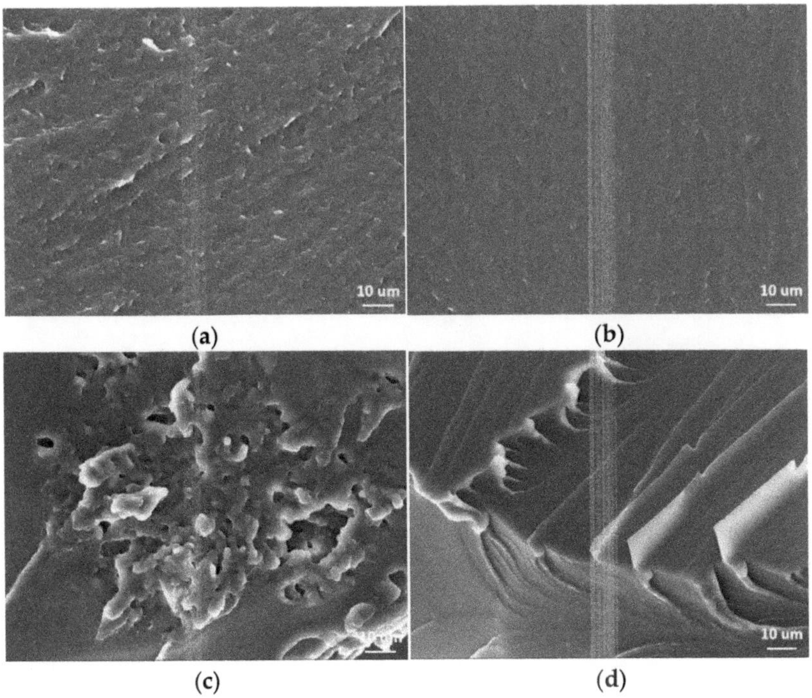

Figure 5. SEM images of tensile fracture surfaces of the four PURs: (**a**) PPG1000-PUR, (**b**) PTMG1000-PUR, (**c**) PCL1000-PUR, (**d**) PC1000-PUR.

3.5. Dynamic Property Analyses

The dynamic properties of the four PURs have been investigated using DMA in strain control. Figure 6a showed the temperature dependence of the loss factor (tan δ) of these PURs at 1 Hz. The peak temperatures marked as the $T_{g\ DMA}$ were extracted in the tan δ curves. Compared to the above $T_{g\ DSC}$ from DSC curves (Figure S4), all $T_{g\ DMA}$ of the four PURs were higher, especially the $T_{g\ DMA}$ of PC1000-PUR, which was up to 34 °C higher than its $T_{g\ DSC}$. The frequency sweep dynamic properties of these PURs are showed in Figure 6b at 25 °C. The storage modulus of the four PURs increased obviously, and the storage modulus of PC1000-PUR gained the most improvement by 90 MPa from 0.1 Hz to 100 Hz when the test frequency increased. It suggested that PC1000-PUR could have better mechanical properties at a high frequency. Moreover, the loss modulus of PTMG1000-PUR and PC1000-PUR increased gradually when the test frequency increased, and the loss modulus of PPG1000-PUR and PCL1000-PUR was not sensitive to the change in frequency. Based on the DMA data, PC1000-PUR could be better at absorbing the energy of high-frequency waves.

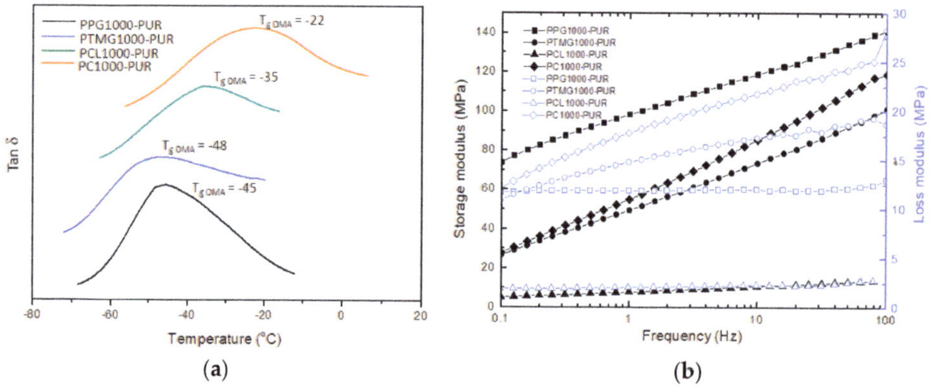

Figure 6. Dynamic properties of the PURs using DMA: (**a**) loss factors depending on temperature, (**b**) storage modulus and loss modulus depending on frequency.

3.6. Fraction-Free Volume Analysis

In an effort to gain further insight into the effect of the soft segment on energy absorption at the molecular level, the molecular models of the four PURs were built in Materials Studio. The fraction-free volume (FFV) analysis reflected the chain packing efficiency of polymer molecules, and a lower FFV indicated tighter atomic packing [50]. The FFV calculations for the four PURs are shown in Table 4. It is worth mentioning that the FFV of the four polymers was calculated by the following equation (Equation (1)) [44]. PTMG1000-PUR had the highest FFV among the four materials, indicating a low dynamic glass transition temperature of the material, which was consistent with the above DMA data:

$$\mathrm{FFV} = \frac{V - 1.3V_W}{V} \quad (1)$$

where V_W was the specific Van der Waals volume of molecules. Also, V was the volume of the molecular cell.

Table 4. Fraction-free volumes of the PURs.

Sample Name	PPG1000-PUR	PTMG1000-PUR	PCL1000-PUR	PC1000-PUR
FFV	18.6	19.8	18.8	18.6

3.7. Binding Energy Analysis

The binding energy (E_{bind}), the negative value of the interaction energy (E_{inter}) of macrodiol and resin segments of PURs, could be evaluated using the interaction strength between macrodiol and resin segments. In this paper, E_{inter} was expressed as the difference between the total energy of the system and the energy of each component. Through MD simulation, the total energy of the system (E_{total}), the energy of macrodiol segments(E_{MS}), and the energy of resin segments (E_{RS}) could be calculated, respectively, and E_{inter} could be calculated through the following formula (Equation (2)).

$$E_{bind} = -E_{inter} = -\left(E_{total} - E_{ss} - E_{Hs} \right) \quad (2)$$

According to the E_{bind} sequence of the four PURs, PPG1000-PUR (656 kcal/mol) < PTMG1000-PUR (798 kcal/mol) < PCL1000-PUR (938 kcal/mol) < PC1000-PUR (942 kcal/mol), there were relatively strong interaction forces between the macrodiol and resin segments of polyester-type PC1000-PUR and PCL1000-PUR because of polar C=O groups of PC1000 and PCL1000. In the Compass force field, the non-bond forces were divided into Van der Waals force and electrostatic force. For the non-bond forces between macrodiol and resin segments of the four PURs, the Van der Waals force was the main interaction force, because the Van der Waals force energy far outweighed the electrostatic force energy (Figure 7). Furthermore, the number of H-Bonds was also an important indicator. The H-Bond number between the macrodiol and resin segments of PCL1000-PUR and PC1000-PUR were also more than that of the other two polyether-type PURs.

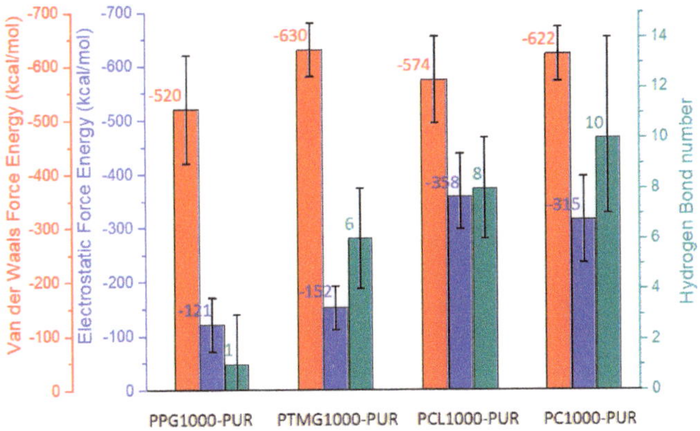

Figure 7. Van der Waals force energy, electrostatic force energy, and H-Bond number on the interface of macrodiol segments and resin segments of the PURs.

3.8. Mean Square Radius of Gyration Analysis

The mean square radius of gyration (R_g) was an important size parameter of polymer chains. For the same molecular weight, the larger R_g represented the looser polymer chains and better polymer chain compatibility. In order to study the distribution of macrodiol and resin segments of PURs, R_g of the macrodiol and resin segments of the PURs was calculated separately with the Analysis module in the MD simulation (Figure 8). The mean square R_g is obtained by statistically averaging the R_g of the macrodiol or resin segments from the four PUR chains. The R_gs of macrodiol segments of PURs were relatively similar in the range of 8.1 to 10.4 Å because of the similar molecular weight of macrodiol segments. However, there was a noticeable difference for the R_gs of resin segments of PURs, especially PC1000-PUR, which showed the biggest R_g of resin segments. This was probably because of an increase in the intermolecular force from PC1000 as macrodiol segments stretch the resin segment chains, resulting in an increase in the mean square rotation of the hard segment chains. The compatibility between these two segments becomes better and better.

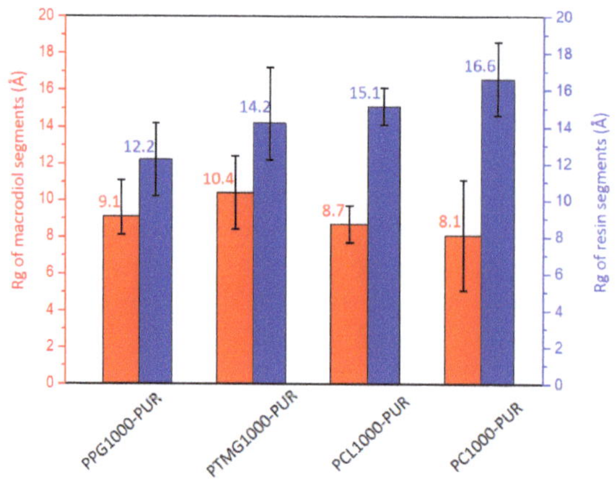

Figure 8. Mean square radius of the gyration of the macrodiol segments and resin segments in the PURs.

3.9. Energy Absorption Mechanism

In order to investigate the energy absorption of the four PURs at the molecular level, a series of simple shear deformation tests were run under different shear rate conditions through MD simulation. In this paper, the total energy of the material system mainly included the internal energy (temperature) and the potential energy ($E_{potential}$), in which the potential energy was the sum of the bond energy (E_{bond}) and non-bond energy ($E_{non\text{-}bond}$) of polymers. An energy minimization procedure was used in these tests to observe the effect of shear rate on the changes in temperature, energy, and H-Bond number in the four PURs.

With shear rate increased, the temperature and $E_{potential}$ of the four PURs gradually rose, indicating that one part of the external energy from simple shear deformation could be absorbed to convert into the internal energy of PURs, and the other part of the energy into $E_{potential}$ (Figure 9). Among the four PURs, the system temperature of PTMG1000-PUR increased the least, probably because of the low atomic density in PTMG1000-PUR and low friction between the atoms, while PC1000-PUR had the highest system temperature change. With the increase in shear rate, the $E_{potential}$ of the four PURs from high to low is as follows: PC1000-PUR > PCL1000-PUR > PTMG1000-PUR > PPG1000-PUR. Classifying the $E_{potential}$ into E_{bond} and $E_{non\text{-}bond}$, it was found that the change in $E_{non\text{-}bond}$ was more substantial, accounting for about 60% of the $E_{potential}$ change, and only 40% of the $E_{potential}$ change for E_{bond} (Figure 9c). Furthermore, the number of H-Bonds changed very significantly in the $E_{non\text{-}bond}$. With the increase in shear rate, the number of H-Bonds decreased rapidly, especially for PC1000-PUR, with a decrease of 15. According to the above simulation calculation, under the dynamic load conditions, the external energy absorption was mainly converted into the $E_{non\text{-}bond}$ of PURs, and the H-Bond deformation in the $E_{non\text{-}bond}$ was the main index of PUR energy absorption (Figure 10).

Based on the combination of experimental results and simulation data, an effective quantification of the performance and energy absorption mechanism of PURs with different macrodiol structural units at the molecular level was provided. Tensile testing revealed that PC1000-PUR exhibited the highest tensile strength and toughness among the four PURs, attributed to the extensive formation of ordered hydrogen bonds and homogeneous phase distribution. PTMG1000 and PCL1000 showed relatively lower mechanical strength and more surface cracks due to the crystallization of their macrodiol segments. In terms of dynamic properties, PC1000-PUR displayed higher storage modulus and loss modulus with increasing frequency, and its glass transition temperature ($T_{g\ DMA}$) was the highest, reaching −22 °C at 1 Hz under dynamic mechanical analysis. MD simulations indicated that PC1000 diols exhibited stronger interaction (approximately 942 kcal/mol) with the resin segments at the molecular level, suggesting their significant contribution to the mechanical properties of the material. The energy absorption mechanism of the PURs was primarily attributed to the change in non-bonded energy, accounting for approximately 60% of the total energy absorbed under dynamic loading conditions. H-Bond deformation was identified as an important factor in energy absorption.

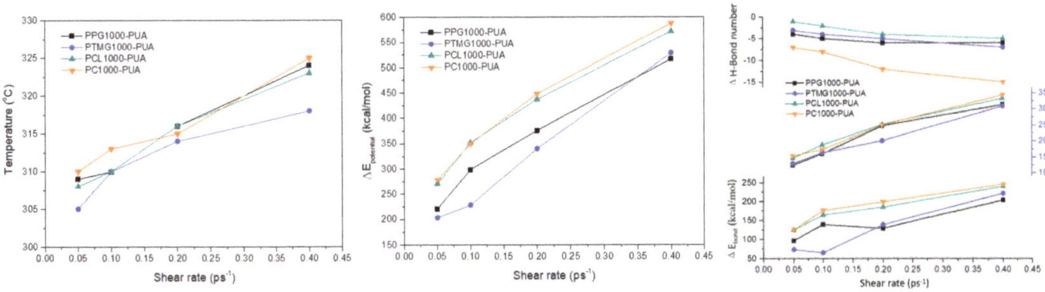

(a) (b) (c)

Figure 9. Changes in the PURs depending on shear rate: (**a**) temperature trends, (**b**) $\Delta E_{potential}$ trends, (**c**) ΔE_{bond}, $\Delta E_{non\text{-}bond}$, and ΔH-Bond.

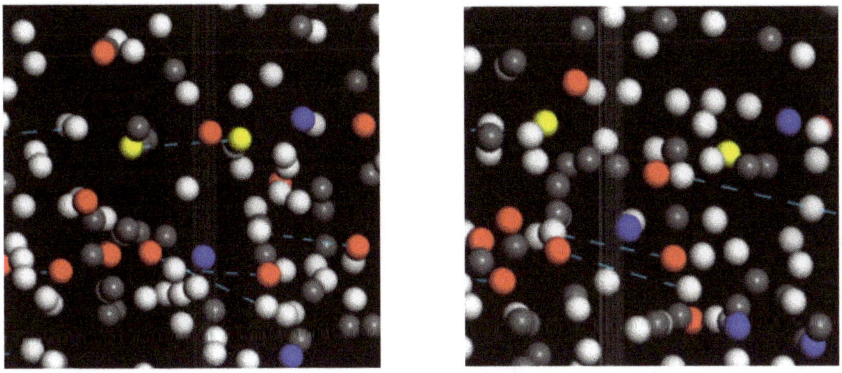

(a) (b)

Figure 10. A zoomed-in view of the PC1000-PUR model (**a**) before and (**b**) after shear deformation. The H-bonds were depicted as dashed lines and the yellow balls were object atoms.

4. Conclusions

The mechanical and dynamic properties of four polyurethane elastomers (PURs) with different macrodiol segments were studied. PC1000-PUR exhibited the highest tensile strength and toughness due to ordered H-bond formation and homogeneous phase distribution. PTMG1000 and PCL1000 showed lower mechanical strength and more surface cracks. PC1000-PUR had higher storage and loss modulus, with the highest $T_{g\ DMA}$. MD simulations revealed a strong interaction between PC1000 diols and resin segments. Energy absorption in PURs was primarily attributed to non-bonded energy change, with H-Bond deformation playing a crucial role. This study provides insights for designing PURs with optimal energy absorption.

Supplementary Materials: The following supporting information can be downloaded at: https://www.mdpi.com/article/10.3390/polym17010107/s1, Figure S1. SEM of the four PUR cross sections. Figure S2. GPC curves of the PURs. Figure S3. DSC curves of the PURs for the first heating scans. Figure S4. DSC curves of the PURs for the second heating scans. Figure S5. TGA curves of the PURs. Figure S6. WAXD profiles of the PURs. Figure S7. AFM images of the PURs. Figure S8. AFM images of the PURs: (a) PPG1000-PUR, (b) PTMG1000-PUR, (c) PCL1000-PUR, (d) PC1000-PUR. Figure S9. SAXS profiles of the PURs. Figure S10. Stress-strain curves of the PURs.

Author Contributions: K.Y.: Conceptualization, investigation, and writing—original draft preparation. S.W.: writing—review. Y.C.: MD simulation and software. H.D.: visualization. Q.W. and Q.C.: supervision. All authors have read and agreed to the published version of the manuscript.

Funding: This research was funded by the National Key R&D Program of China (No. 2021YFC3001204).

Institutional Review Board Statement: Not applicable.

Informed Consent Statement: Not applicable.

Data Availability Statement: The original contributions presented in the study are included in the article/supplementary material, further inquiries can be directed to the corresponding author.

Acknowledgments: We thank Chaoyang Li from Hubei Institute of Aerospace Chemotechnology for the software services of Materials Studio 2020.

Conflicts of Interest: Authors Ke Yang, Shanda Wang, Yanru Chen, Hanhai Dong, Quanguo Wang and Qingli Cheng were employed by the company Sinopec Research Institute of Safety and Engineering Co., Ltd. The authors declare that the research was conducted in the absence of any commercial or financial relationships that could be construed as a potential conflict of interest.

References

1. ur Rehman, S.; Ahmed, R.; Ma, K.; Xu, S.; Aslam, M.A.; Bi, H.; Liu, J.; Wang, J. Ammonium Nitrate Is a Risk for Environment: A Case Study of Beirut (Lebanon) Chemical Explosion and the Effects on Environment. *Ecotoxicol. Environ. Saf.* **2021**, *210*, 111834. [CrossRef] [PubMed]
2. Kundu, B.; Senapati, B.; Matsushita, A.; Heki, K. Atmospheric Wave Energy of the 2020 August 4 Explosion in Beirut, Lebanon, from Ionospheric Disturbances. *Sci. Rep.* **2021**, *11*, 2793. [CrossRef]
3. Yang, X.; Li, Y.; Chen, Y.; Li, Y.; Dai, L.; Feng, R.; Duh, Y.-S. Case Study on the Catastrophic Explosion of a Chemical Plant for Production of M-Phenylenediamine. *J. Loss Prev. Process Ind.* **2020**, *67*, 104232. [CrossRef]
4. Lu, Y.; Wang, T.; Liu, T. Bayesian Network-Based Risk Analysis of Chemical Plant Explosion Accidents. *Int. J. Environ. Res. Public Health* **2020**, *17*, 5364. [CrossRef] [PubMed]
5. Zhang, M.; Kim, R. Occupational Health and Safety in China: From Emergency Response to Jiangsu Chemical Explosion to Long-Term Governance Improvement. *J. Glob. Health* **2020**, *10*, 010315. [CrossRef]
6. Schunck, T.; Bastide, M.; Eckenfels, D.; Legendre, J.F. Explosion Mitigation by Metal Grid with Water Curtain. *Shock Waves* **2021**, *31*, 511–523. [CrossRef]
7. Mazzini, G.; D'Onorio, M.; Caruso, G. Hydrogen Explosion Mitigation in DEMO Vacuum Vessel Pressure Suppression System Using Passive Recombiners. *Fusion Eng. Des.* **2021**, *171*, 112713. [CrossRef]
8. He, H.; Fan, H. Explosion Vibration Mitigation of Meta-Plate with Mass–Spring Metastructures. *Extrem. Mech. Lett.* **2021**, *42*, 101108. [CrossRef]
9. Yang, K.; Wang, S.; Zhou, R.; Cheng, Q.; Wang, Q.; Li, Y. Synthesis of Syndiotactic Polystyrene -Polyisobutylene Graft Copolymers by Cationic Half-sandwich Scandium Complex. *Polym. Eng. Sci.* **2022**, *62*, 3412–3417. [CrossRef]
10. Jajam, K.C.; Sottos, N.R. Energy Absorption Behavior of Polyurea Under Laser-Induced Dynamic Mixed-Mode Loading. *J. Dyn. Behav. Mater.* **2016**, *2*, 379–390. [CrossRef]
11. Iqbal, N.; Tripathi, M.; Parthasarathy, S.; Kumar, D.; Roy, P.K. Polyurea Coatings for Enhanced Blast-Mitigation: A Review. *RSC Adv.* **2016**, *6*, 109706–109717. [CrossRef]
12. Raman, S.N.; Ngo, T.; Mendis, P.; Pham, T. Elastomeric Polymers for Retrofitting of Reinforced Concrete Structures against the Explosive Effects of Blast. *Adv. Mater. Sci. Eng.* **2012**, *2012*, 1–8. [CrossRef]
13. Temizkan, E.; Eroğlu, G.; Ergün, A.; Deligöz, H. Preparation, Characterization, and Influence of Polyurea Coatings on Their Layered Composite Materials Based on Flexible Rebonded Polyurethane. *Polym. Eng. Sci.* **2021**, *61*, 1392–1404. [CrossRef]

14. Wang, T.; Yu, W.-C.; Sun, W.-J.; Jia, L.-C.; Gao, J.-F.; Tang, J.-H.; Su, H.-J.; Yan, D.-X.; Li, Z.-M. Healable Polyurethane/Carbon Nanotube Composite with Segregated Structure for Efficient Electromagnetic Interference Shielding. *Compos. Sci. Technol.* **2020**, *200*, 108446. [CrossRef]
15. Elizalde, F.; Aguirresarobe, R.H.; Gonzalez, A.; Sardon, H. Dynamic Polyurethane Thermosets: Tuning Associative/Dissociative Behavior by Catalyst Selection. *Polym. Chem.* **2020**, *11*, 5386–5396. [CrossRef]
16. Iqbal, N.; Sharma, P.K.; Kumar, D.; Roy, P.K. Protective Polyurea Coatings for Enhanced Blast Survivability of Concrete. *Constr. Build. Mater.* **2018**, *175*, 682–690. [CrossRef]
17. Roland, C.M.; Fragiadakis, D.; Gamache, R.M. Elastomer–Steel Laminate Armor. *Compos. Struct.* **2010**, *92*, 1059–1064. [CrossRef]
18. Bogoslovov, R.B.; Roland, C.M.; Gamache, R.M. Impact-Induced Glass Transition in Elastomeric Coatings. *Appl. Phys. Lett.* **2007**, *90*, 221910. [CrossRef]
19. Das, S.; Yilgor, I.; Yilgor, E.; Inci, B.; Tezgel, O.; Beyer, F.L.; Wilkes, G.L. Structure–Property Relationships and Melt Rheology of Segmented, Non-Chain Extended Polyureas: Effect of Soft Segment Molecular Weight. *Polymer* **2007**, *48*, 290–301. [CrossRef]
20. Grujicic, M.; d'Entremont, B.P.; Pandurangan, B.; Runt, J.; Tarter, J.; Dillon, G. Concept-Level Analysis and Design of Polyurea for Enhanced Blast-Mitigation Performance. *J. Mater. Eng. Perform.* **2012**, *21*, 2024–2037. [CrossRef]
21. Wang, R.; Chen, L.; Lei, Q.; Shi, J.; Chen, C. Preparation and Characterization of Polyethylene Glycol/Polyurea Composites for Shape Stabilized Phase Change Materials. *Polym. Eng. Sci.* **2022**, *62*, 664–676. [CrossRef]
22. He, Y.; Xie, D.; Zhang, X. The Structure, Microphase-Separated Morphology, and Property of Polyurethanes and Polyureas. *J. Mater. Sci.* **2014**, *14*, 7339–7352. [CrossRef]
23. Sheth, J.P.; Klinedinst, D.B.; Wilkes, G.L.; Yilgor, I.; Yilgor, E. Role of Chain Symmetry and Hydrogen Bonding in Segmented Copolymers with Monodisperse Hard Segments. *Polymer* **2005**, *46*, 7317–7322. [CrossRef]
24. Das, S.; Yilgor, I.; Yilgor, E.; Wilkes, G.L. Probing the Urea Hard Domain Connectivity in Segmented, Non-Chain Extended Polyureas Using Hydrogen-Bond Screening Agents. *Polymer* **2008**, *49*, 174–179. [CrossRef]
25. Chen, X.; Shen, Z.; Jia, H.; Gao, Y.; Zhang, M.; Luo, Y.; Luo, Z. Understanding the Self-Healing Mechanism of Polyurethane Elastomer Based on Hydrogen Bonding Interactions through Molecular Dynamics Simulation. *Macromol. Theory Simul.* **2022**, *31*, 2100051. [CrossRef]
26. Wang, X.; Tang, X.-Z.; Hu, X.; Wang, P. Dynamic Tensile Mechanical Behavior of Polyurethane Films Incorporated with Functionalized Graphene Oxide. *Polymer* **2022**, *246*, 124755. [CrossRef]
27. Xia, L.; Tu, H.; Zeng, W.; Yang, X.; Zhou, M.; Li, L.; Guo, X. A Room-Temperature Self-Healing Elastomer with Ultra-High Strength and Toughness Fabricated via Optimized Hierarchical Hydrogen-Bonding Interactions. *J. Mater. Chem. A* **2022**, *10*, 4344–4354. [CrossRef]
28. Sivanesan, D.; Kim, S.; Jang, T.W.; Kim, H.J.; Song, J.; Seo, B.; Lim, C.-S.; Kim, H.-G. Effects of Flexible and Rigid Parts of ε-Caprolactone and Tricyclodecanediol Derived Polyurethane on the Polymer Properties of Epoxy Resin. *Polymer* **2021**, *237*, 124374. [CrossRef]
29. Wang, X.; Xu, J.; Zhang, Y.; Wang, T.; Wang, Q.; Yang, Z.; Zhang, X. High-Strength, High-Toughness, Self-Healing Thermosetting Shape Memory Polyurethane Enabled by Dual Dynamic Covalent Bonds. *Polym. Chem.* **2022**, *13*, 3422–3432. [CrossRef]
30. Santra, S.; Kolay, S.; Sk, S.; Ghosh, D.; Mishra, A.; Roy, L.; Sarkar, K.; Molla, M.R. Supramolecularly Cross-Linked Nanoassemblies of Self-Immolative Polyurethane from Recycled Plastic Waste: High Encapsulation Stability and the Triggered Release of Guest Molecules. *Polym. Chem.* **2022**, *13*, 3294–3303. [CrossRef]
31. Reinecker, M.; Soprunyuk, V.; Fally, M.; Sánchez-Ferrer, A.; Schranz, W. Two Glass Transitions of Polyurea Networks: Effect of the Segmental Molecular Weight. *Soft Matter* **2014**, *10*, 5729–5738. [CrossRef]
32. Castagna, A.M.; Pangon, A.; Choi, T.; Dillon, G.P.; Runt, J. The Role of Soft Segment Molecular Weight on Microphase Separation and Dynamics of Bulk Polymerized Polyureas. *Macromolecules* **2012**, *45*, 8438–8444. [CrossRef]
33. Matsuba, G.; Kobayashi, T.; Chonan, Y. Detailed Analysis of Aliphatic Polyurea Crystals. *J. Fiber Sci. Technol.* **2017**, *73*, 122–125. [CrossRef]
34. Li, S.; Sang, Z.; Zhao, J.; Zhang, Z.; Zhang, J.; Yang, W. Crystallizable and Tough Aliphatic Thermoplastic Polyureas Synthesized through a Nonisocyanate Route. *Ind. Eng. Chem. Res.* **2016**, *55*, 1902–1911. [CrossRef]
35. Choi, T.; Weksler, J.; Padsalgikar, A.; Runt, J. Influence of Soft Segment Composition on Phase-Separated Microstructure of Polydimethylsiloxane-Based Segmented Polyurethane Copolymers. *Polymer* **2009**, *50*, 2320–2327. [CrossRef]
36. Garrett, J.T.; Xu, R.; Cho, J.; Runt, J. Phase Separation of Diamine Chain-Extended Poly(Urethane) Copolymers: FTIR Spectroscopy and Phase Transitions. *Polymer* **2003**, *44*, 2711–2719. [CrossRef]
37. Garrett, J.T.; Siedlecki, C.A.; Runt, J. Microdomain Morphology of Poly(Urethane Urea) Multiblock Copolymers. *Macromolecules* **2001**, *34*, 7066–7070. [CrossRef]
38. Li, F.; Tuinier, R.; van Casteren, I.; Tennebroek, R.; Overbeek, A.; Leermakers, F.A.M. Self-Organization of Polyurethane Pre-Polymers as Studied by Self-Consistent Field Theory: Self-Organization of Polyurethane Pre-Polymers as studied by self-consistent field theory. *Macromol. Theory Simul.* **2016**, *25*, 16–27. [CrossRef]

39. Rong, H.; Xu, M.; Jiang, X.; Lu, X. Synthesis and Molecule Dynamics Study of High Damping Polyurethane Elastomer Based on the Synergistic Effect of Dangling Chains and Dynamic Bonds. *Polym. Chem.* **2022**, *13*, 4260–4272. [CrossRef]
40. Li, C.; Wang, Y.; Yin, Y.; Li, Y.; Li, J.; Sun, D.; Lu, J.; Zhang, G.; Sun, R. A Comprehensive Study of Pyrazine-Contained and Low-Temperature Curable Polyimide. *Polymer* **2021**, *228*, 123963. [CrossRef]
41. Lei, D.; Fu, Y.-Z.; Hu, W.; Li, D.; He, Y.; Gan, L.-H.; Gan, L.; Huang, J. Molecular Dynamics Research on Interfacial Reinforcement between ε-CL-20 and Polymeric Bonding Agents for Humidity-Insensitive Solid Propellant Systems. *J. Polym. Res.* **2021**, *28*, 215. [CrossRef]
42. Grujicic, M.; Snipes, J.S.; Ramaswami, S.; Yavari, R.; Ramasubramanian, M.K. Meso-Scale Computational Investigation of Shock-Wave Attenuation by Trailing Release Wave in Different Grades of Polyurea. *J. Mater. Eng. Perform.* **2014**, *23*, 49–64. [CrossRef]
43. Grujicic, M.; Yavari, R.; Snipes, J.S.; Ramaswami, S.; Runt, J.; Tarter, J.; Dillon, G. Molecular-Level Computational Investigation of Shock-Wave Mitigation Capability of Polyurea. *J. Mater. Sci.* **2012**, *47*, 8197–8215. [CrossRef]
44. Mirhosseini, M.M.; Haddadi-Asl, V.; Khordad, R. Molecular Dynamics Simulation, Synthesis and Characterization of Polyurethane Block Polymers Containing PTHF/PCL Mixture as a Soft Segment. *Polym. Bull.* **2022**, *79*, 643–661. [CrossRef]
45. Yang, Q.; Zheng, H.; Zhang, G.; Dong, B.; Chen, J.; Wang, P.; Li, M.; Hou, D. Evaluating the Thermal and Mechanical Properties of GO-Modified Epoxy/C–S–H Interface: An Atomic Investigation. *J. Mater. Sci.* **2022**, *57*, 20242–20258. [CrossRef]
46. Gu, M.; Ling, X.; Wang, H.; Dang, W.; Teng, X.; Yu, A. Experimental Investigation into Dynamic Mechanical Properties and Explosion Responses of Polyurea Elastomer. *Process Saf. Environ. Prot.* **2022**, *163*, 144–157. [CrossRef]
47. Gu, M.; Ling, X.; Yu, A.; Chen, G.; Wang, H.; Wang, H. Experimental Study of Polyurea-Coated Fiber-Reinforced Cement Boards under Gas Explosions. *Def. Technol.* **2023**, *23*, 201–213. [CrossRef]
48. Yao, K.; Chu, D.; Li, T.; Liu, Z.; Guo, B.-H.; Xu, J.; Zhuang, Z. Atomic-Scale Simulation of Hugoniot Relations and Energy Dissipation of Polyurea under High-Speed Shock. *Eng. Comput.* **2021**, *38*, 1209–1225. [CrossRef]
49. Garrett, J.T.; Runt, J.; Lin, J.S. Microphase Separation of Segmented Poly(Urethane Urea) Block Copolymers. *Macromolecules* **2000**, *33*, 6353–6359. [CrossRef]
50. Yin, C.; Zhao, X.; Zhu, J.; Hu, H.; Song, M.; Wu, S. Experimental and Molecular Dynamics Simulation Study on the Damping Mechanism of C5 Petroleum Resin/Chlorinated Butyl Rubber Composites. *J. Mater. Sci.* **2019**, *54*, 3960–3974. [CrossRef]

polymers

MDPI

Article

The Structural Design of a New Graftable Antioxidant and the Theoretical Study of Its Role in the Cross-Linking Reaction Process of Polyethylene

Yang Du [1], Hui Zhang [1,*], Chi Deng [1], Xia Du [1], Yan Shang [1], Xuan Wang [1,*], Qingguo Chen [1] and Zesheng Li [2]

1 Key Laboratory of Engineering Dielectrics and Its Application of Ministry of Education & School of Material Science and Chemical Engineering, Harbin University of Science and Technology, Harbin 150080, China; duyang950711@163.com (Y.D.); dengchi@potevio.com (C.D.); duxia62@126.com (X.D.); shangyan1972@126.com (Y.S.); qgchen@hrbust.edu.cn (Q.C.)

2 Key Laboratory of Cluster Science of Ministry of Education & School of Chemistry, Beijing Institute of Technology, Beijing 100081, China; zeshengli@bit.edu.cn

* Correspondence: huizhang@hrbust.edu.cn (H.Z.); wangxuan@hrbust.edu.cn (X.W.)

Abstract: Cross-linked polyethylene (XLPE) insulation is used in most advanced power cable technology. However, in traditional cross-linking, the conductivity of the cross-linking system sharply increases due to the presence of additives (antioxidants and cross-linked agents). Therefore, reducing the number of antioxidants to further reduce conductivity is a very promising method. The structural design of a new dual-functional antioxidant 5-allyloxy-2-hydroxyl-3-tert-butylbenzophenone (5ATB) has been established. The antioxidant behavior and grafting reaction of 5ATB after photocatalysis under ultraviolet (UV) conditions were further studied using density functional theory (DFT). The reaction potential energy information of the six reaction channels at the B3LYP/6-311+G(*d,p*) level were obtained. The calculation results indicated that the reaction Gibbs energy barrier of 5ATB with O_2 is approximately 0.48 eV lower than that of the polyethylene chain with O_2 to achieve an anti-oxidative effect. Furthermore, the reaction-active site of 5ATB accepting H is located on the C of CH_2 in a C=C double bond, as demonstrated by an analysis of NBO charge populations. The proposed mechanism has the potential to further expand the design concept of insulation materials for advanced future power cables.

Keywords: power cable; cross-linked polyethylene; antioxidant; grafting reaction; anti-oxidation mechanism

1. Introduction

In order to reduce carbon emissions, there is a necessity to transition from the use of fossil fuels to renewable energy sources, which are considered to be cleaner. However, wind and solar power are typically most abundant in unpopulated areas that are distant from the end user. Reducing power loss during transportation over hundreds to thousands of kilometers is a decisive factor. The necessity for transmission lines arises from the intermittent nature of renewable energy, which necessitates the ability to reverse the direction of energy flow. It is evident that high-voltage direct current (HVDC) transmission lines fulfill the aforementioned requirements and thus represent a pivotal component of future power grids, which are designed to facilitate the integration of renewable energy sources in a seamless manner [1]. In the event of the transmission lines traversing densely populated areas or large bodies of water, it is imperative that they be submerged or buried underground. Therefore, an insulation layer must be employed to encase the cables. The insulation layer must be of a particularly high quality in order to withstand the high transmission voltage of hundreds of kilovolts. The research endeavors in this field have been focused on the augmentation of the transmission voltage to beyond 640 kV, a feat that is presently attainable [2].

Cross-linked polyethylene (XLPE) has become a staple in the field of advanced HVDC cable insulation due to its ability to meet the stringent requirements of very low conductivity. This product is derived from raw materials characterized by exceptional physical and chemical cleanliness, ensuring its reliability and performance in critical applications. The present study demonstrates the necessity of a degassing procedure to eliminate the byproducts of dicumyl peroxide (DCP) cross-linking [3,4]. With the potential for trace residues to augment the electrical conductivity of XLPE, the degassing procedure is imperative for ensuring the quality and reliability of the final product [5–7]. The utilization of ultraviolet (UV) radiation technology devoid of the byproducts of DCP cross-linking has the potential to serve as an effective solution for the fabrication of HVDC cable insulation. This assertion is substantiated by the successful implementation of this technology in the production of low-voltage cable insulation in China [8]. A comparison of the theoretical results obtained from the UV radiation cross-linking process with those of the conventional DCP cross-linking process reveals that the reaction energy barrier of H-abstraction on the polyethylene chain by benzophenone (UV radiation cross-linking) is 0.17 eV, which is 0.08 eV lower than that by cumyl peroxide radicals (DCP cross-linking) [9,10]. This finding suggests that UV radiation cross-linking technology may offer a distinct advantage over conventional DCP cross-linking technology in XLPE production.

Polyethylene is subject to oxidation during the production, storage, and utilization phases, resulting in a decline in its mechanical, physicochemical, and electrical properties. This deterioration contributes to a reduction in the operational stability and in-service life of the cable [11,12]. Antioxidants have been shown to retard the process of oxidation, thereby prolonging the lifespan of XLPE cable insulation [13]. To further reduce conductivity, the most common strategy for reducing the antioxidant content has been proposed. During the production process, oxidation can lead to the formation of radicals in polyethylene molecular chains and increase the cross-linking density, and the migration of antioxidants can reduce service life during the application process. The grafting of antioxidants with low molecular weights onto polyethylene was modified through the application of chemical methods with the objective of averting migration. This process was undertaken to improve the thermal oxidation and aging properties of the material [13–16]. It was found from Li and co-workers that compared with the industrial-produced voltage stabilizer UV-531 (4OB), the voltage stabilizer 4-allyloxy-2-hydroxylbenzophene (4AB) had a better effect on the electrical properties of XLPE [14]. Previous studies have also demonstrated that structures with a 2-hydroxy-benzophenone configuration can scavenge free radicals in the system, serving as antioxidants [15]. Meanwhile, the results showed that the oxidative induction time (OIT) of the polyethylene physically blended with antioxidant 4,4′-thiobis (2-tert-butyl-5-methylphenol) (antioxidant 300) and frequently used in cable insulation was 98.4 min, and the OIT value of polyethylene without any antioxidant was only 0.5 min [16].

Ultraviolet (UV) absorbers, such as 2-hydroxy-benzophenone, have been shown to be capable of absorbing harmful sunlight. The energy absorbed by the absorber is then dissipated through the formation of a reversible six-membered hydrogen bond ring system [15]. This process has been demonstrated to prevent polyethylene from undergoing photo-degradation by UV, thus acting as an antioxidant. Additionally, the absorbers have been found to act as polymer chain-breaking donors [14–17]. The anti-oxidation ability of 2-hydroxy-benzophenone was enhanced by the introduction of electron-donating substituent groups to the *para*-position of the hydroxyl groups, instead of the *ortho*-position [10,17]. The grafting of a UV absorber (4-allyloxy-2-hydroxyl-benzophenone) onto polyethylene has been utilized as a method of enhancing the electrical tree initiation voltage, thereby inhibiting its propagation and augmenting its breakdown strength properties [18].

This work sets out a new graftable antioxidant, which introduces electron-donating allyloxy groups on *para*-position to the hydroxyl groups, thus facilitating oxidation. In addition, the introduction of tertiary butyl groups serves to enhance the production of semi-quinones during the oxidation process. Evidently, it would be highly desirable for the new antioxidant to be grafted to polyethylene, thereby providing long-term protection against oxidative aging and, concomitantly, serving as a voltage stabilizer to resist high voltage. The new antioxidant with dual-functional use in a polyethylene system would further decrease the chemical impurities of the insulation materials.

To the best of our knowledge, no previous theoretical investigation has been performed on the anti-oxidative behavior of antioxidants (5ATB) at molecular and atom levels. We aim to design a new antioxidant and investigate the anti-oxidative reaction mechanisms and the grafting to polyethylene. The elucidation of the reaction mechanism is conducive to the design of highly efficient antioxidant molecules.

2. Computational Methods

In this work, the geometry structure optimization and frequency calculation of the stationary points for the studied six reaction channels on the ground state S_0 (or the triplet state T_1) were carried out by density functional theory (DFT) [18] at the B3LYP/6-311+G(*d,p*) level [19–22]. The B3LYP method was chosen in our present research based on the calculated values of the vibrational frequencies of 4-methylheptane and benzene at the B3LYP/6-311+G(*d,p*) level being in good agreement with the corresponding experimental data. The time-dependent density functional theory (TDDFT) method [23,24] was employed to calculate the excitation energies of the designed antioxidant molecules on the basis of the optimized geometries at the same level. The potential energy surface information was obtained along the minimum energy path (MEP) by the intrinsic reaction coordinate (IRC) theory with a gradient step-size of 0.05 $(amu)^{1/2}$ Bohr [25]. Then, the first and second energy derivatives were obtained to calculate the curvature and the generalized vibrational frequencies along the reaction path. The values of E_g, IP(a), and EA(a) were obtained based on the calculation results of the electronic structure. E_g refers to the energy gaps between the highest occupied molecular orbital (HOMO) and the lowest unoccupied molecular orbital (LUMO). IP(a) and EA(a) refer to the adiabatic ionization potentials and electron affinity energies, respectively. The natural charge population was analyzed by the natural bond orbital (NBO) method [26] based on the optimized geometries. All the electronic structure calculations of the stationary points were completed by the GAUSSIAN09 program package [27].

3. Results and Discussion

Aromatic carbonyl compounds characteristically undergo single electron excitation. The initial excited states are triplets [28,29], and a minimum energy crossing point has been theoretically identified [30]. In this work, the investigation of oxidative behavior during the UV radiation cross-linking process has been completed on the lowest triplet state. The molecular formulae, molecular names, and corresponding abbreviations of the studied molecules are listed in Table 1. 4-methylheptane (Pe) is selected as a model molecule of cross-linkable polyethylene. The newly designed antioxidant is named 5-allyloxy-2-hydroxyl-3-tert-butylbenzophenone (5ATB). In the course of the UV radiation cross-linking process, the ground state (S_0) 5ATB is be excited to the singlet excited state S_1 (n, π^*) (with a calculated value of 3.14 eV) through the absorption of energy emitted from the optical system in the form of UV, and subsequently to the triplet excited state T_1 (n, π^*) (with a calculated value of 0.99 eV) by means of intersystem crossing (ISC).

Table 1. The molecular names, the molecular formulae, and the corresponding abbreviations (ab.) of the studied molecules.

ab.	Molecular Formula	Molecular Name
Pe	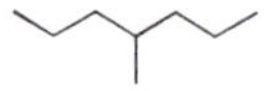	4-methylheptane (model molecule of polyethylene)
4OB	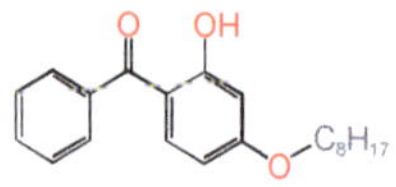 	2-hydroxyl-4-*n*-octoxybenzophenone (ultraviolet absorber UV-531)
4AB	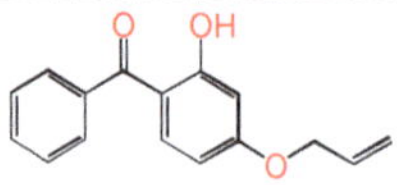 	4-allyloxy-2-hydroxylbenzophenone (graftable UV absorber as voltage stabilizer)
5ATB	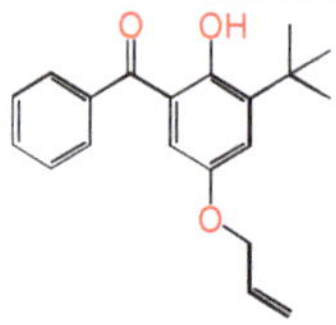 	5-allyloxy-2-hydroxyl-3-tert-butylbenzophenone (designed as a new graftable antioxidant)

Table 2 also lists the Gibbs free energy of the six reaction channels at the B3LYP/6-311+G (*d,p*) level at 298 K (ΔG^0) and the reaction Gibbs energy barrier heights ($\Delta G^{\neq}$) of the transition states (Supplementary Materials). Among them, the reaction channels ①, ①-1, and ①-2 are oxidative processes that occur during the preparation of polyethylene. It is widely acknowledged that polyethylene is susceptible to oxidation during its preparation (①). The free radicals of polyethylene and hydrogen peroxide, formed by the reaction of oxygen and polyethylene after hydrogen extraction, automatically combine (①-1). However, this peroxide is not stable, and the peroxide bond is prone to homolytic cleavage under high temperatures, resulting in the formation of oxide free radicals and hydroxyl free radicals (①-2). To prevent thermal oxidative aging, antioxidants must be introduced. Antioxidants 4AB and 5ATB react with O_2 to exert antioxidant functions (② and ③). At the same time, antioxidant 5ATB also plays a repairing role in the cross-linking process, repairing excess free radicals in the cross-linking system to avoid affecting the cross-linking process of XLPE (④ to ⑥-1).

3.1. Stationary Point Geometries and NBO Charge Population

The optimized geometry structures of the stationary points of the six studied reaction channels, including the reactants, transition states, and products, are presented in Figure 1. The overall reaction process is illustrated in Figure 2, encompassing the oxidative reaction of polyethylene (ab. RH), the antioxidant behavior of the newly designed antioxidant (ab. AH), and the homolytic cleavage of the peroxide bond. The following table (see Table 2) presents a comprehensive listing of the bond lengths of the breaking and forming bonds (b/f) of the six transition states. Additionally, the corresponding bond lengths in equilibrium of the geometry structures of the reactants and products are also given. The labels of the reaction channels and transition states are consistent, with the reactants and products being abbreviated to the corresponding R and P, respectively.

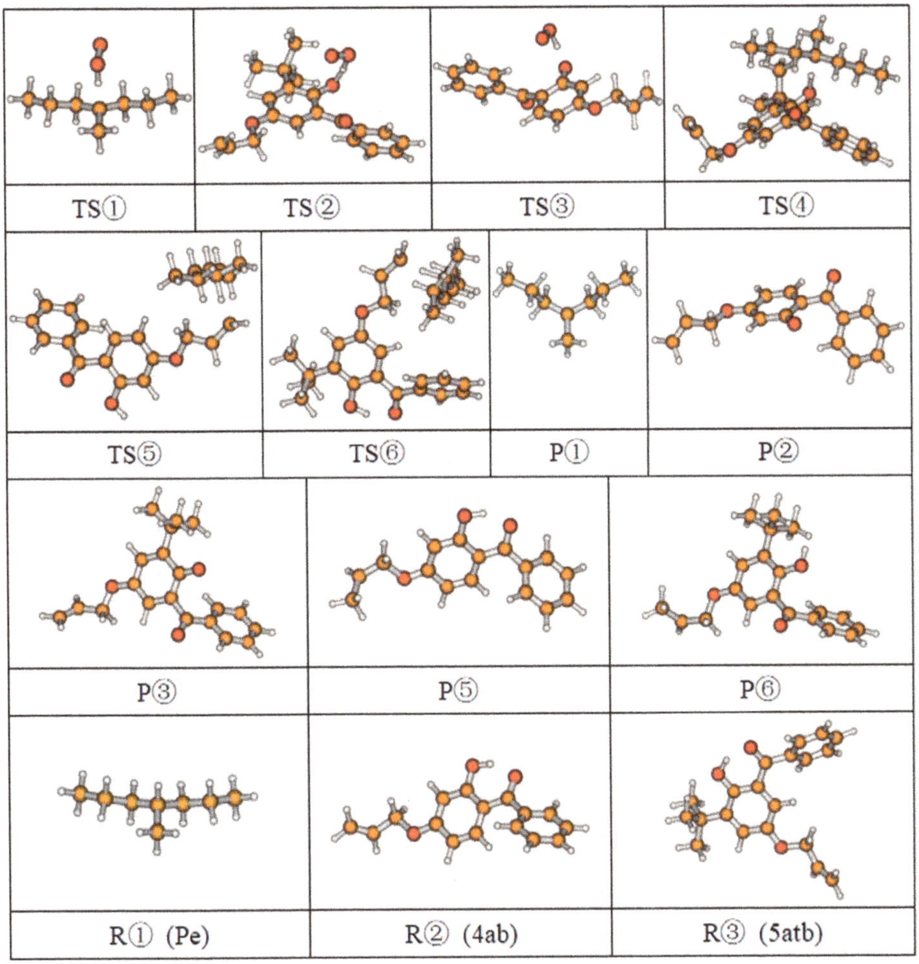

Figure 1. Optimized geometric structures of the six transition states at the B3LYP/6-311+G(*d*,*p*) level.

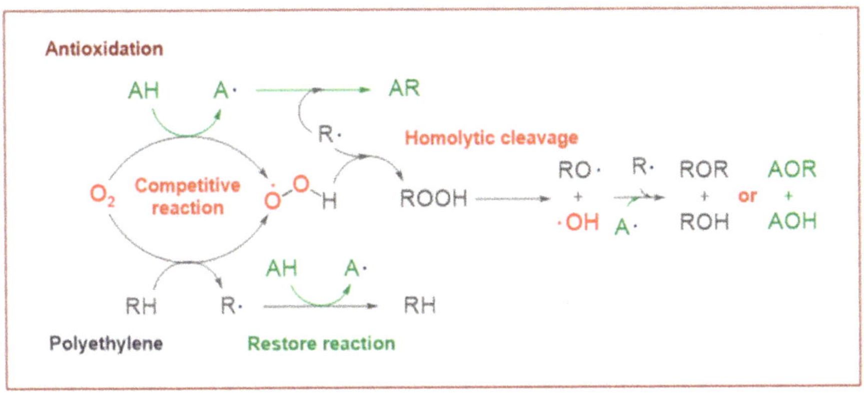

Figure 2. The overall reaction process of the newly designed antioxidant.

Table 2. Optimized geometric structures (in angstrom) of transition states, reaction Gibbs energy barriers ($\Delta G^{\neq}$), and reaction Gibbs free energies, ΔG^0 (in eV).

	Reaction Equation	Reactant	b/f	Product	$\Delta G^{\neq}$	ΔG^0
①	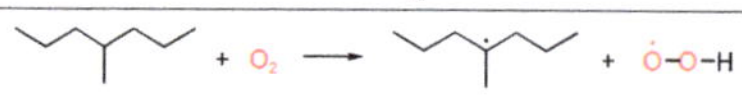	1.100	1.606/1.076	0.977	1.96	1.72
①-1	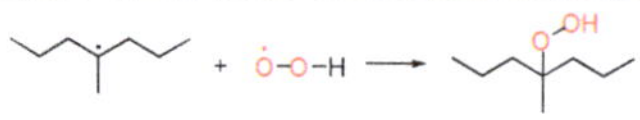	two radicals combined				
①-2	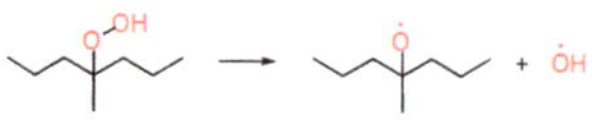	peroxide bond homolytic cleavage				
②	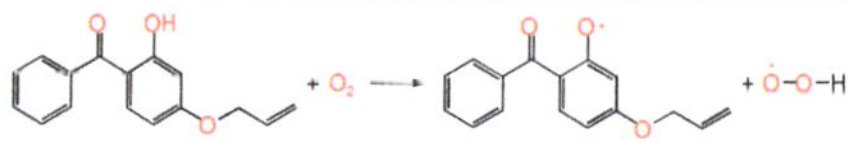	0.963	1.341/1.089	0.977	1.86	1.79
③	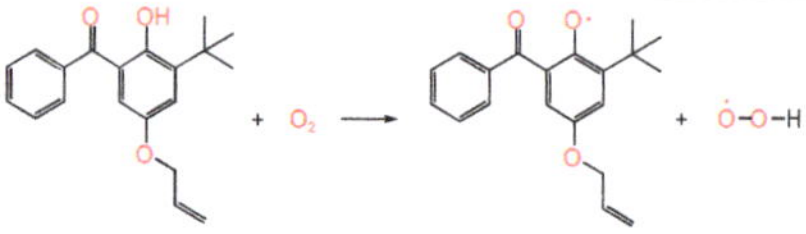	0.991	1.286/1.121	0.977	1.48	1.40
③-1	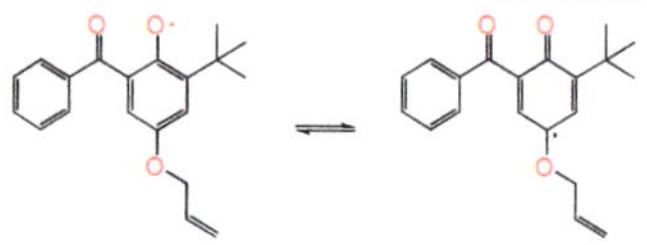	isomerization which is the same as keto-enol tautomerism				
④	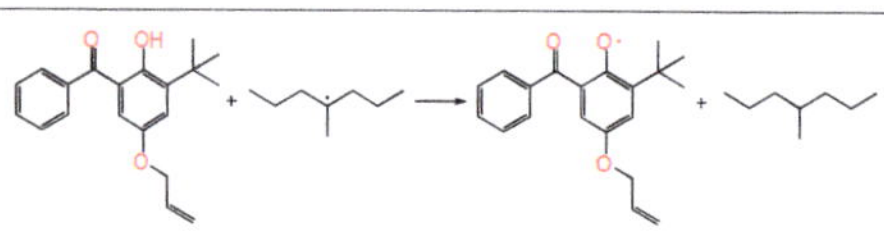	0.963	1.243/1.350	1.100	1.06	−3.31
④-1	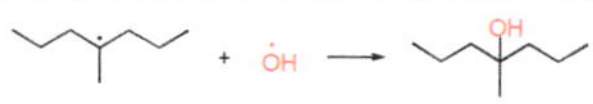	two radicals combined (it can also be Pe radicals or PeO radicals)				
④-2	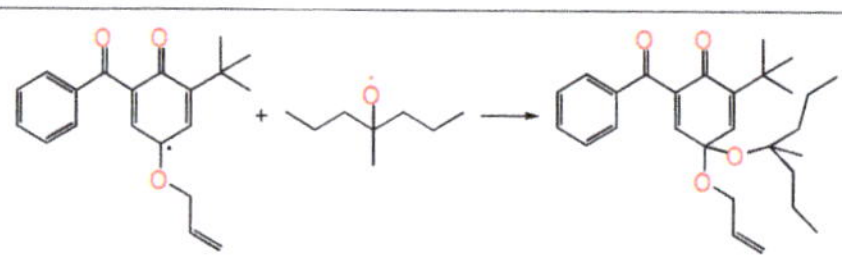	two radicals combined (it can also be OH radicals or Pe radicals)				
⑤	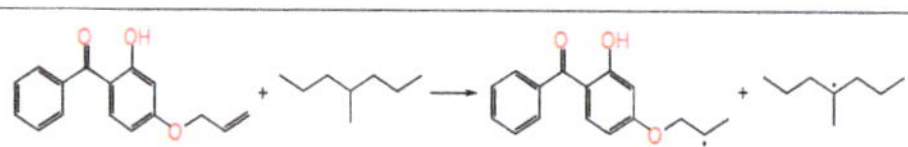	1.100	1.328/1.420	1.101	0.98	−0.96
⑥	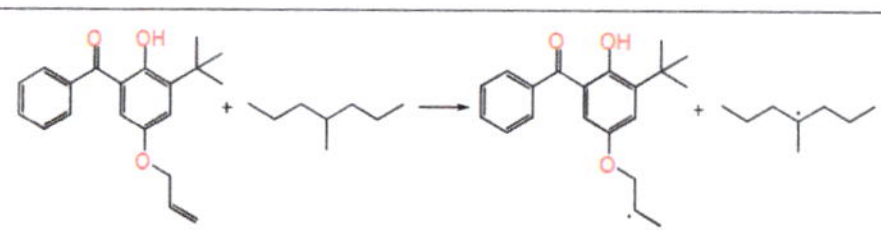	1.100	1.329/1.420	1.101	0.63	−0.46
⑥-1	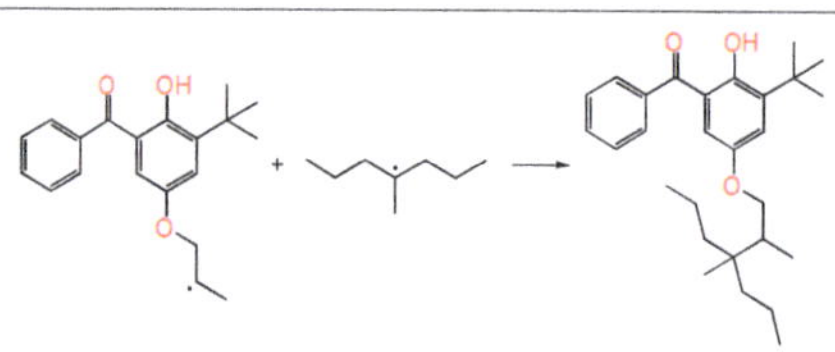	two radicals combined				

As illustrated in Table 2, the transition state structures of the H-abstraction reaction by O_2 channels TS①, TS②, and TS③ exhibit a shared characteristic, namely, the elongation of the breaking bonds exceeds that of the forming bonds in their equilibrium molecules, respectively. This finding indicates that the H-abstraction reactions of Pe, 4AB, and 5ATB by O2 are all produce-like processes. It is evident that the aforementioned reaction channels will process via "later" transition states, which are consistent with Hammond's postulate [31], as applied to endothermic reactions. The NBO calculation results for the six sites of the 4AB and 5ATB molecules are listed in Table 3. As illustrated in Table 3, the natural charge density on the H of the hydroxyl group (OH) in molecules 4AB and 5ATB is higher. This suggests that the covalent bonds between O and H are more susceptible to rupture, resulting in the formation of H radicals. The higher reactivity of the H of the hydroxyl group (OH) is the underlying reason for the antioxidant's ability to function as an anti-oxidative. It has been established that the natural charge density on the C of CH_2 in the C=C double bond of the 4AB and 5ATB molecules is higher than that of the C of CH in the C=C double bond. Furthermore, the negative charge on the C of CH_2 in the C=C double bond is larger, and as a result, its capability of accepting H is also greater. The C of CH_2 in the C=C double bond shows higher reactivity than the C of CH in the C=C double bond. It can be deduced that the grafting reaction of 4AB and 5ATB to the cross-linkable polyethylene molecule chain Pe would be initiated at the C site of CH_2 in the C=C double bond.

Table 3. Natural charge population on the T_1 state of 4AB and 5ATB molecules.

States	Molecular Formula	Natural Charge Population			
		H of OH	O of OH	C of CH on C=C	C of CH_2 on C=C
T_1	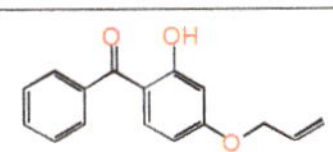	0.472	−0.638	−0.196	−0.348
	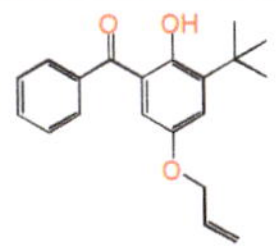	0.479	−0.606	−0.197	−0.348
S_0	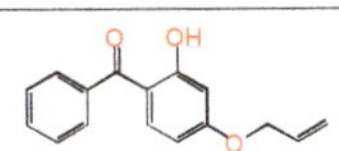	0.504	−0.671	−0.193	−0.353
	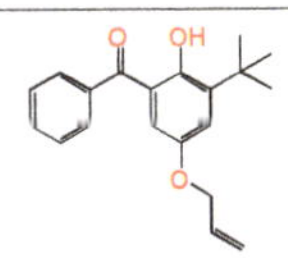	0.503	−0.686	−0.189	−0.359

3.2. Frontier MOs

The calculated values of adiabatic ionization potentials IP(a) and the adiabatic electron affinity energy EA(a), together with the relevant experimental data [32] (in brackets), are listed in Table 4, as well as the values of the HOMO-LUMO gap (E_g).

For the newly designed antioxidant 5ATB (see Table 4), the E_g values of the aryl molecules (Bp, 4OB, 4AB, and 5ATB) are lower than those of the alkyl Pe molecule. It is evident that the introduction of a carbonyl group results in the formation of a π bond with the benzene ring. Conjugation effects occur between the benzene ring and the carbonyl group within the aryl molecules. In comparison with the σ bonds of Pe (E_g = 8.38 eV), it can be observed that the π bonds of aryl molecules possess higher HOMO and lower LUMO. This finding indicates that the E_g value of aryl molecules is relatively low, with 5ATB exhibiting the lowest E_g value of 3.66 eV.

Table 4. The calculated E_g, IP(a), and EA(a) of the studied molecules, as well as the corresponding experimental data in brackets (in eV).

ab.	Molecular Formula	E_g	IP(a)	EA(a)
Pe		8.38	9.41	−1.09
Bp	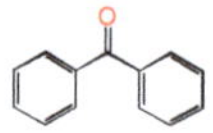	4.90	8.64(9.05)	0.73(0.69 ± 0.05)
4OB	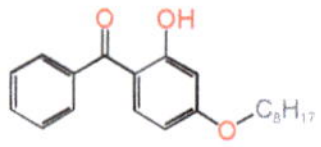	4.32	7.76	0.85
4AB	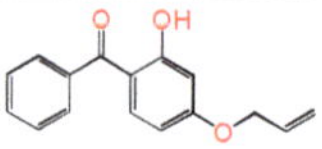	4.34	7.82	0.88
5ATB	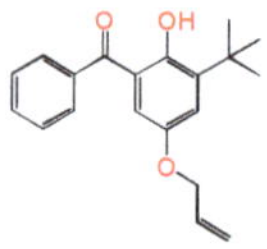	3.66	7.18	1.05

For the newly designed antioxidant 5ATB, as illustrated in Table 4, the IP(a) value is found to be less than that of Pe, which is due to the σ electron of Pe being in the sp^n hybrid orbit, which is more strongly attracted to the atom's nucleus and is not easy to dissociate. It has been established that, in comparison with the σ electron, the π electron of 5ATB is more readily dissociated due to its location in the p orbit, which is a greater distance from the nucleus. The IP(a) values of Bp, 4OB, 4AB, and 5ATB are all found to be less than Pe (9.41 eV), and the IP(a) value of 5ATB (7.18 eV) is the lowest of all.

For the newly designed antioxidant 5ATB in Table 4, the value of EA(a) is larger than Pe, as the heteroatom O is introduced in the aryl conjugate system. It is evident that the heteroatom O has the greater electronegativity, leading to a more potent binding force for the electron. There is the presence of large conjugated π bonds between the benzene ring and the carbonyl group in 5ATB, and tertiary butyl is the electron donating group. This results in an augmentation of the electron cloud density on the benzene ring, thereby facilitating electron acceptance. Furthermore, the distribution of the electronic cloud density of the 4AB molecule is not uniform, with a molecule density primarily at the 3- and 5-position, and the alkoxy is linked to the *meta*-position of the 2-hydroxyl groups. By contrast, the distribution of the electronic cloud density of the 5ATB molecule is uniform, with a molecule density primarily at the 4- and 6-position, and the tertiary butyl and alkoxy being linked to the *ortho*- and *para*-position of the 2-hydroxyl groups, respectively.

It is reported that polycyclic aromatic compounds and benzil-like structures are well suited to the role of voltage stabilizers [33,34]. It has been documented that the ultraviolet absorber (4AB) can be grafted onto polyethylene in order to serve as a voltage stabilizer, thereby enhancing its breakdown strength [17]. It is reported that voltage stabilizers have the capacity to capture high-energy electrons (termed "hot electrons"). These stabilizers then dissipate the energy of these electrons through impact, releasing non-harmful electrons of a lower energy. This process is intended to prevent the degradation of the polymer matrix. The EA(a) value of 5ATB (1.05 eV) is the largest. The results of this study indicate that the designed antioxidant 5ATB has the potential to serve as a voltage stabilizer, thereby providing a high voltage resistance performance.

3.3. Energetics

The reaction Gibbs free energies (ΔG^0) and the reaction Gibbs energy barriers ($\Delta G^{\neq}$) for the six reaction channels are also listed in Table 2. As demonstrated in Figure 2, the reaction process encompasses polyethylene cross-linking, oxidative reactions, anti-oxidative behavior, and the grafting reaction to polyethylene by UV radiation previously mentioned. From Table 2, it can be seen that the reaction process mainly includes four types of reactions, namely the oxidative reaction, the restore reaction of Pe radicals, the isomerization reaction which is the same as keto-enol tautomerism, and the grafting reaction.

3.3.1. Oxidative Reaction

In the course of the research, it was established that the oxidative reaction of the newly designed antioxidant 5ATB (reaction channel ③) is easier than that of Pe (reaction channel ①). This is evidenced by the data in Table 2, which shows that the reaction Gibbs energy barrier $\Delta G^{\neq}{}_{TS③}$ (1.48 eV) is less than that of $\Delta G^{\neq}{}_{TS①}$ (1.96 eV). The newly designed antioxidant 5ATB can undergo a reaction with O_2 forming relatively stable radicals before the polyethylene chain oxidation, which prevents the oxidation of the polyethylene chain. The lower reaction Gibbs energy barrier of the antioxidant is the reason why the antioxidant can play an anti-oxidative role. This is why the OIT value of antioxidant 300, which is frequently used in cable insulation, is 196 times higher than that without any antioxidants [15]. This is consistent with the analytic results of the natural charge density above. The calculated breaking bond dissociation energies (D^{o}_{298}) are 3.55 and 3.91 eV for 5ATB and Pe, respectively, at the B3LYP/6-311+G(*d*,*p*) level. The dissociation energy of the bonds is closely related to the corresponding reaction Gibbs energy barrier heights. The bond dissociation energy of the H atom in the hydroxyl group of the novel antioxidant is lower than that of Pe by approximately 0.36 eV. This means that the reactivity of the H of the hydroxyl groups on 5ATB is comparatively higher than that of Pe. The two radicals of the oxidative products combine to form peroxides (①-1 channel), and the peroxide bond homolytic cleavage to form the other two radicals (①-2 channel). Antioxidants are frequently employed to eliminate free radicals in XLPE insulation materials for high voltage cables. The reaction Gibbs energy barrier $\Delta G^{\neq}{}_{TS③}$ of 1.48 eV is lower than $\Delta G^{\neq}{}_{TS②}$ of 1.86 eV. This finding is consistent with the report that the anti-oxidation ability of 2-hydroxy-benzophenone was improved by the introduction of electron-donating substituent groups to the *para* position of the hydroxyl groups [17].

3.3.2. Restore Reaction

For the reaction channel ④, the aryl 5ATB possess a stronger π-electron delocalization ability than that of alkyl Pe, and the aromatic molecules are chosen as additives added in the XLPE material system, which can restore alkyl radicals through transforming the alkyl radicals to relatively stable aryl radicals. The restore reaction Gibbs energy barrier is 1.06 eV, which can prevent the formation of excessive amounts of polyethylene radicals in the cross-linking system, decrease the cross-linking density, and prevent the breakdown of polyethylene chains during service (the average carbon–carbon single bond energy is 3.60 eV). The two radicals, Pe and OH, combine to form the inactive product alcohol (reaction channel ④-1).

3.3.3. Isomerization Reaction

The reaction channel ③-1 is an isomerization reaction that is analogous to the keto-enol tautomerism, wherein phenolic hydroxyl radicals are transferred to quinone-like structures. This reaction is reversible. The relatively stable 5ATB radicals can combine with the product of the ①-2 channel, forming the inactive product ether (④-2 channel), and can also be combined by OH radicals or Pe radicals to form inactive products.

3.3.4. Grafting Reaction

Compared with the reaction channel ③, we can find from the calculated results in Table 2 that the reaction Gibbs energy barrier $\Delta G^{\neq}{}_{TS⑥}$ (0.63 eV) is lower than $\Delta G^{\neq}{}_{TS③}$ (1.48 eV), and that the reaction channel ⑥ is more advantageous in kinetics. This means that the grafting reaction ⑥ occurs before the oxidative reaction ③. In accordance with the analysis of natural charge density in Table 3, the grafting reaction of 5ATB to Pe can be initiated at the C site of CH_2 in the C=C double bond. A similar case is also seen for the reaction channels ⑤ and ②. The grafting reaction of 4AB to the cross-linkable polyethylene molecule chain Pc can be initiated at the same C site of CH_2 in the C=C double bond. The grafted antioxidants would prevent their own migration and provide long-term protection against oxidative aging resistance.

4. Conclusions

The theoretical calculation results of density functional theory were used to propose the antioxidant mechanism and grafting behavior of the newly designed antioxidant 5ATB after photocatalysis. This is the first time that these mechanisms have been proposed. The reaction potential energy information of the six reaction channels are obtained at the B3LYP/6-311+G(*d,p*) level. The analysis of frontier MOs, NBO charge populations, and the reaction potential energy information also confirmed that the designed 5ATB has dual functions of antioxidation and stabilizing voltage. The underlying mechanism of the designed antioxidant's anti-oxidative effect was elucidated. The Gibbs energy barrier of the reactive antioxidant 5ATB with O_2 was found to be approximately 0.48 eV lower than that of the polyethylene chain with O_2. Additionally, the antioxidant 5ATB exhibited a more pronounced capacity to restore the polyethylene chains. The accepting H reaction of the C-site of CH_2 in the C=C double bond exhibits elevated reactivity, with a lower reaction Gibbs energy barrier height of 0.63 eV, as observed in the grafting of the antioxidant 5ATB onto polyethylene. The use of dual-functional antioxidants has been proven to reduce the introduction of chemical impurities and the electronic conductivity of insulating materials. The designed antioxidant 5ATB has the potential to exhibit excellent electrical performance and meet the requirements of insulation materials for future advanced power cables.

Supplementary Materials: The following supporting information can be downloaded at: https://www.mdpi.com/article/10.3390/polym17040546/s1, Text: The optimized standard orientations of stationary points (reactants, products and transition states) of the six reaction channels at the B3LYP/6-311+G(*d,p*) level.

Author Contributions: All authors contributed to the study conceptualization. Y.D. and C.D. conducted investigation and visualization. H.Z. is responsible for funding acquisition, project administration, and writing—review and editing. X.D. and Y.S. are responsible for formal analysis and methodology. X.W. and Q.C. conducted supervision and validation. Z.L. is responsible for data curation, resources, and software. The writing—original draft of the manuscript was written by Y.D. and all authors commented on previous versions of the manuscript. All authors have read and agreed to the published version of the manuscript.

Funding: This work is supported by the Key Project of the Joint Fund for Regional Innovation and Development of National Natural Science Foundation of China (Grant number U20A20307).

Institutional Review Board Statement: Not applicable.

Data Availability Statement: The original contributions presented in this study are included in the article/Supplementary Materials. Further inquiries can be directed to the corresponding authors.

Conflicts of Interest: The authors declare no conflicts of interest.

References

1. Mazzanti, G.; Marzinotto, M. *Extruded Cables for High-Voltage Direct-Current Transmission*; Wiley: Hoboken, NJ, USA, 2013.
2. Montanari, G.C.; Morshuis, P.H.F.; Zhou, M.; Stevens, G.C.; Vaughan, A.S.; Han, Z.; Li, D. Criteria influencing the selection and design of HV and UHV DC cables in new network applications. *IET High Volt.* **2018**, *3*, 90–95. [CrossRef]
3. Meng, F.B.; Chen, X.R.; Hong, Z.L.; Shi, Y.W.; Zhu, H.S.; Muhammad, A.; Paramane, A.; Huang, R.B.; Han, Z. Insulation Properties and Interfacial Quantum Chemical Analysis of Cross-Linked Polyethylene Under Different Degassing Time for HVDC Cable Factory Joint Applications. *IEEE Trans. Dielectr. Electr. Insul.* **2023**, *30*, 271–278. [CrossRef]
4. Mcnulty, M. Cross-Linked Polyethylene Compound for Cable Systems. In *Wire & Cable Technology International: Serving Manufacturers, Specifiers and Users of Wire and Cable*; Wire & Cable Technology International (WCTI): Akron, OH, USA, 2024; Volume 52.
5. Pourrahimi, A.M.; Pitois, C.; Abbasi, A. XLPE high voltage insulation; A link between DC conductivity and microstructure. *Polym. Test.* **2024**, *131*, 108330. [CrossRef]
6. Ahmed, M.; Lisheng, Z.; Li, F.; Xu, N.; Ren, H. DC Conductivity Fluctuation Due to Temperature Dependence and Cross-linking Byproducts of XLPE Insulation Material in HVDC Cables. In Proceedings of the 2021 IEEE Electrical Insulation Conference (EIC), Denver, CO, USA, 7–28 June 2021; pp. 173–176. [CrossRef]
7. Chen, W.; Yang, Z.; Liu, X.; Wang, Y. Influence of type, content and degassing time of crosslinking agent on electrical tree characteristics of XLPE in high voltage cables. In Proceedings of the 22nd International Symposium on High Voltage Engineering, Xi'an, China, 21–26 November 2021; pp. 1623–1626. [CrossRef]
8. Qu, B.; Bao, W.; Wu, Q.; Shi, W. Recent Developments on photoinitiated crosslinking of polyethylene and its applications for manufacturing insulated wire and cable. In Proceedings of the 2009 IEEE 9th International Conference on the Properties and Applications of Dielectric Materials, Harbin, China, 19–23 July 2009; pp. 33–36. [CrossRef]
9. Zhang, H.; Shang, Y.; Li, M.X.; Zhao, H.; Wang, X.; Han, B.Z. Theoretical study on the radical reaction mechanism in the cross-linking process of polyethylene. *RSC Adv.* **2015**, *5*, 90343–90353. [CrossRef]
10. Zhang, H.; Deng, C.; Du, X.; Shang, Y.; Zhao, H.; Wang, X.; Han, B.Z.; Li, Z.S. Theoretical study on the grafting reaction of maleimide containing 2-hydroxy-benzophenone onto polyethylene. *J. Mol. Model.* **2021**, *27*, 259. [CrossRef] [PubMed]
11. Li, J.C.; Si, Z.C.; Wang, S.H.; Li, S.T.; Zhou, H.; Liu, J.J. Effect of hindered phenolic antioxidants on crosslinking characteristics of low-density polyethylene initiated by peroxide. *Energy Rep.* **2023**, *9*, 159–166. [CrossRef]
12. Sahoo, R.; Karmakar, S. Impact of Accelerated Thermal Aging on Electrical Tree Structure and Physicochemical Characteristics of XLPE Insulation. *IEEE Trans. Dielectr. Electr. Insul.* **2024**, *31*, 429–438. [CrossRef]
13. Deng, C.; Zhang, H.; Du, Y.; Du, X.; Shang, Y.; Yang, H.D.; Wang, X.; Chen, Q.G.; Li, Z.S. Theoretical Study of the Grafting Reaction of a New Antioxidant to Cross-Linked Polyethylene and the Antioxidation Mechanism. *Int. J. Quantum Chem.* **2024**, *124*, e27492. [CrossRef]
14. Li, C.Y.; Zhao, H.; Han, B.Z.; Zhang, H.; Zhang, C.C.; Ai, Y. The Effect of Voltage Stabilizers on the DC Insulation Performance of XLPE. *Chin.S Oc Elec. Eng.* **2018**, *38*, 7071–7079.
15. Allen, N.S. Photostabilising action of ortho-hydroxy aromatic compounds: A critical review. *Polym. Photochem.* **1983**, *3*, 167–187. [CrossRef]
16. Zhang, C.C.; Wang, T.T.; Sun, W.F.; Li, C.Y.; Zhao, H. Grafting of antioxidant onto polyethylene to improve DC dielectric and thermal aging properties. *IEEE Trans. Dielectr. Electr. Insul.* **2021**, *28*, 541–549. [CrossRef]
17. Dobashi, Y.; Kondou, J.; Ohkatsu, Y. Photo-antioxidant abilities of 2-hydroxybenzoyl compounds. *Polym. Degrad. Stab.* **2005**, *89*, 140–144. [CrossRef]
18. Li, C.Y.; Zhang, C.C.; Zhao, H.; Zhang, H.; Wang, X.; Han, B.Z. Grafted UV absorber as voltage stabilizer against electrical degradation and breakdown in cross-linked polyethylene for high voltage cable insulation. *Polym. Degrad. Stab.* **2021**, *185*, 109498. [CrossRef]
19. Truong, T.N.; Duncan, W.T.; Bell, R.L. *Chemical Applications of Density Functional Theory*; American Chemical Society: Washington, DC, USA, 1996. [CrossRef]
20. Lee, C.; Yang, W.; Parr, R.G. Development of the Colle-Salvetti conelation energy formula into a functional of the electron density. *Phys. Rev. B* **1988**, *37*, 785–789. [CrossRef] [PubMed]
21. Miehlich, B.; Savin, A.; Stoll, H.; Preuss, H. Results obtained with the correlation energy density functionals of becke and Lee, Yang and Parr. *Chem. Phys. Lett.* **1989**, *157*, 200–206. [CrossRef]
22. Becke, A.D. Density-functional thermochemistry. III. The role of exact exchange. *J. Chem. Phys.* **1993**, *98*, 5648–5652. [CrossRef]

23. Zangwill, A.; Soven, P. Density-functional approach to local-field effects in finite systems: Photoabsorption in the rare gases. *Phys. Rev. A* **1980**, *21*, 1561–1572. [CrossRef]
24. Levine, Z.H.; Soven, P. Time-dependent local-density theory of dielectric effects in small molecules. *Phys. Rev. A* **1984**, *29*, 625–635. [CrossRef]
25. Hratchian, H.P.; Schlegel, H.B. Accurate reaction paths using a hessian based predictor-corrector integrator. *J. Chem. Phys.* **2004**, *120*, 9918–9924. [CrossRef]
26. Reed, A.E.; Weinstock, R.B.; Weinhold, F. Natural population analysis. *J. Chem. Phys.* **1985**, *83*, 735–746. [CrossRef]
27. Frisch, M.J.; Trucks, G.W.; Schlegel, H.B.; Scuseria, G.E.; Robb, M.A.; Cheeseman, J.R.; Scalmani, G.; Barone, V.; Mennucci, B.; Petersson, G.A.; et al. *Revision A.02*; Gaussian, Inc.: Wallingford, CT, USA, 2009.
28. Rubio-Pons, Ó.; Loboda, O.; Minaev, B.; Schimmelpfennig, B.; Vahtras, O.; Ågren, H. CASSCF calculations of triplet-state properties applications to benzene derivatives. *Mol. Phys.* **2003**, *101*, 2103–2114. [CrossRef]
29. Minaev, B.F.; Knuts, S.; Ågren, H.; Vahtras, O. The vibronically induced phosphorescence in benzene. *Chem. Phys.* **1993**, *175*, 245–254. [CrossRef]
30. Fang, W.H. Ab initio determination of dark structures in radiationless transitions for aromatic carbonyl compounds. *Acc. Chem. Res.* **2008**, *41*, 452–457. [CrossRef]
31. Hammond, G.S. A correlation of reaction rates. *J. Am. Chem. Soc.* **1955**, *77*, 334–338. [CrossRef]
32. Lias, S.G.; Levin, R.D.; Kafafi, S.A.; Bartmess, J.E. *Nist Chemistry Web Book, Nist Standard Reference Database Number 69*; NIST: Gaithersburg, MA, USA, 1998. [CrossRef]
33. Yamano, Y. Roles of polycyclic compounds in increasing breakdown strength of LDPE film. *IEEE Trans. Dielectr. Electr. Insul.* **2006**, *13*, 773–781. [CrossRef]
34. Zhang, H.; Shang, Y.; Zhao, H.; Wang, X.; Han, B.Z.; Li, Z.S. Theoretical study on the tailored side-chain architecture of benzil-like voltage stabilizers for enhanced dielectric strength of cross-linked polyethylene. *RSC Adv.* **2016**, *6*, 11618. [CrossRef]

polymers

MDPI

Article

Dielectric Constant Estimation of Spherical Particle-Filled Nanocomposites Based on the Poon and Shin Model, Considering Interphase Properties

Bin Tang [1,2], Xue Liu [3,*], Shengxiang Deng [4], Wei Zhong [1,*] and Jiang Shao [1,*]

1 School of Mechanical Engineering, Jiangsu University of Science and Technology, Zhenjiang 212100, China; tangbin29@21cn.com
2 The School of Optoelectronic Manufacturing, Zhejiang Industry and Trade Vocational College, Wenzhou 325002, China
3 Henan Key Laboratory of Polyoxometalate Chemistry, College of Chemistry and Chemical Engineering, Henan University, Kaifeng 475004, China
4 Department of Energy and Power Engineering, School of Mechanical and Automotive Engineering, Shanghai University of Engineering Science, Shanghai 201620, China; csdsx@163.com
* Correspondence: xliu@henu.edu.cn (X.L.); zhongwei@just.edu.cn (W.Z.); jiangshao@just.edu.cn (J.S.)

Abstract: A revised version of the Poon and Shin (PS) model, incorporating the effects of the interphase, is introduced to predict the dielectric permittivity of polymer nanocomposites reinforced with spherical nanoparticles. In this modified approach, both the spherical nanoparticle and its surrounding interphase region are treated as an equivalent nanoparticle, modeled as a core–shell structure. This assumption enables a more accurate representation of the composite, where the polymer matrix and the equivalent nanoparticles form a homogeneous mixture. The process of calculating the dielectric permittivity of the composite occurs in two distinct steps. Initially, the dielectric permittivity of the equivalent particle—comprising both the nanoparticle core and its interphase—is computed. Subsequently, the overall dielectric permittivity of the composite material is determined, considering the properties of the polymer substrate and the equivalent nanoparticles, all within the framework of the modified PS model. To verify the validity of the proposed model, experimental data are compared against the predicted values, showing a high level of agreement when the interphase characteristics are appropriately incorporated. Additionally, the influence of various factors, including the properties of the spherical nanoparticles, the interphase, and the polymer matrix, on the dielectric performance of the nanocomposite is thoroughly investigated. This enhanced PS model offers a valuable theoretical framework for designing polymer–spherical nanoparticle composites with superior dielectric properties, paving the way for their potential application in advanced electronic and energy storage devices.

Keywords: modeling; interphase properties; dielectric constant; polymeric nanocomposite

1. Introduction

In recent years, dielectric polymer materials have gained considerable attention due to their potential applications in a wide range of advanced technologies, particularly in energy storage systems, flexible electronics, and wearable devices [1–4]. These materials are widely used as substrates in high-performance capacitors, sensors, and transistors due to their lightweight nature, mechanical flexibility, and ease of processing [5–7]. Some of the most commonly employed polymer matrices include polyimide (PI) [8], poly(vinylidene fluoride) (PVDF) [9], and polydimethylsiloxane (PDMS) [10], among others. Despite their excellent mechanical strength, flexibility, and chemical stability, these polymeric materials typically exhibit low dielectric constants, which significantly hinder their effectiveness in energy storage and electronic applications [11,12]. This limitation has prompted extensive research efforts aimed at enhancing the dielectric properties of polymer-based materials. Strategies such as incorporating high-dielectric nanofillers, modifying polymer structures at the molecular level, and optimizing interfacial interactions have been explored to improve their overall dielectric performance [13–16]. By overcoming these challenges, dielectric polymers with superior properties can be developed, paving the way for next-generation electronic and energy storage technologies.

One effective approach to improving the dielectric performance of polymer nanocomposites is the incorporation of high-dielectric nanoparticles into polymer matrices. Representative nanoparticles commonly used for this purpose include barium titanate (BT) and aluminum oxide (Al_2O_3) [17,18]. For example, Goyal et al. reported the development of BT-filled polymethylmethacrylate (PMMA) composites, which exhibited significantly improved dielectric properties compared to the pure PMMA matrix [19]. The maximum dielectric constant of the BT-PMMA composite reached 69 at a filler content of 65 vol.%. Kim et al. synthesized a high-energy-density poly(vinylidene fluoride-co-hexafluoropropylene) nanocomposite reinforced with BT nanoparticles, achieving a maximum dielectric constant of 35 with 50 vol.% filler content [20]. Zahhaf et al. introduced structured Al_2O_3-PDMS composites which demonstrated superior dielectric properties over their unstructured counterparts [21]. The dielectric constant of the structured Al_2O_3-PDMS composites increased with the Al_2O_3 volume fraction, reaching a dielectric permittivity of 6.25 at an Al_2O_3 content of 40 vol.%.

In the nanocomposites reported in the literature, nanofillers are typically dispersed within the polymer matrix, where they are enveloped by the surrounding polymer material. However, due to their inherently high surface area, nanoparticles tend to aggregate rather than remain uniformly separated, a phenomenon that leads to a significant deterioration in both the mechanical and electrical properties of the composite [22–24]. To counteract this, various techniques, such as ultrasonication and roll milling, are employed to disrupt these agglomerations and facilitate better dispersion [6,25,26]. Furthermore, an interphase region forms at the interface between the nanofillers and the polymer matrix, arising from the interactions between the two phases [27]. This interphase zone, exhibiting unique properties that distinguish it from both the polymer matrix and the nanofillers, is known to have a profound impact on the overall mechanical and electrical performance of the nanocomposite [28–30]. However, the characterization of this interphase region is particularly challenging due to its complex and elusive nature, as well as the limitations of current measurement techniques, which make it difficult to directly assess its physical properties, such as its thickness and dielectric constant. As a result, theoretical methods have become indispensable tools in nanocomposite research, offering invaluable insights for the design and optimization of these materials. Numerous models grounded in distinct theoretical approaches, including the Maxwell–Garnett, Bergman, and Lichtenecker models, have been developed to predict the performance of nanocomposites and guide their fabrication [31–33].

Poon and Shin developed a straightforward analytical formula to predict the dielectric constants of particulate-filled composite materials, assuming that the inclusions are uniformly distributed within the matrix [34,35]. However, it is important to note that the PS model does not account for the interphase effects, which can significantly influence the overall properties of the composite. While various models have been developed in previous studies to address core–shell type inclusions [36–39], research on the intrinsic characteristics of the interphase—such as its thickness and dielectric constant, and their influence on the equivalent particle content and dielectric properties of composites—remains limited. In this study, we extend the PS model to develop a novel mathematical framework that can predict the dielectric constant of spherical particle-filled nanocomposites, while properly incorporating the effects of the interphase. The interphase is modeled as a shell surrounding the spherical particles, as illustrated in Figure 1. The nanocomposite is considered to consist of the polymer matrix, spherical inclusions, and the interphase region. Experimental data are used to validate the accuracy of the modified PS model, with excellent agreement observed between the predicted and measured values. The primary objective of this work is to provide a useful tool for predicting the dielectric properties of polymeric nanocomposites reinforced with spherical inclusions, while simultaneously evaluating the influence of the interphase characteristics.

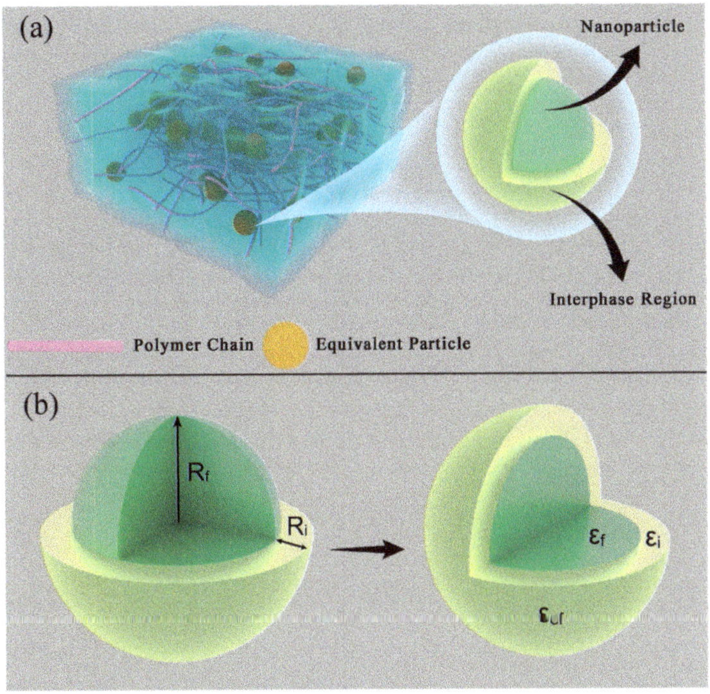

Figure 1. (**a**) Schematic representation of the spherical nanoparticle, its surrounding interphase, and the corresponding equivalent particle model; (**b**) Schematic illustration of the hybrid particle.

2. Model Description

In the PS model (Equation (1)), the displacement field experienced by a single particle is considered to arise from both the surrounding matrix and the polarization of the inclusions embedded within the matrix. The model also accounts for the interactions between these embedded inclusions. However, as previously noted, the PS model does not incorporate the effects of the interphase. Extensive research has demonstrated that an interphase zone develops at the particle surface [40,41], as depicted in Figure 1a,b. This interphase region represents a modified section of the polymer matrix and is confined solely to the outer layer of the nanoparticle. This interphase is primarily a result of the interactions between the filler particles and the polymer matrix. Additionally, the surface area of the embedded inclusions plays a crucial role in the formation and characteristics of the interphase region.

$$\varepsilon = \varepsilon_m + \frac{(\varepsilon_f - \varepsilon_m)V_f}{(1 - V_f)\frac{\varepsilon_f + 2\varepsilon_m - V_f(\varepsilon_f - \varepsilon_m)}{3\varepsilon_m} + V_f} \tag{1}$$

where ε, ε_m, and ε_f are the permittivity of composite, permittivity of matrix, and permittivity of particles, respectively. V_f is the volume fraction of particle.

Previous research has established a connection between the interfacial properties and both the mechanical and electrical performance of composite materials [42,43]. Specifically, the presence of interphase regions can enhance the effective volume fraction of conductive particles, thereby facilitating the development of conductive pathways and lowering the percolation threshold. To simplify the analysis, this study assumes that the particles are idealized as spherical. To account for the interphase effects, a model of equivalent particles with core–shell structures is proposed, where the interphase region is represented as a shell surrounding the spherical core, as shown in Figure 1. The dielectric permittivity of these equivalent particles is initially determined using the Tanaka formula, which is also known as the Maxwell–Garnett formula, as presented in Equation (2) [27,44].

$$\varepsilon_H = \varepsilon_s \frac{\varepsilon_c(1 + 2\rho) + 2\varepsilon_s(1 - \rho)}{\varepsilon_c(1 - \rho) + \varepsilon_s(2 + \rho)} \tag{2}$$

where ε_H, ε_c, and ε_s denote permittivity of the hybrid particle, permittivity of the core part, and permittivity of the shell, respectively. In the Tanaka formula, the parameter ρ is dominated by the size of inner part (radius, R) and shell (thickness, R_s), as defined in Equation (3).

$$\rho = \frac{R^3}{(R + R_s)^3} \tag{3}$$

The Tanaka formula was initially developed to determine the dielectric constant of hybrid particles with core–shell configurations. In the present study, the spherical particle and its surrounding interphase are treated as equivalent particles with a core–shell structure. In this model, the spherical particle and the interphase region correspond to the core and shell components, respectively, as defined in the Tanaka formula. Consequently, the dielectric constant of the equivalent particle, which includes both the spherical particle and the associated interphase, is expressed as follows:

$$\varepsilon_{ef} = \varepsilon_i \frac{\varepsilon_f(1 + 2\rho') + 2\varepsilon_i(1 - \rho')}{\varepsilon_f(1 - \rho') + \varepsilon_i(2 + \rho')} \tag{4}$$

In this expression, ε_{ef} denotes the dielectric permittivity of the equivalent particles, while ε_f refers to the permittivity of the inclusions. The dielectric constant of the shell-like interphase is represented by ε_i. Additionally, ρ' is defined by Equation (5), where R_f and R_i represent the radius of the inclusions and the thickness of the interphase, respectively.

$$\rho' = \frac{{R_f}^3}{(R_f + R_i)^3} \tag{5}$$

The concentration of particles plays a crucial role in determining the dielectric permittivity of the nanocomposite. Given that the interphase region enhances the effective concentration of the filler particles, it is essential to calculate the volume fraction of the equivalent nanoparticles, denoted as V_{ef}. The theoretical filler concentration, represented by V_f, is defined as the ratio of the total volume of the inclusions (V_p) to the volume of the nanocomposite (V_n), as shown in Equation (6).

$$V_f = \frac{V_p}{V_n} \quad (6)$$

where V_p is calculated by the following formula:

$$V_p = n\nu_p \quad (7)$$

In Equation (7), *n* represents the total number of filler particles, while V_p refers to the volume of an individual particle. Since the particles in this study are modeled as spheres, the volume of a single spherical particle V_p is calculated according to Equation (8):

$$\nu_p = \frac{4}{3}\pi {R_f}^3 \quad (8)$$

The volume of an equivalent particle is determined by adding the volume of the filler particle V_p and the volume of the associated interphase region ν_i. Consequently, the total volume of a single equivalent particle ν_{ef} is given by the following expression:

$$\nu_{ef} = \nu_p + \nu_i = \frac{4}{3}\pi(R_f + R_i)^3 \quad (9)$$

According to Equations (5), (8) and (9), the relationship between the volume of single spherical particle ν_p and volume of single equivalent particle ν_{ef} can be established as shown in Equation (10).

$$\nu_{ef} = \frac{\nu_p}{\rho'} \quad (10)$$

The volume fraction of the equivalent inclusions V_{ef} is defined as the ratio between the overall volume of equivalent particles V_{ep} and the volume of nanocomposite, which is expressed in Equation (11).

$$V_{ef} = \frac{V_{ep}}{V_n} = \frac{n\nu_{ef}}{V_n} \quad (11)$$

Consequently, combining Equations (6), (7), (10) and (11), the volume fraction of effective particles V_{ep} is related to the theoretical content of particles V_f, as shown in Equation (12).

$$V_{ef} = \frac{V_f}{\rho'} \quad (12)$$

Therefore, a novel mathematical model that incorporates the influence of the interphase, building upon the PS model, can be formulated as follows:

$$\varepsilon = \varepsilon_m + \frac{(\varepsilon_{ef} - \varepsilon_m)V_{ef}}{(1 - V_{ef})\frac{\varepsilon_{ef} + 2\varepsilon_m - V_{ef}(\varepsilon_{ef} - \varepsilon_m)}{3\varepsilon_m} + V_{ef}} \quad (13)$$

3. Results and Discussion

The refined PS model is applied to estimate the dielectric permittivity of polymer-based nanocomposites containing spherical particles. To assess the reliability of the developed model, its predictions are compared with experimental data, and the key parameters (ε_m, ε_f, R) used in the calculations are summarized in Table 1. Experimental results for PVDF-BT composites [45], poly(arylene ether sulfone) (DPAES)-BT composites [46], and PVDF-BT composites [47] are selected from previously published studies. The dielectric constants (ε_m) of the polymer matrices are taken as 6.5 for PVDF [45], 3.2 for DPAES [46], and 7.5 for PVDF [47], while the permittivity of the incorporated BT nanoparticles (ε_f)

is assumed to be 1240, based on reported values [48]. The average radius (R) of the BT nanoparticles used in the study is set to 50 nm, following experimental findings [49]. The experimental dielectric permittivity data for various polymer nanocomposites are utilized to determine the corresponding interphase thickness and its dielectric constant, with the computed values presented in Table 1. The results indicate that the interphase properties are strongly influenced by the combination of polymer matrix and nanoparticles. Different polymer–nanoparticle pairings lead to variations in both interphase thickness and dielectric permittivity. Despite these differences, the dielectric permittivity of the interphase consistently falls within the range defined by the permittivity of the polymer matrix (ε_m) and that of the BT nanoparticles (ε_f).

Table 1. The parameters of polymer nanocomposites from reported studies and the calculated results based on the developed PS model.

Samples	ε_m	ε_f	R_i(nm)	R(nm)	ε_i
PVDF-BT composite [45]	6.5	1240	20	50	23.7
DPAES-BT composite [46]	3.2	1240	12	50	8
PVDF-BT composite [47]	7.5	1240	5	50	7.8

The predicted dielectric permittivity obtained from the developed model, along with the corresponding experimental data, is presented in Figure 2. As illustrated in the figure, the refined PS model provides highly accurate predictions for all examined samples when appropriate interphase parameters are considered, confirming its reliability in estimating the dielectric properties of $BaTiO_3$–polymer nanocomposites. The strong correlation between experimental results and model predictions highlights the critical role of interphase characteristics in determining the dielectric behavior of these composites. Furthermore, disregarding interphase effects results in significant deviations, leading to inaccurate estimations of the dielectric permittivity for the reported samples. Nevertheless, it is important to acknowledge that the proposed model is developed under the assumption of spherical particles, restricting its applicability to nanocomposites with ball-shaped inclusions. Extending the refined PS model to systems containing nonspherical particles may lead to inaccuracies in prediction results.

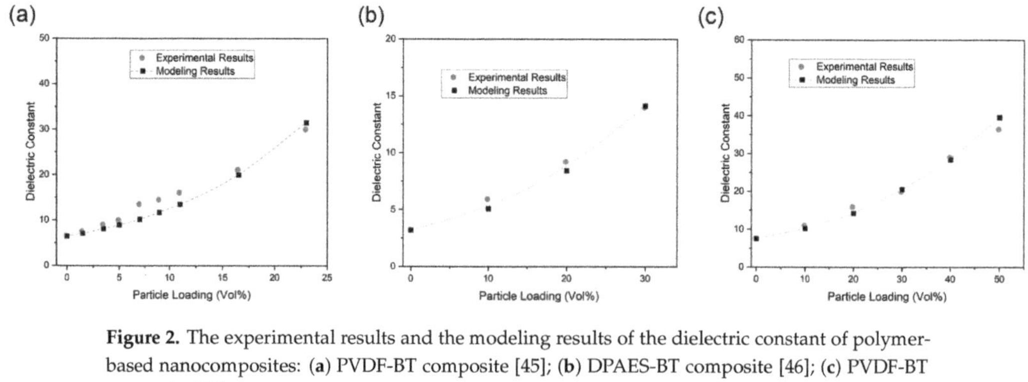

Figure 2. The experimental results and the modeling results of the dielectric constant of polymer-based nanocomposites: (**a**) PVDF-BT composite [45]; (**b**) DPAES-BT composite [46]; (**c**) PVDF-BT composite [47].

To analyze the impact of the interphase on the effective volume concentration of nanoparticles, the volume fraction of the equivalent nanoparticles (V_{ef}) is computed using Equation (12) and compared with the theoretical particle content for the reported samples, as depicted in Figure 3. The results indicate that, across the entire particle loading range, the equivalent volume fraction of $BaTiO_3$ nanoparticles consistently exceeds the theoretical $BaTiO_3$ concentration for all experimental samples. This observation suggests that the interphase is a crucial component of the nanocomposite, significantly enhancing the effective volume fraction of nanoparticles. Consequently, the presence of the interphase plays a fundamental role in shaping the dielectric permittivity of $BaTiO_3$–polymer nanocomposites.

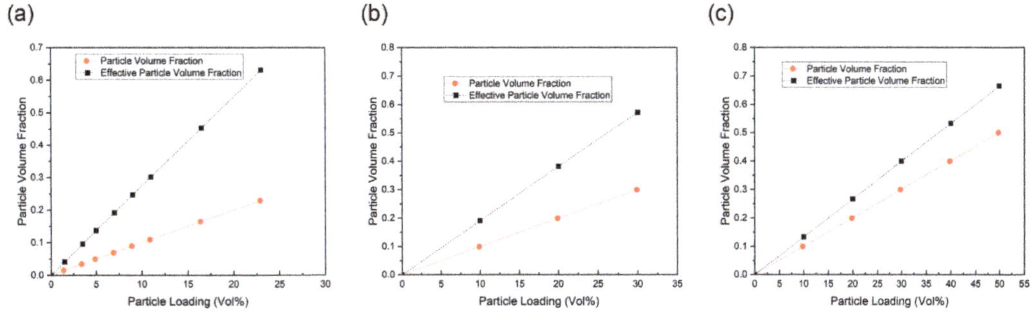

Figure 3. Comparison of particle volume fraction and effective particle volume fraction of polymer-based nanocomposites: (**a**) PVDF-BT composite [45]; (**b**) DPAES-BT composite [46]; (**c**) PVDF-BT composite [47].

The size and concentration of $BaTiO_3$ nanoparticles play a crucial role in determining the dielectric permittivity of $BaTiO_3$–polymer nanocomposites. To systematically evaluate their effects, a parametric study was conducted. Figure 4 illustrates the relationship between the dielectric permittivity of the nanocomposite and the size and volume fraction of $BaTiO_3$ nanoparticles under the conditions $R_i = 20$ nm, $\varepsilon_i = 15$, $\varepsilon_f = 700$, and $\varepsilon_m = 10$. According to the simulation results, these two factors influence the dielectric response in distinct ways. In general, the dielectric constant of the nanocomposite increases with both larger particle size and higher $BaTiO_3$ loading. When the nanoparticle concentration is below 0.43, particle size has a negligible effect on dielectric permittivity. However, at $BaTiO_3$ contents exceeding 0.43, a slight increase in dielectric constant is observed with increasing particle size. The maximum dielectric permittivity of 62.9 is obtained when R = 225 nm and $V_f = 50\%$, whereas the lowest dielectric constant ($\varepsilon_c = 13.9$) is recorded at R = 225 nm with $V_f = 10\%$. These findings suggest that incorporating large $BaTiO_3$ nanoparticles at high concentrations is a viable strategy for enhancing the dielectric performance of $BaTiO_3$–polymer nanocomposites.

In polymer-based nanocomposites, the high interfacial area facilitates exchange coupling, which strengthens polarization, enhances dielectric permittivity, and improves breakdown strength. A high nanoparticle concentration with larger particle sizes contributes to an expanded interfacial area. However, due to the significant surface energy and strong van der Waals interactions, $BaTiO_3$ nanoparticles have a tendency to aggregate within the polymer matrix, reducing the effective interfacial area [50]. This issue is particularly pronounced at high filler loadings and with smaller nanoparticles. Achieving a homogeneous dispersion of nanoparticles—through methods such as ultrasonication or surface modification—can mitigate this problem by maximizing the available interfacial area [51]. Moreover, a well-dispersed nanoparticle distribution leads to a uniform interphase, minimizing the permittivity gradient across the interphase region while enhancing dielectric–dielectric interactions, thereby improving the overall dielectric response.

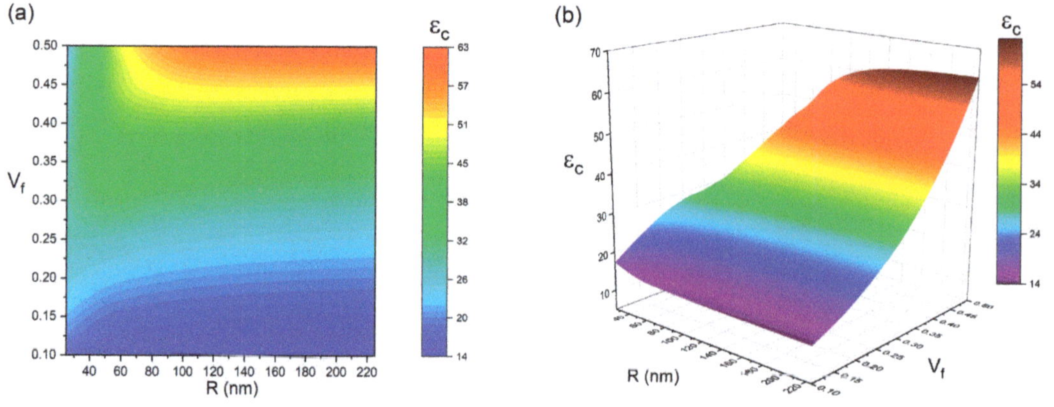

Figure 4. The dependence of ε_c on the R and V_f when $\varepsilon_f = 700$, $\varepsilon_m = 10$, $\varepsilon_i = 15$, $R_i = 20$ nm (**a**) contour plot; (**b**) 3D plot.

Additionally, in nanocomposites containing spherical inclusions, a homogeneous distribution of nanoparticles prevents direct physical contact between them, thereby inhibiting the formation of conductive pathways [52]. This effect reduces the conductivity gradient across the interphase region and lowers dielectric loss while simultaneously improving breakdown strength. Consequently, optimizing nanoparticle dispersion and interfacial interactions is critical for achieving high-performance $BaTiO_3$–polymer nanocomposites with superior dielectric properties.

As previously discussed, the interphase plays a crucial role in determining the dielectric properties of spherical particle-filled nanocomposites. To further explore its impact, a parametric study was conducted to examine how interphase characteristics, including thickness (R_i) and dielectric constant (ε_i), influence the dielectric behavior of the composite. Figure 5 presents the variation in dielectric constant as a function of interphase thickness and permittivity under fixed conditions (R = 50 nm, $\varepsilon_f = 700$, $\varepsilon_m = 10$, and $V_f = 10\%$). The simulation results reveal a direct correlation between interphase attributes and composite permittivity. A thicker interphase layer combined with a higher dielectric constant enhances the dielectric response of the nanocomposite. Conversely, when the interphase is thin and its permittivity is low, the overall dielectric performance deteriorates. The highest dielectric constant ($\varepsilon_c = 82$) is obtained when $R_i = 40$ nm and $\varepsilon_i = 402$, whereas the lowest value ($\varepsilon_c = 5.9$) is recorded at $R_i = 40$ nm and $\varepsilon_i = 2$. These findings suggest that achieving a thick interphase with a high dielectric constant is a key strategy for optimizing the dielectric properties of $BaTiO_3$–polymer nanocomposites.

Beyond improving dielectric performance, the interphase contributes to uniform nanoparticle dispersion within the polymer matrix. It also enhances interfacial interactions between ceramic nanoparticles and the surrounding polymer, reducing the likelihood of particle aggregation [53]. Furthermore, in nanocomposites incorporating conductive fillers, the interphase prevents direct particle-to-particle contact, effectively minimizing tunneling current between adjacent nanoparticles. This suppression of tunneling current results in reduced dielectric loss, enhanced dielectric strength, and an expanded compositional range, thereby improving the overall stability and performance of the composite material.

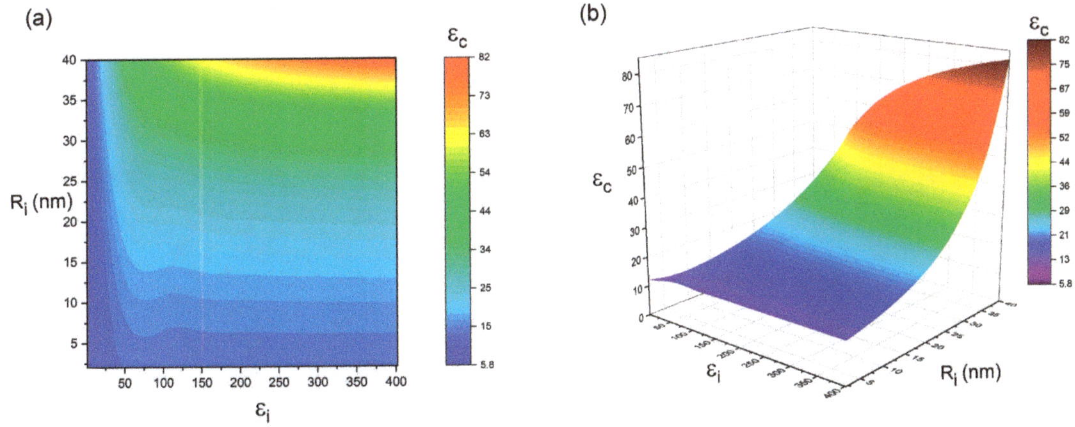

Figure 5. The dependence of ε_c on the ε_i and R_i when ε_f = 700, ε_m = 10, R = 50 nm, and V_f = 0.1 (**a**) contour plot; (**b**) 3D plot.

The formation of the interphase arises from both chemical and physical interactions between the nanoparticles and the surrounding polymer. For instance, the introduction of surfactant species like silane coupling agents allows a better interaction with the polymer matrix, which contributes to the formation of interphase. Due to these interactions, polymer chains adhere strongly to the surfaces of ceramic nanoparticles, leading to the development of interphase layers with distinct properties [54]. Within this region, chain mobility, conformation, and free volume differ significantly from those in the bulk polymer, ultimately influencing the dielectric characteristics of the interphase. The thickness and permittivity of this interphase zone are primarily dictated by the bonding strength between the ceramic nanoparticles and the polymer matrix, which can involve ionic, covalent, hydrogen, or van der Waals interactions, or a combination thereof [55]. Stronger bonding forces, such as ionic and covalent bonds, contribute to a thicker interphase layer. Additionally, robust interfacial interactions within the interphase enhance polarization and charge separation, leading to an interphase region with a high dielectric constant. Since a high-permittivity interphase is crucial for achieving superior dielectric performance in nanocomposites, controlling interphase thickness and properties is essential. A promising approach to tailoring these characteristics involves modifying the surface chemistry of ceramic nanoparticles to manipulate interfacial bonding strength [56].

As the fundamental components of $BaTiO_3$–polymer nanocomposites, both $BaTiO_3$ nanoparticles and the polymer matrix significantly impact the overall dielectric properties of the composite. Therefore, it is essential to examine how the dielectric constants of these two constituents influence the nanocomposite's dielectric response. Based on the modified PS model, Figure 6 illustrates the dependence of the nanocomposite's dielectric permittivity on the dielectric constants of $BaTiO_3$ nanoparticles and the polymer matrix under the parameters R = 50 nm, R_i = 10 nm, V_f = 10%, and ε_i = 20. The results indicate that increasing both ε_m (polymer matrix permittivity) and ε_f ($BaTiO_3$ nanoparticle permittivity) leads to an enhancement in the composite's dielectric properties. Conversely, lower values of ε_m and ε_f result in reduced dielectric performance. Additionally, the modeling results suggest that the polymer matrix has a more pronounced effect on the overall dielectric constant than the $BaTiO_3$ nanoparticles, likely due to the limited particle concentration. The highest dielectric constant (ε_c = 24.8) is achieved when ε_f = 1700 and ε_m = 18, whereas a low-permittivity matrix (ε_m = 2) combined with a small nanoparticle permittivity (ε_f = 100) produces a composite with a significantly reduced dielectric constant (ε_c = 3.28).

These findings highlight that using both a high-permittivity polymer matrix and ceramic fillers is an effective strategy for improving the dielectric performance of $BaTiO_3$–polymer nanocomposites.

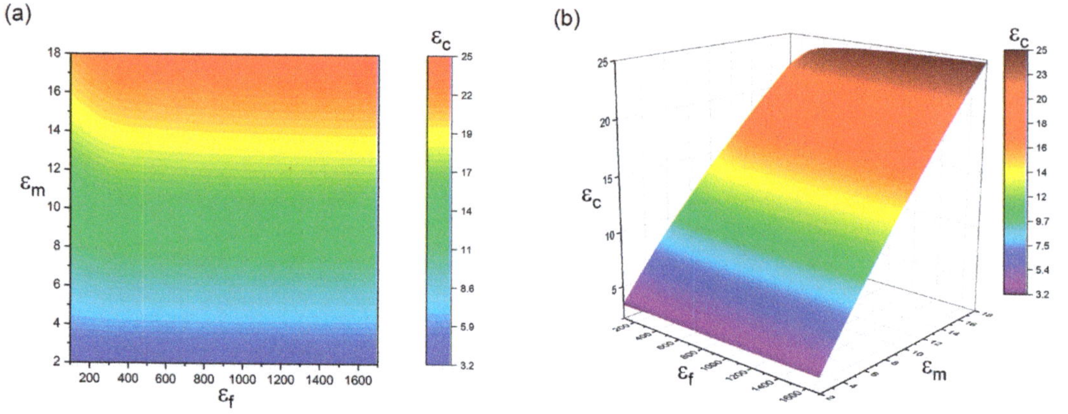

Figure 6. The dependence of ε_c on the ε_m and ε_f when R_i = 10 nm, R = 50 nm, V_f = 0.1, and ε_i = 20 (**a**) contour plot; (**b**) 3D plot.

Polymer-based nanocomposites offer advantages such as cost-effectiveness and ease of processing, making them highly desirable for dielectric applications [57]. The choice of polymer used as the matrix plays a crucial role in determining the overall dielectric permittivity of the composite. Studies have shown that enhancements in the polymer matrix's dielectric constant contribute significantly to the overall improvement of nanoparticle-filled nanocomposites. Specifically, polar polymers such as PVDF exhibit high intrinsic permittivity due to their molecular dipole moments, which facilitate an increase in the dielectric performance of $BaTiO_3$–polymer nanocomposites.

4. Conclusions

A modified PS model was established to predict the dielectric permittivity of polymer nanocomposites filled with spherical particles, incorporating the effects of the interphase. By appropriately selecting interphase parameters, the model demonstrates strong consistency with experimental data. Additionally, the findings reveal that neglecting interphase contributions leads to an underestimation of the effective nanoparticle concentration. Moreover, parameter analysis indicates that the dielectric permittivity of nanocomposites is influenced by multiple factors. To achieve high dielectric performance, both the polymer matrix and embedded nanoparticles should possess high permittivity. A thick interphase layer with a high dielectric constant enhances the dielectric response of the composite, whereas a thin interphase with low permittivity negatively affects its performance. Furthermore, increasing the nanoparticle concentration and using larger particles improves the overall dielectric properties, while a low filler content with small nanoparticles results in reduced permittivity. It is worth noting that this modeling approach is applicable not only to $BaTiO_3$–polymer nanocomposites but also to other polymer-based composites incorporating spherical inclusions. As a result, the developed model is expected to serve as a valuable tool for designing advanced nanocomposites with optimized dielectric properties.

Author Contributions: Conceptualization, X.L. and J.S.; methodology, B.T. and W.Z.; formal analysis, W.Z., S.D. and B.T.; data curation, B.T. and X.L.; writing—original draft preparation, B.T. and W.Z.; writing—review and editing, X.L. and J.S. All authors have read and agreed to the published version of the manuscript.

Funding: This research was funded by the Qing Lan Project of Jiangsu Province.

Data Availability Statement: The original contributions presented in this study are included in the article. Further inquiries can be directed to the corresponding authors.

Conflicts of Interest: The authors declare no conflicts of interest.

References

1. Li, L.; Han, L.; Hu, H.; Zhang, R. A Review on Polymers and Their Composites for Flexible Electronics. *Mater. Adv.* **2023**, *4*, 726–746. [CrossRef]
2. Shao, J.; Zhou, L.; Chen, Y.; Liu, X.; Ji, M. Model-Based Dielectric Constant Estimation of Polymeric Nanocomposite. *Polymers* **2022**, *14*, 1121. [CrossRef] [PubMed]
3. Zha, J.-W.; Zheng, M.-S.; Fan, B.-H.; Dang, Z.-M. Polymer-Based Dielectrics with High Permittivity for Electric Energy Storage: A Review. *Nano Energy* **2021**, *89*, 106438. [CrossRef]
4. Hussain, M.; Zahra, H.; Abbas, S.M.; Zhu, Y. Flexible Dielectric Materials: Potential and Applications in Antennas and RF Sensors. *Adv. Electron. Mater.* **2024**, *10*, 2400240. [CrossRef]
5. Choi, W.; Choi, J.; Han, Y.; Yoo, H.; Yoon, H.-J. Polymer Dielectric-Based Emerging Devices: Advancements in Memory, Field-Effect Transistor, and Nanogenerator Technologies. *Micromachines* **2024**, *15*, 1115. [CrossRef]
6. Shao, J.; Yu, L.; Skov, A.L.; Daugaard, A.E. Highly Stretchable Conductive MWCNT–PDMS Composite with Self-Enhanced Conductivity. *J. Mater. Chem. C Mater.* **2020**, *8*, 13389–13395. [CrossRef]
7. Luo, H.; Wang, F.; Guo, R.; Zhang, D.; He, G.; Chen, S.; Wang, Q. Progress on Polymer Dielectrics for Electrostatic Capacitors Application. *Adv. Sci.* **2022**, *9*, 2202438. [CrossRef]
8. Liu, X.-J.; Zheng, M.-S.; Chen, G.; Dang, Z.-M.; Zha, J.-W. High-Temperature Polyimide Dielectric Materials for Energy Storage: Theory, Design, Preparation and Properties. *Energy Environ. Sci.* **2022**, *15*, 56–81. [CrossRef]
9. Meng, S.; Zhao, T.; Wang, X.; Wang, X.; Zhang, Y. High-dielectric PVDF/MXene Composite Dielectric Materials for Energy Storage Preparation and Performance Study. *Polym. Compos.* **2024**, *45*, 3460–3473. [CrossRef]
10. Wang, W.; Ren, G.; Zhou, M.; Deng, W. Preparation and Characterization of CCTO/PDMS Dielectric Elastomers with High Dielectric Constant and Low Dielectric Loss. *Polymers* **2021**, *13*, 1075. [CrossRef]
11. Shao, J.; Ji, M.; Liu, X. A Developed Rahaman Model for Estimating the Dielectric Permittivity of Polymer–Barium Titanate Nanocomposites. *Ind. Eng. Chem. Res.* **2021**, *60*, 13265–13271. [CrossRef]
12. Liu, X.; Ji, M.; Shao, J. Estimating the Dielectric Constant of BaTiO 3–Polymer Nanocomposites by a Developed Paletto Model. *RSC Adv.* **2021**, *11*, 26056–26062. [CrossRef] [PubMed]
13. Farade, R.A.; Wahab, N.I.A.; Mansour, D.-E.A.; Azis, N.B.; Jasni, J.B.; Veerasamy, V.; Vinayagam, A.; Kotiyal, B.M.; Khan, T.M.Y. The Effect of Interfacial Zone Due to Nanoparticle–Surfactant Interaction on Dielectric Properties of Vegetable Oil Based Nanofluids. *IEEE Access* **2021**, *9*, 107033–107045. [CrossRef]
14. Wu, X.; Chen, X.; Zhang, Q.M.; Tan, D.Q. Advanced Dielectric Polymers for Energy Storage. *Energy Storage Mater.* **2022**, *44*, 29–47. [CrossRef]
15. Likhi, F.H.; Singh, M.; Potdukhe, H.R.; Ajayan, P.M.; Rahman, M.M.; Karim, A. Tuning Dielectric Properties with Nanofiller Dimensionality in Polymer Nanocomposites. *ACS Appl. Mater. Interfaces* **2024**, *16*, 57253–57267. [CrossRef]
16. Hu, J.; Zhang, S.; Tang, B. 2D Filler-Reinforced Polymer Nanocomposite Dielectrics for High-k Dielectric and Energy Storage Applications. *Energy Storage Mater.* **2021**, *34*, 260–281. [CrossRef]
17. Atta, M.R.; Algethami, N.; Farea, M.O.; Alsulami, Q.A.; Rajeh, A. Enhancing the Structural, Thermal, and Dielectric Properties of the Polymer Nanocomposites Based on Polymer Blend and Barium Titanate Nanoparticles for Application in Energy Storage. *Int. J. Energy Res.* **2022**, *46*, 8020–8029. [CrossRef]
18. Shanmugasundram, H.P.P.V.; Jayamani, E.; Soon, K.H. A Comprehensive Review on Dielectric Composites: Classification of Dielectric Composites. *Renew. Sustain. Energy Rev.* **2022**, *157*, 112075. [CrossRef]
19. Goyal, R.K.; Katkade, S.S.; Mule, D.M. Dielectric, Mechanical and Thermal Properties of Polymer/$BaTiO_3$ Composites for Embedded Capacitor. *Compos. B Eng.* **2013**, *44*, 128–132. [CrossRef]
20. Kim, P.; Doss, N.M.; Tillotson, J.P.; Hotchkiss, P.J.; Pan, M.-J.; Marder, S.R.; Li, J.; Calame, J.P.; Perry, J.W. High Energy Density Nanocomposites Based on Surface-Modified BaTiO3 and a Ferroelectric Polymer. *ACS Nano* **2009**, *3*, 2581–2592. [CrossRef]
21. Zahhaf, O.; d'Ambrogio, G.; Le, M.-Q.; Coativy, G.; Grasland, F.; Cottinet, P.-J.; Capsal, J.-F. Dielectrophoretic Alignment of Al2O3/PDMS Composites: Enhancement of Thermal and Dielectric Properties through Structural Sedimentation Analysis. *Mater. Des.* **2021**, *211*, 110134. [CrossRef]
22. Zhang, Z.; He, P.; Ma, W.; Zuo, P.; Liu, X.; Zhuang, Q. Freely Tailorable Yolk-Shell Encapsulation: Versatile Applications in Ultralow-k Dielectric, Drug Delivery Systems, and Catalysts. *Adv. Funct. Mater.* **2023**, *33*, 2302212. [CrossRef]

23. Hornak, J.; Černohous, J.; Prosr, P.; Rous, P.; Trnka, P.; Baran, A.; Hardoň, Š. A Comprehensive Study of Polyurethane Potting Compounds Doped with Magnesium Oxide Nanoparticles. *Polymers* **2023**, *15*, 1532. [CrossRef] [PubMed]
24. Kúdelčík, J.; Hardoň, Š.; Hockicko, P.; Kúdelčíková, M.; Hornak, J.; Prosr, P.; Trnka, P. Study of the Complex Permittivity of a Polyurethane Matrix Modified by Nanoparticles. *IEEE Access* **2021**, *9*, 49547–49556. [CrossRef]
25. Sakthipandi, K.; Sethuraman, B.; Venkatesan, K.; Alhashmi, B.; Purushothaman, G.; Ansari, I.A. Ultrasound-Based Sonochemical Synthesis of Nanomaterials: A Comprehensive Review. In *Handbook of Vibroacoustics, Noise and Harshness*; Springer: Singapore, 2024; pp. 1117–1162.
26. Azani, M.; Hassanpour, A. Nanotechnology in the Fabrication of Advanced Paints and Coatings: Dispersion and Stabilization Mechanisms for Enhanced Performance. *ChemistrySelect* **2024**, *9*, e202400844. [CrossRef]
27. Shao, J.; Liu, X.; Ji, M. Effect of Interfacial Properties of Filled Carbon Black Nanoparticles on the Conductivity of Nanocomposite. *J. Appl. Polym. Sci.* **2022**, *139*, 51604. [CrossRef]
28. Idumah, C.I.; Obele, C.M. Understanding Interfacial Influence on Properties of Polymer Nanocomposites. *Surf. Interfaces* **2021**, *22*, 100879.
29. Zhang, X.; Li, B.; Dong, L.; Liu, H.; Chen, W.; Shen, Y.; Nan, C. Superior Energy Storage Performances of Polymer Nanocomposites via Modification of Filler/Polymer Interfaces. *Adv. Mater. Interfaces* **2018**, *5*, 1800096. [CrossRef]
30. Wang, S.; Luo, Z.; Liang, J.; Hu, J.; Jiang, N.; He, J.; Li, Q. Polymer Nanocomposite Dielectrics: Understanding the Matrix/Particle Interface. *ACS Nano* **2022**, *16*, 13612–13656. [CrossRef]
31. Han, R.; Ren, J.; Quan, H.; Wang, X.; Qi, W.; Gao, D.; Zhou, Z.; Li, Q.; Chen, G.-X. Construction of Parallel-Structured Composites Based on a Theoretical Model to Achieve an Ultra-High Dielectric Constant. *J. Mater. Chem. A Mater.* **2024**, *12*, 22785–22796. [CrossRef]
32. Koledintseva, M.; DuBroff, R.; Schwartz, R. A Maxwell Garnett Model for Dielectric Mixtures Containing Conducting Particles at Optical Frequencies. *Prog. Electromagn. Res.* **2006**, *63*, 223–242. [CrossRef]
33. Mandal, A.; Biswas, D.; Mondal, R.; Lalwani, S.K.; Kabi, S.; Modak, N. Electrical and Dielectric Enhancements in Lithium-Ion Doped Vanadium-Zinc-Phosphate Glass for Energy Storage Applications. *Mater. Chem. Phys.* **2025**, *333*, 130389. [CrossRef]
34. Araújo, M.C.; Costa, C.M.; Lanceros-Méndez, S. Evaluation of Dielectric Models for Ceramic/Polymer Composites: Effect of Filler Size and Concentration. *J. Non-Cryst. Solids* **2014**, *387*, 6–15. [CrossRef]
35. Poon, Y.M.; Shin, F.G. A Simple Explicit Formula for the Effective Dielectric Constant of Binary 0–3 Composites. *J. Mater. Sci.* **2004**, *39*, 1277–1281. [CrossRef]
36. Gutiérrez, Y.; Ortiz, D.; Alcaraz de la Osa, R.; Saiz, J.M.; González, F.; Moreno, F. Electromagnetic Effective Medium Modelling of Composites with Metal-Semiconductor Core-Shell Type Inclusions. *Catalysts* **2019**, *9*, 626. [CrossRef]
37. Chettiar, U.K.; Engheta, N. Internal Homogenization: Effective Permittivity of a Coated Sphere. *Opt. Express* **2012**, *20*, 22976–22986. [CrossRef]
38. Vo, H.T.; Shi, F.G. Towards Model-Based Engineering of Optoelectronic Packaging Materials: Dielectric Constant Modeling. *Microelectron. J.* **2002**, *33*, 409–415. [CrossRef]
39. Odegard, G.M.; Clancy, T.C.; Gates, T.S. Modeling of the Mechanical Properties of Nanoparticle/Polymer Composites. In *Characterization of Nanocomposites*; Jenny Stanford Publishing: Singapore, 2017; pp. 319–342.
40. Noein, L.; Razzaghi-Kashani, M. Improvement in Dielectric Properties of Polyacrylate Copolymers by Engineering the Interphase Region with Polymer-Grafted Reduced Graphene Oxide. *Mater. Res. Bull.* **2024**, *180*, 112999. [CrossRef]
41. Amraei, J.; Jam, J.E.; Arab, B.; Firouz-Abadi, R.D. Effect of Interphase Zone on the Overall Elastic Properties of Nanoparticle-Reinforced Polymer Nanocomposites. *J. Compos. Mater.* **2019**, *53*, 1261–1274. [CrossRef]
42. Ali, Z.; Yaqoob, S.; Yu, J.; D'Amore, A. Critical Review on the Characterization, Preparation, and Enhanced Mechanical, Thermal, and Electrical Properties of Carbon Nanotubes and Their Hybrid Filler Polymer Composites for Various Applications. *Compos. Part C Open Access* **2024**, *13*, 100434.
43. Zha, J.-W.; Wang, F.; Wan, B. Polymer Composites with High Thermal Conductivity: Theory, Simulation, Structure and Interfacial Regulation. *Prog. Mater. Sci.* **2024**, *148*, 101362. [CrossRef]
44. Tanaka, T. A Quantum Dot Model for Permittivity of Polymer Nanocomposites. In Proceedings of the 2016 IEEE Conference on Electrical Insulation and Dielectric Phenomena (CEIDP), Toronto, ON, Canada, 16–19 October 2016; IEEE: Piscataway, NJ, USA, 2016; pp. 40–43.
45. Dou, X.; Liu, X.; Zhang, Y.; Feng, H.; Chen, J.-F.; Du, S. Improved Dielectric Strength of Barium Titanate-Polyvinylidene Fluoride Nanocomposite. *Appl. Phys. Lett.* **2009**, *95*, 132904. [CrossRef]
46. Liu, J.; Li, X.; Ma, S.; Zhang, J.; Jiang, Z.; Zhang, Y. Enhanced High-temperature Dielectric Properties of Poly (Aryl Ether Sulfone)/BaTiO3 Nanocomposites via Constructing Chemical Crosslinked Networks. *Macromol. Rapid Commun.* **2020**, *41*, 2000012. [CrossRef] [PubMed]

47. Xie, L.; Huang, X.; Yang, K.; Li, S.; Jiang, P. "Grafting to" Route to PVDF-HFP-GMA/$BaTiO_3$ Nanocomposites with High Dielectric Constant and High Thermal Conductivity for Energy Storage and Thermal Management Applications. *J. Mater. Chem. A Mater.* **2014**, *2*, 5244–5251. [CrossRef]
48. Fan, B.-H.; Zha, J.-W.; Wang, D.-R.; Zhao, J.; Dang, Z.-M. Experimental Study and Theoretical Prediction of Dielectric Permittivity in BaTiO3/Polyimide Nanocomposite Films. *Appl. Phys. Lett.* **2012**, *100*, 092903. [CrossRef]
49. Tong, Y.; Liu, Y.; Zhang, D.; Wu, J.; Lai, J.; Ju, S.; Li, J.; Li, Y.; Liu, J. Effect of Interfacial Area and Particle Size on the Microstructure and Dielectric Properties of BaTiO3-SiO2 Nanocomposites. *J. Mater. Sci. Mater. Electron.* **2024**, *35*, 843. [CrossRef]
50. Zamperlin, N.; Bottacini, A.; Callone, E.; Pegoretti, A.; Fontana, M.; Dirè, S. Barium Titanate Functionalization with Organosilanes: Effect on Particle Compatibility and Permittivity in Nanocomposites. *Molecules* **2022**, *27*, 6499. [CrossRef]
51. Low, Z.L.; Low, D.Y.S.; Tang, S.Y.; Manickam, S.; Tan, K.W.; Ban, Z.H. Ultrasonic Cavitation: An Effective Cleaner and Greener Intensification Technology in the Extraction and Surface Modification of Nanocellulose. *Ultrason. Sonochemistry* **2022**, *90*, 106176.
52. Chen, X.; Zhou, W.; Meng, X.; Lv, Y.; Kong, F.; Lin, N.; Zheng, J.; Han, C. Synergistically Enhanced Dielectric Properties and Thermal Conductivity in Percolating Flaky Copper/Poly (Vinylidene Fluoride) Nanocomposites via Engineering Magnesium Oxide as a Buffer Interlayer. *J. Appl. Polym. Sci.* **2024**, *141*, e56070. [CrossRef]
53. Trembecka-Wójciga, K.; Ortyl, J. Enhancing 3D Printed Ceramic Components: The Function of Dispersants, Adhesion Promoters, and Surface-Active Agents in Photopolymerization-Based Additive Manufacturing. *Adv. Colloid Interface Sci.* **2024**, *332*, 103251. [CrossRef]
54. Huang, J.; Zhou, J.; Liu, M. Interphase in Polymer Nanocomposites. *JACS Au* **2022**, *2*, 280–291. [CrossRef] [PubMed]
55. Alasfar, R.H.; Ahzi, S.; Barth, N.; Kochkodan, V.; Khraisheh, M.; Koç, M. A Review on the Modeling of the Elastic Modulus and Yield Stress of Polymers and Polymer Nanocomposites: Effect of Temperature, Loading Rate and Porosity. *Polymers* **2022**, *14*, 360. [CrossRef] [PubMed]
56. Osterberg, M.; Henn, K.A.; Farooq, M.; Valle-Delgado, J.J. Biobased Nanomaterials—The Role of Interfacial Interactions for Advanced Materials. *Chem. Rev.* **2023**, *123*, 2200–2241. [CrossRef] [PubMed]
57. Suthar, V.; Joseph, J.; Kadam, S.; Bhatt, R.; Shah, B. Functionalized Polymer Nanocomposites for Dielectric Applications. In *Advances in Functionalized Polymer Nanocomposites*; Elsevier: Amsterdam, The Netherlands, 2024; pp. 915–958.

polymers

MDPI

Article

A Tale of Two Chains: Geometries of a Chain Model and Protein Native State Structures

Tatjana Škrbić [1,2,*], Achille Giacometti [1,3], Trinh X. Hoang [4], Amos Maritan [5] and Jayanth R. Banavar [2]

1 Department of Molecular Sciences and Nanosystems, Ca' Foscari University of Venice, 30170 Venice, Italy; achille.giacometti@unive.it
2 Department of Physics and Institute for Fundamental Science, University of Oregon, Eugene, OR 97403, USA; banavar@uoregon.edu
3 European Centre for Living Technology (ECLT), Ca' Bottacin, Dorsoduro 3911, Calle Crosera, 30123 Venice, Italy
4 Institute of Physics, Vietnam Academy of Science and Technology, Hanoi 11108, Vietnam; txhoang@iop.vast.ac.vn
5 Department of Physics and Astronomy, University of Padua, 35122 Padua, Italy; amos.maritan@unipd.it
* Correspondence: tatjana.skrbic@unive.it or tskrbic@uoregon.edu; Tel.: +39-324-6365131

Abstract: Linear chain molecules play a central role in polymer physics with innumerable industrial applications. They are also ubiquitous constituents of living cells. Here, we highlight the similarities and differences between two distinct ways of viewing a linear chain. We do this, on the one hand, through the lens of simulations for a standard polymer chain of tethered spheres at low and high temperatures and, on the other hand, through published experimental data on an important class of biopolymers, proteins. We present detailed analyses of their local and non-local structures as well as the maps of their closest contacts. We seek to reconcile the startlingly different behaviors of the two types of chains based on symmetry considerations.

Keywords: symmetry; geometry; local and non-local structure; contact maps

1. Introduction

Polymer science [1–4], the study of chain molecules, including linear polymers, is a flourishing subject that has led to life-changing progress in several technologies, including plastics, textiles, and the design of novel materials. At the same time, linear chain molecules form the very basis of life, including both the DNA molecule, whose information is translated into the sequence of amino acids, and proteins, which serve as amazing molecular machines in living cells. While conventional polymer models with stiffness have proved to be adequate for describing the relevant physics of the DNA molecule [5–7], the physical behavior of canonical polymers and proteins is strikingly different. In contrast to DNA, the structural changes during protein folding occur on multiple length scales at once, making it difficult to separate the relative contributions of the myriad interactions [8–13].

A linear chain is composed of many interacting monomers that are tethered together in a railway train topology. If the only interaction is self-avoidance, a single chain is in a coil phase whose large-scale behavior is in the same class as a self-avoiding walk. Upon adding an attractive interaction between pairs of non-adjacent monomers, the chain undergoes compaction into a highly degenerate compact phase at low temperatures [14–44]. While the notion of phases and phase transitions for a polymeric chain strictly refers to a chain with an infinite number of monomers, proteins are modest length chains, which yet exhibit several common characteristics. Notably, the compact state of proteins is modular and made up of two kinds of secondary building blocks: topologically one-dimensional helices [45] and two-dimensional sheets made up of zig-zag strands [46]. The helices and the strands are connected by turns or loops [47–50]. The nature of the ground states of compact polymers is qualitatively distinct from that of proteins and ordinarily does not exhibit any secondary motifs. The common characteristics of proteins are believed to derive from the shared backbone of distinct amino acid sequences.

Here, we present an analysis of these two distinct classes of behaviors to understand their similarities and distinctions. We do this in two complementary ways. For conventional polymers, we study the simplest model of tethered hard spheres of diameter σ and a bond length equal to the sphere diameter. Following the standard nomenclature, we call this the tangent sphere model. We impose an attractive interaction between all pairs of non-adjacent monomers through a square well of range 1.6 σ and depth -1, which sets the energy scale without loss of generality.

A sphere is isotropic and looks the same when viewed from any direction. There is, nevertheless, a preferred axis at the location of each main chain sphere corresponding to the tangent along the chain or the direction along which the chain is oriented at that location. Replacing the spheres with objects such as unidirectional coins and uniaxial discs, allowing neighboring spheres to overlap, or adding side chains to the spheres along the main chain are all steps that break the spherical symmetry and yield ground state structures, which resemble protein structures to varying degrees [51–61]. To avoid clouding the issue by studying an approximate model for proteins, we resort instead to a careful analysis of experimental data of over 4000 protein native state structures (see Section 2). A side-by-side comparison of the local structures and non-local contacts of proteins to those of the tangent sphere model at both low and high temperatures provides a vivid picture of the different views of a chain molecule. Our goal here is to assess how well the tangent sphere model describes the protein backbone.

2. Materials and Methods

2.1. Our Protein Dataset

Our protein data set consists of 4391 globular protein structures from the Protein Data Bank (PDB), a subset of Richardsons' Top 8000 set [62] of high-resolution, quality-filtered protein chains (resolution < 2Å, 70% PDB homology level) that we further distilled out to exclude structures with missing backbone atoms, as well as amyloid-like structures (for the full list of the PDB identifiers of protein structures in our database see Table S1 in the Supplementary Materials of Refs. [58,63]). The program DSSP (CMBI version 2.0) [64] has been used to determine the backbone hydrogen bonding pattern and thus place each protein residue in context within a protein chain: within an α-helix, it is labeled an 'α-residue'; within a β-strand, it is labeled a 'β-residue'; or elsewhere, it is tagged as a 'loop-residue'.

2.2. Numerical Simulations of a Chain of Tethered Tangent Spheres

To obtain a set of independent equilibrium configurations of a chain of tethered tangent spheres comprised of $n = 80$ spherical beads, subject to an attractive potential for a wide range of temperatures, we have employed standard replica exchange (RE) (or parallel tempering) canonical simulations [65,66]. Most of our simulations were carried out with a chain of 80 hard spheres of diameter σ. The model is a tangent sphere model because the bond length is also constrained to be equal to σ. At any temperature, the spheres do not self-intersect and are hard. We introduce a generic attractive square-well attraction of range $R_{att} = 1.6\sigma$ between all pairs of spheres and magnitude ε, which sets the characteristic energy scale. The attractive interaction causes the chain to become compact at low temperatures.

The RE calculation [65,66] relies on a set of canonical simulations run in parallel at a set of M carefully chosen different temperatures, T_i, i = 1, 2, ... M. Each simulation represents a replica or a system copy in thermal equilibrium. The key advantage is the possibility of swapping replicas at different temperatures without affecting the equilibrium condition at each temperature. This permits rapid equilibration even when there is a rugged free energy landscape. In each Monte Carlo (MC) simulation of one replica, new moves are accepted with the standard Metropolis acceptance probabilities [67]. We ensure that the number of swaps that entail the exchange of replicas is large enough to ensure the fidelity of the statistics. The efficiency of the RE scheme depends on the number of replicas, the selected set of temperatures as well as of the swap moves frequency. For best performance, the acceptance rate of swaps is tuned to be around 20% [68]. The RE simulation results are conveniently analyzed using the weighted histogram analysis method [69]. We employed 30 replicas with a finer temperature mesh at lower values of the reduced temperatures (k_BT/ε) in the range k_BT/ε = 0.3–0.5 with a separation of neighboring temperatures of 0.02. In the k_BT/ε = 0.5–1 interval, the separation of neighboring temperatures was 0.05, and for k_BT/ε = 1–4, the separation interval was 0.2. We allowed for the RE swaps only between neighboring temperatures. The exchange moves were attempted every 100 MC steps per monomer. The length of the simulations was 10^9 MC steps per monomer and per replica.

For sampling chain configurations at infinite temperature, we have used the standard Metropolis MC algorithm [67], in which all proposed updates of the chain configuration that respect its self-avoidance are accepted. In both simulation protocols, standard local moves, including crankshaft, reptation (or slithering-snake) moves, endpoint moves, and the non-local pivot move [70], are employed with equal probabilities.

The n = 80 beads tangent sphere polymer exhibits two continuous (second order) 'transitions' (see inset of Figure 5b). At a temperature $k_BT/\varepsilon \sim 3$ (ε denotes the energy scale of attraction), there is a coil-to-globule transition (signaled by a kink in the specific heat per bead C_V/Nk_B). This is a finite-size counterpart of the θ-point for this system. At low temperatures $k_BT/\varepsilon \sim 0.4$, there is a second 'transition' into a compact globule phase, signaled by the maxima of the specific heat per bead C_V/Nk_B. The results we will present for the tangent sphere model are at infinite temperature and k_BT/ε = 0.3.

In the next section, we will present some definitions and general observations pertaining to the local structure of a chain, especially in the continuum limit. We then go on to the Results Section, (Section 4) which is divided into two parts. First, we depict some similarities and some critical differences between the polymer model and protein structures in terms of the power law behavior of certain geometrical measures of local structure. In the second part, we will highlight some striking differences between the geometries of the model polymer and protein native state structures. We then conclude with a brief discussion.

3. General Considerations

Three-Body Interactions and the Self-Avoidance of a Continuum Tube

We will begin our analysis with some definitions and general observations. For fixed bond length, a local chain conformation is specified by two angles, θ and μ. θ is a measure of bond bending; a straight conformation has $\theta = \pi$. The angle between successive binormals, μ, is the second angular coordinate and is the dihedral or torsional angle [63].

The standard model of a linear chain in polymer science, represented by tethered spheres, does not lend itself, in a natural manner, to housing helices, which are recurrent motifs in biomolecules [71–74]. In contrast, a tube, such as a garden hose, which can be thought of as a chain of discs, can be wound readily into a helix. The protein α-helix has a geometry akin to that of a tube wound tightly into a space-filling helix [75].

Consider a collection of uniform, untethered hard spheres. The standard prescription for ensuring that spheres do not overlap is to ensure that the distance between the centers of every pair of spheres is no smaller than the sphere diameter. Pairwise interactions often capture the essence of interacting systems. The simplest generalization to a chain topology is tethered hard spheres (with the same self-avoidance constraint), which, as we have noted, is the simplest conventional model of polymer physics. In contrast, the alternative description of a chain as a tube leads to an unusual condition of self-avoidance.

It has been shown that the correct way to determine whether a tube *in the continuum limit* is self-avoiding or not entails discarding pairwise interactions and invoking appropriate many-body interactions [75,76]. This is illustrated by considering the self-avoidance of a tube of non-zero thickness (see Figure 1). Knowledge of the distance between a pair of points on the tube axis (say A and B or B and D) does not discriminate between the two contexts of nearby points along the axis or in different parts along a chain. In the continuum limit, points A, B, and C, locally positioned along the axis, can become infinitesimally close to each other. This cannot be the case, however, for non-local points, where self-avoidance is a prime consideration, and one must ensure that the two points do not approach each other too closely. Knowledge of the coordinates of a pair of points does not inform you about the context that the two points are in—are they points locally along the axis (which can of course be arbitrarily close to each other) or are they non-local points (that may have come too close to each other possibly signaling an intersection)? This inability to discriminate between local and non-local pairs of points is at the root of the problem.

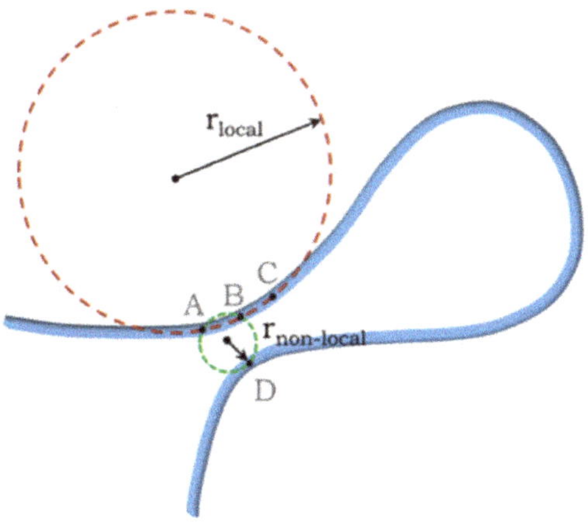

Figure 1. Sketch of the axis of a self-avoiding continuum tube (depicted in blue). The points A, B, and C lie alongside each other on the tube axis whereas point D is a nearby point from another part of the tube. The three-body prescription is to draw circles through all triplets of points on the tube axis and ensure that none of the radii is smaller than the tube radius. For a local triplet of points, one obtains the local radius of curvature whereas the non-local radius is a measure of the distance of closest approach of two parts of the tube.

The standard method in polymer physics of ensuring self-avoidance in the continuum limit is to first make the tube infinitesimally thin (a natural limiting case for a continuum chain of tethered spheres) and then use a singular δ-function potential [77] interaction: there is no energy cost as long as two points on the axis do not overlap exactly but there is an infinite energy cost when there is in fact an overlap. There are at least three problems with this [78]. First, the δ-function potential is singular (unlike say a familiar Lennard-Jones potential) and one needs to use renormalization group theory to introduce an artificial cut-off length scale to carry out the calculations and then demonstrate that this length scale can safely approach zero and yet retain the validity of the results. Second, the δ-function pairwise potential does not preserve the topology of a closed string—the number of knots is not necessarily conserved. Finally, in standard polymer physics, there is no description of the self-avoidance of a continuum self-avoiding tube (or surface) of non-zero thickness.

These problems are deftly averted by discarding pairwise interactions and working with a suitable many-body potential. In the continuum limit, there is a simple geometrical condition [75] to ascertain both whether a tube (and by extension a surface) is self-avoiding non-locally and is not too tightly wound locally. One can draw a circle through any triplet of points along the tube axis and measure the radius. The prescription for self-avoidance of a continuum tube is to consider all possible triplets, local or otherwise, and ensure that every one of the three body radii is greater than or equal to the tube radius. The radius of the circle passing through the triplet of points (A,B,C) in Figure 1 is the local radius of curvature as the points approach each other and results in a kink in the tube when the radius of curvature becomes smaller than the tube radius. On the other hand, the radius of the circle drawn through (A,B,D) is a measure of the distance of approach of two parts of the tube and must not be smaller than the tube thickness in order to respect self-avoidance. Likewise, the self-avoidance of a surface or layer (or a sheet of paper) of non-zero thickness necessarily entails discarding both pairwise and three-body interactions and working with a suitable four-body potential. One considers all quartets of points on the symmetry plane of a surface and draws spheres through each quartet. A surface is self-avoiding if the sphere radius of each quartet (local or non-local) is larger than the thickness of the surface. We note that this many-body prescription is strictly needed only in the continuum limit [76].

For a discrete chain, such as the ones we focus on in this paper, the pairwise distance between a local pair of points is the bond length, and there is no singular behavior at the local level because of a natural cut-off length scale. Also, a minimum threshold of the non-local three-body radius can readily be measured by directly accessing the non-local pairwise distance. We will make use of these simplifications for a discrete chain in our analysis below. It is important to note that there are myriad local interactions, including hydrogen bonds, as well as entropic effects that will play a key role in governing the local behavior of a real chain. Despite the absence of an imperative need to invoke many-body interactions for a discrete chain, the three-body and four-body radii do yield interesting information. An infinitely large three-body radius signals co-linearity of the three points (bond bending angle θ equal to 180°; $\theta = 0°$ is excluded because of steric overlap), whereas an infinite four-body radius is associated with planarity of the quartet of points (μ equal to 0°, 180°, or −180°).

4. Results

4.1. Power Law Scaling

Power laws often signify scale invariance and are a signature of an absence of a characteristic scale [79]. A liquid–vapor system at its critical point exhibits critical opalescence. The system appears milky white because light of all wavelengths scatter from the droplets and bubbles of liquid and vapor of all sizes thoroughly interspersed among each other. Another example of a non-trivial power law is the fractal dimension of a self-avoiding walk in three dimensions of around 5/3 [2]. Here, we discuss a somewhat trivial but surprising realization of 'universal' power law behavior arising in the statistics of the local conformation of a discrete chain molecule. We alert the reader that, unlike in critical phenomena, here, there is neither any many-body emergent behavior nor the need to invoke a system in the thermodynamic limit.

The procedure that we follow is simple. For a set of three (four) points, one can readily draw a circle (sphere) passing through them. The center of a circle (sphere) is the point equally distant from all three (four) points and can be determined as a solution of a suitable system of linear equations. In the case of three points, the solution is simple, and the radius of the circle R is related to the area A of the triangle passing through the three points and its sides a, b, and c: $R = abc/4A$. For four points, we merely solve the equations on a computer to obtain the radius R of the sphere.

Our goal here is to measure the radii R associated with many realizations of these points and obtain cumulative probability distribution functions of the inverse radius $X = 1/R$. Figure 2 shows plots of the cumulative distribution function (CDF) of the inverse radii ($X = 1/R$) (the probability $P(1/r < 1/R)$ as a function of R that a given radius r is larger than R). The circles (spheres) in question are drawn through three (four) points (chosen consecutively along a chain and randomly in some instances) in three dimensions.

We have studied the following: (a) Consecutive triplets (quartets) of C_α atoms along the backbones of globular proteins when the Ramachandran ω angles characterizing a consecutive triplet have canonical values of $|\omega| \sim 180°$. Case (a) occurs in around 99.7% of the cases in globular proteins yielding the *trans* isomeric conformation of a peptide backbone, where the two neighboring C_α atoms are on opposite sides of the peptide bond, with a bond length approximately equal to 3.81 Å [63]. (b) Consecutive triplets (quartets) of C_α atoms in globular proteins in which at least one of the two Ramachandran ω angles has a rare non-canonical value of $|\omega| \approx 0°$ that occurs in ~0.3% of cases. This happens when two neighboring C_α atoms are on the same side of the peptide bond, resulting in a shorter bond length of around ~2.95 Å [63]. This corresponds to the so-called *cis*-conformation of a protein backbone [80]. Cases (a) and (b) are combined for the quartets because they show very similar behavior. (c) Three (four) points selected from a two-step (three-step) self-avoiding random walk of a tangent sphere model (hard spheres of diameter 3.81 Å and bond length 3.81 Å with no other interaction besides the non-overlapping of hard spheres (polymer at infinite temperature). (d) Three (four) points selected from a two-step (three-step) self-avoiding random walk of a tangent sphere model (spheres of diameter 3.81 Å and bond length 3.81 Å) subject to an attractive square-well interaction of range $R_{att} = 1.6\sigma \approx 6$Å (polymer at low temperature). (e) Three (four) points chosen randomly within a three-dimensional sphere of unit radius. (f) Three (four) points selected as points on a two-step (three-step) random walk (no self-avoidance or steric constraints) in three dimensions with a fixed bond length of 3.81 Å (corresponding to the distance between consecutive C_α atoms in proteins [63]).

In the case of triplets ('3-body' case) shown in Figure 2a, we see that the last five systems (cases b–f described in the previous paragraph) exhibit power law behavior with approximately the same exponent of 2, but the canonical protein triplet does *not* ('Proteins $|\omega|$ ~180°' class in Figure 2a).

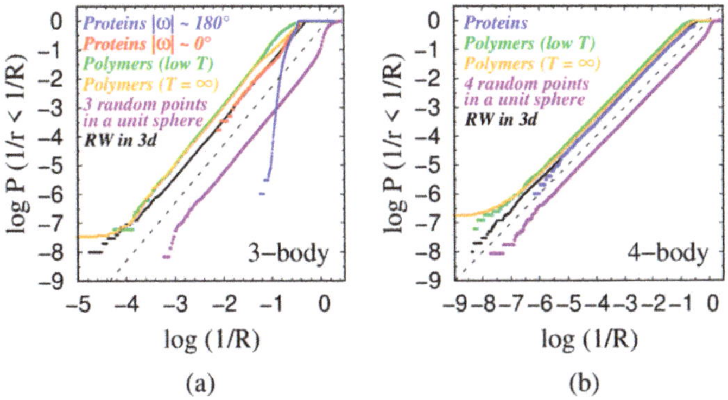

Figure 2. (**a**) Cumulative probability distributions of the inverse radii $X = 1/R$ of circles drawn through three consecutive points along different classes of chains. (blue) 965,122 triplets of the backbones of globular proteins in our data set (defined by C_α atoms) when both the Ramachandran ω angles characterizing a consecutive triplet have canonical values of $|\omega| \approx 180°$; (red) 5774 consecutive triplets of C_α atoms in globular proteins in which at least one of the two Ramachandran ω angles is $|\omega| \approx 0°$; (green) 16,391,622 triplets taken from ≈200,000 low temperature ($k_BT/\varepsilon = 0.3$) configurations, obtained using replica-exchange (RE) simulations, of a chain of 80 tangent spheres of diameter σ with an attractive square well potential of range $R_{att} = 1.6\sigma \approx 6$ Å; (orange) 25,598,976 triplets obtained from MC simulations at $T = \infty$; (purple) 143,557,206 triplets of points chosen uniformly from within a unit sphere in three dimensions; and (black) 100,000,000 two-step random walks in three dimensions with fixed bond length of 3.81 Å. The gray dashed line has a slope of 2 and is a guide to the eye. (**b**) Cumulative probability distributions of the inverse radii $X = 1/R$ of spheres, whose surface passes through four consecutive points in different classes of chains: (blue) 957,723 local quartets along the backbones of globular proteins (employing C_α atoms); (green) 16,181,473 local quartets selected from ≈200,000 chain configurations, obtained from RE simulations, of a chain of 80 tangent spheres of diameter σ subject to an attractive square-well potential of range $R_{att} = 1.6\sigma \approx 6$ Å in the low-temperature phase ($k_BT/\varepsilon = 0.3$); (orange) 25,270,784 local quartets obtained from MC simulations at $T = \infty$; (purple) 112,754,340 quartets of points chosen uniformly within a unit sphere in three dimensions; and (black) 100,000,000 three-step random walks in three dimensions with a fixed bond length of 3.81 Å. The gray dashed line is a guide to the eye and has a

slope of 1. In all simulations, the bond lengths have been chosen to be 3.81 Å, equal to the mean value of the distance between the two consecutive C_α atoms along the protein chain. The distinctive behaviors of the purple curves (corresponding to the random points cases) occur because one can obtain circles of arbitrarily small radii, a situation precluded in the other cases due to steric considerations. The behaviors of the tangent polymer model at high and low temperatures are essentially the same. The local behavior is governed by the same steric constraints in both cases, and the CDF does not change. In contrast, for real polymers, recent experimental studies [81,82] have shown the importance of mechanical properties in determining the local curvature in the context of super-lubricity at the single-molecule level.

For a canonical protein backbone in its *trans* conformation, quantum chemistry does not allow the bond bending angle θ to be greater than ≈150°, thereby preventing too large a value of R. In contrast, for a non-canonical protein backbone in its *cis* conformation [80], two consecutive C_α atoms along the protein chain are much closer to one another, and the backbone in many of these cases has PRO residues. This stiffens up the protein backbone with respect to the canonical case, permitting large bond bending angles, θ, that can almost reach ≈180°.

The second panel in Figure 2, Figure 2b shows similar 'universal' power law behavior for the CDF in the case of quartets ('4-body' case) of the inverse radius (X = 1/R) of a sphere for various cases, detailed in the caption, this time with an exponent 1. We provide a simple rationalization of these findings in the next section for a triplet of points.

4.2. Rationalization of the Power-Law Exponent

To illustrate the origin of the power law behavior, we present here a simple derivation of the probability distribution P(X = 1/R) of the inverse radii of circles drawn through three points of a two-step d-dimensional random walk with a fixed bond length b, which provides a natural length scale. We do not present a similar derivation for the radius of the sphere passing through four points because it is more complex and is best handled numerically (which is what we have done). The quantity Xb is a dimensionless quantity, which enters in the derivation below. The input is p(θ), the probability distribution of the bond bending angle θ. For a d-dimensional random walk, the probability distribution p(θ) scales as [83],

$$p(\theta) \sim (\sin(\theta))^{d-2} \tag{1}$$

R and θ are related by

$$R = b/(2\cos(\theta/2)) \tag{2}$$

or equivalently:

$$X = 1/R = (2\cos(\theta/2))/b \tag{3}$$

Noting that

$$(X)dX = p(\theta)d\theta \tag{4}$$

one obtains

$$P(X) \sim (Xb)^{d-2}\,[1-(Xb/2)^2]^{(d-3)/2} \tag{5}$$

One thus obtains asymptotically (when Xb << 1 or in the large radius R limit) a power law behavior of the probability distribution P(1/R) with an exponent (d − 2) for d > 2 with a power law correction. This means that the cumulative probability distribution P(1/r < 1/R), being an integral of the probability P(1/R), displays a power law behavior with an exponent (d − 1) (=2 in three dimensions) with a power law correction. This correction is relatively small when $(Xb)^2$ is much smaller than 1 and yields good power law behavior, as observed in Figures 2a and 3a. The pivotal quantity that determines the asymptotic exponent is the behavior of p(θ) when θ approaches 180° and the three points become co-linear. The difference in behavior in 2 and 3 dimensions is shown in Figure 3b. The numerical simulations are in good accord with the prediction of Equation (1).

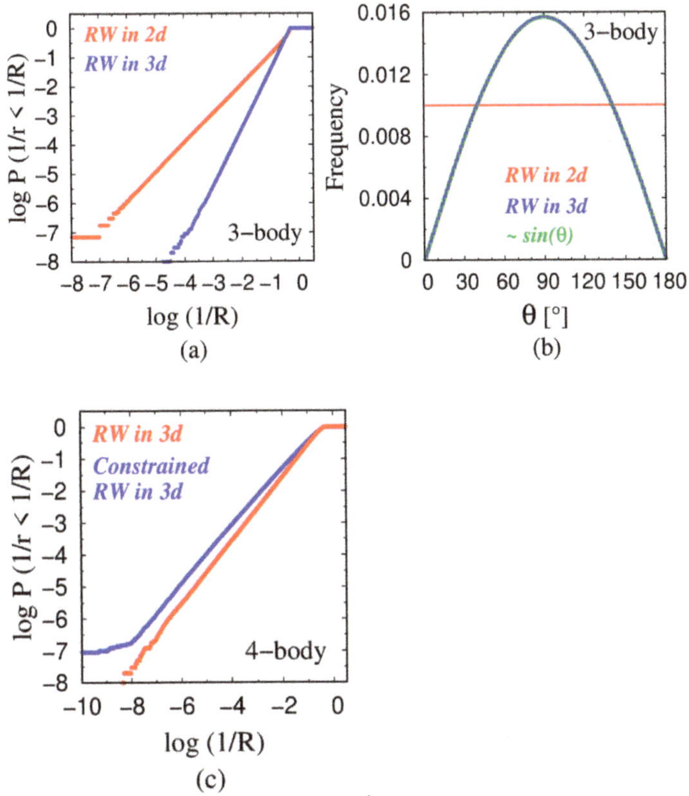

Figure 3. (**a**) Cumulative probability distributions of the inverse radii $X = 1/R$ of the circles drawn through three points of a two-step random walk with a fixed bond length of 3.81 Å, when the walk is performed in two dimensions (red) and in three dimensions (blue). (**b**) The distribution of the bond bending angle θ in the two-step random walk in two dimensions is uniform $p(\theta)$ = const. (red histogram), while, in three dimensions, $p(\theta) = \sin\theta$ (blue histogram and the green line). (**c**) Cumulative probability distributions of the inverse radii $X = 1/R$ of the spheres drawn through four points of a three-step random walk with a fixed bond length of 3.81 Å, when the walk is performed in three dimensions (red) and for a constrained random walk. The constraint arises because the first three points (first two steps of the random walk) are sampled from a plane (sampling in two dimensions—after all, any three points do lie in a plane), while the final step (fourth point) is sampled in full three-dimensional space (blue). The constrained random walk corresponds to a fractal dimension $2 < d < 3$ for the effective sampling of the variables controlling $1/R$ and reduces the steepness of the power law.

The '4-body' case works in a similar manner to the '3-body' case, except that the radius now depends on two independent variables. An unexpected sensitivity of the power law exponent to the choice of the 4 points is demonstrated in Figure 3c. Quartets derived from a plain 3-step random walk in three dimensions (3D) exhibit power law behavior with an exponent of 1 (in accord with the results shown in Figure 2b). We have also considered a simple variant of the plain random walk that we call a constrained random walk. Here, we define the first two points to lie along the x-axis. We then place the third point randomly on a pre-determined x–y plane. Superficially, this may not seem to be an onerous constraint because any three points will necessarily lie in a plane. Nevertheless, a constrained random walk still shows power law behavior but with a distinct exponent, behaving as though it is in a fractal dimension regime. This is because the sampling of phase space is now altered in a relevant manner. We note that this regime may occur when a polymer system happens to be in the vicinity of a surface or a solid wall.

4.3. Chain Geometries

We begin with an analysis of distinct local chain geometries. We compare the behaviors of the tangent sphere model at low and high temperatures, on one hand, and that of the native states of globular proteins, on the other. The local structures of these cases are shown in Figure 4 through their characteristic (θ,μ) plots [63]. These plots are drawn by measuring local pairs of bond-bending and dihedral angles for nearly a million monomers (for the polymer models) and residues (for protein native state structures). In contrast to the plots of model polymers (Figure 4a,b), the (θ,μ) plot for globular protein native state structures (Figure 4c) exhibits significant structure (distinct from the features present in the low-temperature tangent sphere model) and signal the presence of the building blocks of helices and sheets.

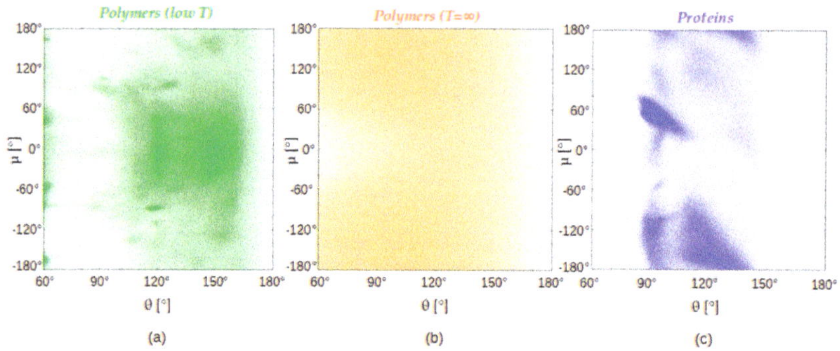

Figure 4. (θ,μ) cross plots of the bond bending angle θ versus dihedral angle μ for 966,505 randomly chosen monomers belonging to two different polymer classes and for the same number of residues in protein native state structures. (**a**) (light green) Polymer chain consisting of 80 tangent spheres of diameter σ subject to an attractive square well potential of range $R_{att} = 1.6\sigma \approx 6$ Å at low temperature ($k_BT/\varepsilon = 0.3$) studied using RE simulations. (**b**) (orange) Polymer chain consisting of 80 tangent spheres of diameter σ at infinite temperature accessed by means of MC simulations with the only interaction being steric avoidance of all pairs of spheres. (**c**) (blue) For 966,505 residues of the 4391 globular proteins in our data set.

The trends in Figure 4 are analyzed in Figure 5, which depicts a histogram of how far the nearest non-local monomer or residue is along the chain sequence. A non-local contact is defined to be one that is separated by at least three positions along the chain with the two beads nearest to each other (i, j) satisfying $|i-j| \geq 3$. For both helices and turns, there are sharp maxima of close-by neighbors at a sequence separation of 3.

There are at least two characteristic length scales associated with the conformation of a chain. One is the local behavior as measured by the local radius of curvature of a triplet of contiguous points. The second is the distance to the nearest non-local contact. For the space-filling conformation of a continuum tube of non-zero thickness, these two length scales become equal [75]. Figure 6 shows a histogram of the two scales, local and non-local, for several situations. The local radius is $R = b/(2\cos(\theta/2))$, where θ is the bond bending angle associated with a local triplet. Structures in the histogram of local radii denote a preference for certain angles of θ. The α-residues exhibit a sharp peak corresponding to $\theta_\alpha \approx 92°$, the β-residues for $\theta_\beta \approx 120°$, and the loop-residues have a pronounced maximum close to the helical value and a less prominent maximum around 111°, see Figure 6a. These peaks are also reflected in the high-density regions in the (θ,μ) cross plot of protein native state structures, shown in Figure 4c. Figure 6b shows the histograms of the relevant non-local length scales for all five cases. All five curves exhibit a single peak denoting a relevant non-local length scale. Table 1 is a compilation of these characteristic local and non-local length scales.

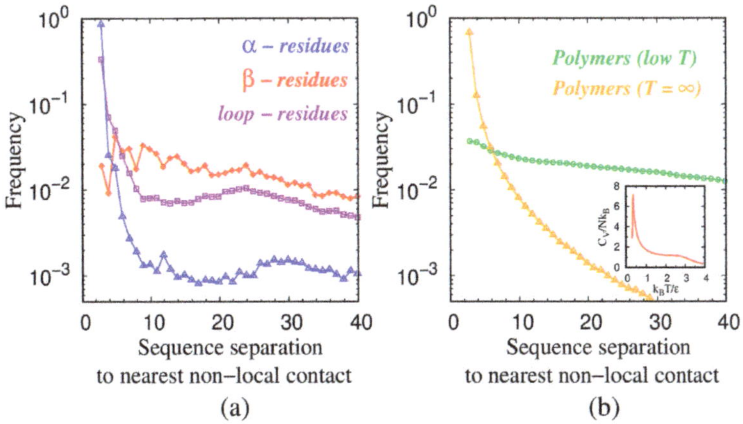

Figure 5. Frequency distribution of the sequence separation $|i-j|$ along the chain of the nearest non-local contact of bead i found at location j. In panel (**a**), the blue points indicate 313,574 α-residues out of a total of 975,287 residues in our data set from 4391 globular proteins; the red points are 214,501 β-beads; and the purple points denote an analysis of 442,821 loop-residues. (**b**) the green and yellow points show the distinct smooth behaviors of a tangent polymer model at low ($k_BT/\varepsilon = 0.3$) and infinite temperatures. The behavior is monotonic, with the closest non-local contacts always being close along the sequence. The inset shows the specific heat per bead C_V/Nk_B for an 80-bead long tangent sphere polymer as a function of the reduced temperature k_BT/ε. There are two continuous (second order) 'transitions': at a high temperature ~3, there is a coil-to-globule transition and at ~0.4, there is a transition into a compact globule.

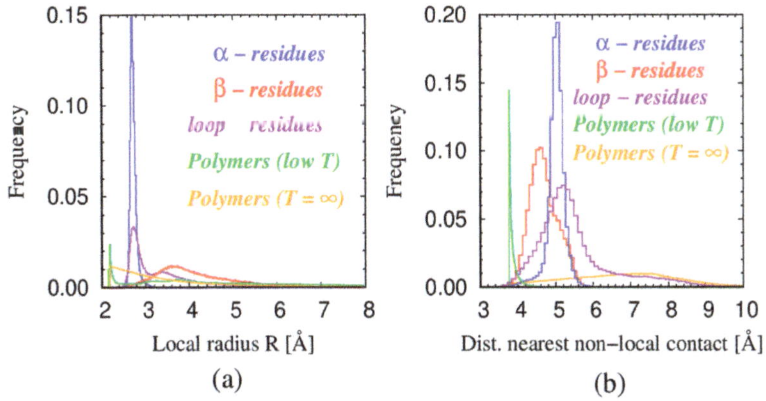

Figure 6. (**a**) Frequency distribution of the local radius of curvature R for five different classes of consecutive triplets. We consider only pure protein triplets in which all three residues are in the *same* structural class. There are 256,154 α triplets (blue) comprising ~26% of all 970,896 triplets. There are 134,643 β triplets (red) and 313,923 loop triplets (purple). There are 16,391,622 low-temperature polymer triplets (green) and 25,270,784 infinite-temperature triplets (orange). (**b**) Frequency distributions of the distances to the closest non-local contact, defined as $|j-i| \geq 3$, of monomers belonging to different classes with the same color code as in panel (**a**).

Table 1. Characteristic local and non-local length scales for three residue types in proteins and for monomers of a tangent polymer chain in the low and high-temperature phases.

Monomers	Most Frequent Value of Local Radius [Å]	Relevant Non-Local Length Scale [Å]
α—residues in proteins	2.73	5.06
β—residues in proteins	3.6	4.67
loop—residues in proteins	2.75	5.26
Spheres in tangent polymers at low T	2.21	3.81
Spheres in tangent polymers at T = ∞	2.47	7.72

Figure 7 shows five representative chain conformations (denoted A–E), each having 80 monomers. We will present the key characteristics of these conformations to highlight similarities and especially the differences. Figure 7 also presents the contact maps for each of these five conformations that show, for each monomer (labeled 1–80), its nearest in distance monomer (separated by at least three positions along the chain). The points in red indicate the contacts for which the pairwise distance is greater than 6 Å. Such distant contacts are rare in the polymer models, unlike the three protein chains. There is little structure in the globular polymer structure A. The infinite temperature polymer B is characterized by the nearest contacts being nearby in sequence, as evidenced by the points in its contact map being close to the diagonal.

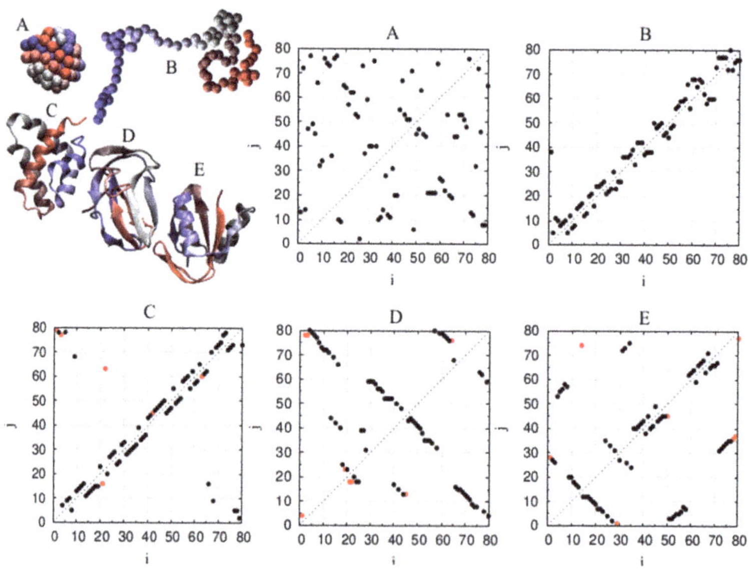

Figure 7. Representative conformations of chains of length 80: (**A**) a low-temperature globule and (**B**) an infinite-temperature coil of a generic tangent polymer chain; (**C**) all-α protein [PDB code: 3bqp, chain B]; (**D**) all-β protein [PDB code: 1bdo, chain A]; and (**E**) α/β protein [PDB code: 3l9, chain X]. The color coding of the conformation does not represent secondary motifs but rather depicts how far a monomer is along the sequence. The chain beginning is depicted in red color that morphs towards blue at the chain end. The contact maps shown in the other panels are those indicating the closest non-local contact j of a given monomer. The black points indicate those that are found within 6 Å, and the red points are those that are found further away than 6 Å of these configurations. The choice of 6 Å is dictated by the fact that the radial distribution function of proteins exhibits a pronounced minimum at this value [61].

This feature of the locality of the closest contacts is also seen in protein α-helices C, where coordinated closest contacts are of the (i, i + 3) type. Furthermore, a distinctive pattern is seen for β sheets D, which display coordinated (i, j) contacts in which bead index j ≥ i + 4 and takes on a coordinated pattern that is consistent with the situation of two strands coming together and forming parallel β sheets (when points in the contact map are parallel to the principal diagonal but shifted away from it) or anti-parallel β-sheets (when points in the contact map are placed along the directions that are perpendicular to the diagonal). The mixed α/β protein E has the features present in both α-helices and two types of β-sheets (parallel and anti-parallel). The similarity of the α helix contact map and that of the infinite temperature polymer is in accord with expectations that helices are more prone to nucleate from the coil phase than β sheets because of the prevalence of short-range contacts.

Figure 8 shows the relative importance of the local radius of curvature and the relevant non-local distance in determining the nature of compact conformations in the five cases. In all five panels, we have scaled the quantities with the appropriate characteristic length scale defined in Table 1. Even though the scaling factors are clearly different, the behavior of the polymer chain is superficially similar at high and low temperatures. The scaled value of the closest non-local distance is substantially flat around a value of 1, with the fluctuations being a bit larger at infinite temperatures. In contrast, the local radius exhibits bigger swings in the low-temperature conformation compared to that at infinite temperature. The α helix is special in that both the local and closest non-local scaled distances are close to each other and to the value 1. In the continuum limit, this equality would result in a space-filling helix [75]. The β regions and the loops do not exhibit this kind of behavior, with the non-local scaled distance often being smaller than the scaled local radius of curvature.

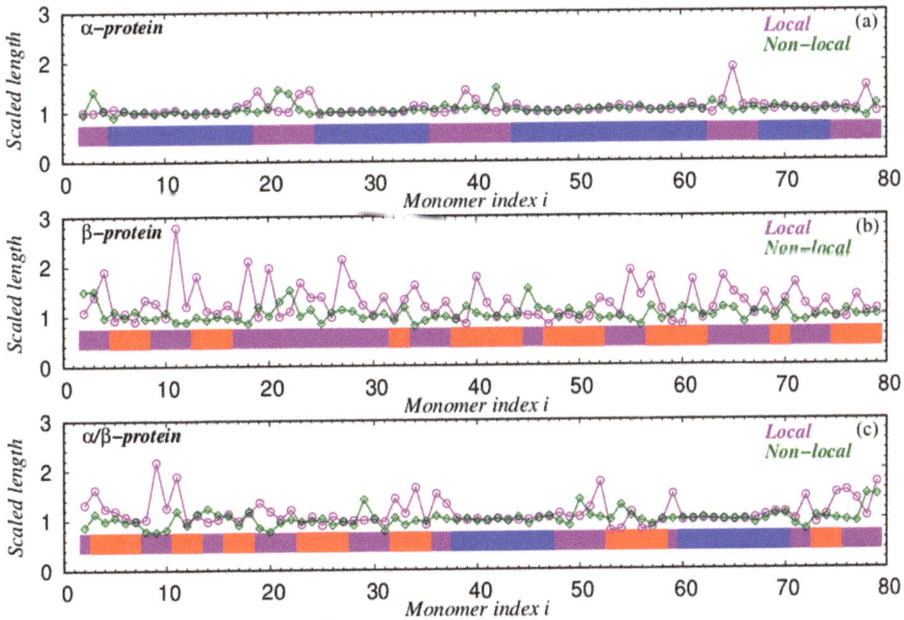

Figure 8. *Cont.*

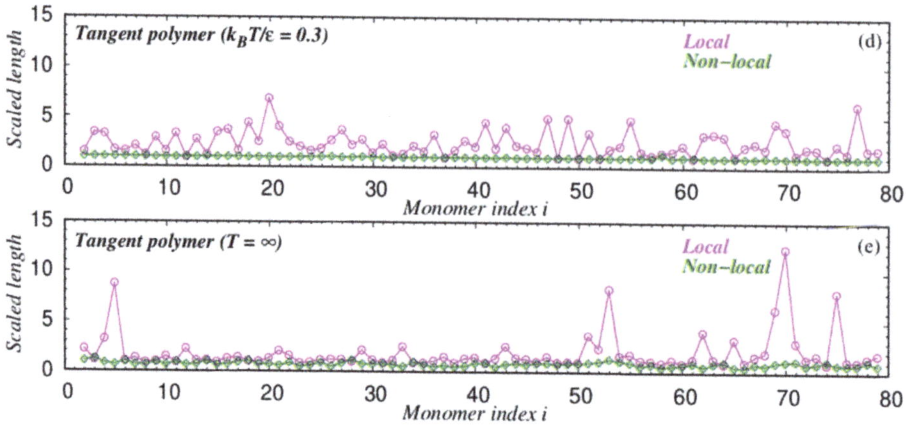

Figure 8. Scaled values of the local radius and minimal non-local distance for the five conformations shown in Figure 7. For each residue of the three proteins, depending on its type ('α', 'β', or 'loop'), these quantities are appropriately scaled with the value of the corresponding characteristic length scale presented in Table 1. (**a**) All-α protein 80 residues long [PDB code: 3bqp, chain B]. (**b**) All-β protein 80 residues long [PDB code: 1bdo, chain A]. (**c**) α/β protein 80 residues long [PDB code: 3l9, chain X]. In the top three panels, the blue ribbons indicate α-helical parts of the sequence, the red ribbons indicate β-sheets, and the purple ribbons indicate loop regions. (**d**) A tangent polymer conformation at $k_BT/\varepsilon = 0.3$. (**e**) A tangent polymer conformation at $T = \infty$.

5. Discussion

Our principal focus in this paper was to study the similarities and differences between chains viewed in two separate ways. The first, a baby model in polymer science, is a tangent sphere model subject to a square-well attraction. We have carried out extensive simulations of the model in the high- and low-temperature phases. We compare the behaviors with those of experimentally determined (and presented in the Protein Data Bank) structures of more than 4000 proteins. The relevant local behavior can be measured by determining the radius of a sphere passing through a local triplet or the radius of a sphere whose surface passes through a set of four consecutive monomers. Surprisingly, we found power law behavior of the probability distribution functions of the radii (with the notable exception of amino acid triplets in proteins). We presented a simple rationalization of this behavior.

We then went on to underscore the numerous distinctions between the model results and protein data. Proteins are complex molecules which follow the rules of quantum chemistry [45,46,84,85] and are influenced by the interactions with the surrounding solvent molecules [86–98]. A protein is a distinct sequence of amino acids. Yet, proteins show remarkable common characteristics. They generally fold reproducibly and rapidly into their native state structures. These structures are directly implicated in protein function. The native state conformations are modular and made up of building blocks, notably helices, and sheets comprised of zig-zag strands. These common characteristics arise because proteins share the same backbone despite having distinct sequences. An important open challenge is the determination of the simplest chain model that would suitably describe the striking features of the common backbones of protein native state structures.

Author Contributions: Conceptualization, J.R.B., A.M. and T.Š.; methodology, J.R.B., A.M. and T.Š.; software, T.X.H. and T.Š.; validation, J.R.B., A.M. and T.Š.; formal analysis, T.Š.; investigation, J.R.B., A.M. and T.Š.; resources, J.R.B., A.G., T.X.H., A.M. and T.Š.; data curation, T.Š.; writing—original draft preparation, J.R.B.; writing—review and editing, J.R.B., A.G., T.X.H., A.M. and T.Š.; visualization, J.R.B., A.M. and T.Š.; supervision, J.R.B.; project administration, J.R.B.; funding acquisition, J.R.B., A.G., T.X.H., A.M. and T.Š. All authors have read and agreed to the published version of the manuscript.

Funding: This project received funding from AdiR (Assegnazioni Dipartimentali per la Ricerca, at Ca' Foscari University of Venice) and European Union's Horizon 2020 research and innovation program under Marie Skłodowska-Curie Grant Agreement No. 894784 (T.Š.). J.R.B. was supported by a Knight Chair at the University of Oregon. A.G. acknowledges support from the Grant PRIN-COFIN 2022JWAF7Y. T.X.H. was supported by the Vietnam Academy of Science and Technology under grant No. NVCC05.02/24-25.

Institutional Review Board Statement: Not applicable.

Data Availability Statement: The raw data supporting the conclusions of this article will be made available by the corresponding author on request.

Conflicts of Interest: The authors declare no conflicts of interest.

References

1. Flory, P. *Statistical Mechanics of Chain Molecules;* Inter-Science Publishers: New York, NY, USA, 1969.
2. de Gennes, P.G. *Scaling Concepts in Polymer Physics;* Cornell University Press: Ithaca, NY, USA, 1979.
3. Khokhlov, A.R.; Grosberg, A.Y.; Pande, V.S. *Statistical Physics of Macromolecules (Polymers and Complex Materials)*, 4th ed.; American Institute of Physics: New York, NY, USA, 2002.
4. Rubinstein, M.; Colby, R.H. *Polymer Physics (Chemistry), 1st ed;* Oxford University Press: London, UK, 2003.
5. Kratky, O.; Porod, G. Röntgenuntersushung gelöster Fagenmoleküle. *Rec. Trav. Chim. Pays Bas* **1949**, *68*, 1106–1123. [CrossRef]
6. Schellman, J.A. Flexibility of DNA. *Biopolymers* **1974**, *13*, 217–226. [CrossRef]
7. Bustamante, C.; Bryant, Z.; Smith, S.B. Ten years of tension: Single-molecule DNA mechanics. *Nature* **2003**, *421*, 423–427. [CrossRef]
8. Kauzmann, W. Some factors in interpretation of protein denaturation. *Adv. Protein Chem.* **1959**, *14*, 1–63. [CrossRef]
9. Privalov, P.L.; Gill, S.J. Stability of protein structure and hydrophobic interaction. *Adv. Protein Chem.* **1988**, *39*, 191–234. [CrossRef]
10. Dill, K.A. Dominant forces in protein folding. *Biochemistry* **1990**, *29*, 7133–7155. [CrossRef] [PubMed]
11. Stickle, D.F.; Presta, L.G.; Dill, K.A.; Rose, G.D. Hydrogen bonding in globular proteins. *J. Mol. Biol.* **1992**, *226*, 1143–1159. [CrossRef]
12. Horinek, D.; Serr, A.; Geisler, M.; Pirzer, T.; Slotta, U.; Lud, S.Q.; Garrido, J.A.; Scheibel, T.; Hugel, T. Peptide absorption on a hydrophobic surface results from interplay of solvation, surface, and intrapeptide forces. *Proc. Natl. Acad. Sci. USA* **2008**, *105*, 2842–2847. [CrossRef]
13. Newberry, R.W.; Raines, R.T. Secondary Forces in Protein Folding. *ACS Chem. Biol.* **2019**, *14*, 1677–1686. [CrossRef] [PubMed]
14. Khokhlov, A.R. On the θ-behaviour of a polymer chain. *J. Phys.* **1977**, *38*, 845–849. [CrossRef]
15. de Gennes, P.G. Collapse of a flexible polymer chain II. *J. Phys. Lett.* **1978**, *39*, 299–301. [CrossRef]
16. Ceperley, D.; Kalos, M.H.; Lebowitz, J.L. Computer Simulation of the Dynamics of a Single Polymer Chain. *Phys. Rev. Lett.* **1978**, *41*, 313. [CrossRef]
17. Fixman, M. Simulation of polymer dynamics. I. General theory. *J. Chem. Phys.* **1978**, *69*, 1527–1537. [CrossRef]
18. Williams, C.; Brochard, F.; Frisch, H.L. Polymer Collapse. *Annu. Rev. Phys. Chem.* **1981**, *32*, 433–451. [CrossRef]
19. Webman, I.; Lebowitz, J.L.; Kalos, H. A Monte Carlo Study of the Collapse of a Polymer Chain. *Macromolecules* **1981**, *14*, 1495–1501. [CrossRef]
20. Bruns, W.; Bansal, R. Molecular dynamics study of a single polymer chain in solution. *J. Chem. Phys.* **1981**, *74*, 2064–2072. [CrossRef]
21. Dünweg, B.; Kremer, K. Molecular dynamics simulation of a polymer chain in solution. *J. Chem. Phys.* **1993**, *99*, 6983–6997. [CrossRef]
22. Tanaka, G.; Wayne, L. Chain Collapse by Atomistic Simulation. *Macromolecules* **1995**, *28*, 1049–1059. [CrossRef]
23. Wittkop, M.; Kreitmeier, S.; Göritz, D. The collapse transition of a single polymer chain in two and three dimensions: A Monte Carlo study. *J. Chem. Phys.* **1996**, *104*, 3373–3385. [CrossRef]
24. Binder, K.; Paul, W. Monte Carlo Simulations of Polymer Dynamics: Recent Advances. *J. Polym. Sci. Part B Polym. Phys.* **1997**, *35*, 1–31. [CrossRef]
25. Michel, A.; Kreitmeier, S. Molecular dynamics simulation of the collapse of a single polymer chain. *Comput. Theor. Polym. Sci.* **1997**, *7*, 113–120. [CrossRef]
26. Honeycutt, J.D. A general simulation method for computing conformational properties of single polymer chains. *Comput. Theor. Polym. Sci.* **1998**, *8*, 1–8. [CrossRef]
27. Aksimentiev, A.; Hołyst, R. Single-chain statistics in polymer systems. *Prog. Polym. Sci.* **1999**, *24*, 1045–1093. [CrossRef]
28. Mavrantzas, V.G.; Boone, T.D.; Zervopoulou, E.; Theodorou, D.N. End-Bridging Monte Carlo: A Fast Algorithm for Atomistic Simulation of Condensed Phases of Long Polymer Chains. *Macromolecules* **1999**, *32*, 5072–5096. [CrossRef]
29. Sikorski, A. Computer Simulation of Adsorbed Polymer Chains with a Different Molecular Architecture. *Macromol. Theory. Simul.* **2001**, *10*, 38–45. [CrossRef]

30. Kreer, T.; Baschnagel, J.; Müller, M.; Binder, K. Monte Carlo Simulation of Long Chain Polymer Melts: Crossover from Rouse to Reptation Dynamics. *Macromolecules* **2001**, *34*, 1105–1117. [CrossRef]
31. Fujiwara, S.; Sato, T. Structure formation of a single polymer chain. I. Growth of trans domains. *J. Chem. Phys.* **2001**, *114*, 6455–6463. [CrossRef]
32. Müller-Plathe, F. Coarse-Graining in Polymer Simulation: From the Atomistic to the Mesoscopic Scale and Back. *Chem. Phys. Phys. Chem.* **2002**, *3*, 754–769. [CrossRef]
33. Abrams, C.F.; Lee, N.-K.; Obukhov, S.P. Collapse dynamics of a polymer chain: Theory and simulation. *Europhys. Lett.* **2002**, *59*, 391. [CrossRef]
34. Auhl, R.; Everaers, R.; Grest, G.S.; Kremer, K.; Plimpton, S.J. Equilibration of long chain polymer melts in computer simulations. *J. Chem. Phys.* **2003**, *119*, 12718–12728. [CrossRef]
35. Rampf, F.; Binder, K.; Paul, W. The Phase Diagram of a Single Polymer Chain: New Insights From a New Simulation Method. *J. Polym. Sci. Part B Polym. Phys.* **2006**, *44*, 2542–2555. [CrossRef]
36. Binder, K.; Baschnagel, J.; Müller, M.; Paul, W.; Rampf, F. Simulation of Phase Transitions of Single Polymer Chains: Recent Advances. *Macromol. Symp.* **2006**, *237*, 128–138. [CrossRef]
37. Binder, K.; Paul, W.; Strauch, T.; Rampf, F.; Ivanov, V.; Luettmer-Strathmann, J. Phase transitions of single polymer chains and of polymer solutions: Insights from Monte Carlo simulations. *J. Phys. Condens. Matter* **2008**, *20*, 494215. [CrossRef]
38. Taylor, M.P.; Paul, W.; Binder, K. All-or-none proteinlike folding transition of a flexible homopolymer chain. *Phys. Rev. E* **2009**, *79*, 050801. [CrossRef] [PubMed]
39. Taylor, M.P.; Paul, W.; Binder, K. Phase transitions of a single polymer chain: A Wang–Landau simulation study. *J. Chem. Phys.* **2009**, *131*, 114907. [CrossRef] [PubMed]
40. Rossi, G.; Monticelli, L.; Puisto, S.R.; Vattulainen, I.; Alla-Nissilla, T. Coarse-graining polymers with the MARTINI force-field: Polystyrene as a benchmark case. *Soft Matter.* **2011**, *7*, 698–708. [CrossRef]
41. Taylor, M.P.; Paul, W.; Binder, K. Applications of the Wang-Landau Algorithm to Phase Transitions of a Single Polymer Chain. *Polym. Sci. Ser. C* **2013**, *55*, 23–28. [CrossRef]
42. Jiang, Z.; Dou, W.; Sun, T.; Shen, Y.; Cao, D. Effects of chain flexibility on the conformational behavior of a single polymer chain. *J. Polym. Res.* **2015**, *22*, 236. [CrossRef]
43. Tzounis, P.-N.; Anogiannakis, S.D.; Theodorou, D.N. General Methodology for Estimating the Stiffness of Polymer Chains from Their Chemical Constitution: A Single Unperturbed Chain Monte Carlo Algorithm. *Macromolecules* **2017**, *50*, 4575–4587. [CrossRef]
44. Gartner, T.E., III; Jayaraman, A. Modeling and Simulations of Polymers: A Roadmap. *Macromolecules* **2019**, *52*, 755–786. [CrossRef]
45. Pauling, L.; Corey, R.B.; Branson, H.R. The structure of proteins: Two hydrogen-bonded helical configurations of the polypeptide chain. *Proc. Natl. Acad. Sci. USA* **1951**, *37*, 205. [CrossRef]
46. Pauling, L.; Corey, R.B. The pleated sheet, a new layer configuration of polypeptide chains. *Proc. Natl. Acad. Sci. USA* **1951**, *37*, 251. [CrossRef]
47. Richardson, J.S. The anatomy and taxonomy of protein structure. *Adv. Prot. Chem.* **1981**, *34*, 167. [CrossRef]
48. Rose, G.D.; Young, W.B.; Gierasch, L.M. Interior turns in globular proteins. *Nature* **1983**, *304*, 654. [CrossRef]
49. Rose, G.D.; Gierasch, L.M.; Smith, J.A. Turns in peptides and proteins. *Adv. Protein Chem.* **1985**, *37*, 1–109. [CrossRef]
50. Škrbić, T.; Giacometti, A.; Hoang, T.X.; Maritan, A.; Banavar, J.B. III. Geometrical framework for thinking about globular proteins: Turns in proteins. *Proteins*, 2024; *online version of record before inclusion in an issue*. [CrossRef]
51. Škrbić, T.; Badasyan, A.; Hoang, T.X.; Podgornik, R.; Giacometti, A. From polymers to proteins: The effect of side chains and broken symmetry on the formation of secondary structures within a Wang-Landau approach. *Soft Matter* **2016**, *12*, 4783–4793. [CrossRef]
52. Škrbić, T.; Hoang, T.X.; Giacometti, A. Effective stiffness and formation of secondary structures in a protein-like model. *J. Chem. Phys.* **2016**, *145*, 0849041–08490410. [CrossRef]
53. Škrbić, T.; Hoang, T.X.; Maritan, A.; Banavar, J.R.; Giacometti, A. The elixir phase of chain molecules. *Proteins* **2019**, *87*, 176–187. [CrossRef] [PubMed]
54. Škrbić, T.; Hoang, T.X.; Maritan, A.; Banavar, J.R.; Giacometti, A. Local symmetry determines the phases of linear chains: A simple model for the self-assembly of peptides. *Soft Matter* **2019**, *15*, 5596–5613. [CrossRef] [PubMed]
55. Škrbić, T.; Banavar, J.R.; Giacometti, A. Chain stiffness bridges conventional polymer and bio-molecular phases. *J. Chem. Phys.* **2019**, *151*, 174901. [CrossRef] [PubMed]
56. Škrbić, T.; Hoang, T.X.; Giacometti, A.; Maritan, A.; Banavar, J.R. Spontaneous dimensional reduction and ground state degeneracy in a simple chain model. *Phys. Rev. E* **2021**, *104*, L0121011–L0121017. [CrossRef]
57. Škrbić, T.; Hoang, T.X.; Giacometti, A.; Maritan, A.; Banavar, J.R. Marginally compact phase and ordered ground states in a model polymer with side spheres. *Phys. Rev. E* **2021**, *104*, L0125011–L0125017. [CrossRef]
58. Škrbić, T.; Maritan, A.; Giacometti, A.; Rose, G.D.; Banavar, J.R. Building blocks of protein structures: Physics meets biology. *Phys. Rev. E* **2021**, *104*, 0144021–0144027. [CrossRef]
59. Škrbić, T.; Hoang, T.X.; Giacometti, A.; Maritan, A.; Banavar, J.R. Proteins–A celebration of consilience. *Intl. J. Mod. Phys. B* **2022**, *36*, 2140051. [CrossRef]

60. Banavar, J.R.; Giacometti, A.; Hoang, T.X.; Maritan, A.; Škrbić, T. A geometrical framework for thinking about proteins. *Proteins*, 2023, *online version of record before inclusion in an issue.* [CrossRef]
61. Škrbić, T.; Giacometti, A.; Hoang, T.X.; Maritan, A.; Banavar, J.B. II. Geometrical framework for thinking about globular proteins: The power of poking. *Proteins*, 2023, *online version of record before inclusion in an issue.* [CrossRef]
62. 3D Macromolecule Analysis & Kinemage Home Page at Richardson Laboratory. Available online: http://kinemage.biochem.duke.edu/research/top8000/ (accessed on 1 June 2020).
63. Škrbić, T.; Maritan, A.; Giacometti, A.; Banavar, J.R. Local sequence-structure relationships in proteins. *Protein Sci.* **2021**, *30*, 818–829. [CrossRef]
64. Kabsch, W.; Sander, C. Dictionary of protein secondary structure: Pattern recognition of hydrogen-bonded and geometrical features. *Biopolymers* **1983**, *22*, 2577–2637. [CrossRef] [PubMed]
65. Swendsen, R.H.; Wang, J.S. Replica Monte Carlo Simulation of Spin Glasses. *Phys. Rev. Lett.* **1986**, *57*, 2607. [CrossRef] [PubMed]
66. Geyer, C.J. Computing Science and Statistics. In Proceedings of the 23rd Symposium on the Interface; American Statistical Association: New York, NY, USA, 1991; p. 156.
67. Metropolis, N.; Rosenbluth, A.W.; Rosenbluth, M.N.; Teller, A.H. Equation of State Calculations by Fast Computing Machines. *J. Chem. Phys.* **1953**, *21*, 1087–1092. [CrossRef]
68. Rathore, N.; Chopra, M.; de Pablo, J.J. Optimal allocation of replicas in parallel tempering simulations. *J. Chem. Phys.* **2005**, *122*, 024111. [CrossRef] [PubMed]
69. Ferrenberg, A.M.; Swendsen, R.H. Optimized Monte Carlo data analysis. *Phys. Rev. Lett.* **1989**, *63*, 1195–1198. [CrossRef] [PubMed]
70. Madras, N.; Sokal, A.D. The pivot algorithm: A highly efficient Monte Carlo method for the self-avoiding walk. *J. Stat. Phys.* **1988**, *50*, 109–186. [CrossRef]
71. Creighton, T.E. *Proteins: Structures and Molecular Properties*; W. H. Freeman: New York, NY, USA, 1983.
72. Lesk, A.M. *Introduction to Protein Science: Architecture, Function and Genomics*; Oxford University Press: Oxford, UK, 2004.
73. Bahar, I.; Jernigan, R.L.; Dill, K.A. *Protein Actions*; Garland Science: New York, NY, USA, 2017.
74. Berg, J.M.; Tymoczko, J.L.; Gatto, G.J., Jr.; Stryer, L. *Biochemistry*; Macmillan Learning: New York, NY, USA, 2019.
75. Maritan, A.; Micheletti, C.; Trovato, A.; Banavar, J.R. Optimal shapes of compact strings. *Nature* **2000**, *406*, 287–290. [CrossRef] [PubMed]
76. Banavar, J.R.; Gonzalez, O.; Maddocks, J.H.; Maritan, A. Self-interactions of strands and sheets. *J. Stat. Phys.* **2003**, *110*, 35–50. [CrossRef]
77. Doi, M.; Edwards, S.F. *The Theory of Polymer Dynamics*; Clarendon Press: New York, NY, USA, 1993.
78. Marenduzzo, D.; Flammini, A.; Trovato, A.; Banavar, J.R.; Maritan, A. Physics of thick polymers. *J. Polym. Sci. Part B Polym. Phys.* **2005**, *43*, 650–679. [CrossRef]
79. Stanley, H.E. Scaling, universality, and renormalization: Three pillars of modern critical phenomena. *Rev. Mod. Phys.* **1999**, *71*, S358. [CrossRef]
80. Ramachandran, G.N.; Mitra, A.K. An explanation for the rare occurrence of cis peptide units in proteins and polypeptides. *J. Mol. Biol.* **1976**, *107*, 85. [CrossRef] [PubMed]
81. Vilhena, J.G.; Pawlak, R.; D'Astolfo, P.; Liu, X.; Gnecco, E.; Kisiel, M.; Glatzel, T.; Pérez, R.; Häner, R.; Decurtins, S.; et al. Flexible Superlubricity Unveiled in Sidewinding Motion of Individual Polymeric Chains. *Phys. Rev. Lett.* **2022**, *128*, 216102. [CrossRef]
82. Cai, W.; Trefs, J.L.; Hugel, T.; Balzer, B.N. Anisotropy of $\pi-\pi$ Stacking as Basis for Superlubricity. *ACS Mater. Lett.* **2023**, *5*, 172–179. [CrossRef]
83. Blumenson, L.E. A Derivation of n-Dimensional Spherical Coordinates. *Am. Math. Mon.* **1960**, *67*, 63–66. [CrossRef]
84. Rose, G.D. Protein folding-seeing is deceiving. *Protein Sci.* **2021**, *30*, 1606–1616. [CrossRef]
85. Rose, G.D. From propensities to patterns to principles in protein folding. *Proteins* 2023, *ahead of print.* [CrossRef]
86. Lesk, A.M.; Chothia, C. Solvent Accessibility, Protein Surfaces, and Protein Folding. *Biophys. J.* **1980**, *32*, 35–57. [CrossRef]
87. Kyte, J.; Doolittle, R.F. A simple method for displaying the hydropathic character of a protein. *J. Mol. Biol.* **1982**, *157*, 105–132. [CrossRef]
88. Rose, G.D.; Geselowitz, A.R.; Lesser, G.J.; Lee, R.H.; Zehfus, M.H. Hydrophobicity of Amino Acid Residues in Globular Proteins. *Science* **1985**, *229*, 834–838. [CrossRef]
89. Ben-Naim, A. Solvent effects on protein association and protein folding. *Biopolymers* **1990**, *29*, 567–596. [CrossRef] [PubMed]
90. Alonso, D.O.V.; Dill, K.A. Solvent Denaturation and Stabilization of Globular Proteins. *Biochemistry* **1991**, *30*, 5974–5985. [CrossRef] [PubMed]
91. Southall, N.T.; Dill, K.A.; Haymet, A.D.J. A View of the Hydrophobic Effect. *J. Phys. Chem. B* **2002**, *106*, 521–533. [CrossRef]
92. Baldwin, R.L.; Rose, G.D. How the hydrophobic factor drives protein folding. *Proc. Natl. Acad. Sci. USA* **2016**, *113*, 12462–12466. [CrossRef] [PubMed]
93. Hayashi, T.; Yasuda, S.; Škrbić, T.; Giacometti, A.; Kinoshita, M. Unraveling protein folding mechanism by analyzing the hierarchy of models with increasing level of detail. *J. Chem. Phys.* **2017**, *147*, 125102. [CrossRef] [PubMed]
94. Davis, C.M.; Gruebele, M.; Sukenik, S. How does solvation in the cell affect protein folding and binding? *Curr. Opin. Struct. Biol.* **2018**, *48*, 23–29. [CrossRef] [PubMed]

95. Hayashi, T.; Inoue, M.; Yasuda, S.; Petretto, E.; Škrbić, T.; Giacometti, A.; Kinoshita, M. Universal effects of solvent species on the stabilized structure of a protein. *J. Chem. Phys.* **2018**, *149*, 0451051–04510517. [CrossRef]
96. Škrbić, T.; Zamuner, S.; Hong, R.; Seno, F.; Laio, A.; Trovato, A. Vibrational entropy estimation can improve binding affinity prediction for non-obligatory protein complexes. *Proteins* **2018**, *86*, 393–404. [CrossRef] [PubMed]
97. Dongmo Foumthuim, C.J.; Carrer, M.; Houvet, M.; Škrbić, T.; Graziano, G.; Giacometti, A. Can the roles of polar and non-polar moieties be reversed in non-polar solvents? *Phys. Chem. Chem. Phys.* **2020**, *22*, 25848–25858. [CrossRef] [PubMed]
98. Carrer, M.; Škrbić, T.; Bore, S.L.; Milano, G.; Cascella, M.; Giacometti, A. Can Polarity-Inverted Surfactants Self-Assemble in Nonpolar Solvents? *J. Phys. Chem. B* **2020**, *124*, 6448–6458. [CrossRef] [PubMed]

polymers

MDPI

Article

Critical Adsorption of Polyelectrolytes onto Patchy Particles with a Low-Dielectric Interior

Dante A. Anhesini [1], Daniel L. Z. Caetano [1], Icaro P. Caruso [1], Andrey G. Cherstvy [2] and Sidney J. de Carvalho [1,*]

[1] Institute of Biosciences, Humanities and Exact Sciences, São Paulo State University (UNESP), São José do Rio Preto 15054-000, Brazil; dante.anhesini@unesp.br (D.A.A.); daniel.caetano@unesp.br (D.L.Z.C.); icaro.caruso@unesp.br (I.P.C.)

[2] Institute for Physics & Astronomy, University of Potsdam, D-14476 Potsdam-Golm, Germany; a.cherstvy@gmail.com

* Correspondence: sidney.carvalho@unesp.br

Abstract: A polyelectrolyte (PE) chain in the vicinity of an oppositely charged surface can exhibit a discontinuous transition from the adsorbed to the desorbed state once the electrostatic attractive interactions are not strong enough to overcome the entropic losses caused by the PE-surface adsorption. In the context of PE–protein interactions, the heterogeneity of the charge distribution and the effects of a low dielectric permittivity underneath the surface are crucial. Studies of the combined effects of these two properties are very sparse, especially in the spherical geometry; we thus fill this gap here. We study the adsorption of PE chains onto spherical particles with heterogeneously charged surfaces, with the main focus on the critical-adsorption conditions and the effects of a low-dielectric core. Metropolis Monte Carlo simulations are employed, with the PE exploring the phase-space around the binding particle in the canonical ensemble. Two adsorption–desorption transitions are observed when the particle possesses a net charge of the *same sign* as that of the PE, resulting in *nonmonotonic* behavior of the critical charge density required for the PE–particle electrostatically driven adsorption. An increased affinity between the PEs and low-dielectric particles with variable heterogeneous charge distributions is observed, in contrast to the behavior detected for homogeneous low-dielectric particles. This higher affinity occurs when the Debye screening length in the solution becomes comparable to the dimensions of a patch of the opposite sign to the PE. A number of real-life applications of the considered PE–particle system is presented in the discussion, in particular regarding the properties of the complex formation between various PEs and globular proteins featuring a dipolar-type distribution of electric charges on their surfaces, such as insulin and bovine serum albumin.

Keywords: adsorption of polyelectrolytes; electrostatics; electrostatic potential; low-dielectric interfaces; patchy interfaces; conditions of critical adsorption; polyelectrolyte–protein complexation

1. Introduction

1.1. Adsorption, Critical Adsorption, and Complex Formation

The adsorption of flexible polyelectrolytes (PEs) [1–4] onto oppositely charged surfaces has been studied extensively; see the theoretical and computational [5–65] and experimental [66–82] studies. Often, it exhibits a phase-transition-like behavior when the ionic strength of the solution or the charge density of the surface electric charges reach the values at which the screened electrostatic (ES) attractive interaction energy can no longer overcome the translational and conformational entropic losses caused by the polymer adsorption [41]. In a series of studies over several decades on the PE complexation with various micelles and dendrimers developed and conducted by a group of experts around Dubin [66–70,72–81], it was shown that the critical conditions for the PE adsorption obey a power law that relates the *critical* (indicated by the subscript "c" hereafter) surface charge density σ and the reciprocal Debye screening length κ, namely [41]

$$\sigma_c(\kappa) \sim \kappa^{\alpha}, \tag{1}$$

with a geometry-dependent exponent α.

Several theoretical approaches have been proposed to evaluate the statistical properties of the adsorbed PE chains and the nature of the PE-surface adsorption–desorption transition [5–10,12,14,16,18,24,28,29,36,41–43,45,48,53,57,59,62,64,83]. For instance, Cherstvy and Winkler presented in ref. [41] a unified approach for the three basic geometries using the well-known quantum mechanical Wentzel–Kramers–Brillouin (WKB) method [84] to approximately solve the Edwards [85] equation for the PE density profiles and to quantify the PE adsorption onto *homogeneously* charged planar, cylindrical, and spherical interfaces. Note that the properties of the strong PE–surface adsorption—when the PE chain assumes a specific pattern on the oppositely and homogeneously charged surfaces—was also investigated [10,11,23,25,28,36,45]. The results were shown to have a good agreement with the experimental data for the conditions when the surface can be considered to be homogeneously charged. In these three situations, the exponent α assumes the values $\alpha = \{1,2,3\}$ for the spherical, cylindrical, and planar surfaces, respectively. The conditions of the critical adsorption of flexible and worm-like PEs inside the oppositely charged confining cylindrical and spherical cavities were also considered [28,36,42,53,86], with some applications, e.g., to the charge stoichiometry of the nucleic acids and of the capsid proteins for a number of spherical single-stranded DNA and RNA viruses [27,40].

Several computational approaches have been adopted over the last few decades (i) to investigate the effects of important and realistic features not considered in the simplest Gaussian-chain model of PEs, such as a finite length of the PE, a nonuniformity of its charge distribution, a stiffness of the chain, etc., as well as (ii) to verify the description of the PE-surface adsorption provided by the analytically solvable approximate models; see refs. [15,39,46,50–52,54,63].

1.2. *Effects of Charge Heterogeneities and of Dielectric Boundaries*

PE adsorption onto *heterogeneously* charged surfaces was also studied (see, i.a., refs. [13,16,17,31,34,35,44,47,56]), and the effects of (i) the charge patterning on the adsorbed PE conformations, (ii) the spatial distribution of monomers near the surface, and (iii) the strength of the adsorption energy onto the critical-adsorption conditions were examined. Particularly interesting are the applications to protein–PE complexes [20,21,26,30,38,49,55,58,79,80], which can occur even between *similarly charged* macromolecules. This phenomenon is often referred to as adsorption on the "wrong side" of the isoelectric point [82,87,88]. Two mechanisms have been proposed to explain such a complexation: (i) the presence of certain protein regions with a charge opposite to that of the PE (so-called "charge patches"), which promote the adsorption for the situations when the local attractive ES interactions [68,89] overcome the repulsive interactions caused by the net protein charge, and (ii) the charge-regulation effects with charge fluctuations of certain titratable groups can lead to the appearance of an induced charge that locally favors the PE–particle adsorption [90,91].

As a consequence of the charge-patch mechanism, some experiments on the interaction between heparin (a polyanion) and insulin (a small globular protein) have showed [76], e.g., a *nonmonotonic* behavior of the critical pH as a function of the ionic strength, in complete distinction to what is observed for a homogeneously charged case. This nonmonotonicity physically occurs because at low salt concentrations—when the Debye length is long compared to the protein dimensions—an increasing ionic strength favors the adsorption of a polyanion due to ES screening of the ES *repulsions* caused by the net charge of the protein. In contrast, when the Debye length is comparable to the protein dimensions, the local *attractive* ES interactions of the PE to the charged patches of the protein become dominant, and the homogeneous-charge behavior is restored [92,93].

To investigate the critical conditions of the PE ES-driven adsorption onto patchy particles more generally, in the past, we have used the Metropolis Monte Carlo method to simulate a PE chain in the presence of a net-neutral dipolar Janus nanosphere [48]. The latter can be seen as a simplified model for a globular protein with two large oppositely charged patches. These simulations [48] have revealed significant differences in the conditions of critical adsorption, as compared to those of the PEs onto a homogeneously charged sphere [41]. Generally, the critical-adsorption studies between charged protein-like patchy particles and PE chains are rather scarce; we try to fill this gap in the current study.

For the protein–PE interactions, the macromolecular dielectric permittivity becomes important due to a low polarizability of the protein interior, in comparison to reorientations of the water molecules in an aqueous environment [94]. Some studies have examined the PE adsorption in the *presence* of low-dielectric substrates [37,43,60,61], a relevant scenario not only for the PE–protein complexation, but also for the PE deposition onto materials such as mica and silica [71]. In the presence of dielectric boundaries, the PE adsorption onto an uncharged planar [61] and spherical surface [37] or onto surfaces with homogeneous charge distributions [22,43] was considered. In this last scenario, the image-charge repulsion [95] causes an *exclusion* of PEs from the vicinity of homogeneously charged surfaces, as predicted from the solution of the Edwards equation in the so-called Debye-Hückel approximation [96] of the linear ES [43,97] and by the Monte Carlo simulations in the absence of added salt [22].

The effects of both a variable surface-charge distribution and a different dielectric-permittivity value was recently studied by Wang and collaborators, who adopted a planar surface with a discrete-charge distribution [60]. However, the adsorption–desorption critical conditions were not explicitly addressed in ref. [60]. Here, the dielectric discontinuities were also shown to *enhance* the PE adsorption onto surfaces with discrete—rather than uniform—charge distributions [60]. Despite of these efforts, systematic studies addressing the combined and synergistic effects of heterogeneous/arbitrary charge distributions and of variable low-dielectric constant in the interior of spherical particles on the properties of PE adsorption in general and—most importantly from the perspective of critical phenomena—onto the PE-surface adsorbed–desorbed transition are rare in the literature.

Therefore, to obtain a more complete picture of PE adsorption onto protein-like patchy spherical particles, it is important to consider the effects of variable and jump-wise changing dielectric permittivity on a macromolecular scale. The main goal of the current work is, therefore, to generalize the critical conditions for the PE adsorption onto spherical particles by considering (as a concrete example) a heterogeneous charge distribution formed by two oppositely charged symmetric regions/caps with an arbitrary charge density and variable dimensions.

For the current study, the problem is formulated such that a spherical particle with an *arbitrary* surface-charge distribution and with a *low-dielectric* core is considered as an attractive nucleus for adsorption of a PE chain, the critical-adsorption conditions of which need to be examined. To the best of our knowledge, this critical-adsorption problem has never been posed and solved previously in full, and the solution to it regarding the most important critical-adsorption demarcation curve as a function of the added simple salt is not known. This is the main novelty aspect of the current study.

Prior to performing the computer simulations, some nontrivial analytical calculations of the ES potential for an *arbitrary* distribution of the surface charges and in the presence of an *arbitrary* dielectric constant inside the core of a spherical particle need first to be performed. Despite certain similarities with the previous work of the authors on PE-Janus-particle adsorption [48], these cap-type charge distributions are chosen here for illustrative purposes and also as a point of reference (to be able to quantify the differences in the obtained results). Therefore, the novelty lies in the *simultaneous* consideration of the combined effects of an arbitrary charge distribution, a low dielectric constant inside the particle, and the spherical geometry of the adsorbing particle on the statistical properties of the adsorbed PE chain in general and on the location of the adsorption–desorption transition for a realistic choice of the model parameters (relevant to some real-world-conditions) in particular. The latter are, e.g., the critical conditions of adsorption of PEs onto multiple kinds of PE-binding proteins; see, e.g., Ref. [76].

1.3. Plan of This Paper

In the current study, we consider the properties of critical PE adsorption onto a surface of a low-dielectric spherical particle/protein with a *potentially arbitrary* distribution of electric charges on its surface. As the initial approximation for a patchy distribution of charges, we employ the model of a spherical particle with dielectric discontinuity on the surface and with two oppositely charged hemispheres, which might be separated by a neutral region (a modified Janus-type [48] particle; see Figure 1). Namely, we divide the sphere into three regions with the (correspondingly) positive, zero, and negative charge densities, controlling the size and charge density of these patchy regions in simulations. Similar to our previous studies [48,50], we employ extensive Monte Carlo computer simulations to uncover the critical-adsorption conditions of a PE as functions of (i) the charge density of the patches, (ii) the size of the patches, and (iii) the salinity of the surrounding solution. A comparison of these new findings with the critical-adsorption conditions of a single PE onto a uniformly charged sphere [41] with no dielectric jump is presented too.

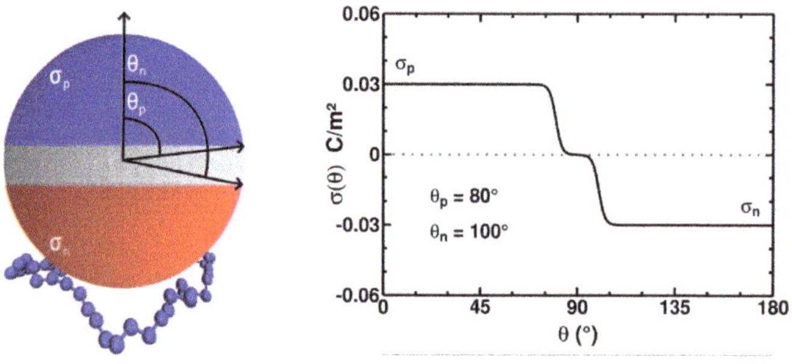

(a) (b)

Figure 1. (**a**) Illustrative representation of a spherical particle, with the blue, red, and white colors indicating the positive, negative, and neutral regions, respectively. The positively charged PE chain is illustrated as being adsorbed on the negatively charged cap of this Janus-type [48] particle. (**b**) A plot of Equation (2) representing the surface charge density of the particle with the following model parameters: the angles are $\theta_p = 80^\circ$ and $\theta_n = 100^\circ$, and the surface charge densities are $\sigma_p = \sigma_n = 0.03\,\mathrm{C/m^2}$.

This paper is organized as follows. In Section 2, we outline in detail the physical model used in the simulations and discuss the approximations employed, we present the main equations of the model, and we outline the calculations for the distribution of the ES potential. In Section 3, we present the results of the simulations and some underlying data trends of the statistical analysis. Finally, we conclude in Section 4 by providing an overview of possible experimental data where the current results are relevant and by mentioning possible future developments of the proposed model.

2. Model and Simulations

2.1. Main Equations for the ES Potential

To study the PE adsorption onto a spherical patchy particle, we adopt a coarse-grained model based on a continuum description of the solvent, the particle interior, and on the Debye–Hückel approximation to describe the ES interactions at a given ionic strength [48]. We consider a spherical simulation box containing a free PE and a patchy particle of a fixed radius $a = 70$ Å permanently located at its center. The simulation box is large enough not to affect the polymer conformations and the adsorption properties to be extracted from the simulation data.

We adapt this model for a particle with the charge density depending on the polar coordinate $0 \leq \theta \leq \pi$ as follows:

$$\sigma(\theta) = \frac{\sigma_p}{2}\left[1 - \tanh\left(n(\theta - \theta_p)\right)\right] - \frac{\sigma_n}{2}\left[1 + \tanh\left(n(\theta - \theta_n)\right)\right]. \quad (2)$$

The surface of the particle for the angles (i) $0 \leq \theta \leq \theta_p$ has the charge density σ_p, (ii) $\theta_n \leq \theta \leq \pi$ has the charge density $-\sigma_n$, and (iii) $\theta_p \leq \theta \leq \theta_n$ has a neutral charge density; see Figure 1. Here, the indices "p" and "n" in $\sigma_{p,n}$ (which are both positive based on the construction of Equation (2)) correspond to the properties of the positive and negative spherical caps, respectively. The special case of $\theta_p = \theta_n = \pi/2$ produces the so-called "Janus particle"; we studied the conditions of the critical PE adsorption onto such net-neutral particles without a dielectric-constant jump previously in ref. [48].

The *dimensionless* ES potential Ψ is defined via the standard ES potential ϕ as

$$\Psi = e_0\phi/(k_B T). \quad (3)$$

Here, e_0 is the elementary electric charge and $k_B T$ is the thermal energy. The potential Ψ_σ of a patchy particle is given via an infinite series in terms of the modified spherical Bessel functions of the second kind $k_l(x)$ and of the Legendre polynomials $P_l(\cos\theta)$ [98]; see Appendix A for the details (and also ref. [28] for similar expansions for other heterogeneous patterns of charges on a sphere). Below, we consider the first eleven terms of this series (sufficient for all situations and conditions considered below), i.e.,

$$\Psi_\sigma(r,\theta) = \frac{e_0}{2\epsilon_0 k_B T}\sum_{l=0}^{10} \alpha_l k_l(\kappa r) P_l(\cos\theta), \quad (4)$$

where ϵ_0 is the dielectric permittivity of the vacuum, and the coefficients α_l are expressed as follows:

$$\alpha_l = \frac{(2l+1)a}{\epsilon_{\text{part}} l k_l(\kappa a) - \epsilon_{\text{water}} \kappa a k_l'(\kappa a)} \int_0^\pi \sigma(\theta) P_l(\cos\theta)\sin\theta d\theta. \quad (5)$$

Here, ϵ_{part} and ϵ_{water} are the values of the dielectric constant of the particle interior and of the electrolyte solvent, respectively; $k_l'(x) = dk_l(x)/dx$ denotes the standard derivative. The indices "part" and "water" are related to the interior of the particle and to the aqueous solution, respectively. We also use the standard notations for the inverse Debye screening length,

$$\kappa = \sqrt{8\pi l_B n_0}, \tag{6}$$

and for the Bjerrum length in the electrolyte solution,

$$l_B = e_0^2/(4\pi\epsilon_0\epsilon_{\text{water}}k_B T). \tag{7}$$

We thus consider an aqueous solution containing the monovalent anions and cations with the number density n_0. For simplicity, below, we neglect the dependence of ϵ_{water} on n_0; typically, $\epsilon_{\text{water}}(n_0)$ is a decreasing function of n_0 because of a progressively disrupted ability of interconnecting hydrogen bonds to be created in salty water at high concentrations of the added salt [99–101].

The model parameter $n = 20$ specifies the "abruptness" of the charge-density variation in the patch boundary (see Figure 1b); its value was chosen so that the integral of Equation (5) does not measurably change as n increases. Note that smooth variations in the charge-density profiles eliminate some problems possibly occurring for the jumps in the charge-density profiles when computing the coefficients in the corresponding eingenfunction expansions.

The PE chain is modeled as a set of $N = 50$ spherical beads of radius $b = 2$ Å (the PE length is fixed hereafter), each with a monovalent positive charge. The beads are linearly connected by the harmonic potential $u_r(r) = k_r(r - r_0)^2$, where the equilibrium distance is set to $r_0 = 7$ Å and the spring constant is $k_r = 0.5$ N/m (see also our previous simulations-based studies [48,50,57]). The ES interaction energy between the ith and jth beads on the PE separated by a distance $r_{ij} > 4$ Å is given by a screened Coulomb or Yukawa-type [102] potential (computed along the polymer chain):

$$u_{\text{PE-PE}}(r_{ij}) = \frac{l_B}{r_{ij}} \exp(-\kappa r_{ij}). \tag{8}$$

This summation of the Debye–Hückel intercharge potentials thus properly accounts for the ES persistence length of the PE chain in this approximation of exponentially screened ES interactions [25,103]. This level of polymer coarse-graining is rather standard in the PE–surface adsorption studies; for a PE in front of a oppositely uniformly charged surface, it does produce the correct scaling relation for $\sigma_c(\kappa a)$ at $\kappa a \gg 1$ [48,57,63]. We thus expect this polymer model to also work well here. In the following, we thus employ the theory of a continuum linear-PB ES and of a continuum dielectric medium. Note also that the performance of the linear- versus nonlinear-PB equations regarding the calculation of the ES potential and the subsequent effects of the potential differences on the location of the adsorption–desorption curve were carefully investigated by the authors previously [50].

Although the Debye–Hückel approximation enables us to calculate the above-mentioned ES potentials analytically, this benefit comes at a certain "cost". It is well known that this approach is only appropriate at low ES potentials (with a magnitude considerably smaller than about 25 mV). This is valid close to the critical-adsorption conditions, where the adsorbed and desorbed states coexist. For higher magnitudes of the ES potentials, the Debye–Hückel approximation leads to *stronger* adsorption compared to the adsorption based on the accurate ES potential obtained, e.g., via the approximate solutions of the nonlinear PB equation [50]. Furthermore, the nonlinear dependence of the surface charge density on the ES potential at the surface leads to a higher limiting value of κ, beyond which the adsorption is not observed at all [50].

Beyond the mean-field approaches, the incorporation of explicit ions into the simulations would make it possible to evaluate the entropic contributions related, i.a., to the release of condensed/bound counterions [49], which are not being considered here. The effects of ion–ion correlations beyond the mean field are known to be significant, especially for multivalent ions and highly charged surfaces [104–108].

The ES interaction energy between the kth bead of the PE at position $\vec{r_k} = (r_k, \theta_k)$ and the patchy particle is given as follows [109] (expressed in units of $k_B T$):

$$u_{\text{PE-part}}(r_k, \theta_k) = \Psi_\sigma(r_k, \theta_k) + \frac{1}{2}\Psi_{k,\text{polar}}(r_k, \theta_k) + \sum_{i=1, i\neq k}^{N} \Psi_{i,\text{polar}}(r_k, \theta_k). \tag{9}$$

Here, the first term describes the direct ES interactions via the potential (4), the second term refers to the polarization charges generated by the kth bead close to the dielectric interface, and the third term refers to the polarization charges generated by the presence of all others beads of the PE. The potential $\Psi_{j,\text{polar}}(r, \theta)$ of the polarization charges (image charges, reaction field) caused by the presence of the jth monomer located at position $\vec{r_j} = (r_j, \theta_j)$ is derived in Appendix B and is given as follows:

$$\Psi_{j,\text{polar}}(r, \theta) = \sum_{l=0}^{10} \frac{e_0^2 \kappa}{4\pi\epsilon_0 \epsilon_{\text{water}} k_B T} \beta_l(r_j) k_l(\kappa r) P_l(\cos\theta_{rj}), \tag{10}$$

where the coefficients β_l are given in terms of the modified spherical Bessel functions of the first kind $i_l(x)$ as follows:

$$\beta_l(r_j) = \left[\frac{\epsilon_{\text{water}} \kappa a i_l'(\kappa a) - \epsilon_{\text{part}} l i_l(\kappa a)}{\epsilon_{\text{part}} l k_l(\kappa a) - \epsilon_{\text{water}} \kappa a k_l'(\kappa a)}\right] (2l+1) k_l(\kappa r_j). \tag{11}$$

The ES-potential calculations above are developed to provide a general description applicable to other setups, including others definitions of $\sigma(\theta)$ based on a specific system. In all the results presented here, we assume $\theta_p = \theta_n$, meaning that no neutral region between the positive and negative patches exists. This is the first approximation to generalize the PE adsorption onto the Janus particles, studied previously [48]. The current model enables us to describe the situations ranging from a homogeneously charged surface to a spherical interface via controlling the geometry of the oppositely charged regions by varying only a few model parameters. This allows us also to evaluate the critical-adsorption conditions for the situation when the particle has a net charge similar to that of the adsorbing PE chain, when the PE is locally attracted to the particle via attractive ES contributions.

2.2. Details of Computer Simulations

The computer simulations are performed using the Metropolis Monte Carlo method to sample the phase space, generating a myriad of different configurations of the PE chain through all allowed monomer movements. Specifically, the pivot and crankshaft movements (see ref. [110]), along with translational and rotational motions of the entire chain, as well as arbitrary monomer displacements, are employed in this sampling process.

The simulations consist of two steps. First, various configurations of the polymer are generated to bring the system to a state of equilibrium. The system is considered to be in equilibrium when the total energy fluctuates around its mean value; this state is achieved after $\sim 10^6$ moves in the simulations. In the second step, a large number of configurations are generated to estimate all relevant statistical averages. To ensure the independence between consecutive configurations, they are separated by 10^3 moves in the simulations. All simulations are performed at a room temperature of $T = 298.15$ K and $\sim 10^7$ PE configurations were used to estimate the statistical averages.

In the course of the equilibration process, the computer simulations begin with a PE chain in a random configuration located at a random position inside the simulation box. The amplitude of the movements is large to ensure no prolonged localization of the polymer in one of possible metastable states. After the equilibration is complete, the simulations are performed using the pivot and crankshaft movements, and monomer displacements in order to sample all available configurations of the PE in space. The amplitude of these movements is defined by an acceptance of the randomly generated polymer configurations of about 50%. These implemented movements are broadly recognized in the simulation community as well-suited to guarantee a good sampling of the available polymer configurations. Along with these movements, the translational motion of the entire PE chain with a high amplitude is used to promote the "take off" the PE from possible metastable states or to promote a transition between different states of this kind (in particular, between the adsorbed and desorbed states).

As mentioned after Equation (8), the ES contribution to the PE's length of persistence $l_p = l_{p,\text{ES}} + l_{p,\text{non-ES}}$ is automatically accounted for in our current simulations. The non-ES contribution to the PE persistence is, however, neglected, as in a number of our previous studies on the critical PE-surface adsorption, both simulations- and theory-based. For the current setup, such a supposition is quite realistic: the PE chains involved in the formation of polymer–proteins complexes are usually very soft [76], with $l_p \sim 10\ldots 40$ Å. If necessary—e.g., for some particular PEs and specific adsorption conditions—the effects of the "structural" mechanical non-ES persistence can be easily incorporated into our future specialized studies of the critical adsorption of flexible-to-semiflexible PE chains.

The statistical deviations from the mean values of the binding energy $\langle E_B \rangle$ and of the radius of gyration $\sqrt{\langle R_g^2 \rangle}$ were calculated by dividing the full set of 10^7 simulated polymer configurations into 10 sections, each with 10^6 configurations. The mean binding energies are then calculated for all these sections separately and then used to compute the standard deviation $\sqrt{\langle E_B^2 \rangle - \langle E_B \rangle^2}$. For most of the results presented below, these deviations result in the error bars around $\langle E_B \rangle$, which are comparable to or smaller than the symbol size used (also for uncertainties of determination of $\sqrt{\langle R_g^2 \rangle}$; thus not shown).

To calculate the values of $\sigma_{n,c}$, we carry out computer simulations for a number of systematically separated values of the charge density σ_n in a range that ensures that the fraction of polymer configurations in the adsorbed state varies between 0 and 1. We then fit the dataset of the fraction of PE configurations in the adsorbed state versus σ, denoted as $f_{\text{ads}}(\sigma)$, with a sigmoidal function $f_{\text{ads}}(\sigma) = \frac{1}{1+e^{\alpha(\sigma-\sigma_{1/2})}}$ with the fit parameters α and $\sigma_{1/2}$. The parameter $\sigma_{1/2}$ then corresponds to the critical values $\sigma_{n,c}$ presented in the plots below for each set of the model parameters. The same procedure is adopted for calculating the values of κ_c. For all critical properties calculated in this study, the correlation coefficient assumes the values $\gtrsim$0.9.

3. Main Results

3.1. PE–Particle Binding Energy

Firstly, we analyze the influence of the charge distribution on a particle on the properties of the PE adsorption *without* considering a dielectric interface (no jump of the dielectric constant). We examine the mean binding energy

$$\langle E_B \rangle = \left\langle \sum_{k=1}^{N} u_{\text{PE-part}}(r_k, \theta_k) \right\rangle \tag{12}$$

as a function of the salt concentration to understand how the range of ES interactions determines the results of the competition between the net- and local-charge density of our Janus-type particle.

Figure 2a presents $\langle E_B \rangle$ as a function of κa for three different charge densities and for a patch size defined by $\theta_p = \theta_n = 120°$. The light-blue curve in Figure 2a shows the results for the particle with a neutral net charge, while the blue and dark-blue ones refer to the particles with a positive net charge (the same charge as that of the PE chain). The light-blue curve exhibits a behavior analogous to that of the PE adsorption on a homogeneously oppositely charged surface [39] due to the attractive PE–particle interactions of a dipolar nature. At low salt concentrations, the ES attraction is strong enough to overcome the entropic loss due to the reduction of translational and conformational degrees of freedom of the PE chain caused by the adsorption process. As the salt concentration increases, the ES screening progressively weakens the attractive ES interactions that favors the emergence of configurations with the PE chain being desorbed, as already discussed in ref. [39]; see Figure 2a. At high enough ionic strengths, the mean binding energy nearly vanishes, $\langle E_B \rangle \approx 0$, and the adsorption is completely suppressed by strong screening.

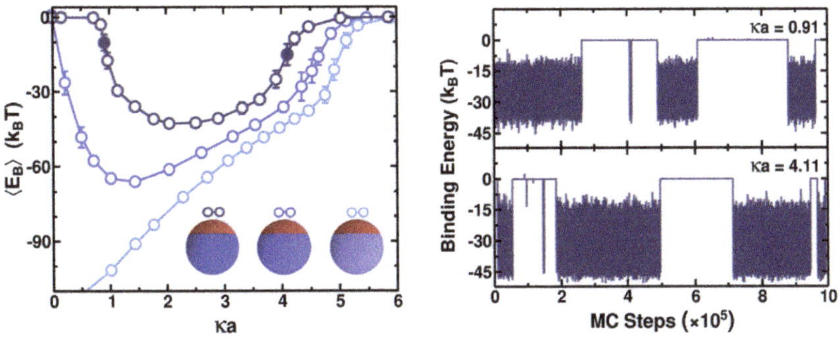

(a) (b)

Figure 2. (**a**) Mean binding energy $\langle E_B \rangle$ as a function of κa for a PE chain containing $N = 50$ monomers onto a sphere of radius $a = 70$ Å and with the charge density given in Equation (2), with the patch sizes defined by the angles $\theta_p = \theta_n = 120°$. The parameters are $\sigma_n = 0.03$ C/m^2 and $\sigma_p = 0.01$, 0.02, and 0.03 C/m^2 for the light-blue, blue, and dark-blue curves, respectively. In the legend of panel (**a**) and hereafter, the intensity of the blue color for the spherical Janus-type particle is correlated with the magnitude of the respective surface charge density. (**b**) Trajectories of the variation in the binding energy E_B(number of MC steps) with time recorded throughout the course of the simulations, plotted for the parameters corresponding to those of the filled blue symbols in panel (**a**).

In the situations when the charge density of the positive patch increases—see the blue and dark-blue curves in Figure 2a—the mean binding energy exhibits a nontrivial behavior. At low salt concentrations, $\langle E_B \rangle \approx 0$, characterizing the desorbed state of the PE, caused by the ES repulsion due to a nonzero net charge of the particle. In the regimes of higher salt concentrations, local attractive interactions become more and more important, leading to the PE–particle adsorption and to an increase in the magnitude of the mean binding energy. Consequently, after $\langle E_B \rangle$ reaches its minimal value (corresponding to the strongest PE–particle ES interactions because $E_B < 0$), the behavior becomes analogous to that for PE adsorption onto a particle with an opposite net electric charge; see refs. [39,48].

The dark-blue curve in Figure 2a follows the free-PE-chain behavior until the data approach the boundary $\kappa a \approx 1$, whereas the blue curve deviates at smaller κa values. This difference arises because the dark-blue curve in Figure 2a corresponds to a particle with a higher density of positive charges, requiring stronger ES screening in order to ensure the dominance of the local interactions between the positively charged PE and the negatively charged cap of the spherical particle.

Figure 2b illustrates the binding energy E_B(number of MC steps) at each step of the Monte Carlo simulations at regimes of low and high ionic strength. These conditions are indicated by the filled points in the dark-blue curve in Figure 2a, at $\kappa a = 0.91$ and at $\kappa a = 4.11$, respectively. These points are situated within the range of κa where there is a coexistence of the adsorbed and desorbed states of the PE chain. The desorbed state is characterized by a binding energy nearly equal to zero, whereas in the adsorbed state, the binding energy fluctuates between $\approx -45\ k_BT$ and $\approx -10\ k_BT$. In the same way as that observed at high salt concentrations, we observe a desorption $\rightarrow$ adsorption first-order-like transition when the ES screening reaches the regime where the local attractive interactions become dominant over the repulsion stemming from the particle's net charge. Figure 2b shows that the PE adsorption onto a more positively charged particle with $\sigma_p = 0.03\ \mathrm{C/m^2}$ (the dark-blue curve) is observed only when the value of κa reaches about unity, unlike for the particle that is less positively charged (see the blue curve in Figure 2a), in which a very small increase in the salt concentration from $\kappa = 0$ leads to an increase in $|E_B|$ and, as a result, to a significant PE–particle adsorption.

3.2. Radius of Gyration of the Polymer

Figure 3 presents the root mean square radius of gyration of the PE chain $\sqrt{\langle R_g^2 \rangle}$ as a function of κa for PE adsorption onto a particle with the same charge densities as in Figure 2a. We also use the same symbol and color schemes in these two figures for clarity of presentation. In the case of a neutral particle, at low salt concentrations, the PE chain has a radius of gyration much smaller than that expected for a free chain (see the gray dashed line in Figure 3), highlighting the dramatic reduction in the chain dimensions caused by adsorption onto the particle. With increasing salt concentration, the radius of gyration decreases initially because the screening of the monomer–monomer ES interactions makes fewer extended configurations of the PE being occupied in the space of configurations. This behavior is analogous to the PE adsorption onto the homogeneously charged spherical [39] and cylindrical [46] macroions as well as onto the Janus nanospheres [48].

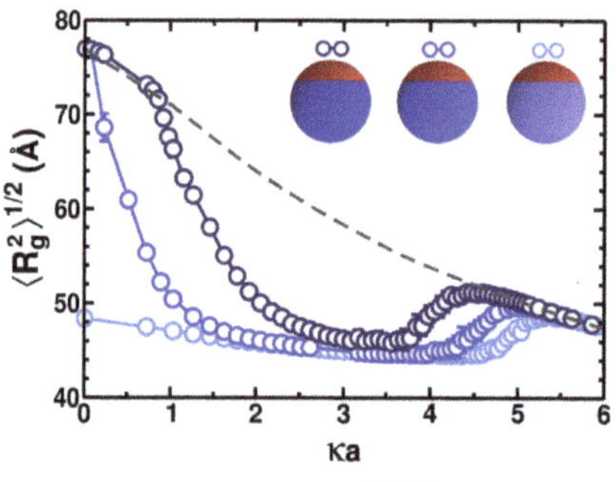

Figure 3. Radius of gyration $\sqrt{\langle R_g^2 \rangle}$ as a function of κa for a PE chain of $N = 50$ monomers. The color and symbol schemes are identical to those in Figure 2a. The gray dashed curve illustrates the results for a free PE chain in the absence of adsorbing spherical particle. As κ increases, the PE–PE ES repulsions (8) decrease, making the chain less extended, as shown. The particle has a patch size defined by the angles $\theta_p = \theta_n = 120°$ and the sphere radius is $a = 70$ Å.

The increase in the charge density of the positive patch—which leads to a more positive total charge of the particle—causes the positive PE chain to follow the free-chain behavior because the like-charge PE–particle ES repulsions prevent the PE–particle adsorption. This occurs at $\kappa a \approx 0$ for the spherical particles with $\sigma_p = 0.02\ \mathrm{C/m^2}$ (the blue curve in Figure 3) or $\kappa a \sim 1$ for the particle with $\sigma_p = 0.01\ \mathrm{C/m^2}$ (the dark-blue curve). This takes place, logically, in the same regimes of the ionic strengths in which the mean binding energy is nearly vanishing in Figure 2a. Increasing the salt concentration in Figure 3, the transition from the desorbed to the adsorbed states is accompanied by a decrease in the average radius of gyration due to a spatial confinement of the PE chain. Additionally, stronger screening of the ES interactions reduces the monomer–monomer repulsion along the PE, leading to less extended configurations of the PE chain.

The overall *nonmonotonic* dependence of $\sqrt{\langle R_g^2(n_0)\rangle}$ is a clear manifestation of a compactification transition at relatively low salinities and of a decompactification of the PE chain at considerably higher salt concentrations. The latter PE adsorption–desorption transition takes place in the range of $\kappa a \approx 4\ldots5$. Despite strong ES screening, which leads to overall more compact PE chain configurations, the increase in $\langle R_g^2\rangle^{1/2}$ reported at these κa values in Figure 3 is caused by the contribution of configurations in the desorbed state, having larger values of the radius of gyration.

3.3. *Adsorption–Desorption Curve and Critical Charge Density*

The discontinuous transitions shown in Figure 2b enable us to define a *critical* charge density of the negative patch, $\sigma_{n,c}$. The latter is defined when the fraction of the number of configurations in which the PE is in the adsorbed state is exactly 1/2. This criterion was also adopted in several of our previous studies, see, i.a., refs. [48,54,57] for the details on the methodology of determining the value of σ_c based on the analysis of the simulated PE configurations.

Figure 4 shows $\sigma_{n,c}$ as a function of κa for two values of the positive charge densities, namely $\sigma_p = 0.03\ \mathrm{C/m^2}$ (the dark-blue curve) and $\sigma_p = 0.01\ \mathrm{C/m^2}$ (the light-blue curve), computed with the same patch size as in Figure 2. As indicated in Figure 4, the region above the curves comprises the values of σ_n and κa at which the PE chain is adsorbed to the patchy particle, while in the regions below the respective critical curves, stable adsorption is not observed. As seen in Figure 4, for the region of small solution salinities of $\kappa a \lesssim 1\ldots3$, when the PE chain mostly experiences the charge of the entire particle, its net charge is particularly important as a limiting factor of adsorption; in contrast, for $\kappa a \gg 1$, the PE interacts mainly *locally* and the long-range ES interactions are strongly screened. Thus, the charge density of the negative patch needed to have adsorption is large, in order to compensate for the repulsions stemming from the positive portion of the spherical particle.

At low ionic strengths, the critical surface charge density $\sigma_{n,c}$ of the ES–attractive negative cap reaches a value of $\sigma_{n,c} \approx 0.05\ \mathrm{C/m^2}$ for $\sigma_p = 0.03\ \mathrm{C/m^2}$ and of $\sigma_{n,c} \approx 0.02\ \mathrm{C/m^2}$ for the situation $\sigma_p = 0.01\ \mathrm{C/m^2}$, respectively. These situations for the positive cap charge densities are considered in Figure 4. As expected, smaller ES–attractive negative spherical caps should have a larger magnitude of the surface charge density in order to adsorb the positively charged PE chain, which is being ES repelled from the larger positive cap on the opposite side of the particle. For both values of σ_p considered, the computed critical values of $\sigma_{n,c}$ are *smaller* than the negative surface charge densities needed to fully neutralize the particle (see the respective arrows for the densities that would produce net-neutral particles on the right side of Figure 4). We thus observe the complexation of the PE onto a *likely charged* particle. A similar physical phenomenon is observed for the PE–protein complexation on the "wrong side" of the isoelectric point [82,87].

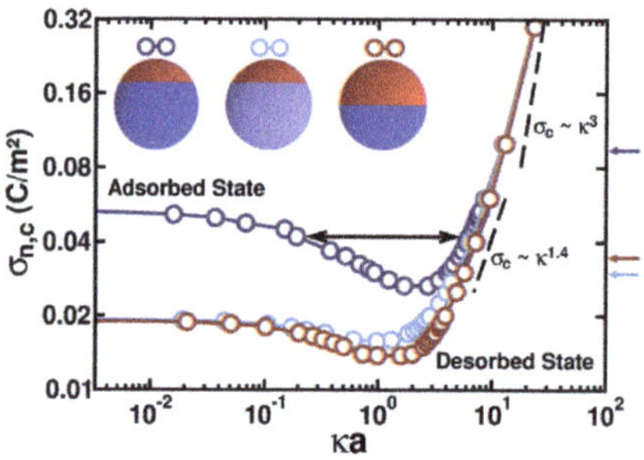

Figure 4. Magnitude of the critical surface charge density $\sigma_{n,c}$ as a function of κa for a PE chain with $N = 50$ monomers. The particles represented by the dark-blue and light-blue curves have starting densities of $\sigma_p = 0.03\ \mathrm{C/m^2}$ and $0.01\ \mathrm{C/m^2}$, respectively, with $\theta_p = \theta_n = 120°$, whereas the particle represented by the dark-red curve has $\sigma_p = 0.03\ \mathrm{C/m^2}$ and $\theta_p = \theta_n = 90°$ (see the legend in the plot). The arrows of the respective colors on the right axis indicate (for the $\sigma_c(\kappa a)$-data presented) the densities where the net charge of the respective particles (without the adsorbed PE chain) is equal to zero; that is, net-neutral Janus-type particles. The scaling relations of the type (1) at intermediate and large κa values are indicated as the dashed lines. For the black-circle data, the region of κa indicated by the double arrow shows the critical charge densities at which—upon increasing the value of κa — the desorbed-to-adsorbed and the adsorbed-to-desorbed transitions take place (see also the description in the text).

The effect of a variable patch size is now examined; see the dark-red curve in Figure 4 plotted for the parameters $\theta_n = 90°$ and $\sigma_p = 0.03\ \mathrm{C/m^2}$. The critical charge density $\sigma_{n,c}$ needed to trigger the PE adsorption onto a particle with these properties is similar to what we have obtained for the situation $\sigma_p = 0.01\ \mathrm{C/m^2}$ and $\theta_n = 120°$. The decrease in the positively charged area of the particle is thus compensated by the increasing of the surface charge density σ_p, as intuitively expected. In particular, for small κa values, the particles with a *large positively charged cap*—which is ES repelling the positively charged PE chain—require much *larger* values of surface charge density on the *smaller negatively charged* cap. The latter is the one eventually attracting the PE and triggering the ES-driven adsorption.

As a support of this statement, the reader should compare the low-salt behavior of the dark-blue and dark-red curves in Figure 4. In these cases, the value of $\sigma_{n,c}$ decreases with increasing κa, and $\sigma_{n,c}$ reaches a *minimum* at $\kappa a \approx 1.5$; see Figure 4. Under these conditions, the Debye screening length becomes comparable to the size of the negative patch of the particle, i.e., $\kappa^{-1} \sim a$, thus reducing the influence of particle's net charge.

All of the curves in Figure 4 converge (at a fixed a value) for $\kappa a \gg 1$ to the state dominated by the *local* ES interactions; this is the regime of strongly screened ES. In this regime, we recover the behavior observed for the positively charged PE adsorbed onto a homogeneously negatively charged planar surface, with the known scaling relation $\sigma_{n,c}(\kappa) \sim \kappa^3$ [5,41]. For an experimental range of salinities and particle dimensions, with a typical range of $5 \lesssim \kappa a \lesssim 15$ [33,74,75], the scaling relations predicted from our simulations are in agreement with the ones observed for the PE complexation with several types of spherical colloids, namely $\sigma_c(\kappa) \sim \kappa^{1.4}$, as indicated in Figure 4 by the dashed line.

We stress here that the nonmonotonic behavior of $\sigma_{n,c}$ versus κa presented in Figure 4 gives rise to the existence of *two phase transitions* in our computer simulations. Namely, with an increasing κa value, the PE goes from the desorbed to the adsorbed state at some critical surface charge density $\sigma_{n,c}$, then at a larger κa value and at the same $\sigma_{n,c}$ value, a transition from the adsorbed to the desorbed state takes place. Note that this nonmonotonic $\sigma_{n,c}(\kappa a)$–behavior emerges for the adsorption of only *similarly charged* PE onto a spherical particle (for the *oppositely charged* PEs and surfaces of any geometries, the critical surface charge density always *increases* with κa [41]). For the PE adsorption onto a net-neutral Janus particle treated in ref. [48], a monotonic behavior of σ_c versus κa was also observed.

Also note here that the so-called resolubilization or decondensation at comparatively high concentrations of the added salt is a well-studied phenomenon for the complex formation of DNA molecules with a number of DNA-condensing cations and (at least locally) positively charged proteins [25,111–113].

3.4. Implications of Low-Dielectric Interfaces

Now, we discuss the main new aspect of the current study, namely the effects of a low-dielectric interior of patchy heterogeneously charged particles on the characteristics of critical PE–particle adsorption.

The low-ϵ_{part} interior of homogeneously charged surfaces causes a *weaker* PE adsorption due to the ES-induced repulsion of the PE monomers from their image charges [114] being formed underneath the adsorbing interface; see refs. [43,61] for the investigation of these effects on the PE adsorption. For the PE adsorption onto a patchy particle, in addition to the ES potential of the polarization charges given in Equation (10), the potential due to the particle-charge distribution described in Equation (4) is also affected by a different/low value of the dielectric constant under the adsorbing boundary; see Equation (4). We incorporated these effects of a low-dielectric interior in the model and performed the respective computer simulations of PE adsorption onto such spherical low-dielectric patchy particles.

Figure 5 presents the monomer-density distribution as a function of the distance from the particle center for four different patch sizes with $\theta_p = \theta_n$ (i.e., no neutral region on the spherical particle, see Figure 1a). In Figure 5a, we show the case without a dielectric interface; that is, $\epsilon_{part} = \epsilon_{water} = 78.7$, while in Figure 5b, the patchy spherical particle has a low internal dielectric constant of $\epsilon_{part} = 4$. Thus, it is possible to separately analyze the influence of the charge-patch size and of the internal dielectric constant on the PE–particle adsorption.

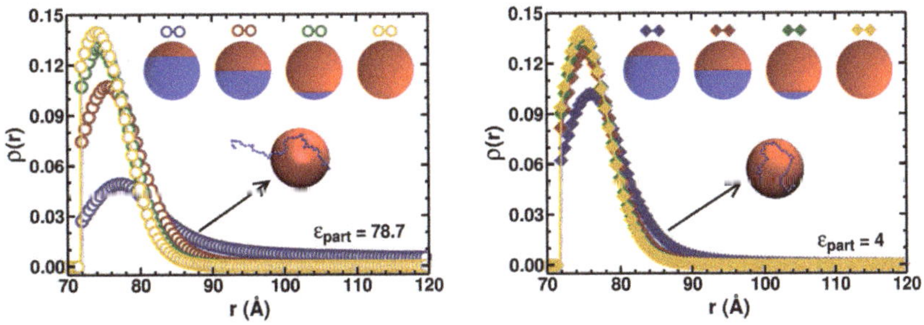

(a) (b)

Figure 5. Density distribution of the PE monomers under the conditions *above* the adsorption–desorption transition as a function of the distance from the center of a particle with a dielectric constant of (**a**) $\epsilon_{part} = 78.7$ (empty circles) and (**b**) $\epsilon_{part} = 4$ (filled diamonds). The particle radius is $a = 70$ Å, the surface charge densities are $\sigma_n = \sigma_p = 0.03\ \mathrm{C/m^2}$. For the dark-blue, dark-red, dark-green, and dark-yellow curves, the patch sizes are defined by $\theta_p = \theta_n = \{120°, 90°, 45°, 0°\}$, respectively, as indicated in the legends. The salt concentration in the solution is fixed at 5 mM (corresponding to $\kappa a = 1.625$, valid for the curves of all colors presented in these plots).

Starting with the case of the homogeneously charged sphere—i.e., $\theta_p = \theta_n = 0$, see the dark-yellow curves in Figure 5a,b—the distribution of the PE monomers in the presence of a dielectric interface/boundary is slightly less confined compared to the case with a uniform-dielectric medium (compare the dark-yellow curves in panels (a) and (b) of

Figure 5). As mentioned above, this is caused by the repulsive potentials of the polarization charges, similar to the trends predicted in ref. [43] based on the analytical WKB-type computations of the adsorption onto low-dielectric interfaces. The leading-order term of the ES potential of the particle (see the $l = 0$ term in Equation (4)) does not change with variations in the internal dielectric constant. The potential due to the polarization charges (given in Equation (10)) is responsible for a *decrease* in the binding affinity, causing the PE configurations be *less confined*, as compared to those for the PE–particle interactions for the situation without a dielectric interface.

As a positively charged region on the surface of the patchy spherical particle *increases* in size (i.e., when the values of $\theta_p = \theta_n$ increase; see the caption of Figure 5), the PE-monomer distribution becomes *less* confined close to the surface due to the ES-driven repulsion from a positively charged surface patch. When the dielectric constant of the particle is much smaller than that of the solvent, $\epsilon_{\text{part}} \ll \epsilon_{\text{water}}$, a more attractive ES potential generated by the particle overcomes the repulsion of the PE monomers from their image charges. This results in a more confined adsorbed PE chain than in the scenario with no dielectric interfaces.

Figure 6 presents the critical negative charge density $\sigma_{n,c}$ as a function of κa for a particle with the patch sizes defined by $\theta_p = \theta_n = 120°$. The computed adsorption–desorption demarcation boundaries demonstrate that it is possible to also see a significant attraction of the PE chain to the particle with a low-dielectric core; compare the empty circles and filled diamonds in Figure 6. Despite a repulsive ES contribution of the induced polarization charges in the scenario $\epsilon_{\text{part}} \ll \epsilon_{\text{water}}$, a stronger attractive contribution emerges due to the existence of a region of the particle with a *similar* charge to that of the PE chain. Therefore, we do find from the simulations that somewhat *smaller* negative surface charge densities are needed to promote the ES-driven adsorption of the positively charged PE chain in this case. The effect is not dramatic, but the differences we detect in the respective $\sigma_{n,c}$ values are clear and systematic; these differences are in the same direction in the entire range of the salt concentrations considered in our simulations (the empty circles are above the filled diamonds in Figure 6).

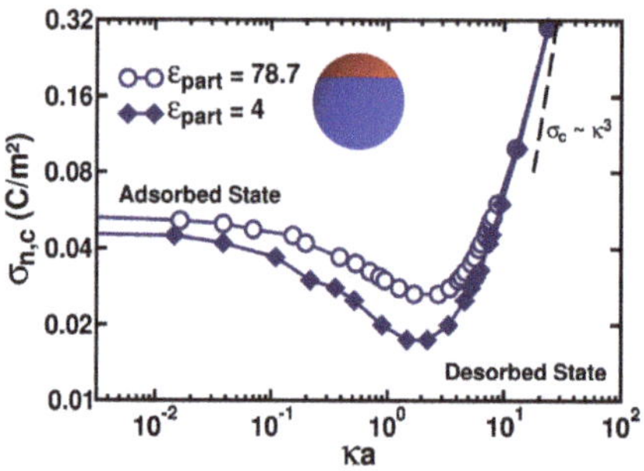

Figure 6. Critical surface charge density of the negative patch $\sigma_{n,c}$ plotted as a function of κa for a PE chain with $N = 50$ monomers. The values of the parameters are as follows: the positive charge density is $\sigma_p = 0.03$ C/m², the patch sizes are defined by $\theta_p = \theta_n = 120°$, and the sphere radius is $a = 70$ Å. The empty dark-blue circles represent a particle with a dielectric constant equal to that of the medium, while the filled dark-blue diamonds are for the situation $\epsilon_{\text{part}} \ll \epsilon_{\text{water}}$; see the legend.

Figure 7 presents the critical values of $\kappa_c a$ as a function of the patch-controlling angles $\theta_p = \theta_n$ for the cases with and without the dielectric interface. This time, for the *fixed* $\sigma_{n,p}$ values of the charge densities of the spherical caps and for *varying* patch angles, we conducted systematic computer simulations to rationalize the critical-adsorption conditions. From the simulation data, however, the values of critical $\kappa_c a$ are now being extracted (for the situation of fixed surface charge densities of the caps). The criterion to calculate the critical κ_c values here is identical to that used to enumerate the $\sigma_{n,c}$ values in Figure 4. Namely, it is a 50%–50% distribution between the adsorbed and desorbed conformations of the polymer, as recorded during the simulation time. These cases are shown by the filled diamonds and empty circles in Figure 7, respectively. We remind the reader here that the adsorbed state exists for the conditions below the shown adsorption–desorption boundary curves, while the desorbed state is above them.

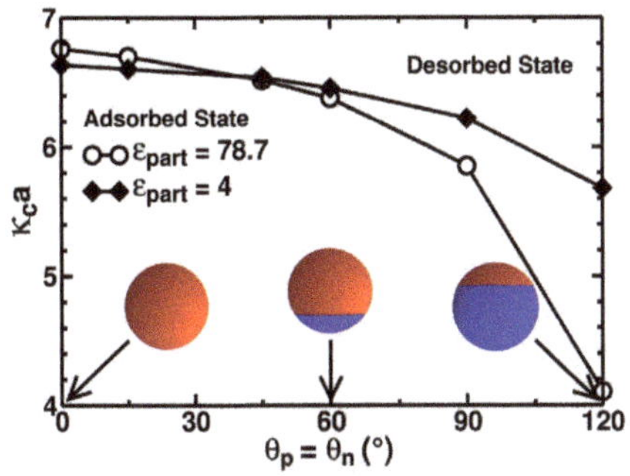

Figure 7. Critical dimensionless solution-salinity parameter $\kappa_c a$ plotted as a function of the patch dimensions on the spherical particle. The charge densities of the spherical caps are *fixed* at $\sigma_p = \sigma_n = 0.03$ C/m². The empty black circles represent a particle with a dielectric constant equal to that of the medium, while the filled black diamonds are for the situation $\epsilon_{\text{part}} \ll \epsilon_{\text{water}}$.

We observe that as $\theta_p = \theta_n$ decreases and the Janus-type patchy particle effectively approaches a homogeneously charged sphere, the PE chain experiences a weaker attraction to the particles with a *low* dielectric constant inside them due to the repulsive contribution of the polarization charges, as given in Equation (10). This is evidenced by a somewhat lower ionic concentration needed to promote the PE desorption for the case with $\epsilon_{\text{part}} = 4$ (compare the values of $\kappa_c a$ on the left side of Figure 7). As the size of the positive patch increases, this behavior gets reverted at $\theta_p = \theta_n \sim 45°$, and the critical $\kappa_c a$ value becomes substantially larger for the case of the particle with the low-dielectric interior (compare for this the values of $\kappa_c a$ on the right side of Figure 7).

This can be understood as follows: the ES potential of the polarization charges Ψ_{polar} given by Expression (10) generated by the presence of a monomer close to the particle with a low dielectric constant is always repulsive and is independent of the particle charge distribution. When the particle is homogeneously charged, carrying a charge opposite to that of the adsorbing PE chain, the ES potential Ψ_σ is attractive (largely independent of the internal dielectric constant of the particle). The sum of these two contributions for the case when the particle has a low dielectric constant leads to *weaker* adsorption compared to the case when the dielectric interface is absent, i.e., when $\Psi_{\text{polar}} = 0$, as can be seen in Figure 7 for $\theta_p = \theta_n = 0$. With an increase of $\theta_p = \theta_n$, the emergence of a patch similarly charged to the PE on the particle surface causes Ψ_σ to increase (i.e., for the convention of the signs chosen, the potential Ψ_σ becomes less negative). Importantly, this variation in the potential is more intense for the case when the particle has a high dielectric constant. This variation for a high-dielectric case *compensates* for the summation of the repulsive contribution of the polarization charges in the low-dielectric case, and for $\theta_p = \theta_n > 45^o$ (see Figure 7), the total potential in the low-dielectric case becomes *more attractive* to the PE chain.

4. Discussion, Conclusions, and Perspectives

In this work, we investigated the characteristics of the adsorption of a positively charged PE chain onto spherical Janus-type particles featuring patches of both negative and positive charges on their surface. The particular focus was on the effect of a dielectric interface between such a patchy particle and the surrounding solvent. A key finding was the observation of two distinct adsorption–desorption transitions for PE interacting with particles bearing a net charge of the same sign. This phenomenon can be discussed in terms of either an energetic or a conformational transition. These transitions were found to be characterized by abrupt changes in the mean binding energy, reflecting the dynamical transitions between the adsorbed (more compact) and desorbed (more extended) states of the PE. The maximum of the PE–particle binding affinity was found to occur at $\kappa a \approx 1.5$, i.e., when the Debye length is comparable to the size of the negative patch, $\lambda_D = 1/\kappa \approx 46.7$ Å for the particle radius $a = 70$ Å used in the simulations.

The presence of a dielectric interface leads to *enhanced* PE adsorption onto patchy particles. Note that earlier theoretical results for the PE adsorption onto *homogeneously* charged particles with a low-dielectric interior have predicted a *reduction* in PE adsorption [43]. However, as the particle approaches a uniformly charged sphere, this trend of the enhanced adsorption reverses, in agreement with the previous findings in the literature [28,61]. The enhanced adsorption observed here in the situation of heterogeneously charged Janus-type spherical particles can be attributed to a competition between the more attractive effects generated by a patchy particle with a low internal dielectric constant and the repulsion of the PE monomers from their image charges. Regarding the effects of the dielectric-constant mismatch, a recent study [60] has also demonstrated an increased binding affinity of the PEs to oppositely charged surfaces bearing discrete charge patterns. In the study [60], the localized charged sites acted as "bridges" between the monomers and their image charges; an effect further amplified by a dielectric mismatch.

In general, any distribution of patches of electric charges in terms of their size and geometric arrangement is possible on the spherical surface within the current model. This variability from the ES viewpoint, plus the low-dielectric particle interior—one of the main novelty of this study—is ultimately needed to apply such a model *quantitatively* to the real-world scenarios of PE–protein ES-driven complexation and solubilization (as in the works of Dubin mentioned in the introduction). As any charge distribution can be treated in terms of the resulting ES potential, the PDB data for a given globular protein can be directly imported into the ES model of this type. This is, evidently, our next step in this direction. Further steps into a "terra incognita" include the issues of accommodating more generalized/sophisticated conditions, such as some particular multivalent ions, the impact of a pH-regulated charged groups, implications of the charge regulation, and more realistic (non-ideally spherical) protein geometries.

From the perspective of applications, one can highlight the agreement of our observed trends with the experimental results of the *nonmonotonic* behavior for the binding affinity as a function of the salt concentration observed for heparin–antithrombin [81] and heparin–insulin [76] complexes. As mentioned in Section 1.2, the latter study examined the properties of the interactions between the polyanion heparin and insulin protein. Insulin is a type of globular protein and contains clearly defined positively and negatively charged caps on its surface; see Figure 8. The authors of ref. [76] have demonstrated the existence of transitions between the complexed/compact and noncomplexed/decondensed states at a given critical value of the pH, namely at pH_c [76].

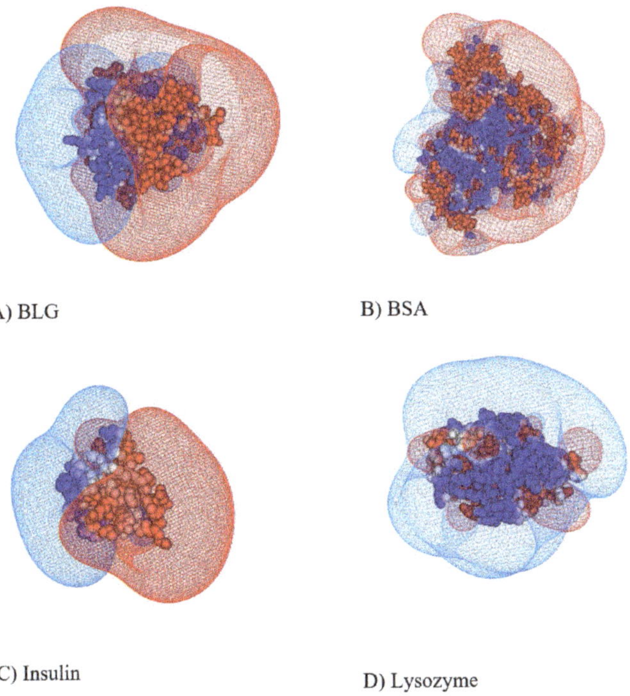

Figure 8. Distribution of the ES potential on the surface of four abundant dipolar-type protein molecules at pH 7 and ionic strength of 0.15 M for bovine serum albumin (BSA), lysozyme, insulin, and β-lactoglobulin (BLG). Reprinted with permission from ref. [76]. Copyright 2003 American Chemical Society.

The variation in the ionic strength was shown to exhibit *nonmonotonic* behavior of the pH_c as a function of κ; see Figure 2 in ref. [76]. The maximal pH value was shown to be attained when the Debye length $1/\kappa$ became of the order of the protein radius ($\kappa a \approx 1$). We observed a very similar trend for the critical adsorption condition of $\sigma_{n,c}$, as reported in Figures 4 and 6. The current study thus enables one to extend this analytical and simulational approach to more realistic systems of, e.g., numerous important PE–protein complexes, as mentioned in Section 1.1.

One more example of the role of the ES interactions in certain biological functions of some nucleic acid-binding proteins is the M2-1 protein of the human Respiratory Syncytial Virus (hRSV) [115]. M2-1 is a transcriptional anti-termination factor of the viral polymerase and its globular domain displays a heterogeneous ES-potential surface, including a positively charged patch, playing a key role in the interactions that M2-1 establishes with RNA [115].

Another example involves the so-called cold-shock proteins (Csp), acting to preserve cell viability at low temperatures by binding to nucleic acids and thus modulating the gene expression. Caruso et al. [116] recently demonstrated that the CspA of *Corynebacterium pseudotuberculosis* also exhibits an asymmetric surface of the ES potential, containing a positively charged patch identified as the primary site of interactions with the single-stranded DNA (Y-box sequence).

The role of image charges quantified here can also enable a more detailed analysis and, hopefully, serve for a deeper interpretation of the data on PE–protein complexation, i.a., at varying solution salinities. The effects of image charges are of importance both for wrapping and complexation of very flexible PEs such as single-stranded DNA/RNA (i.a., upon their wrapping [23,117] around single-wall carbon nanotubes [118,119]) and PAA [28] molecules, for intermediately flexible PEs such as Na-PSS, PDADMAC, and hyaluronic acid [28,74,77], as well as for semiflexible or rigid (depending of the external salinity) PEs such as the double-stranded DNA in its canonical high-humidity B-form.

Finally, the current model can include two types of specificity to study specific systems. The first one is regarding the charge-regulation effects, recognized to be important for PE–protein interactions [26,38]. These effects can result in an induced net charge or in a modification of the charge distribution of a protein due to the presence of the vicinal PE chain. To achieve this, a specific spatial distribution of the protein residues and of their respective pK_a values must be considered. Another specificity is the "structural" persistence length of the PE, in addition to the ES PE persistence already considered in the present simulations. This can be achieved by including a force field with the parameters dependent on a specific PE molecular structure.

Author Contributions: Conceptualization, S.J.d.C.; methodology, S.J.d.C. and D.L.Z.C.; software, D.A.A. and D.L.Z.C.; formal analysis, S.J.d.C. and D.L.Z.C.; investigation, D.A.A.; writing—original draft preparation, D.A.A. and I.P.C.; writing—review and editing, A.G.C., S.J.d.C., and D.L.Z.C.; supervision, S.J.d.C.; funding acquisition, S.J.d.C. and I.P.C. All authors have read and agreed to the published version of the manuscript.

Funding: This study was financed in part by the Coordenação de Aperfeiçoamento de Pessoal de Nível Superior–Brasil (CAPES)–Finance Code 001, and by the São Paulo Research Foundation (FAPESP), Brasil, Processes Numbers #2022/07231-7 and #2025/03288-2.

Data Availability Statement: All data used in this study appear in the submitted article.

Acknowledgments: This research was supported by resources supplied by the Center for Scientific Computing (NCC/GridUNESP) of the São Paulo State University (UNESP). I.P.C. and D.L.Z.C. acknowledge the support provided by PROPG/PROPe UNESP. S.J.C. and D.L.Z.C. acknowledge Guilherme Volpe Bossa for fruitful discussions. The authors thank E. Seyrek for a high-resolution image of Figure 8.

Conflicts of Interest: The authors declare no conflicts of interest.

Abbreviations

The following abbreviations are used in this manuscript:

PE	polyelectrolyte
ES	electrostatic / electrostatics
WKB	Wentzel–Kramers–Brillouin

Appendix A. ES Potential of a Patchy Particle

The dimensionless ES potential Ψ obeys the Laplace equation inside the particle ($\Psi = \Psi_{\text{part}}$ for $r \leq a$) and the linearized Poisson–Boltzmann equation in the external electrolyte solution ($\Psi = \Psi_{\text{water}}$ for $r > a$), i.e.,

$$\nabla^2\Psi_{\text{part}}(r,\theta) = 0 \quad \text{and} \quad \nabla^2\Psi_{\text{water}}(r,\theta) = \kappa^2\Psi_{\text{water}}. \tag{A1}$$

The Equation (A1) consist of a set of two second-order differential equations to be solved, subject to an appropriate set of four boundary conditions. Two of these conditions demand a finite potential Ψ_{part} at $r = 0$ and a vanishing Ψ_{water} at infinity. Then, we can write the following expansions [114]:

$$\Psi_{\text{part}}(r,\theta) = \sum_{l=0}^{\infty} A_l r^l P_l(\cos\theta) \quad \text{and} \quad \Psi_{\text{water}}(r,\theta) = \sum_{l=0}^{\infty} B_l k_l(\kappa r) P_l(\cos\theta). \tag{A2}$$

The remaining two conditions are related to the continuity of both the potentials and the electric displacements at the interface between the particle interior and the electrolyte,

$$\Psi_{\text{water}}(r = a,\theta) = \Psi_{\text{part}}(r = a,\theta) \tag{A3}$$

and

$$\epsilon_{\text{water}} \frac{\partial \Psi_{\text{water}}}{\partial r}\bigg|_{r=a} - \epsilon_{\text{part}} \frac{\partial \Psi_{\text{part}}}{\partial r}\bigg|_{r=a} = -\frac{e_0 \sigma(\theta)}{\epsilon_0 k_B T}. \tag{A4}$$

Upon applying (A3) and (A4), we obtain $A_l = B_l k_l(\kappa a)/a^l$ and

$$\sum_{l=0}^{\infty} B_l P_l(\cos\theta) \left[\epsilon_{\text{water}} \kappa k_l'(\kappa a) - \epsilon_{\text{part}} \frac{l k_l(\kappa a)}{a} \right] = -\frac{e_0 \sigma(\theta)}{\epsilon_0 k_B T}. \tag{A5}$$

In Equation (A1), we neglect the effects of high salt concentrations onto the form of the right-hand side of the Poisson–Boltzmann equation; see refs. [120,121]. We can find the solution to Equation (A5) by using the orthogonality conditions of the Legendre polynomials [98,114], i.e., $\int_{-1}^{1} P_m(x) P_n(x) dx = 2\delta_{mn}/(2m+1)$, where δ_{mn} denotes the Kronecker δ-symbol. Then, for an arbitrary $\sigma(\theta)$ distribution (2), the coefficients are as follows:

$$B_l = \frac{e_0}{2\epsilon_0 k_B T} \frac{(2l+1)a}{\epsilon_{\text{part}} l k_l(\kappa a) - \epsilon_{\text{water}} \kappa a k_l'(\kappa a)} \int_0^{\pi} \sigma(\theta) P_l(\cos\theta) \sin\theta d\theta. \tag{A6}$$

The ES potential $\Psi_{\text{water}}(r,\theta)$ in Equation (A2) with the expansion coefficients (A6) is exactly the surface-charge-induced ES potential $\Psi_\sigma(r,\theta)$ in Equation (4) of the main text.

Appendix B. ES Potential of the Polarization Charges at the Particle Surface

The ES potential of the polarization charges—emerging due to the presence of the jth-monomer of the PE close to the particle—is obtained from the ES potential of a charge q_j in an electrolyte solution in the presence of a dielectric sphere. In this case, the ES potentials inside and outside of the particle are given by (A1), with the addition of the ES potential from the monomer itself at the external region. As the potential should be finite at $r = 0$, and to vanish at infinity, we can write the expansions

$$\Psi_{\text{part}}(r,\theta,\phi) = \sum_{l=0}^{\infty} \sum_{m=-l}^{+l} A_{lm} r^l Y_{lm}(\theta,\phi) \tag{A7}$$

and

$$\Psi_{\text{water}}(r,\theta,\phi) = \sum_{l=0}^{\infty} \sum_{m=-l}^{+l} B_{lm} k_l(\kappa r) Y_{lm}(\theta,\phi) + \frac{l_B z_j}{e_0} \frac{e^{-\kappa|\vec{r}-\vec{r}_j|}}{|\vec{r}-\vec{r}_j|}, \tag{A8}$$

where $Y_{lm}(\theta, \phi)$ are the standard spherical harmonics [114]. The screened Coulomb potential can be written as the following expansion [122]

$$\frac{e^{-\kappa|\vec{r}-\vec{r}_j|}}{|\vec{r}-\vec{r}_j|} = \sum_{l=0}^{\infty} \sum_{m=-l}^{+l} 4\pi\kappa k_l(\kappa r_j) i_l(\kappa r) Y_{lm}(\theta, \phi) Y^*_{lm}(\theta_j, \phi_j), \tag{A9}$$

where $i_l(x)$ is the modified spherical Bessel functions of the first kind.

The potentials (A7) and (A8) are subject to the continuity of both the potentials themselves and of the electric displacements at the interface between the particle interior and the electrolyte solution. Applying these boundary conditions,

$$\Psi_{\rm water}(r = a, \theta) = \Psi_{\rm part}(r = a, \theta) \quad \text{and} \quad \epsilon_{\rm water} \frac{\partial \Psi_{\rm water}}{\partial r}\bigg|_{r=a} = \epsilon_{\rm part} \frac{\partial \Psi_{\rm part}}{\partial r}\bigg|_{r=a}, \tag{A10}$$

yields for the coefficients B_{lm} the result

$$B_{lm} = \frac{4\pi l_B z_j \kappa k_l(\kappa r_j)}{e_0} \left[\frac{\epsilon_{\rm water} \kappa a i'_l(\kappa a) - \epsilon_{\rm part} l i_l(\kappa a)}{\epsilon_{\rm part} l k_l(\kappa a) - \epsilon_{\rm water} \kappa a k'_l(\kappa a)} \right] Y^*_{lm}(\theta_j, \phi_j). \tag{A11}$$

The ES potential of the polarization charges is given by the first term in Equation (A8). Their final form is obtained by substituting (A11) into expression (A8) and by employing the so-called "addition theorem" for the spherical harmonics [98,114]

$$P_l(\cos\theta_{rj}) = \frac{4\pi}{2l+1} \sum_{m=-l}^{+l} Y_{lm}(\theta, \phi) Y^*_{lm}(\theta_j, \phi_j), \tag{A12}$$

which results in Equation (10) of the main text.

References

1. Oosawa, F. *Polyelectrolytes*; Marcel Dekker, Inc.: New York, NY, USA, 1971.
2. Grosberg, A.Y.; Khokhlov, A.R. *Statistical Physics of Macromolecules*; AIP Press: Melville, NY, USA, 1994.
3. Rubinstein, M.; Colby, R.H. *Polymer Physics*; OUP Oxford: Oxford, UK, 2003.
4. Muthukumar, M. 50th anniversary perspective: A perspective on polyelectrolyte solutions. *Macromolecules* **2017**, *50*, 9528–9560. [CrossRef] [PubMed]
5. Wiegel, F.W. Adsorption of a macromolecule to a charged surface. *J. Phys. A* **1977**, *10*, 299–303. [CrossRef]
6. Muthukumar, M. Adsorption of a polyelectrolyte chain to a charged surface. *J. Chem. Phys.* **1987**, *86*, 7230–7235. [CrossRef]
7. Von Goeler, F.; Muthukumar, M. Adsorption of polyelectrolytes onto curved surfaces. *J. Chem. Phys.* **1994**, *100*, 7796–7803. [CrossRef]
8. Eisenriegler, E.; Hanke, A.; Dietrich, S. Polymers interacting with spherical and rodlike particles. *Phys. Rev. E* **1996**, *54*, 1134–1152. [CrossRef]
9. Kong, C.Y.; Muthukumar, M. Monte Carlo study of adsorption of a polyelectrolyte onto charged surfaces. *J. Chem. Phys.* **1998**, *109*, 1522–1527. [CrossRef]
10. Netz, R.R.; Joanny, J.F. Complexation between a semiflexible polyelectrolyte and an oppositely charged sphere. *Macromolecules* **1999**, *32*, 9026–9040. [CrossRef]
11. Netz, R.R.; Joanny, J.F. Adsorption of semiflexible polyelectrolytes on charged planar surfaces: Charge compensation, charge reversal, and multilayer formation. *Macromolecules* **1999**, *32*, 9013–9025. [CrossRef]
12. Gurovitch, E.; Sens, P. Adsorption of polyelectrolyte onto a colloid of opposite charge. *Phys. Rev. Lett.* **1999**, *82*, 339–342. [CrossRef]
13. Ellis, M.; Kong, C.Y.; Muthukumar, M. Polyelectrolyte adsorption on heterogeneously charged surfaces. *J. Chem. Phys.* **2000**, *112*, 8723–8729. [CrossRef]
14. Dobrynin, A.V.; Deshkovski, A.; Rubinstein, M. Adsorption of polyelectrolytes at opppositely charged surfaces. *Macromolecules* **2001**, *34*, 3421–3436. [CrossRef]

15. Stoll, S.; Chodanowski, P. Polyelectrolyte adsorption on an oppositely charged spherical particle. Chain rigidity effects. *Macromolecules* **2002**, *35*, 9556–9562. [CrossRef]
16. McNamara, J.; Kong, C.Y.; Muthukumar, M. Monte Carlo studies of adsorption of a sequenced polyelectrolyte to patterned surfaces. *J. Chem. Phys.* **2002**, *117*, 5354–5360. [CrossRef]
17. Huang, Y.W.; Chun, K.Y.; Gupta, V.K. Adsorption of a polyelectrolyte on surfaces with manometer sized chemical patchiness. *Langmuir* **2003**, *19*, 2175–2180. [CrossRef]
18. Netz, R.R.; Andelman, D. Neutral and charged polymers at interfaces. *Phys. Rep.* **2003**, *380*, 1–95. [CrossRef]
19. Dobrynin, A.V.; Rubinstein, M. Effect of short-range interactions on polyelectrolyte adsorption at charged surfaces. *J. Phys. Chem. B* **2003**, *107*, 8260–8269. [CrossRef]
20. de Vries, R.; Weinbreck, F.; de Kruif, C.G. Theory of polyelectrolyte adsorption on heterogeneously charged surfaces applied to soluble protein–polyelectrolyte complexes. *J. Chem. Phys.* **2003**, *118*, 4649–4659. [CrossRef]
21. de Vries, R. Monte Carlo simulations of flexible polyanions complexing with whey proteins at their isoelectric point. *J. Chem. Phys.* **2004**, *120*, 3475–3481. [CrossRef]
22. Messina, R. Effect of image forces on polyelectrolyte adsorption at a charged surface. *Phys. Rev. E* **2004**, *70*, 051802. [CrossRef]
23. Cherstvy, A.G.; Winkler, R.G. Complexation of semiflexible chains with oppositely charged cylinder. *J. Chem. Phys.* **2004**, *120*, 9394–9400. [CrossRef]
24. Dobrynin, A.V.; Rubinstein, M. Theory of polyelectrolytes in solutions and at surfaces. *Progr. Pol. Sci.* **2005**, *30*, 1049–1118. [CrossRef]
25. Cherstvy, A.G.; Winkler, R.G. Simple model for overcharging of a sphere by a wrapped oppositely charged asymmetrically neutralized polyelectrolyte: Possible effects of helical charge distribution. *J. Phys. Chem. B* **2005**, *109*, 2962–2969. [CrossRef]
26. Barroso da Silva, F.L.; Lund, M.; Jönsson, B.; Åkesson, T. On the complexation of proteins and polyelectrolytes. *J. Phys. Chem. B* **2006**, *110*, 4459–4464. [CrossRef]
27. Belyi, V.A.; Muthukumar, M. Electrostatic origin of the genome packing in viruses. *Proc. Natl. Acad. Sci. USA* **2006**, 103, 17174–17178. [CrossRef]
28. Cherstvy, A.G.; Winkler, R.G. Strong and weak adsorptions of polyelectrolyte chains onto oppositely charged spheres. *J. Chem. Phys.* **2006**, *125*, 064904. [CrossRef] [PubMed]
29. Winkler, R.G.; Cherstvy, A.G. Critical adsorption of polyelectrolytes onto charged spherical colloids. *Phys. Rev. Lett.* **2006**, *96*, 066103. [CrossRef] [PubMed]
30. Ulrich, S.; Seijo, M.; Stoll, S. The many facets of polyelectrolytes and oppositely charged macroions complex formation. *Curr. Opin. Colloid Interface Sci.* **2006**, *11*, 268–272. [CrossRef]
31. Hoda, N.; Kumar, S. Brownian dynamics simulations of polyelectrolyte adsorption onto charged patterned surfaces. *Langmuir* **2007**, *23*, 1741–1751. [CrossRef]
32. Carrillo, J.M.Y.; Dobrynin, A.V. Molecular dynamics simulations of polyelectrolyte adsorption. *Langmuir* **2007**, *23*, 2472–2482. [CrossRef]
33. Winkler, R.G.; Cherstvy, A.G. Adsorption of weakly charged polyelectrolytes onto oppositely charged spherical colloids. *J. Phys. Chem. B* **2007**, *111*, 8486–8493. [CrossRef] [PubMed]
34. Hoda, N.; Kumar, S. Theory of polyelectrolyte adsorption onto surfaces patterned with charge and topography. *J. Chem. Phys.* **2008**, *128*, 124907. [CrossRef]
35. Hoda, N.; Kumar, S. Brownian dynamics simulations of polyelectrolyte adsorption in shear flow: Effects of solvent quality and charge patterning. *J. Chem. Phys.* **2008**, *128*, 164907. [CrossRef]
36. Man, X.; Yang, S.; Yan, D.; Shi, A.C. Adsorption and depletion of polyelectrolytes in charged cylindrical system within self-consistent field theory. *Macromolecules* **2008**, *41*, 5451–5456. [CrossRef]
37. Seijo, M.; Pohl, M.; Ulrich, S.; Stoll, S. Dielectric discontinuity effects on the adsorption of a linear polyelectrolyte at the surface of a neutral nanoparticle. *J. Chem. Phys.* **2009**, *131*, 174704. [CrossRef]
38. da Silva, F.L.B.; Jönsson, B. Polyelectrolyte–protein complexation driven by charge regulation. *Soft Matter* **2009**, *5*, 2862–2868. [CrossRef]
39. de Carvalho, S.J. First-order–like transition in salt-induced macroion-polyelectrolyte desorption. *Europhys. Lett.* **2010**, *92*, 18001. [CrossRef]
40. Wang, J.; Muthukumar, M. Encapsulation of a polyelectrolyte chain by an oppositely charged spherical surface. *J. Chem. Phys.* **2011**, *135*, 194901. [CrossRef] [PubMed]
41. Cherstvy, A.G.; Winkler, R.G. Polyelectrolyte adsorption onto oppositely charged interfaces: Unified approach for plane, cylinder, and sphere. *Phys. Chem. Chem. Phys.* **2011**, *13*, 11686–11693. [CrossRef] [PubMed]
42. Cherstvy, A.G. Critical polyelectrolyte adsorption under confinement: Planar slit, cylindrical pore, and spherical cavity. *Biopolymers* **2012**, *97*, 311–317. [CrossRef]

43. Cherstvy, A.G.; Winkler, R.G. Polyelectrolyte adsorption onto oppositely charged interfaces: Image-charge repulsion and surface curvature. *J. Phys. Chem. B* **2012**, *116*, 9838–9845. [CrossRef]
44. Nunes, S.C.C.; Pinto, P.; Pais, A.A.C.C. Nonrandom adsorption of polyelectrolyte chains on finite regularly charged surfaces. *J. Comput. Chem.* **2013**, *34*, 1198–1209. [CrossRef]
45. Winkler, R.G.; Cherstvy, A.G. Strong and weak polyelectrolyte adsorption onto oppositely charged curved surfaces. In *Polyelectrolyte Complexes in the Dispersed and Solid State I: Principles and Theory*; Springer: Berlin/Heidelberg, Germany, 2013; pp. 1–56.
46. de Carvalho, S.J.; Caetano, D.L.Z. Adsorption of polyelectrolytes onto oppositely charged cylindrical macroions. *J. Chem. Phys.* **2013**, *138*, 244909. [CrossRef]
47. Kampmann, T.A.; Boltz, H.H.; Kierfeld, J. Controlling adsorption of semiflexible polymers on planar and curved substrates. *J. Chem. Phys* **2013**, *139*, 034903. [CrossRef] [PubMed]
48. de Carvalho, S.J.; Metzler, R.; Cherstvy, A.G. Critical adsorption of polyelectrolytes onto charged Janus nanospheres. *Phys. Chem. Chem. Phys.* **2014**, *16*, 15539–15550. [CrossRef]
49. Yigit, C.; Heyda, J.; Ballauff, M.; Dzubiella, J. Like-charged protein-polyelectrolyte complexation driven by charge patches. *J. Chem. Phys.* **2015**, *143*, 064905. [CrossRef] [PubMed]
50. de Carvalho, S.J.; Metzler, R.; Cherstvy, A.G. Critical adsorption of polyelectrolytes onto planar and convex highly charged surfaces: The nonlinear Poisson-Boltzmann approach. *New J. Phys.* **2016**, *18*, 083037. [CrossRef]
51. Caetano, D.L.Z.; de Carvalho, S.J.; Metzler, R.; Cherstvy, A.G. Critical adsorption of periodic and random polyampholytes onto charged surfaces. *Phys. Chem. Chem. Phys.* **2017**, *19*, 23397–23413. [CrossRef] [PubMed]
52. Stornes, M.; Linse, P.; Dias, R.S. Monte Carlo simulations of complexation between weak polyelectrolytes and a charged nanoparticle. Influence of polyelectrolyte chain length and concentration. *Macromolecules* **2017**, *50*, 5978–5988. [CrossRef]
53. Shojaei, H.R.; Muthukumar, M. Adsorption and encapsulation of flexible polyelectrolytes in charged spherical vesicles. *J. Chem. Phys.* **2017**, *146*, 244901. [CrossRef]
54. Caetano, D.L.; de Carvalho, S.J. Conformational properties of block-polyampholytes adsorbed on charged cylindrical surfaces. *Euro. Phys. J. E* **2017**, *40*, 33. [CrossRef]
55. Xu, X.; Angioletti-Uberti, S.; Lu, Y.; Dzubiella, J.; Ballauff, M. Interaction of proteins with polyelectrolytes: Comparison of theory to experiment. *Langmuir* **2019**, *35*, 5373–5391. [CrossRef]
56. Bakhshandeh, A.; Segala, M. Adsorption of polyelectrolytes on charged microscopically patterned surfaces. *J. Mol. Liq.* **2019**, *294*, 111673. [CrossRef]
57. Caetano, D.L.Z.; de Carvalho, S.J.; Metzler, R.; Cherstvy, A.G. Critical adsorption of multiple polyelectrolytes onto a nanosphere: Splitting the adsorption–desorption transition boundary. *J. R. Soc. Interface* **2020**, *17*, 20200199. [CrossRef]
58. Kim, S.; Sureka, H.V.; Kayitmazer, A.B.; Wang, G.; Swan, J.W.; Olsen, B.D. Effect of protein surface charge distribution on protein–polyelectrolyte complexation. *Biomacromolecules* **2020**, *21*, 3026–3037. [CrossRef]
59. Chang, Q.; Jiang, J. Adsorption of block-polyelectrolytes on an oppositely charged surface. *Macromolecules* **2021**, *54*, 4145–4153. [CrossRef]
60. Wang, R.; Ginzburg, V.V.; Jiang, J.; Wang, Z.G. Adsorption of a polyelectrolyte chain at dielectric surfaces: Effects of surface charge distribution and relative dielectric permittivity. *Macromolecules* **2023**, *56*, 7653–7662. [CrossRef]
61. Vahid, H.; Scacchi, A.; Sammalkorpi, M.; Ala-Nissila, T. Adsorption of polyelectrolytes in the presence of varying dielectric discontinuity between solution and substrate. *J. Chem. Phys.* **2024**, *161*, 134907. [CrossRef] [PubMed]
62. Morrison, G.; Thirumalai, D. Scaling regimes for wormlike chains confined to cylindrical surfaces under tension. *Eur. Phys. J. E* **2024**, *47*, 6. [CrossRef]
63. Africo, F.A.; Cherstvy, A.G.; Jurado de Carvalho, S. Critical adsorption of polyelectrolytes onto highly oppositely charged surfaces: Effects of charge renormalization. *J. Chem. Phys.* **2024**, *161*, 194905. [CrossRef] [PubMed]
64. Martens, C.M. Critical adsorption and charge reversal in polyelectrolyte solutions: Analytical mean-field theory. *J. Chem. Phys* **2024**, *161*, 054901. [CrossRef]
65. Yadav, V.; Jaiswal, P.K.; Podgornik, R.; Kumari, S. Charge regulation effect on nanoparticles interaction mediated by polyelectrolyte. *arXiv* **2025**, arXiv:2507.00549. [CrossRef]
66. Dubin, P.L.; Rigsbee, D.R.; McQuigg, D.W. Turbidimetric and dynamic light scattering studies of mixtures of cationic polymers and anionic mixed micelles. *J. Colloid Interface Sci.* **1985**, *105*, 509–515. [CrossRef]
67. Dubin, P.L.; Curran, M.E.; Hua, J. Critical linear charge density for binding of a weak polycation to an anionic/nonionic mixed micelle. *Langmuir* **1990**, *6*, 707–709. [CrossRef]
68. Park, J.M.; Muhoberac, B.B.; Dubin, P.L.; Xia, J. Effects of protein charge heterogeneity in protein-polyelectrolyte complexation. *Macromolecules* **1992**, *25*, 290–295. [CrossRef]
69. McQuigg, D.W.; Kaplan, J.I.; Dubin, P.L. Critical conditions for the binding of polyelectrolytes to small oppositely charged micelles. *J. Phys. Chem.* **1992**, *96*, 1973–1978. [CrossRef]

70. Li, Y.; Dubin, P.L.; Spindler, R.; Tomalia, D.A. Complex formation between poly (dimethyldiallylammonium chloride) and carboxylated starburst dendrimers. *Macromolecules* **1995**, *28*, 8426–8428. [CrossRef]
71. Shubin, V.; Samoshina, Y.; Menshikova, A.; Evseeva, T. Adsorption of cationic polyelectrolyte onto a model carboxylic latex and the influence of adsorbed polycation on the charge regulation at the latex surface. *Colloid Polym. Sci.* **1997**, *275*, 655–660. [CrossRef]
72. Yoshida, K.; Sokhakian, S.; Dubin, P.L. Binding of polycarboxylic acids to cationic mixed micelles: Effects of polymer counterion binding and polyion charge distribution. *J. Colloid Interface Sci.* **1998**, *205*, 257–264. [CrossRef]
73. Miura, N.; Dubin, P.L.; Moorefield, C.; Newkome, G. Complex formation by electrostatic interaction between carboxyl-terminated dendrimers and oppositely charged polyelectrolytes. *Langmuir* **1999**, *15*, 4245–4250. [CrossRef]
74. Zhang, H.; Ohbu, K.; Dubin, P.L. Binding of carboxy-terminated anionic/nonionic mixed micelles to a strong polycation: Critical conditions for complex formation. *Langmuir* **2000**, *16*, 9082–9086. [CrossRef]
75. Feng, X.H.; Dubin, P.L.; Zhang, H.W.; Kirton, G.F.; Bahadur, P.; Parotte, J. Critical conditions for binding of dimethyldodecylamine oxide micelles to polyanions of variable charge density. *Macromolecules* **2001**, *34*, 6373–6379. [CrossRef]
76. Seyrek, E.; Dubin, P.L.; Tribet, C.; Gamble, E.A. Ionic strength dependence of protein-polyelectrolyte interactions. *Biomacromolecules* **2003**, *4*, 273–282. [CrossRef]
77. Kayitmazer, A.B.; Seyrek, E.; Dubin, P.L.; Staggemeier, B.A. Influence of chain stiffness on the interaction of polyelectrolytes with oppositely charged micelles and proteins. *J. Phys. Chem. B* **2003**, *107*, 8158–8165. [CrossRef]
78. Kayitmazer, A.B.; Shaw, D.; Dubin, P.L. Role of polyelectrolyte persistence length in the binding of oppositely charged micelles, dendrimers, and protein to chitosan and poly(dimethyldiallyammonium chloride). *Macromolecules* **2005**, *38*, 5198–5204. [CrossRef]
79. Cooper, C.L.; Dubin, P.L.; Kayitmazer, A.B.; Turksen, S. Polyelectrolyte–protein complexes. *Curr. Opin. Colloid Interface Sci.* **2005**, *10*, 52–78. [CrossRef]
80. Kayitmazer, A.B.; Seeman, D.; Minsky, B.B.; Dubin, P.L.; Xu, Y. Protein–polyelectrolyte interactions. *Soft Matter* **2013**, *9*, 2553–2583. [CrossRef]
81. Minsky, B.B.; Zheng, B.; Dubin, P.L. Inhibition of antithrombin and bovine serum albumin native state aggregation by heparin. *Langmuir* **2014**, *30*, 278–287. [CrossRef] [PubMed]
82. Lunkad, R.; Barroso da Silva, F.L.; Košovan, P. Both charge-regulation and charge-patch distribution can drive adsorption on the wrong side of the isoelectric point. *J. Am. Chem. Soc.* **2022**, *144*, 1813–1825. [CrossRef]
83. Boroudjerdi, H.; Kim, Y.W.; Naji, A.; Netz, R.R.; Schlagberger, X.; Serr, A. Statics and dynamics of strongly charged soft matter. *Phys. Rep.* **2005**, *416*, 129–199. [CrossRef]
84. Langer, R.E. On the connection formulas and the solutions of the wave equation. *Phys. Rev.* **1937**, *51*, 669–676. [CrossRef]
85. Doi, M.; Edwards, S.F. *The Theory of Polymer Dynamics*; International Series of Monographs on Physics; Clarendon Press: Oxford, UK, 1988.
86. Muthukumar, M. Polymers under Confinement. In *Advances in Chemical Physics*; John Wiley & Sons, Ltd.: Hoboken, NJ, USA, 2012; Chapter 4, pp. 129–196.
87. Blanco, P.M.; Achetoni, M.M.; Garcés, J.L.; Madurga, S.; Mas, F.; Baieli, M.F.; Narambuena, C.F. Adsorption of flexible proteins in the 'wrong side' of the isoelectric point: Casein macropeptide as a model system. *Colloids Surf. B Biointerfaces* **2022**, *217*, 112617. [CrossRef]
88. Simončič, M.; Hritz, J.; Lukšič, M. Biomolecular complexation on the "wrong side": A case study of the influence of salts and sugars on the interactions between bovine serum albumin and sodium polystyrene sulfonate. *Biomacromolecules* **2022**, *23*, 4412–4426. [CrossRef] [PubMed]
89. Warshel, A.; Russell, S.T. Calculations of electrostatic interactions in biological systems and in solutions. *Q. Rev. Biophys.* **1984**, *17*, 283–422. [CrossRef]
90. Lund, M.; Åkesson, T.; Jönsson, B. Enhanced protein adsorption due to charge regulation. *Langmuir* **2005**, *21*, 8385–8388. [CrossRef]
91. Lund, M.; Jönsson, B. Charge regulation in biomolecular solution. *Q. Rev. Biophys.* **2013**, *46*, 265–281. [CrossRef]
92. Li, W.; Persson, B.A.; Morin, M.; Behrens, M.A.; Lund, M.; Zackrisson-Oskolkova, M. Charge-induced patchy attractions between proteins. *J. Phys. Chem. B* **2015**, *119*, 503–508. [CrossRef]
93. Lund, M. Anisotropic protein–protein interactions due to ion binding. *Colloids Surf. B Biointerfaces* **2016**, *137*, 17–21. [CrossRef] [PubMed]
94. Simonson, T.; Brooks, C.L. Charge screening and the dielectric constant of proteins: Insights from molecular dynamics. *J. Am. Chem. Soc.* **1996**, *118*, 8452–8458. [CrossRef]
95. Cherstvy, A.G. Electrostatics of DNA complexes with cationic lipid membranes. *J. Phys. Chem. B* **2007**, *111*, 7914–7927. [CrossRef] [PubMed]
96. Debye, P.; Hückel, E. Zur Theorie der Elektrolyte. I. Gefrierpunktserniedrigung und verwandte Erscheinungen. *Phys. Z.* **1923**, *XXIV*, 185–206.

97. Cherstvy, A.G. Electrostatic interactions in biological DNA-related systems. *Phys. Chem. Chem. Phys.* **2011**, *13*, 9942–9968. [CrossRef]
98. Morse, P.M.C.; Feshbach, H. *Methods of Theoretical Physics*; International Series in Pure and Applied Physics; McGraw-Hill: New York, NY, USA, 1953.
99. Hasted, J.B.; Ritson, D.M.; Collie, C.H. Dielectric properties of aqueous ionic solutions. Parts I and II. *J. Chem. Phys.* **1948**, *16*, 1–21. [CrossRef]
100. Levy, A.; Andelman, D.; Orland, H. Dielectric constant of ionic solutions: A field-theory approach. *Phys. Rev. Lett.* **2012**, *108*, 227801. [CrossRef] [PubMed]
101. Gavish, N.; Promislow, K. Dependence of the dielectric constant of electrolyte solutions on ionic concentration: A microfield approach. *Phys. Rev. E* **2016**, *94*, 012611. [CrossRef] [PubMed]
102. Yukawa, H. On the interaction of elementary particles. I. *Proc. Phys.-Math. Soc. Japan. 3rd Ser.* **1935**, *17*, 48–57.
103. Cherstvy, A.G. Effect of a low-dielectric interior on DNA electrostatic response to twisting and bending. *J. Phys. Chem. B* **2007**, *111*, 12933–12937. [CrossRef]
104. Netz, R.; Orland, H. Beyond Poisson-Boltzmann: Fluctuation effects and correlation functions. *Eur. Phys. J. E* **2000**, *1*, 203–214. [CrossRef]
105. Nguyen, T.T.; Grosberg, A.Y.; Shklovskii, B.I. Macroions in salty water with multivalent ions: Giant inversion of charge. *Phys. Rev. Lett.* **2000**, *85*, 1568–1571. [CrossRef]
106. Grosberg, A.Y.; Nguyen, T.T.; Shklovskii, B.I. Colloquium: The physics of charge inversion in chemical and biological systems. *Rev. Mod. Phys.* **2002**, *74*, 329–345. [CrossRef]
107. Levin, Y. Electrostatic correlations: From plasma to biology. *Rep. Progr. Phys.* **2002**, *65*, 1577–1632. [CrossRef]
108. dos Santos, A.P.; Diehl, A.; Levin, Y. Electrostatic correlations in colloidal suspensions: Density profiles and effective charges beyond the Poisson–Boltzmann theory. *J. Chem. Phys.* **2009**, *130*, 124110. [CrossRef]
109. Böttcher, C.J.F.; van Belle, O.C.; Bordewijk, P.; Rip, A.; Yue, D.D. Theory of electric polarization. *J. Electrochem. Soc.* **1974**, *121*, 211Ca. [CrossRef]
110. Frenkel, D.; Smit, B. *Understanding Molecular Simulation: From Algorithms to Applications*; Elsevier: London, UK, 2023.
111. Teif, V.B.; Bohinc, K. Condensed DNA: Condensing the concepts. *Progr. Biophys. Mol. Biol.* **2011**, *105*, 208–222. [CrossRef]
112. Cherstvy, A.G.; Teif, V.B. Electrostatic effect of H1-histone protein binding on nucleosome repeat length. *Phys. Biol.* **2014**, *11*, 044001. [CrossRef]
113. Collette, D.; Dunlap, D.; Finzi, L. Macromolecular crowding and DNA: Bridging the gap between In Vitro and In Vivo. *Intl. J. Mol. Sci.* **2023**, *24*, 17502. [CrossRef]
114. Jackson, J.D. *Classical Electrodynamics*, 3rd ed.; Wiley: New York, NY, USA, 1999.
115. Caruso, I.P.; Guimaraes, G.C.; Machado, V.B.; Fossey, M.A.; Willbold, D.; Almeida, F.C.L.; Souza, F.P. Biophysical and dynamic characterization of fine-funed binding of the human Respiratory Syncytial Virus M2-1 core domain to long RNAs. *J. Virol.* **2020**, *94*, jvi.01505. [CrossRef]
116. Caruso, I.P.; Panwalkar, V.; Coronado, M.A.; Dingley, A.J.; Cornélio, M.L.; Willbold, D.; Arni, R.K.; Eberle, R.J. Structure and interaction of *Corynebacterium pseudotuberculosis* cold shock protein A with Y-box single-stranded DNA fragment. *FEBS J.* **2018**, *285*, 372–390. [CrossRef]
117. Tani, M.; Wada, H. How a soft rod wraps around a rotating cylinder. *Phys. Rev. Lett.* **2024**, *132*, 058204. [CrossRef]
118. Zheng, M.; Jagota, A.; Semke, E.D.; Diner, B.A.; Mclean, R.S.; Lustig, S.R.; Richardson, R.E.; Tassi, N.G. DNA-assisted dispersion and separation of carbon nanotubes. *Nat. Mat.* **2003**, *2*, 338–342. [CrossRef] [PubMed]
119. Li, Y.; Wen, Y.; Beltrán, L.C.; Zhu, L.; Tian, S.; Liu, J.; Zhou, X.; Chen, P.; Egelman, E.H.; Zheng, M.; et al. Understanding DNA-encoded carbon nanotube sorting and sensing via sub-nm-resolution structural determination. *Scie. Adv.* **2025**, *11*, eadt9844. [CrossRef] [PubMed]
120. Eigen, M.; Wicke, E. The thermodynamics of electrolytes at higher concentration. *J. Phys. Chem.* **1954**, *58*, 702–714. [CrossRef]
121. Borukhov, I.; Andelman, D.; Orland, H. Steric effects in electrolytes: A modified Poisson-Boltzmann equation. *Phys. Rev. Lett.* **1997**, *79*, 435–438. [CrossRef]
122. Abramowitz, M.; Stegun, I. *Handbook of Mathematical Functions: With Formulas, Graphs, and Mathematical Tables*; Applied Mathematics Series; Dover Publications: Garden City, NY, USA, 1965.

polymers

MDPI

Review

Polymers in Physics, Chemistry and Biology: Behavior of Linear Polymers in Fractal Structures

Hector Eduardo Roman

Department of Physics, University of Milano-Bicocca, Piazza della Scienza 3, 20126 Milano, Italy; hector.roman@unimib.it

Abstract: We start presenting an overview on recent applications of linear polymers and networks in condensed matter physics, chemistry and biology by briefly discussing selected papers (published within 2022–2024) in some detail. They are organized into three main subsections: polymers in physics (further subdivided into simulations of coarse-grained models and structural properties of materials), chemistry (quantum mechanical calculations, environmental issues and rheological properties of viscoelastic composites) and biology (macromolecules, proteins and biomedical applications). The core of the work is devoted to a review of theoretical aspects of linear polymers, with emphasis on self-avoiding walk (SAW) chains, in regular lattices and in both deterministic and random fractal structures. Values of critical exponents describing the structure of SAWs in different environments are updated whenever available. The case of random fractal structures is modeled by percolation clusters at criticality, and the issue of multifractality, which is typical of these complex systems, is illustrated. Applications of these models are suggested, and references to known results in the literature are provided. A detailed discussion of the reptation method and its many interesting applications are provided. The problem of protein folding and protein evolution are also considered, and the key issues and open questions are highlighted. We include an experimental section on polymers which introduces the most relevant aspects of linear polymers relevant to this work. The last two sections are dedicated to applications, one in materials science, such as fractal features of plasma-treated polymeric materials surfaces and the growth of polymer thin films, and a second one in biology, by considering among others long linear polymers, such as DNA, confined within a finite domain.

Keywords: linear polymers; self-avoiding random walks; fractals; reptation; proteins and evolution; deposition of thin polymeric films; packing of DNA chains within the cell

1. Introduction

The study of polymers is tantamount to new scientific and technological developments during the last decades. Among them, linear polymers play an important role in this review, which focuses on the theoretical and simulation approaches of such structures embedded in complex environments, either in the form of single chains or a many polymer system such as a melt. The core of this work concerns statistical properties of single chains, modeled by self-avoiding walks (SAWs), confined within scale-invariant structures such as deterministic and random fractals. Accurate methods are discussed to generate long SAW chains in arbitrary geometries, of which the reptation algorithm plays a major role. The interest in the latter is justified by the many potential applications it can deliver to deal with both static as well as dynamic chain configurations. This review on SAWs complements the existing literature on the subject and should be useful to polymer scientists involved in complex modeling for the interpretation of experimental results.

Linear polymers are constituted by units which may display a broad type of structures, ranging from quite simple repeated ones (homopolymers) to different complex molecular units (heteropolymers) as in the case of proteins. For many applications, simplifications are introduced consisting of what is called coarse-grained modeling. These simplified models are still quite difficult to study theoretically and numerically due the long-range connectivity of the single monomers and their repulsive interactions at short distances.

To alert the reader to recent contributions from the vast polymer community on modeling and simulations aspects, we have selected papers published during 2022–2024 with the aim of illustrating as clearly as possible the impact and relevance of polymers in physics, chemistry and biology. The papers collection is confined to Section 2 aimed at illustrating the 'state-of-the-art' in these research areas, and each paper is discussed in sufficient detail so that the non-expert reader can obtain basic knowledge on the subject at first sight without the need to rely on external sources. Also, the subjects are clearly catalogued so that the reader can rapidly move on to the point of particular interest. We expect this format to be useful to obtain a broad idea of the issues of interest in polymer science today.

In addition to the state of the art (Section 2) and the theory of SAWs in fractal structures (Section 3), the overview includes for completeness three more sections: experimental aspects of polymers (Section 4), applications in materials science (Section 5) focusing on the use of plasma techniques, and applications in biology (Section 6) reviewing very recent results (from 2024) on the genome and aminocyanines. The experimental section is included to provide a basic description on linear polymers, such as, e.g., polyethylene, polypropylene, and plasma-treated polymeric surfaces, which is necessary to understand the experimental results to be discussed in Sections 5 and 6.

Applications in Section 5 cover recent results on the wetting properties of plasma-treated polymeric materials surfaces, which is a subject of current interest in the plasma community. This section brings in the concept of fractal scaling, which is essential to understand the hydrophilicity/hydrophobicity properties of these materials, making contact with many of the issues discussed in Section 3. A further intriguing application is the modeling of thin films formation by attaching linear polymers on plasma-treated surfaces, which is essentially based on an implementation of the reptation algorithm considered in Section 3. Last but not least are the applications in biology dealing with long polymer chains, such as the double-helix of the genome. Issues related to modeling the transcription of DNA and how to deal with many linear chains in confined geometries, such as the nucleus of a cell, are also discussed. The overview ends with the concluding remarks in Section 7.

2. State of the Art: Polymers in Physics, Chemistry and Biology

In the following, we discuss results published during 2022–2024 on polymers with the aim of covering, as thoroughly as possible, the theoretical aspects of current interest on these complex systems in the fields of physics, chemistry and biology. The selected papers are organized for convenience into eight main subsections with a brief description of the content in each case.

- Polymers in Physics:
    - (Section 2.1) Coarse-grained models and MD simulations: single polymer, melt, porous and fibrous materials, bottlebrushes, copolymers, surfactant, crystals, anomalous dynamics, translocation, polymerization, rubbers, polyelectrolytes, polymer through capillaries, charged polymers, bioplastics, branched polymers, synthetic hydrogels.
    - (Section 2.2) Structural properties, self-assembly, smart materials: block polymers, amphiphilic diblock copolymers, chains on graphene, knotted chains, additive manufacturing, conjugated polymers, plasma techniques.
- Polymers in Chemistry:
    - (Section 2.3) Structural properties: polymer electrolyte and fuel cells, rubber, thermoplastic polymers, star-branched molecules, dendrimers.
    - (Section 2.4) Quantum mechanical calculations and MD simulations: solubility of polymers, pyrolysis for polystyrene, insulating materials, expoxy resins, oilfield scales inhibitors.

- (Section 2.5) Environmental issues, batteries, multifunctional materials: aging, gas desorption, biodegradable polymers, gas separation membranes, polymers electrolytes for Li batteries, polymer actuators, 3D printing.
- (Section 2.6) Rheological properties, viscoelasticity: fluid dynamics, epoxy-matrix composites, entangled polymers, thermoplastics, viscoelasticity, orientational relaxation, polymers rheology.

- Polymers in Biology:
  - (Section 2.7) Macromolecules, proteins, structures and dynamics: large-scale molecular simulations, actin filaments, electric field effects on GLP-2 peptide, soft vesicle, 2D tethered semiflexible polymers, entangled polymer networks, chirality of polymer knots, model polymers vs. globular proteins, polyelectrolyte pore translocation, protein adsorption dynamics, nucleosome dynamics.
  - (Section 2.8) Biomedical applications: 3D polymer hydrophilic networks, skin repair issues, material models for limb orthoses.

2.1. Polymers in Physics: Coarse-Grained Models and MD Simulations

Single polymer: A simple linear polymer model is a self-avoiding walk (SAW) defined on a lattice (see Section 3 for more details). The lattice helps us to count the non-local contacts (between non-consecutive nearest neighbors along the chain) for a single chain so that one can study the effects of interactions between monomers if they become close in space [1–4]. If the interaction is attractive, the chain will eventually collapse below some critical temperature, which is called T_θ. In [5], the authors compute all possible SAWs up to 27 steps ($N = 28$ monomers), requiring about 4×10^{17} different configurations on the simple cubic lattice. They find $T_\theta \simeq 3.713(2)$, in units of the contact energy, which is comparable to accurate very long-chains calculations. They suggest a systematic approach to improve further on these results.

Polymer melt: If we move from a single-chain problem to a many-chain scenario, like in a polymer melt, the situation becomes even more intricate. These systems possess complex viscoelactic behavior due to interchain interactions. Rouse (1953) proposed a tractable model to describe such melts. Despite its popularity and widespread use [6–10], it has many drawbacks. In [11], the author reviews several aspects of the Rouse model, suggesting its inadequacy to describe realistically such complex polymer solutions. Other competing models, such as the Kirkwood–Riseman model, for concentrated solutions are reviewed together with a selection of computer simulation results to test the theoretical predictions. In particular, the so-called reptation model is discussed in association with the highly entangled chains regime. The Kirkwood–Riseman model is extended to include interchain hydrodynamic interactions, yielding pseudoviral series for the dependence on the concentration and molecular weight of the self-diffusion coefficient and shear viscosity [12].

Porous materials: Viscoelastic phase separation and the mechanical properties of a coarse-grained polymer melt model have been studied using MD simulations [13], applying an arrested phase separation strategy by quenching the melt at different times (Figure 1). The resulting systems consist of interpenetrating bicontinuous gas–solid phases, yielding a type of glassy structure characterized by a power-law relaxation decay. In the elastic regime, both the Young and shear moduli of the emerging porous structure attain low values compared to dense glassy systems. For large deformations, the stress–strain curves display a highly intermittent behavior characterized by repeated avalanches of plastic events. The work examines for the first time the role that polymer chain connectivity plays in determining the elastic and plastic responses of these glassy materials (see also [14–17]).

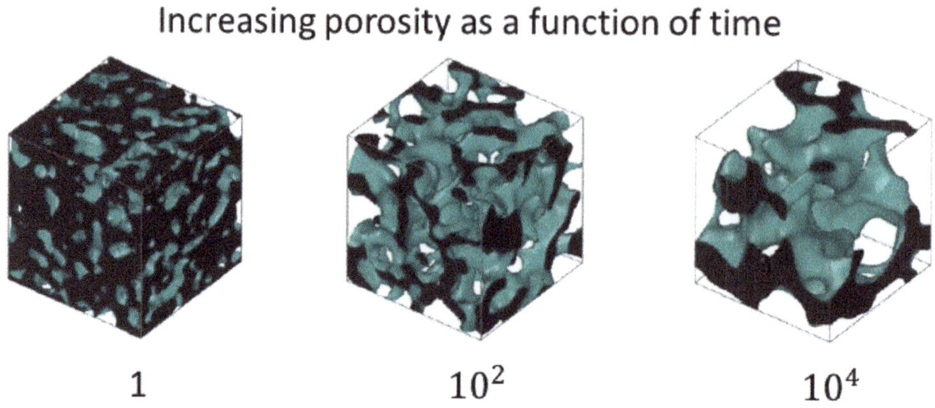

Figure 1. Molecular dynamics simulations of the quenching at temperature T of a melt of $N = 10^3$ linear polymer chains with $M = 20$ monomers each. Along the chain, nearest-neighbor monomers interact via a harmonic potential, and the remaining intra- and inter-chain monomers interact via a Lennard–Jones potential. Here, $T \simeq (1/4)\,T_g$ of the glass transition temperature T_g, and time increases from left to right (1, 10^2, 10^4) a.u. (adapted from [13]).

Bottlebrushes: Moving onto different polymer shapes, we consider next the case of molecular bottlebrushes (MBBs), which possess a well-defined linear backbone upon which side chains are densely grafted, constituting a type of branched macromolecules [18–21]. In [22], the authors aim to study the aggregation of MBBs in solvents of various qualities using dissipative particle dynamics (see Figure 2). They study the number of MBB clusters and their size distribution, allowing them to define a mean aggregation number. A low volume fraction of bottlebrushes is considered for different types of solvents and comparing different cutoffs criteria to define which bottlebrushes belong to a cluster. It is found that both short-lived transients and long-lived agglomerates can be present in the system. This work can be extended to study agglomerates of more complex macromolecules as well.

Amphiphilic copolymers: Going a bit up on polymer structure complexity, one finds the large family of amphiphilic copolymers, which comprise both hydrophilic and hydrophobic parts. The copolymers can be built into blocks, i.e., strings of one type of monomers, chemically linked in a linear or branched fashion, which in addition can carry a net charge as in the case of block polyelectrolytes [23–28]. In [29], the coarse-grained dissipative particle dynamics (DPD) simulation method is reviewed and its applications to the field of self-organization and self-assembly of amphiphilic copolymers are discussed. Of particular interest are polyelectrolytes with special attention to implementing explicit electrostatics in DPD numerical calculations. The importance and suitability of this coarse-grained method for understanding these complex polymer systems are stressed, and several relevant works are critically discussed.

Surfactant at water–oil interface: The physical phenomenology taking place at the interface between immiscible fluids is essential to understand the behavior of the systems in the presence of oil–water or oil–gas flows. Applications range from multilayer coating and phase separation to industrial oil mobilization. Coarse-grained modeling methods, such as DPD, are suitable to simulate multifluid systems on large spatial and temporal scales, which are much larger than those achieved using MD simulations [30–32]. In [33], DPD methods are employed to study the flow of oil and water in a narrow slit under different flow conditions, in which large surfactant molecules can be included in the simulations (see Figure 3). In the case of Poiseuille flow, a critical shear rate occurs beyond which the surfactants desorbed, forming micelles and thus destabilizing the interface, while under a Couette flow, they remain stable over the whole range of shear rates studied.

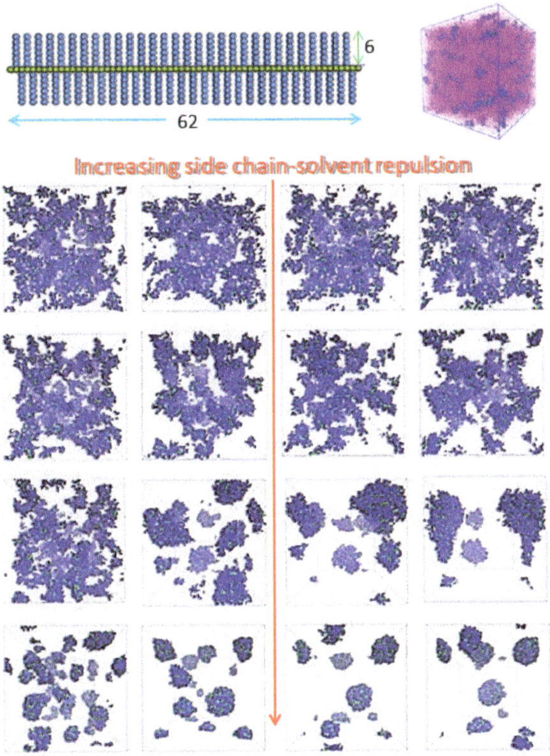

Figure 2. Mesoscale modeling of agglomeration of molecular bottlebrushes. (**Upper panel**) Bottlebrush architecture displaying the backbone beads (green circles) and the side chain ones (blue circles). In this example, the former are 62 and the side chain length is 6. On the right side, a snapshot of the equilibrated system of 77 bottlebrushes in the case of the lowest repulsion parameter between side chain beads and solvent molecules, the latter considered to be a good solvent. (**Lower panel**) Bottlebrushes structures at increasing repulsion parameter (poorer solvents) at four simulations time steps (from left to right): T = (10^5, 10^6, 2×10^6, 3×10^6) (adapted from [22]).

Polymer crystals: Arrays of homopolymers ordered in a crystalline structure are relevant to a number of physical and chemical systems such as solar cells, semiconductors, biological materials, or conventional plastics. Understanding polymer crystallization is still controversial despite the large amount of research performed over the years [34–37]. In [38], the authors present the results of large-scale Monte Carlo (MC) simulations based on a system of 54 chains of 1000 monomers (hard spheres) each by developing an MC method that is able to correctly sample polymer conformational space in an efficient fashion. This is particularly hard to achieve for very long chains at very high concentrations. The simulations start from a purely random configuration, which rapidly develops into an hexagonal closed packed (HCP) polymorph and is followed by a random hexagonal close packed (rHCP) morphology. The latter finally goes into a stable face-centered cubic (FCC) crystal of very high perfection. In [39], the authors also studied local and global order in dense packed linear, semiflexible polymers of hard spheres by extensive MC simulations as a function of volume fractions (see also [40–42]). They analyze the way in which packing density and chain stiffness affect the ability of the chains to self-organize at short and long length scales. The critical volume fraction for the phase transition is determined, and the different crystal morphologies are obtained, yielding a complete phase diagram as a function of packing density and equilibrium bending angle.

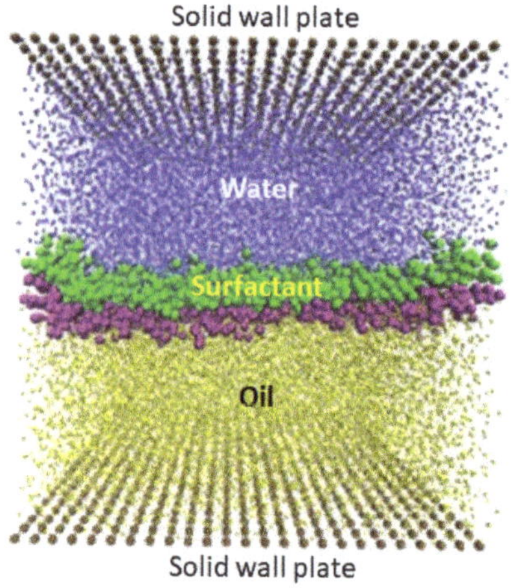

Figure 3. Simulations of a surfactant at the water–oil interface with 50% of water and oil between two solid walls under Poiseuille flow. As is apparent, all the surfactant molecules, representing sodium dodecylsulfate and octaethylene glycol monododecyl ether, remained at the water/oil interface with the hydrophilic head (green) beads facing toward the aqueous phase, whereas the hydrophobic tail (purple) beads tended to face the oil phase (adapted from [33]).

Fibrous materials: Modeling fibrous materials is based on the available statistical databases of their structure, often obtained in the form of 2D distribution functions, to predict actual 3D fiber orientations (see, e.g., [43–46]). In [47], the authors suggest new methods to generate more accurate 3D fiber orientation probability distributions from 2D data and identify a way to determine suitable parameters entering the Mises–Fisher random walks approach (Figure 4). The new method succesfully accounts for the observed projection distortion and orientation distributions in real fibrous materials. The problem of dealing with fiber–fiber overlaps, emerging when the fibers are randomly generated in the simulation domain, is considered, and efficient solutions are suggested to alleviate it.

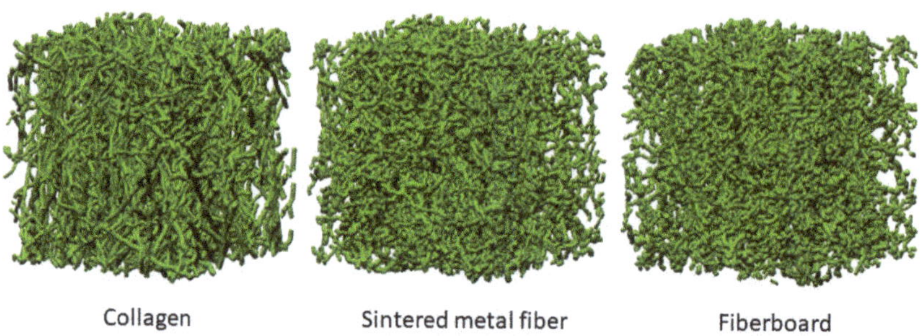

Figure 4. Simulations of fibrous materials. Shown are three cases with increasing (from left to right) tortuosity, using von Mises–Fisher directional probability distribution functions of random walks, aimed at modeling: Collagen, sintered metal fiber, and fiberboard, respectively (adapted from [47]).

Anomalous dynamics: Polymer melts often display short-time anomalous dynamics behavior in obvious contrast to their single-molecule counterparts (see, e.g., [48–51]). In [52], short homopolymer chains are considered by studying MD simulation chain trajectories in a melt. Microscopic molecular mechanisms yielding the observed anomalous behavior are discussed in terms of subdiffusive dynamics obtained from the analysis of van Hove distribution functions and other structural properties.

Translocation through a nanomembrane: Linear polymers can cross nanometer-sized pores when an external force is applied [53–55]. This translocation problem is relevant for understanding transport through nanomembranes, where typically the polymer moves under the presence of an electrostatic potential (see Figure 5). In [56], an analytical expression describing polymer flux through a nanopore is derived by assuming a linear dependence of the friction coefficient on the number of chain monomers N attraversing the pore and a parabolic behavior for the open-free (Gibbs) energy.

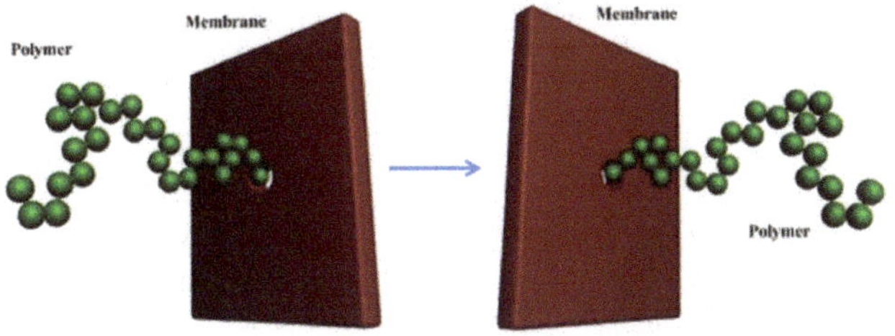

Figure 5. A linear polymer attempts to cross a membrane (from left to right) through a small pore, typically of nanometer size (adapted from [56]).

Polymerization (HDDMA): The radical polymerization process of acrylate compounds is studied in [57] to investigate the gel-point transition of 1,6-hexanediol dimethacrylate (HDDMA) [58–60] using a coarse-grained force field that reduces significantly the computer time. This model allows the simulation of much larger system sizes and smaller radical concentrations than in previous works (see Figure 6). Polymerization is modeled using reactive classical MD extended upon the implementation of a non-equilibrium MD scheme. A full cluster analysis of the resulting network substructures is performed, which is in good agreement with previously published all-atom calculations.

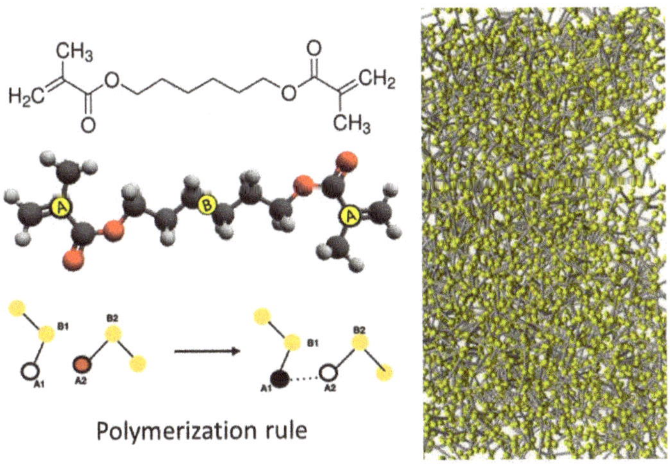

Figure 6. Polymerization of 1,6-hexanediol dimethacrylate (HDDMA) aimed at studying the gel-point transition. (**Left panel**) Structural formula of HDDMA (top image), molecular structure (middle image), and polymerization rule (bottom image). In the latter, a new bond is created between the atoms (A1, A2) if a reaction occurs, and two angle potentials, (B1-A1-A2) and (A1-A2-B2), are added to the chain, where the reactive atom is drawn in red and the potentially reactive one in white, while the inactive one is drawn in black. (**Right panel**) Snapshot of the polymeric structure for the initial radical concentration of 3% (adapted from [57]).

Nitrile butadiene rubber: Nitrile butadiene rubber (NBR) has excellent oil resistance and mechanical properties, being used to seal hydraulic systems. The problem arises when NBR deteriorates after undergoing thermal cycles, the action of oxygen and other chemicals, in addition to mechanical stress, leading to a loss of elasticity with deleterious consequences to the container working conditions. To gain insight into NBR aging (see, e.g., [61–64]), MD simulations have been implemented [65], using well-known force field (COMPASS) potentials [66] (see also [67]), where hydroxyl groups and carbonyl groups are introduced into the rubber molecular chains that mimic oxidative aging (see Figure 7). It becomes clear that elevated temperatures can dramatically increase the mobility of rubber molecular chains and fractional free volume, while strain compression results in further packing of the chains. A comprehensive picture thus emerges that helps understanding the effects of oxidative aging on rubber materials at the very molecular level.

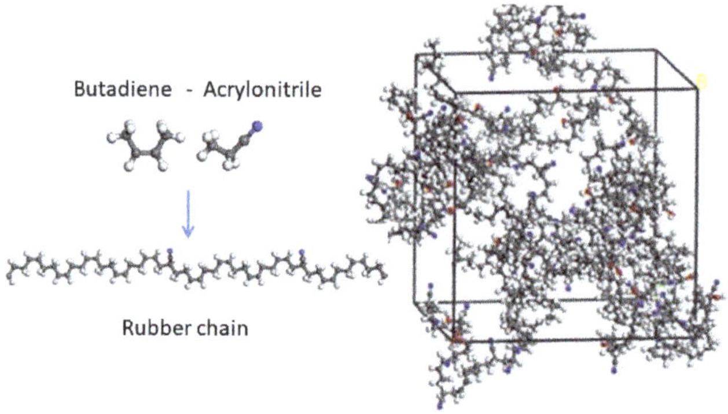

Figure 7. Molecular dynamics simulations of the formation of rubber. (**left**) The two molecular units, butadiene (**left**) (C brown sphere, H white sphere), and acrylonitrile (**right**) (N blue sphere), which are allowed to polymerize to yield a rubber chain. (**right**) A full optimization of a system of 5 chains, containing 65 % of butadiene molecules of 50 units each, modified with hydroxyl groups (OH-) (O red sphere). The cubic system obeys periodic boundary conditions (PBC) and has a cell size of about 30 Å (adapted from [65]).

Liquid wood: One of the major players to environmental pollution comes from the construction industry, mainly due to its fossil-fuel consumption and in loco production of particulate pollution. A possible solution would be using new materials adhering to environmental requirements, such as being of natural origin and therefore biodegradable [68–70]. In [71], the authors study the polymeric biocomposite commonly known as 'liquid wood' (LW), which is a good candidate for that purpose, and in addition, it is expected to be used for the restoration of degraded wooden moieties in civil structures, such as historical buildings, monuments, etc. Computer simulations are performed to study the mechanical behavior of LW by assuming that structure and functionality can be well described by a multifractal object, possessing different scaling properties at different spatial scales. A comparison with experimental data is discussed in support of the theoretical approach.

Aramid/SiO_2 mixtures: Aramid, meta-aramid, and aramid-mica papers play a role in building insulating materials [72,73]. In [74], mixtures of poly-m-phenyleneisophthalamide (PMIA) with SiO_2 (silica) nanoparticles (NP) have been studied by MD simulations (see Figure 8). Among these papers, PMIA is commonly used due to its suitable characteristics, but the range of applicability is limited by its rather poor thermal conductivity (see also [75]). It is indeed found that the thermal expansion coefficient of the SiO_2/PMIA mixture decreases, both in the glass and rubber states, while the thermal stability is increased, suggesting that the addition of silica NP improves the thermal and structural properties of PMIA by apparently hindering the chain's mobility.

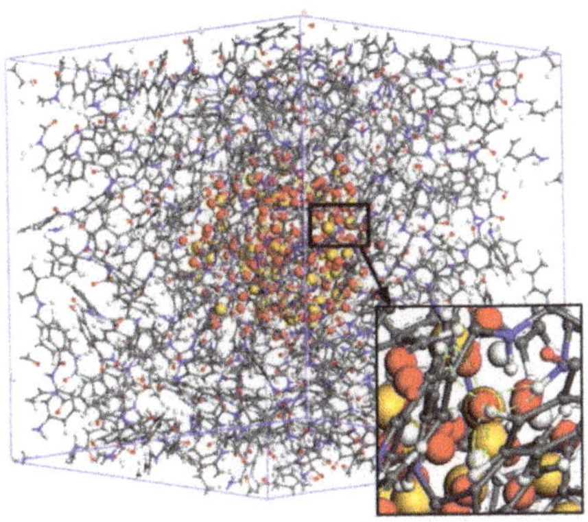

Figure 8. Snapshot of the simulations of poly-m-phenyleneisophthalamide (PMIA) network (C brown dots, H white dots, O blue dots), containing nano-silica particles (SiO_2) (Si yellow spheres). The inset shows a local zoom of hydrogen bonding in the SiO_2/PMIA model (adapted from [74]).

Charged polymers: The mechanical properties of charged polymers can vary quite significantly depending on the way they become charged, but the microscopic mechanism responsible for the observed correlations between structural and dynamical properties remains largely unknown [76–78]. In [79], the authors study by means of all-atom MD simulations the structure and diffusion behavior of both randomly and end-functionalized ionic poly(dimethylsiloxane) (PDMS) melts in the presence of bromide (Br^-) counterions dispersed in the melt (see Figure 9), finding that the density of the ionic PDMS melts is larger than for neutral melts. The counterions' main role is to slow down PDMS diffusion, while the counterion structure has a weak temperature dependence displaying an Arrhenius type of diffusion coefficient. The results contribute to our theoretical understanding of the correlations between the structure and dynamical properties of ion-containing polymers (see also [80]).

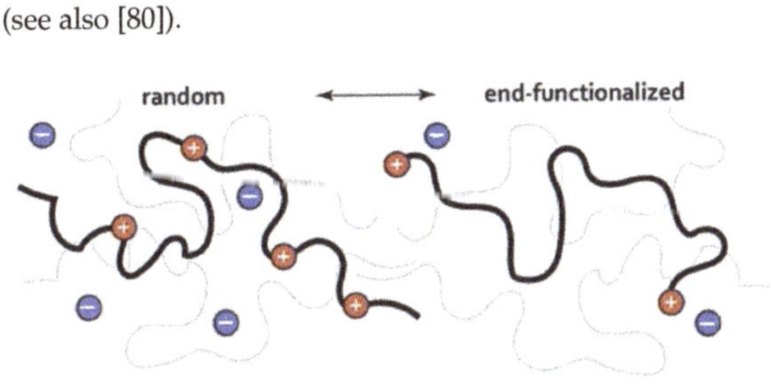

Figure 9. Mixtures of linear positively charged (red circles) polymers (linear segments) and mobile negative counterions (blue circles). The **left panel** illustrates the case of randomly charged polymers, while in the case depicted on the **right panel**, the polymers are charged only at their ends. The gray thin lines represent similar linear polymers present in the mixture (adapted from [79]).

Bioplastics: The large accumulation of plastics waste in the global environment is endangering both wildlife and human health worldwide [81–84]. The biocompatible polymer polyethylene vanillate (PEV) is considered a good candidate to replace the widely used polyethylene terephthalate (PET), reaching us in our daily lives. In [85], the authors compare the structural dynamics and physical data of both plastics using all-atom MD simulations of a single polymer chain made of $N = 100$ monomers. It is found that PEV flexibility is larger than for PET, predicting glass transition temperatures T_g of about 353 K and 345 K, respectively, which is consistent with current experimental evidence. In addition, the work suggests that both plastics share similar physical and chemical properties enabling the replacement of PET by its bio-counterpart.

Branched polymers: The molecular weight distribution (MWD) and configurational architecture are two important indicators of polymer characteristics. Indeed, they have a strong impact on material properties such as the dynamic modulus, fracture resilience, gel and glass transition temperatures, and viscosity [86–88]. MC simulations have been implemented [89] to efficiently deal with the challenging problem of obtaining the MWD of the branched polymer polyetherimide (PEI), which was formed via a step-growth polymerization (see Figure 10). The MC simulation results are compared with Flory–Stockmayer theory predictions, showing that the latter works well only below the gel point. The MC method can be used to accurately predict MWD in all cases, becoming a promising technique to deal with polymeric materials at the macroscopic level.

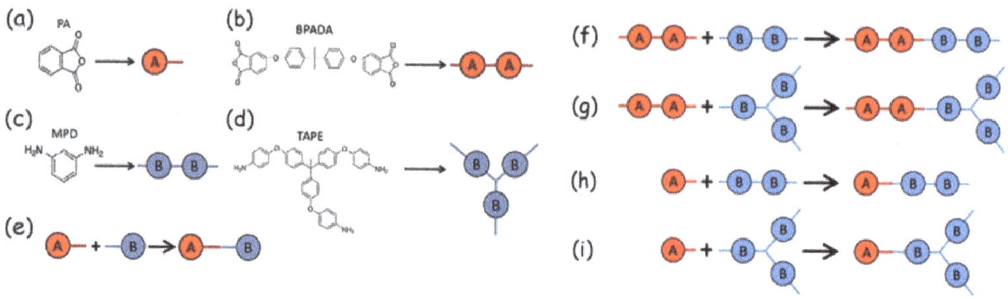

Figure 10. (**a–d**) The four types of monomers constituting a branched polyetherimide (PEI), used as models in the MC simulations, include: phthalic anhydride (PA), 4,4′-bisphenol A dianhydride (BPADA), m-phenylenediamine (MPD), and tris[4-(4-aminophenoxy)phenyl] ethane (TAPE) (the latter containing three NH_2 terminals). The B bead represents a full functional group containing one amine, and the A bead represents a functional group containing a carboxylic anhydride. (**e**) The condensation reaction between an amine group and a carboxylic anhydride one, taking place in the polymerization of PEIs, is represented by the formation of a bond between beads A and B. (**f–i**) The polymerization of branched PEIs is represented by the following four reactions: (**f**) BPADA + MPD (see (b,c) above); (**g**) BPADA + TAPE (see (b,d) above); (**h**) PA + MDP (see (a,c) above); (**i**) PA + TAPE (see (a,d) above) (adapted from [89]).

Synthetic hydrogels: The production of synthetic hydrogels is mainly based on the use of cross-linked polyacrylamide (cPAM) materials, keeping with their simple aqueous synthesis, great chemical stability and reproducible hydrogels formation (cf. Figure 11).

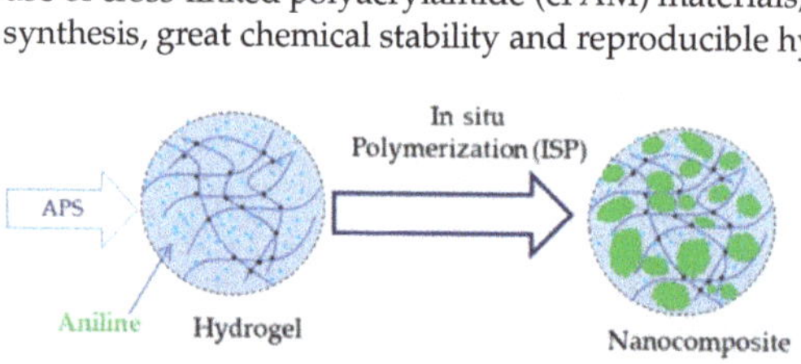

Figure 11. Illustration of the polymerization of aniline. Aniline molecules (green dots) are located inside a porous hydrogel, and by application of an oxidant (e.g., APS [90]), they start aggregating (larger green particles) by the process known as in situ polymerization (ISP), yielding a nanocomposite (adapted from [90]).

Used in the oil industry as a chemical method for water control, cPAM enhances crude oil flow by acting on water, which would otherwise block oil flow; thus, oil recovery can be significantly improved [91–93]. In [90], the authors consider hydrogels made of cPAM and conducting materials obtained from polyanilines (PANIs). Combinations of these two materials produce composites showing enhanced properties, intermediate between those of cPAM and PANIs. Typically, a gel composite is formed by radical polymerization, which is followed by incorporating the PANIs into the network by the oxidative polymerization of anilines (Figure 11). Such composites find a variety of technological applications, e.g., in photothermal/electromechanical actuators, supercapacitors, and motion detection sensors.

2.2. Polymers in Physics: Structural Properties, Self-Assembly, Smart Materials

Block copolymers: Block copolymers (BCPs) are attracting a great deal of attention due to their self-assembling characteristics, with potential use in the realm of nanopatterning, as they can yield a wide range of nanostructures well below the optical lithography resolution limit [94–96]. Self-assembly is a sort of microphase separation taking place due to the amphiphilic [28,29] nature of BCPs, which in combination with liquid phase infiltration (LPI) processes can open a path to the creation of novel nanomaterials with particular functional features. In [97], the authors provide an overview on the mechanisms underlying LPI processes and the role the associated infiltration parameters play in determining the nanomaterial properties. A bunch of applications in the field of photonics, plasmonics and electronics are comprehensibly highlighted.

Bis-salphen chains on graphene: Molecular self-assembly is becoming a powerful tool to easily and accurately obtaining nanopatterns on very large length scales [98–100]. In [101], the authors apply density functional theory (DFT) and MD simulations to study the self-assembly of bis-(Zn)salphen compounds which interact with both pristine and functionalized graphene (see Figure 12). They demonstrate how graphene sheets, and also carbon nanotubes (CNTs), can act as condensation centers of bis-salphen chains, which intially stick to the carbon-based large exposed surface, followed by a translational motion along the surface favored by their weak interactions with π–π carbon dangling bonds. The process ends when they become entangled with other chains, leading to their self-assembly. For other applications of graphene-related material interfaces, see the review in [102].

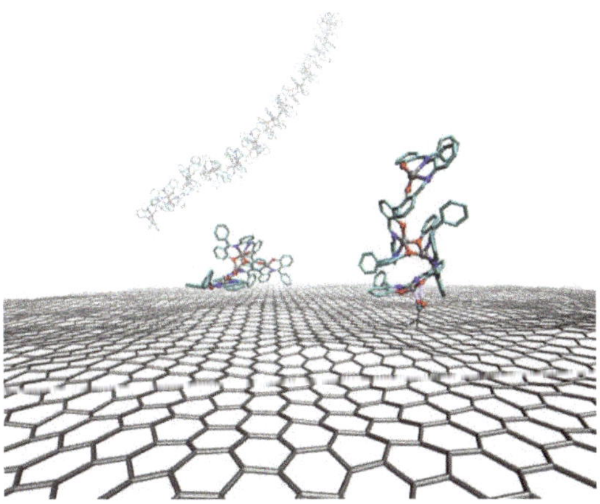

Figure 12. Complex polymer chains interacting with graphene (brown network). The colors in the polymer chains represent: N (Blue), O (red), C (green), Zn (brown) (adapted from [101]).

Polymer/graphene interfaces: In [103], the authors use MD simulations to study the stability of thermoplastic polymer/graphene oxide interfaces. For the former, they consider polyethersulfone (PES) and PEI, while graphene is oxidized in different ways, finding that PES becomes strongly stabilized in the opposite fashion as PEI does (see Figure 13). It is concluded that the observed different orders of stability are due to a balance between electrostatic interactions, in which atoms attract each other under a strong charge bias, thus stabilizing the interface, while excluded-volume effects between functional groups destabilize the interface as they prevent the $\pi - \pi$ stacking of aromatic rings [104–106].

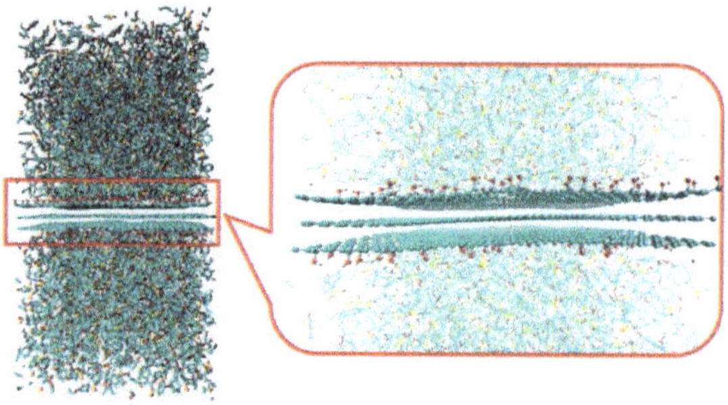

Figure 13. Snapshot of an ensemble of PEI chains (green dots) near a functionalized graphene sheet (dark green) containing -OH dangling bonds (red dots) fixed at random locations on the carbon structure (adapted from [103]).

Knoted polymer chains: In [107], the authors consider the issue of knoted polymer chains. This phenomenon is actually quite common in our daily life, but it is also relevant to polymer chains at the microscale, in particular in the case of biopolymers such as DNA and proteins, as well as for synthetic polymers. They combine MD simulations for a single polymer chain with multiparticle collision dynamics (MPCD) for solvent molecules to study the interesting problem of untying a trefoil knot polymer chain, which translocates through a narrow channel. This study may shed light into the translocation behavior of knotted polymer chains through a capillary (see also [108–112]).

Porous polyimide: In [113], the authors review properties of porous polyimide (PI), known for its applications as an interlayer insulating material displaying low dielectric constant, κ, aimed at improving transmission speeds between chips in large-scale integrated circuits. Several physical and chemical methods for reducing the dielectric constant of PI-based materials are mentioned and discussed. The work emphasizes key technical problems and challenges encountered in optimizing porous polyimide materials aimed at pushing the lower limit for κ even further (see also [114–116]).

Additive manufacturing: Thanks to recent advances in material synthesis and additive manufacturing (AM) technologies (see, e.g., [117]), the advent of metamaterials, known also as lattice materials since they are based on cellular architectures, has become possible. Inspired by nature, they possess a variety of multifunctional attributes [118–120]. In [121], the authors review a collection of recent developments on promising polymeric metamaterials (PMMs). The review covers different areas such as the design, fabrication and testing of PMM, stressing the need for their timely developments similarly to the ones carried on triply periodic minimal surfaces (TPMSs) lattices. This point of view extends to future research on the fabrication of PMMs under complex loading scenarios in order to understand their behavior. The achievement of this goal can be greatly facilitated by relying on machine learning algorithms to alleviate many of the current difficulties encountered in the fabrication of PMMs using AM techniques.

Conjugated polymers: Conjugated polymers have undergone large developments as organic semiconductors, competing with silicon-based inorganic materials which have dominated the semiconductor industry over the years [122–124]. In [125], the authors review several properties of conjugated polymers related to their structures, glass transition phenomena at T_g, and crystallization scenarios, which is aimed at understanding their relationships for optimizing their performances. For instance, it is known that in conjugated polymers, side-chain interactions affect T_g. Also, their growth processes and crystallization paths have an impact on mechanical and electrical properties. The review provides a thorough discussion on advanced characterization techniques such as X-ray diffraction, atomic force microscopy, and thermal analysis, helping to understand how molecular ordering and polymer–crystal interfaces take place. It lays a firm conceptual foundation for further research and development on conjugated polymer-based materials.

Block copolymers and brushes: Self-consistent field theory (SCFT) is a powerful theoretical technique to deal with a many-body system in which the complex interactions scenario is treated within a mean-field approach [126–128]. In [129], the authors introduce a simple 3D SCFT algorithm based on real-space methods with adaptive discretization, which increases the accuracy and efficiency of the numerical calculations, and apply it to the problem of polymeric materials surfaces. The algorithm performance is tested on two very distinct polymeric systems: block copolymer (BCP) [97] films and polymer brushes (see also [22]). The key point is the use of a finer contour discretization at grafting chain ends, thereby increasing spatial resolution in regions influenced by external forces, yielding one of the most accurate SCFT treatments of 3D polymeric systems so far.

Plasma techniques: We close with a brief discussion on plasma techniques, specially in view of their applications to polymeric materials [130–134]. In [135], the author presents a historical overview of the development and achievements of cold plasma technology (see Figure 14). The study includes a variety of plasma sources such as low-pressure glow discharges and atmospheric pressure plasmas used for dielectric barrier devices. Key operational parameters such as pressure, input gas type and flow rate, applied electric field, and type of discharges are critically examined. The review also discusses applications of cold plasmas to polymeric materials surface modifications and associated properties such as wettability and adhesion. A number of plasma techniques are considered which are relevant to polymeric materials and films. Further applications cover the biomedical sector and the crucial role of plasmas for water purification, gas separation, and energy production. It includes applications to bioplastics and the possibility of developing self-healing materials using this useful technology. A recent book summarizes some of the most important applications of low energy plasmas and the more advanced plasma-assisted supersonic jet deposition (PASJD) technique [136].

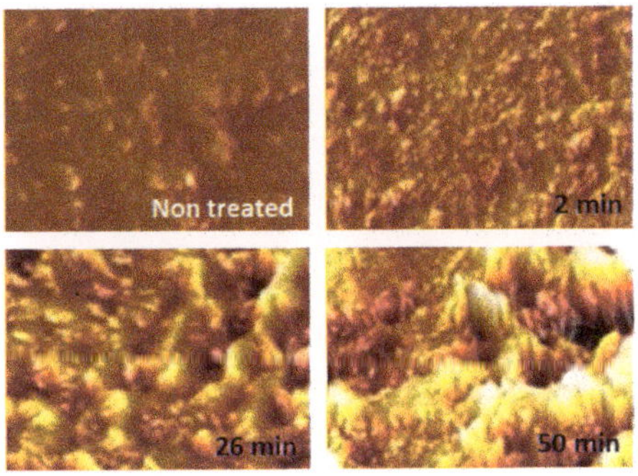

Figure 14. Images of polytetrafluoroethylene (PTFE) surfaces obtained using atomic force microscopy (AFM). The PTFE surfaces have been treated with a radio frequency plasma torch for different exposure times as indicated in the figure: 0 min (original surface), 2 min, 26 min, and 50 min. The resulting root-mean-square roughnesses are 22 nm, 33 nm, 68 nm, and 150 nm, respectively (adapted from [135]).

2.3. *Polymers in Chemistry: Structural Properties and Solubility Issues*

Polymer electrolyte for fuel cells: Market emerging electric vehicles partially rely on polymer electrolyte membrane fuel cells (PEMFCs) with an increased use of alternative sulfonated hydrocarbon-based proton exchange membranes (PEMs) [137–140], as they largely overcome many of the drawbacks associated with perfluorosulfonic acid polymers. In addition, PEMs have higher chemical and mechanical stability and significantly lower production costs. In [141], the authors employ MD simulations to study the effects of side chains length in sulfonated polystyrene grafted poly(arylene ether sulfone)s (SPAES), which are used as proton exchange membranes, where side chains length improve proton transport in hydrophilic environments.

Nitrile butadiene rubber: NBR is a non-natural rubber obtained from the synthesis of acrylonitrile (ACN) and butadiene (C_4H_6). It is mainly used for fabricating seals in general and hoses to transport fuel/oil in the aeronautical/automotive industry, since it has a wide temperature range of stability (−40 °C to 108 °C) (see also Figure 7 and [65]). The development of new bio-gasoline/diesel fuels requires a careful examination of NBR response to different structural and chemical stress factors [142–144]. In [145], the authors investigate the swelling response of NBR to different acrylonitrile (ACN) contents to validate a prediction model of its oil-resistant features. They base their analysis on a modified Flory–Huggins interaction parameter χ_{HSP}, being a function of the Hansen solubility parameter (HSP). The NBR swelling response can be evaluated quite accurately, suggesting that the new model can be used to keep the oil resistance under control even in extreme application conditions.

Dendrimers: Dendrimers consist of a central core molecule from which many branches can emerge yielding a hierarchical topology, and they constitute a prominent category of synthetic materials [146–149]. They possess a structural organization with a globular morphology at the nanoscale which reminds us of proteins (see [150] and Figure 15a).

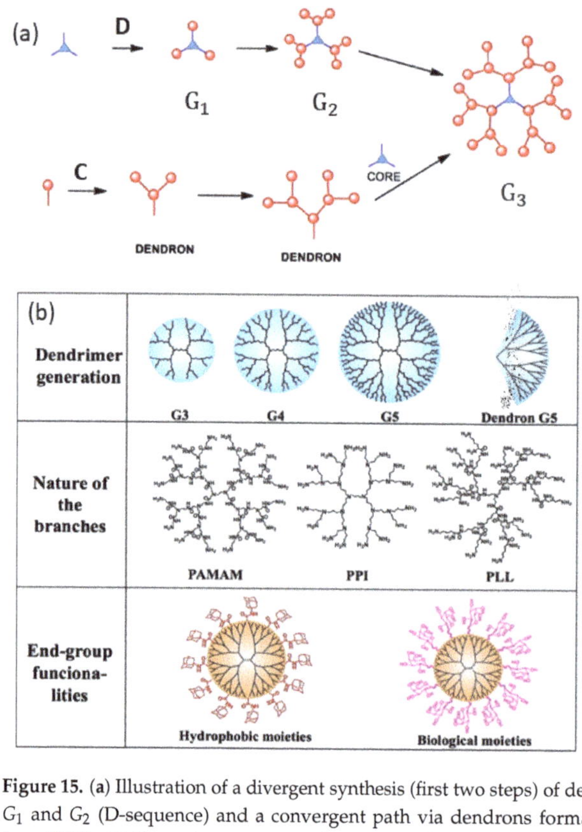

Figure 15. (**a**) Illustration of a divergent synthesis (first two steps) of dendrimer G_3 from dendrimers G_1 and G_2 (D-sequence) and a convergent path via dendrons formation (C-sequence) (adapted from [150]). (**b**) The main parameters determining the structural properties of dendrimers [151].

In [151], the authors review the properties and applications of quintessential dendrimer varieties such as polyamidoamine (PAMAM), poly(propylene imine) (PPI), polylysine (PLL), and polyester dendrimers (see Figure 15b). The focus of the work lies in the discussion of the essential properties and applications of these types of dendrimers, which are widely used today.

2.4. Polymers in Chemistry: Quantum Mechanical Calculations and MD Simulations

Pyrolysis for polystyrene: The exuberant production and uncontrolled consumption of plastics are causing severe, apparently irreparable damages to all forms of life on Earth. Pyrolysis turns out to be a possible solution for waste management, including recycling, landfill disposal and combustion [152–154]. In [155], the authors report the experimental development, and mathematical modeling, of a new thermal pyrolysis process for polystyrene (PS). The theoretical model significantly improves on previously discussed approaches by solving the full MWD without ad hoc assumptions, yielding a more complete understanding of the pyrolysis process.

Oilfield scales inhibitors: The unsolicited aggregation of different materials within the main working parts in an oil or gas production system, commonly referred to as scale deposition, is a major issue in the industry of energy commodities. Typically, oilfield scales contain calcium carbonates and sulfates, including barium sulfate and iron sulfides, among which calcium carbonate ($CaCO_3$) is one of the main components [156–158]. In [159], the authors employ DFT and ab initio molecular dynamics (AIMD) techniques to study the performance of scales inhibitors, such as polyacrylamide (PAM) and its silica functionalized counterpart (PAM-Silica). The latter was found to be a suitable calcium scale inhibitor due to its high binding affinity with Ca^{2+}, suggesting that it could be used as an efficient scale inhibitor.

Insulating materials: As discussed in Section 2.2, PI is widely used as an insulating material (see also [113]). In many applications, high temperatures are involved which require the material to maintain a robust thermal stability and good thermal conductivity [160–163]. In [164], the authors show that certain arrangements of boron nitride nanosheets (BNNSs) with polyimide (PI) can yield insulating materials with high thermal conductivity. They support their conclusions with extensive MD simulations showing the inhibitory role played by BNNSs on the decomposition process of PI (see Figure 16), which can be caused by different molecular entities.

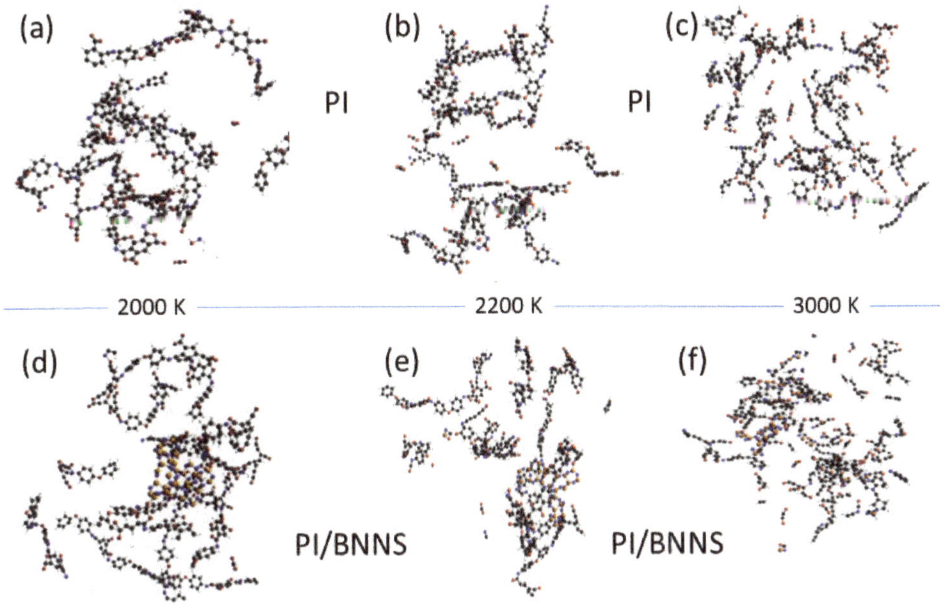

Figure 16. Molecular dynamics simulations of pyrolysis. (**Upper panel**) PI neat: (**a**) 2000 K, (**b**) 2200 K, (**c**) 3000 K. (**Lower panel**) PI/BNNS: (**d**) 2000 K, (**e**) 2200 K, (**f**) 3000 K (adapted from [164]).

2.5. Polymers in Chemistry: Environmental Issues, Batteries, Multifunctional Materials

Gas desorption of H_2 from polymer media: Hydrogen gas is attracting a great deal of interest nowadays as an environment-friendly energy source, because it can be produced from clean energy suppliers such as solar, wind, hydro, and nuclear technologies. The ever-increasing need for the use of hydrogen fuel requires the development of newly designed infrastructure for its handling, storage and transportation, and polymer materials are widely employed for these purposes. Indeed, a hydrogen tank liner provides a safe storage space to contain and protect the hydrogen gas stored inside a cylinder, ensuring a high level of containment preventing hydrogen to leak or escape out in the environment. Polyamide, polyethylene, and/or cross-polyethylene, build the innermost part of the liner. All these polymers display minimal permeability for the highly volatile hydrogen gas, thus providing the container with a very good barrier that minimizes possible gas losses during its storage. Since the operating conditions require high-pressure hydrogen gas, a rapid decompression event outside the liner can yield to gas desorption during which hydrogen diffuses rapidly out of the container, producing blisters or cracking that may cause a permanent damage to the polymer [165–167]. In [168], the authors study numerically the diffusion of hydrogen through a polymer material for storages of different geometries such as cylindrical, spherical and sheet-shaped containers. They quantify their results in terms of the desorption equilibrium time, representing the time for which H_2 diffusion loss attains its saturation level, displaying an exponentially growing trend as a function of the square of both thickness and diameter of the cylinder-shaped specimen, while it was found to be proportional to the diameter square for the sphere-shaped specimen and to the thickness square for the sheet-shaped specimen. The model calculations based on solving the diffusion equation are in good agreement with the experimental results, thus providing a useful tool for predicting the time-dependence of H_2 desorption in polymer materials.

Biodegradable polymer materials: To mitigate the burden of waste management due to the exponential growth of anthropogenic causes of environmental pollution, new regulating policies are encouraging the more extensive use of biodegradable materials [169–172]. In [173], the authors review recent developments within the topic of biomaterials. They consider in particular the prominent role played by polycaprolactone (PCL) as one of the best candidates, whose use has increased at a rapid pace in recent years. Indeed, PCL-based biocomposites, reinforced by adding NP, display interesting mechanical and thermal properties relevant to many applications.

Bio-synthesizable, biodegradable polymers: Petroleum-based synthetic plastics are designed to possess good performance and durability. Unfortunately, these properties are in hard conflict with the new requirements of biodegradability and recyclability [174–177]. In [178], the authors determine the mechanical properties of polyhydroxyalkanoates (PHAs) and their relation to chemical behavior using MD simulations. PHAs have lately emerged as very promising bio-synthesizable, biocompatible, and bio-degradable polymers expected to replace petroleum-based plastics. The work suggests rules for tailoring the mechanical properties of PHAs, opening the door to future theoretical studies aimed at identifying functional PHA polymer candidates for targeted applications.

Gas separation membranes: Membrane technology has seen an upward trend in recent years due to its simplicity and lower costs than its peers, particularly regarding gas separation. Certainly, polymers occupy a prevailing role among the different materials constituting the membranes. Gas separations involving CO_2 are an essential requirement to cope with the greenhouse effect and with the aim of pursuing a model of sustainable growth [179–182]. In [183], the authors review the use of Pebax polymers, i.e., block copolymers (see also Section 2.2 and [97,129]) made up of rigid-polyamide and soft-polyether blocks, for developing Pebax-based mixed-matrix membranes (MMMs), in which inorganic fillers are typically added in different contents (see Figure 17, and also the similar case of NP dispersed in ionic conductors [184], which may be relevant to MMMs). These promising fillers may yield new innovative membranes with improved performances.

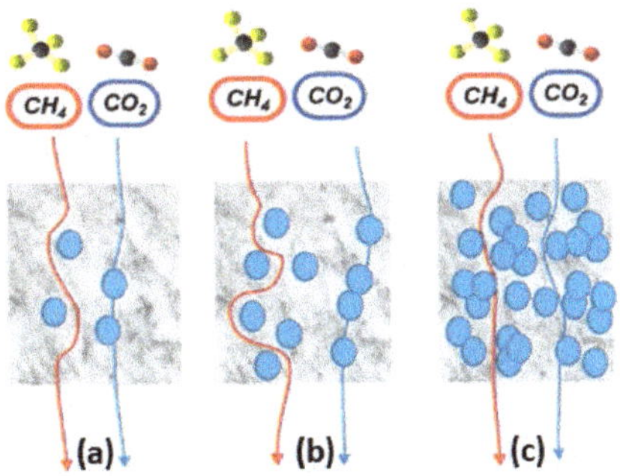

Figure 17. Fillers, represented by the full blue circles, dispersed in MMMs. (**a**) Very low filler contents may not achieve the desired selection effects. (**b**) At intermediate filler concentrations, one expects to find an optimal filler behavior. In many cases, very low filler contents are actually sufficient to sharply modify the polymer matrix characteristics. This aspect is rather encouraging for large-scale separations at still competitive costs. (**c**) At too high filler concentrations, one reaches a threshold value leading to particle agglomeration, where the MMMs display a reduced permselectivity [185] (adapted from [183]).

Polymer electrolytes in proton-based batteries: Polymer electrolytes (Section 2.1 and [28]) are expected to have further impact on the development of advanced electrochemical devices like batteries, supercapacitors, fuel cells, solar cells, sensors, etc. [186–189]. In [190], the authors review the ever-expanding use of polymer electrolytes in proton-conducting batteries, which has become a prominent research field yielding the creation of new and more complex polymer-electrolyte materials, stressing the need for additional structural characterizations to further optimize these novel materials. The work reviews recent advancements in proton-conducting polymer electrolytes characterization for their use in solid-state batteries.

Polymer electrolytes for Li-based batteries: Lithium-ion batteries (LIBs) play such a dominant role in electric vehicles and portable devices applications due to their high energy density and long lifespan, but safety concerns remain, and their extremely high energy density is a limiting factor for developing new more efficient devices [191–194]. In [195], the authors review polymer electrolytes and the possibility of combining them with ionic salts, which has indeed produced significant advances in battery technology in addition to dealing with safety issues, increased capacity, and longer life cycles. The review also provides future perspectives for developing novel polymer electrolytes to be applied for high performance Li-based batteries.

Epoxy resins: In [196], the authors review properties of epoxy resins related to their structural characteristics, chemical and electrical properties, and their modifications (see also [197–200]). Indeed, due to their multifunctional characteristics, they have become fundamental material components in different industrial sectors. The work discusses the mechanisms determining their dielectric breakdown behavior and new strategies to enhance the dielectric strength as well as the way fillers and additives have an impact on their insulation properties.

Polymer actuators: Nature-inspired polymer actuators, developed to mimic the real movements and functions of natural organisms, are expected to find applications in biomedical engineering, soft robotics, and energy harvesting, since they can efficiently convert electrical, thermal, or chemical energy into mechanical motion [201–204]. In [205], the authors review recent progress in the development of electrospun actuators, commonly based on polymers such as stimuli-sensitive hydrogels, shape-memory polymers (SMPs), and electroactive polymers. The surveyed future polymer actuators scenarios can contribute to the search for more advanced and multifunctional systems with improved performance and new elaborated applications.

Three-dimensional (3D) printing for protheses: AM, commonly known as 3D printing (see Section 2.2 and [117,121]), has the potential to revolutionize future achievements in medical fields such as healthcare [206–208]. Indeed, this technology is able to reshape the manufacturing of prosthetics and prostheses, no matter how complicated, for a wide range of polymeric materials. In [209], the authors review applications of polymer and its composites in the medical sector using 3D printing technology, discussing the creation of tailored prostheses, anatomical models for surgical planning and training. The article also discusses applications for drug delivery systems (DDSs) and tissue engineering, in addition to stereo-lithography, fused deposition modeling (FDM), and selective laser sintering (SLS) techniques. It is expected that manufacturing new medical devices will lead to better healthcare services and improved patient outcomes.

2.6. Polymers in Chemistry: Rheological Properties, Viscoelasticity

Computational fluid dynamics: Fluids displaying viscoelastic behavior show both viscous and elastic responses to external stimuli [210–212]. Many industrial applications actually deal with viscous fluids, so the question arises of how to accurate characterize the latter in order to optimize the associated industrial operations. In [213], the authors implement a variety of theoretical models to describe complex rheological behavior using a recently developed computational fluid dynamics (CFD) software, which is able to simulate Newtonian, generalized-Newtonian and viscoelastic flows, based on finite differences on suitably defined hierarchical grids (HiGs). It is found that the new methodology implemented in a HiG system can successfully reproduce the rheological behavior observed in a variety of fluids of interest to the broad rheology community, in particular typical non-monotonic flow curves of micellar solutions and plug-flow velocity profiles of viscoelastic fluids with yield-stress.

Epoxy–matrix composites: Composite materials can be coveniently reinforced by adding suitable amounts of natural fibers, aimed at lowering production costs, material toxicity, and specific weight [214–218]. In [219], the authors study theoretically epoxy–matrix composites reinforced with randomly oriented Cuban henequen (also known as sisal or agave fourcroydes) long fibers to determine essential parameters describing uniaxial three-element viscoelasticity. Their analysis is based on a constitutive model, with both integer and fractional index derivatives, to understand the mechanical evolution of the composite at different fiber/matrix compositions, discussing several methods to estimate the parameters accurately. The predictions are consistent with the experimental data, yielding a new feasible and accurate strategy to describe the viscoelastic behavior of fiber/matrix composites.

Elongational rheology: The nonlinear elongational rheology of polymers behaves differently than in the case of linear viscoelasticity and nonlinear shear rheology. According to tube theory (see Section 3), the elongational viscosity of entangled polymers should decrease with applied strain rate under specific elongation rate conditions [220–222]. In [223], the authors analyze the elongational viscosity of polypropylene carbonate (PPC) melts, using reptation dynamics for the single chains [224] (see Figure 18), and that of a polystyrene (PS) melt to estimate effects of chemistry, since the latter has an entanglement number per chain and a polydispersity index similar to PPC. The model uses slip-link networks consisting of strands, nodes, and dangling ends. Each path connecting two dangling ends corresponds to a polymer, which can go through strands and connected nodes. At each network node, one slip-link is located to connect two polymer chains in accord with the assumed entanglement rule. A 3D motion of the slip-links, the chain sliding through them, and their creation and destruction at the chain ends are considered, where the slip-link position is determined by solving a Langevin-type of equation. A constant monomeric friction overestimates elongational viscosity, therefore friction is allowed to diminish with increasing segment orientation. The effect of chemistry is studied by simulating PS, suggesting that PPC and PS behave similarly in terms of the reduction of friction under fast deformations.

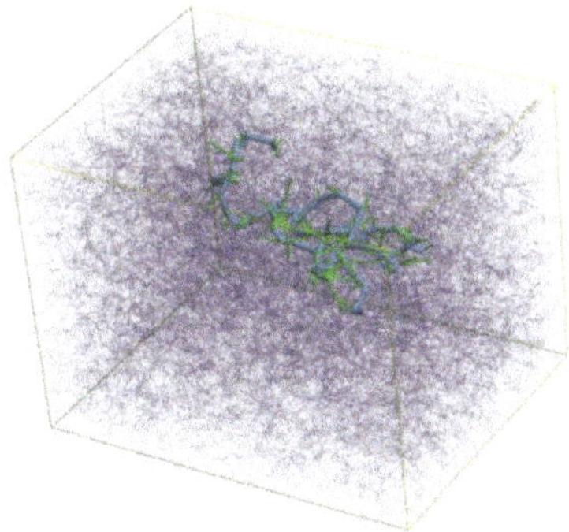

Figure 18. A multi-chain simulation based on the primitive chain network model to study the elongational rheology of polymers, such as polypropylene carbonate. A typical snapshot of a chain with $N = 88$ units (green lines) is shown, while the thin black lines are the other chains. The thick green lines are segments entangled to the test chain. Periodic B.C. are used (adapted from [223]).

Polymers rheology theories: The question how polymer chains do interact in a melt, and the ways they respond to shear deformations, specially at high temperatures, are some of the main aspects that need to be answered for understanding the behavior of polymer viscous liquids [225–227]. It is generally accepted that there is a characteristic molecular weight separating two different scenarios: the low polymer mass one consisting of unentangled chains described by the Rouse model and the heavier polymer mass regime, where chains become strongly entangled and the de Gennes reptation model plays a major role (see, e.g., [228]). In [229], the authors examine both failures and challenges of this current paradigm, and they suggest possible theoretical concepts to move forward our understanding on polymer rheology.

Thermoplastic polyimides on C-nanotubes: It is known that the mechanical properties of thermoplastic semi-crystalline polyimides can be enhanced by the presence of single-walled carbon nanotubes (SWCNTs) (see, e.g., [230–232]). In [233], the authors perform all-atom MD simulations up to microseconds time scales to understand the behavior of thermoplastic semi-crystalline 4,4′-Bis(4-aminophenoxy)biphenyl (BAPB)-based compound polyimide R-BAPB near an SWCNT. They find an enhanced viscosity of the polymer melt in agreement with experiments. More importantly, the viscosity increase is related to the conspicuous structural ordering of the chains near the carbon nanotube, thus excluding the possible formation of new interchain links. As a result, a strong anisotropy in the rheological properties of the R-BAPB near the carbon surface emerges, which is attributed to the existence of the polyimide chains' preferred orientations.

Orientational relaxation: Orientational relaxations of linear one-component polymer melts are well understood on the basis of Doi–Edwards–de Gennes tube-like models. The situation is, however, still controversial regarding relaxation phenomena in the case of a long linear polymer immersed within an ensemble of shorter chains (see, e.g., [234–236]). In [237], the authors discuss the relaxation of a self-unentangled long chain dispersed within a one-component polymer melt made of shorter chains, finding that while the long-chain relaxation is well approximated by a constraint release Rouse (CRR) mechanism, there is no unique prediction on the CRR relaxation time scale, τ_{obs}, representing the entanglement's mean release time. The aim of the work is to discuss different approaches yielding τ_{obs} and compare them to a large set of experimental viscoelastic data, including poly(methyl-)methacrylate and 1,4-polybutadiene blends, polystyrene and polyisoprene blends (see also [238]). It is found that τ_{obs} grows as a power-law of the short chains molar mass with an exponent of about 2.5 instead of 3 as predicted by the CRR mechanism. They suggest a new description of τ_{obs} implemented within a tube-like model consistent with the experimental observations.

2.7. Polymers in Biology: Macromolecules, Proteins, Structures and Dynamics

Large-scale molecular simulations: In [239], the authors explore the possibility of simulating molecular systems on very large scales. They remark the fact that whole micelles, bigger than 20 nm in size, formed by the self-assembly of hundreds of copolymers each containing as many as 50 units, have not been studied numerically so far. To overcome this lack of theoretical information, extensive MD simulations are performed on 900 flexible amphiphilic triblock copolymers made of 80 units each, aimed at describing both ribonucleic acid (RNA) and small interfering RNA (siRNA), the latter causing gene silencing through repression of translation in the cell [240–242]. Typically, siRNA is carried by a protein in the cell forming an RNA-induced silencing complex (RISC). The siRNA then unbinds from RISC and binds to its complementary messenger RNA (mRNA), silencing the gene that encodes it by inducing mRNA cleavage. It is concluded that such detailed MD simulations are possible and useful for the optimization of physicochemical characterizations of siRNA micelle complexes for improving the design of future drug delivery systems.

Actin filaments: Actin is the most abundant highly conserved protein in eukaryotic cells, participating in the majority of protein–protein interactions. More specifically, actin filaments (F-actins), resulting from the polymerization of the monomeric G-actin protein, are highly charged double-stranded rod-like polyelectrolytes, and they can be organized into higher-order structures, forming bundles and networks thus providing mechanical support [243], cell shape, and motion to the cell surface, which is essential for enabling the cell diverse functions [244,245]. In [246], the authors introduce a unique approach that combines dynamics and electrophoresis light-scattering experiments, with an extended semiflexible worm-like chain model built upon an assumed asymmetric distribution of polymer lengths, to characterize the polyelectrolyte and hydrodynamic properties of actin filaments in aqueous electrolyte solutions. The optimized models can be used to calculate other actin filaments properties such as stability, intrinsic viscosity, axial tension, elastic stretch modulus, and the force associated with the increase and compression of filaments length.

Electric field effects on GLP-2 peptide: Incretins are hormones produced in the gut in response to food, stimulating glucose-dependent insulin release, such as glucagon-like peptide-1 (GLP-1) and glucose-dependent insulinotropic polypeptide (GIP). They are secreted by the so-called enteroendocrine K- and L-cells of the intestinal mucosal epithelium. The L-cells secrete glucagon-like peptide-2 (GLP-2) [247–249], which is a single-chain protein of 33 amino acids of 3.8 kDa total weight containing 264 atoms. In [250], the authors perform extensive MD simulations of GLP-2 to study the effects of electric fields on its conformation, finding that its stability is reduced by the electric field, transforming it into unstable turn and coil structural shapes (see Figure 19), but returning to the linear shape at higher fields. It is expected that MD simulations of the effects of electric fields on the GLP-2 structure can provide a useful theoretical basis for understanding the biological function of this important hormone in vivo.

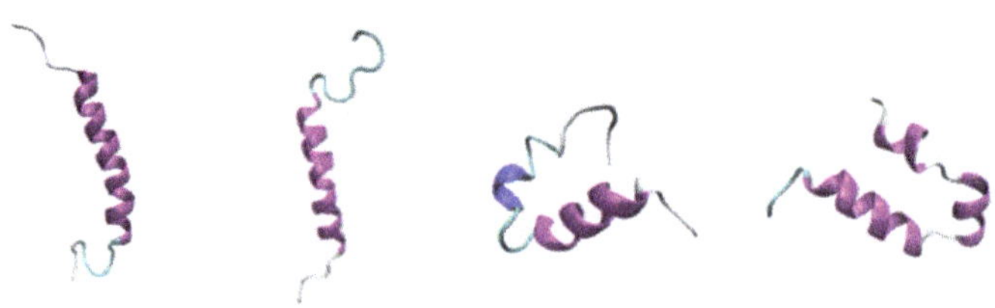

Figure 19. Three-dimensional structures of GLP-2 after MD simulations of 100 ns duration. Structures obtained by application of a uniform electric along the z-direction of intensity: (0, 0.4, 0.5, 0.6) V/nm, while for larger fields, the linear structure of the α-helix is restored (adapted from [250]).

Soft vesicle–polymer chain conformations: A vesicle is a complex biological macromolecule consisting of a liquid, or cytoplasm, enclosed by a lipid bilayer, and it is important for both biological and materials science processes, such as endocytosis via transformations of cell membrane shape, drug delivery, nanochemistry, and micro-reactors [251–253]. In [254], the authors employ MC simulations to investigate the interaction between a soft vesicle with a linear polymer (see Figure 20), yielding different types of induced vesicle conformational changes (see also [255]). The effects of polymer bending stiffness and strength of the assumed attractive interaction between vesicle and polymer are discussed. The results are expected to be useful for engineering future applications of such vesicle shape transformations.

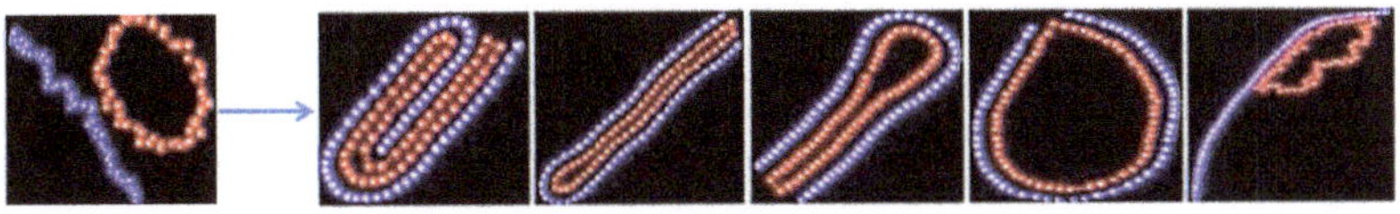

Figure 20. Interaction of a linear polymer chain (blue beads) with a vesicle, represented here as a polymer ring (red beads) in two dimensions, for different values of the bending stiffness (adapted from [254]).

Charged polymers: Our current understanding of the different thermodynamic phases of charged polymers in solution is still incomplete essentially due to the interacting multi-component nature of the system, such as long-range electrostatic and excluded volume interactions, the different components' translational entropies, chain connectivity [256–259], etc. In [260], the authors study the different phases undergone by charged polymers in solutions containing polar molecules by developing a novel liquid-state (LS) theory with short-range interactions. Specifically, interactions between charged polymer groups with external counterions, in addition to those between neutral polymer parts and charged intrapolymer segments, are taken into account by means of an extra Helmholtz free energy within the framework of perturbed-chain statistical associating fluid theory (PC-SAFT). A computational scheme used for salt-free polymer solutions is developed, allowing easy computations of the binodal curve and critical points for different model parameters.

Chirality of polymer knots: Polymer knots are formed when a long polymer chain is looped on itself. They occur quite naturally in both synthetic and biological polymers, such as DNA and proteins. Knotted polymers are attracting increasing interest due to progress achieved in macromolecular synthesis, biology, mathematics, and molecular simulations [261–263]. In [264], the authors study coarse-grained DNA polymers pushed inside chiral and achiral open channels using MD simulations, and they investigate the polymer metrics in terms of their span, monomer distributions and topological changes in the channels. The compression affects the polymer topology differently in chiral channels than in achiral ones. The former yields equichiral knots with the same handedness as the channels one.

Entangled polymer networks: The tube model for polymer melts is widely seen as the reference scheme for understanding the complex behavior of flexible polymers solutions, as it reduces the many-body problem into a much simpler mean field-type approach, in which a test chain is constrained to move within a tube virtually constructed by the surrounding polymers. It has, however, a number of shortcomings such as the lack of thermal fluctuations of the tube itself, which are responsible for the observed disentanglement effects [265–268]. In [269], the authors report experimental evidence that correlated constraints release does occur in entangled polymer networks but not in their crosslinked semiflexible counterparts (see Figure 21). In their experiments, they track single semiflexible DNA nanotubes, which are very similar to F-actin filaments in both thickness and length, embedded both in entangled and crosslinked F-actin networks, suggesting that very different reptation dynamics take place in these systems. This work is of primary importance for future theoretical work on the disentanglement issue.

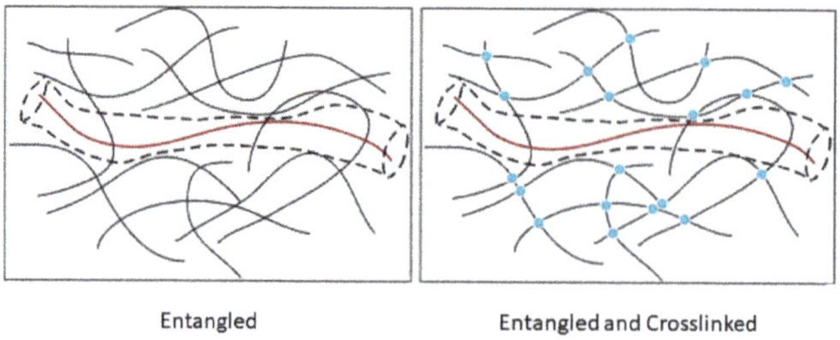

Figure 21. Schematic illustrations of (**left panel**) an entangled semiflexible polymer network of filaments (black lines) with an embedded tracer filament (red line). (**right panel**) In this case, the filaments (black lines) are either intertwined or connected by a crosslinker (blue circles). The dashed tubular structure indicates the space available to the tracer (red) filament, which is generally denoted as the reptation tube (adapted from [269]).

Polyelectrolyte pore translocation: In [270], the authors study the translocation of a polyelectrolyte chain through a pore driven by an electric field using MD simulations based on a coarse-grained model of hydrophobic-polar (HP) monomers, where the former represent neutral monomers, and the latter charged ones uniformly distributed along the hydrophobic backbone [271] (see Figure 22).

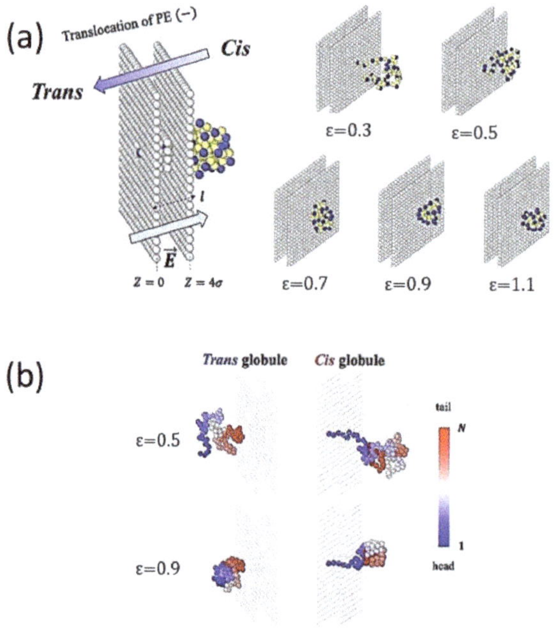

Figure 22. Illustration of a translocation process through a bilayer membrane (5σ wall thickness, where σ is the bead diameter), separating the cis-side from the trans-side of the membrane. (**a**) The negatively charged polymer is driven by the applied electric field from the cis- to the trans-region. Dark-colored beads represent negatively charged monomers and are considered as hydrophilic (polar), while the white ones carry no charge, representing the hydrophobic ones. Different solvent conditions can be described using different values of the attractive LJ potential strength, $\varepsilon \geq 0$ (i.e., $\varepsilon = 0$ for a good solvent), among hydrophobic sites. (**b**) Examples showing the translocation from the cis to the trans side for a polymer of $N = 90$ monomers and two different LJ potential strengths $\varepsilon = 0.5$ and $\varepsilon = 0.9$. The colors range from blue (head) to red (tail) of the chains (adapted from [270]).

The hydrophobic chain is initially found in the cis-side (Figure 22) in a globular form, where H- and P-type monomers are partially segregated. In order to translocate, the globule must first partially unfold, where interplay between translocation through a realistic pore and globule unraveling is discussed. The translocation dynamics of the chain is studied as a function of the solvent conditions, showing that short translocation times occur for slightly poor solvents (see also [272–276]). The time distribution near the minimum is rather shallow and the translocation time is roughly constant for intermediate hydrophobicities. The chain dynamics depends on the channel- and chain-internal frictions related to globule uncoiling. The results are compared with the predictions of a simplified Fokker–Planck equation describing the head monomer location in space.

Protein adsorption dynamics: In [277], the authors study universal aspects of protein adsorption dynamics by polymerized surfaces, using both a coarse-grained model for hydrophobic polar proteins (see Figure 23) and a polymer brush representing the binding surface. The proteins are uniformly placed on top of the brush whose multibead–spring chains are tethered to a solid implicit wall. They find that the most important factor determining the adsorption efficiency is the brush grafting density, while protein size and related hydrophobicity also play a role [278–281]. The rate of protein adsorption, density profiles and protein shapes, together with the respective mean binding force are compared for the different protein adsorption scenarios.

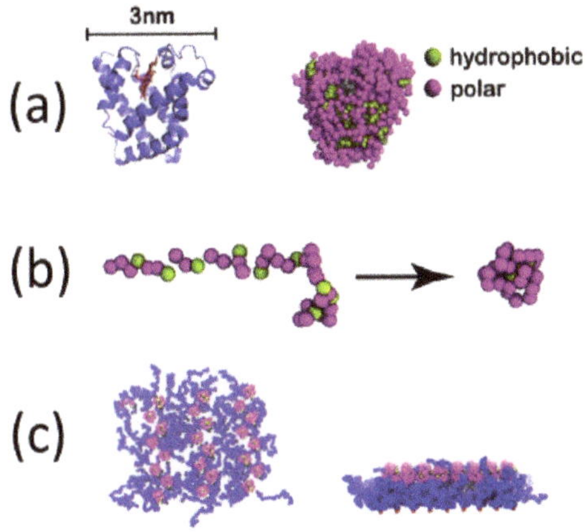

Figure 23. (**a**) Myoglobin protein: (a-left image) Native structure showing the α-helices. (a-right image) Coarse-grained model consisting of 151 amino acids, each one represented by a sphere. Polar amino acids are in red color, while hydrophobic (non-polar) ones are in green color. (**b**) Coarse-grained model of S25 protein consisting of 40 polar amino acids and 25 hydrophobic ones: (b-left image) Starting elongated configuration. (b-right image) Collapsed structure, where the hydrophobic sites have been brought inside the structure, leaving essentially all polar ones on its surface. (**c**) Planar polymer 'mushroom' in the case of 24 S25 proteins: (c-left image) Top view perpendicular to the surface. (c-right image) Side view (adapted from [277]).

Nucleosome dynamics: A nucleosome is the basic repeating subunit of chromatin packaged inside the cell's nucleus [282,283]. In humans, DNA is typically 1.8 m long and it must be packed into a cell nucleus of about 10 µm diameter, where nucleosomes play a key role in packing DNA. A single nucleosome consists of about 150 base pairs of DNA sequence wrapped around a core of eight histone proteins (see Figure 24). To allow for DNA transcription, chromatin must first be brought into its open form called euchromatin (right part of the figure). In contrast, to yield the compact chromosome structure, the nucleosomes must repeatedly fold onto themselves. However, much less clear is the picture of nucleosome regulatory mechanism, thus requiring a more systemic discussion (see, e.g., [284–286]). In [287], the authors review the regulatory factors of nucleosome dynamics by considering histone modification, DNA methylation, nucleosome-transcription factors and nucleosome-remodeling proteins and cations, from both computational and experimental approaches. It is suggested that integrating the latter with suitable nucleosomes packing modes, and the application of deep learning techniques, can yield promising opportunities for further discoveries.

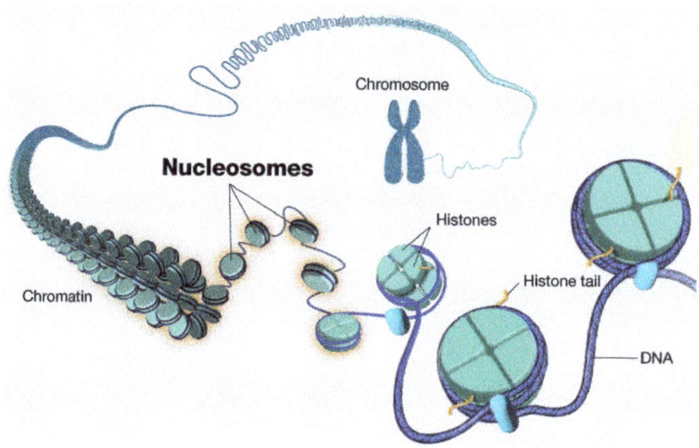

Figure 24. Schematic structure of chromatin showing the nucleosome units, inside which DNA pairs are wrapped around histone proteins. Close chromatin (called heterochromatin) is densely packed (**left** side of the figure) and transcription cannot occur. For transcription to occur, chromatin must be open, yielding the so-called euchromatin (**right** part of the figure). Condensing DNA into chromosomes (**top right** part of the figure) prevents DNA tangling and damage during cell division. Courtesy: National Human Genome Research Institute (https://www.genome.gov/genetics-glossary/Nucleosome (accessed on 27 November 2024)).

Model polymers vs. globular proteins: Recently, the issue of linear chain molecules has been addressed using MC simulations of standard polymer chains of tethered spheres, at both low and high temperatures, and their behavior compared with experimental data on globular proteins taken from the Protein Data Bank (PDB) [288]. They discuss detailed analyses of both local and non-local structures (see also [289–291]), and the associated maps of their closest contacts, as an attempt to reconcile using symmetry considerations the apparently different behaviors observed for the model and real chains.

2.8. Polymers in Biology: Biomedical Applications

Biomedical applications of polymers: The search for materials having applications in human body-related issues is a rapidly evolving field in the so-called biomedical sector. Such biomaterials can be fully synthetic or based on natural components fulfilling strict requirements of biocompatibility. Their range of applications is broad, being used for the development of prostheses, replacement of bones or tissues, and even artificial organs. Interestingly, the use of polymers in biomedical applications has significantly increased during the last few decades [292–294], despite some intrinsic limitations such as their higher costs than standard materials, lower strength compared to metallic components, lack of advanced design optimization and insufficient incorporation of reinforcement components. In [295], the authors review biomedical applications of polymers related to important advances in the synthesis and modification techniques for obtaining novel biomaterials, e.g., aimed as replacements of both hard and soft tissues, and to be used in limb prostheses, dentistry and bone fracture issues. The emergence of less invasive surgery techniques, together with faster surgical sutures, can contribute to the further expansion of the use of polymers in biomedicine.

Three-dimensional (3D) polymer hydrophilic networks: Among polymer biomaterials, 3D polymer hydrophilic networks, made of crosslinked macromolecular chains known as polymeric hydrogels (Section 2.1 and [90]), have attracted a great deal of attention [296–298]. They exhibit high swelling in water and water retention, suitable response to external stimuli, and adaptable mechanical features. In addition to the biomedical sector, polymeric hydrogels have found applications also as hydrogel electrolytes in flexible aqueous energy storage devices. In [299], the authors discuss synthetic processes for developing polymer hydrogels using polymerization techniques, such as frontal polymerization (FP) based on a propagating self-sustained reaction across an ensemble of monomeric units in a batch reactor. A summary is presented of the main recent developments achieved related to the design, preparation, and application of FP-derived polymeric hydrogels.

Skin repair issues: The outermost layer of the human body, called the integumentary system, includes skin, hair, scales, feathers, hooves, and nails, of which the skin constitutes the largest organ. The skin has up to seven layers of ectodermal tissue, roughly described in terms of the superficial epidermis, the subcutaneous hypodermis and the deeper dermis, and it is embodied with a complex self-regulatory function. The epidermis is composed largely of keratinocytes, undergoing constant renewal due to the action of epidermal stem cells that keep producing either new or translocation expansion cells. The skin protects the body against pathogens and excessive water loss, while it is devoted to heal itself in case of damage by forming a scar tissue [300–302]. It has recently been shown that it can also absorb man-made per- and polyfluoroalkyl substances (PFAS), known as forever chemicals, which are highly persistent chemicals not present in nature. In [303], the authors deal with the issue of wounds unable to effectively heal via normal repair mechanisms by reviewing different healing strategies such as the possibility of inducing changes in physicochemical and biological properties. The latter are obtained by incorporating different polymers and fillers into specifically designed polyurethane dressings, discussing their applications in wound repair and regeneration. Several polymers are considered, of which natural-based ones include collagen, chitosan, and hyaluronic acid, while synthetic-based polymers are constituted by polyethylene glycol, polyvinyl alcohol, and polyacrylamide, in addition to other active ones such as the antimicrobial trifluoroacetate salt (LL37 peptide), platelet lysate, and exosomes. They conclude with a discussion of future developments and applications of novel polyurethane dressings.

Material models for limb orthoses: FDM is a widely used technique in the biomedical sector (see also Sections 2.2 and 2.5, and [117,121,209]) due to its versatility in dealing with various materials. However, the appropriate material selection for a specific application can become an issue, and the understanding of the mechanical behavior, in particular of polymeric materials, becomes essential to obtain the desired result [304–306]. In [307], using the finite elements numerical approach (see examples in Figure 25), the authors analyze three material models: The Bergström–Boyce (BB), the three-network (TN), and the three-network viscoplastic (TNV) models, in order to determine the mechanical behaviors of the polymeric materials, acrylonitrile butadiene styrene (ABS), polylactic acid (PLA), and polyethylene terephthalate glycol (PETG), aimed to be used in personalized upper limb orthoses. The new approach thus relies on the combination of both theoretical and experimental investigations to predict the suitability of the mechanical characteristics required for obtaining accurate personalized orthoses.

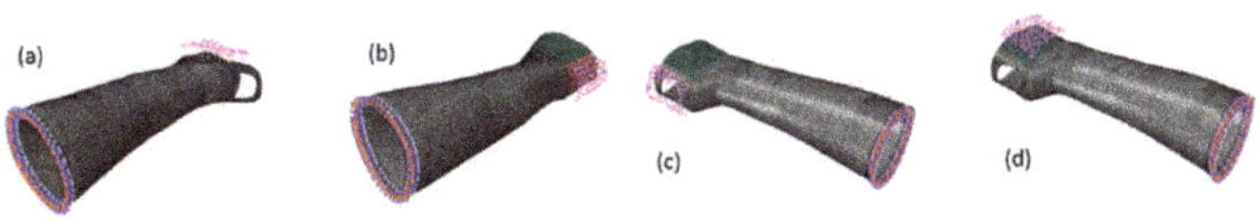

Figure 25. Finite elements (FE) mesh, boundary conditions, and load directions for four crucial wrist movements: (**a**) flexion, (**b**) extension, (**c**) radial and (**d**) ulnar. The Abaqus software (from Dassault Systems [307]) was employed for the FE analysis of orthoses, where the splint exhibited 11,481 node points and 37,225 linear four-node tetrahedral elements (adapted from [307]).

3. Theoretical Aspects of Polymers

The simplest polymer chains are those constituted by N (monomers) units, each one representing a molecule of the same type. They are typically considered to be immersed in a liquid solution (made of single molecules), which allows the chains to move around, displaying in general a quite challenging dynamical behavior. This complexity arises because of intermolecular interactions, in which the monomers are covalently bonded to their neighboring partners having bond strengths much larger than the other interactions present in the system, thus providing the chain with a good stability even at relatively high temperatures. In addition to these purely local interactions, the molecules exhibit a strong repulsion to other monomers, located beyond their nearest-neighbor ones, if they become close enough in space. In many cases, however, there is a medium-long-range attractive interaction present between such non-local partners which may become important at low temperatures. Typically, they represent van der Waals types of attractions, which may dominate the whole interaction behavior in the case of neutrally charged monomers, leading to a collapse of the chain below some critical temperature, T_θ.

3.1. Deterministic and Random Fractals: Percolation Clusters and SAWs

For temperatures $T \gg T_\theta$, highly diluted chains attain an open structure displaying a universal scale-invariant shape whenever the single molecules in solution behave as a 'good solvent'; i.e., they fully screen non-local interactions between monomers. In these circumstances, the so-called SAWs, which mimic the internal connectivity of a chain, can be used to model diluted polymer chains. Effects of non-local, not fully screened interactions emerge in a poor solvent and can still be treated within the SAWs models at finite temperatures. The simplest case of an SAW in a good solvent is the one in which only hardcore interactions are considered between non-local monomers. However, one can imagine situations in which SAW chains can move either on top of a solid surface or in a solution immersed within a porous medium. The most interesting materials for us are those which display some type of fractal scaling in space, since in this case the standard universality class of the SAW can be modified. Ideal fractal materials can be modeled by relying on specific construction rules yielding mathematically exact fractal shapes. Examples of such deterministic fractal objects are displayed in Figure 26, in which the lack of disorder allows us to concentrate on the effects of the fractal medium alone on the scaling properties of SAWs.

2D 3D

Figure 26. Illustrations of the Sierpinski fractal carpet in 2D and the fractal sponge in 3D [308]. Both fractals are generated according to the rule (n, k), where n is the length of the initiator and k refers to the number of subunits deleted from the generator (c.f. the red square ('seed') shown at the bottom-left part of the figure). The seed is made of $n \times n$ (here $n = 3$) squares (cubes), of which the central square ($k = 1$) is deleted for the 2D case, and $k = 7$ cubes are deleted for the 3D fractal. The process is continued deterministically afterwards. The 2D fractal is shown up to the third iteration, and the 3D sponge up to the fourth one. The sites available to the SAW are shown as white squares in 2D and white cube faces in 3D. The fractal dimensions, $d_f(d)$, can be calculated exactly, $d_f(d) = \ln(n^d - k)/\ln(n)$, yielding $d_f(2) \simeq 1.89278$ and $d_f(3) \simeq 2.72683$.

The ideal fractals shown in Figure 26 may still be built experimentally as follows. The 2D carpet could be created by deposition techniques using a fractal 'mask', in which the black zones are grown along the vertical direction creating a sort of barriers for the SAWs, preventing them from occupying those locations and thus forcing the chains to wonder over the white available places. The 3D case is a bit more cumbersome, since the building of the porous material would require 3D printing techniques. To study fractal effects, two possibilities can be envisaged regarding which part of the structure represents the polymer conducting medium. To be noted is that both the white cubes and the empty pores constitute connected 3D structures by their own. In the case shown in Figure 26, the polymers can be considered to slide along the white surface areas, thus effectively moving through a space of reduced dimension, $d_f < 3$. The second possibility would be to consider the complementary set to the one shown in the figure, in which the pores are 3D printed and becoming the solid part of the object. The remaining empty part of the object would build a conducting medium of dimension d_f once filled with a good solvent, allowing the polymers to diffuse through it. The connected solid part could eventually be made of a transparent material in order to monitor the polymer(s) motions inside the structure.

More realistic fractal media, closer to the case of natural shapes [309], can be modeled by different types of random fractals (see, e.g., [309–312]), among which percolation clusters are very useful, since they describe a quite general type of structural disorder [310,312–314] (see Figure 27).

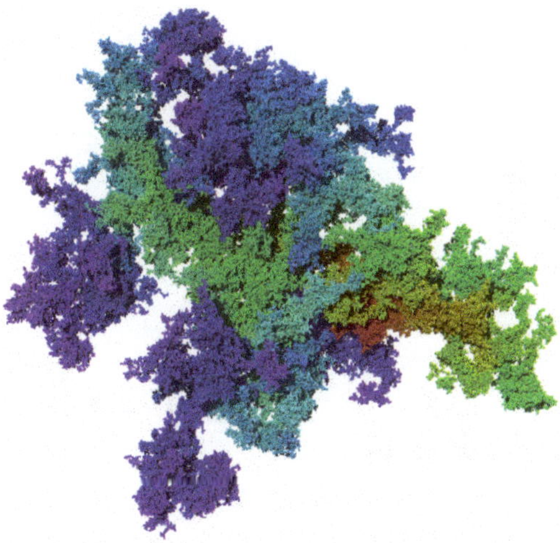

Figure 27. A large percolation cluster at the critical site concentration $p \simeq 0.3116$ in 3D on a simple cubic lattice. The cluster has been grown using standard techniques [315,316] (see also [310,317] for a generalization of the growth rules). The fractal dimension of percolation clusters is $d_f \simeq 2.53 \pm 0.02$ (see, e.g., [318]). The different colors represent the length of the shortest (linear) path connecting a site to the seed of the cluster, the latter is located within the red sites and the blue sites have the longest paths (or chemical distances [310]) from the seed. The cluster contains over a million sites, and the longest chemical distance is 2000 in lattice units (courtesy of Francesco Marini). The fractal dimension of the shortest path is $d_{min} = 1.13077(2)$ in 2D and $d_{min} = 1.3756(6)$ in 3D [319]. The shortest paths are essential when dealing with the transport of matter along the connected sites of the cluster.

As is apparent from Figure 27, large percolation clusters exhibit a beautiful complex topology which necessitates more than one fractal dimension to fully describe it, in this case the mass fractal dimension, d_f, and the one of the shortest paths, d_{min}. The latter brings us to the concept of linear structures displaying self-similarity or fractal scaling. Among them, SAWs play a very important role, since they can be used to study the statistical properties of linear polymers [310,320–323]. An example of an SAW in 2D is shown in Figure 28.

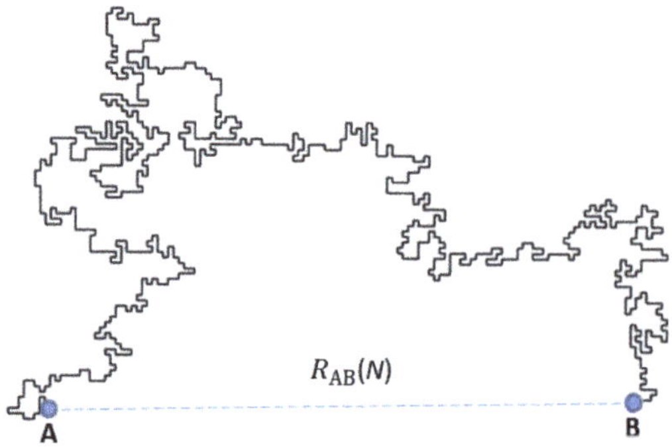

Figure 28. A linear polymer in a good solvent, modelled by a 2D SAW of 1000 monomers on the square lattice, generated using the reptation method discussed in Section 3.2.4. SAWs are linear fractals with a fractal dimension $d_f = 4/3$ in 2D (see e.g. [324]). Interestingly, the shortest paths in percolation clusters are less compact than SAWs in both 2D and 3D. In the figure, we have highlighted the ends points of the trace of the SAW, denoting them with the letters A and B. The dashed line has a length $R_{AB}(N)$, which upon a configurational average depends only on the number of SAW steps, N, between the endpoints (here $N = 1000$). Self-trail crossing are forbidden for SAWs.

3.2. Structural Properties of SAWs in Regular Lattices and in Fractal Structures

Next, we review the structural properties of SAWs in Section 3.2.1 and their relation to fractal percolation clusters. Then, we consider the cases in which SAWs are embedded in deterministic (Section 3.2.2) and random (Section 3.2.3) fractals as well as the way their structural properties change with respect to their standard behaviors in regular lattices. Finally, in Section 3.2.4, we describe a simple method for generating SAWs of arbitrary length using the reptation method valid in any substrate.

3.2.1. Structural Properties of SAWs in Regular Lattices

SAWs display a characteristic spatial extension as a function of the number of steps (monomers), N. To obtain it accurately, one considers an SAW trail for fixed N, such as in Figure 28, and perform an average of the end-to-end distance square, $R^2_{AB}(N)$, over fully distinct configurations. The result can be written in the scaling form,

$$R \equiv \langle R^2_{AB} \rangle_N^{1/2} \simeq a\, N^{\nu}, \quad \text{for } N \gg 1, \tag{1}$$

where the scaling exponent is given by the Flory [320] celebrated relation,

$$\nu(d) = \frac{3}{d+2}, \quad \text{with } \nu(d) = \frac{1}{2} \text{ for } d \geq 4. \tag{2}$$

Note that $\nu(d)$ depends on the spatial dimension d, yielding the exact result for $d \neq 3$, except for logarithmic corrections at the critical dimension, $d_c = 4$, above which they become similar to random walks (RWs). For convenience, one often considers a lattice to generate the SAWs configurations, but the same result for $\nu(d)$ holds in the continuum.

Efficient algorithms have been developed over time to generate long SAWs chains, of which pivot strategies have become popular (see, e.g., [325–327]), including MC methods [328]. An improved pivot algorithm has been recently introduced, yielding the most accurate estimation of ν in $d = 3$ so far, $\nu(3) = 0.587597(7)$ [329], which is smaller but still very close to the value $\nu(3) = 0.6$ predicted by Flory.

To be noted is that Equation (1) can be inverted to obtain the number of monomers (or mass of the SAW) confined within a distance R between the end ones, yielding

$$N(R) \simeq c\, R^{d_f}, \quad \text{with } d_f = \frac{1}{\nu(d)}, \tag{3}$$

where d_f is the fractal dimension of the SAWs in a space of dimension d. Therefore, one finds $d_f(2) = 4/3$ (see Figure 28), $d_f(3) \simeq 1.70184(2)$ (see [329]), and $d_f = 2$ for $d \geq 4$, corresponding to the fractal dimension of standard RWs. We say that SAWs are self-similar objects with fractal dimension $d_f(d)$.

A relevant quantity for SAWs is the total number of distinct chain configurations of N steps, C_N, which is expected to obey the mixed-scaling behavior,

$$C_N \simeq b\,\mu^N\,N^{\gamma-1}, \tag{4}$$

where μ is the effective coordination number of the lattice considered, and $\gamma \geq 1$ is the second critical exponent for SAWs which depends solely on the dimensionality d of the space, and it is denoted as the enhancement exponent. Indeed, $\gamma = 1$ for RWs. The scaling relation in Equation (4) can be rewritten in the more convenient form for numerical estimations of the exponents,

$$\frac{\ln C_N}{N} = \frac{\ln b}{N} + \ln \mu + (\gamma - 1)\,\frac{\ln N}{N}. \tag{5}$$

This relation has been used in [330] to estimate values of μ and γ on square lattices using the exact enumeration data of all SAWs walks up to $N = 50$ [331–333], yielding $\mu = 2.641 \pm 0.005$ and $\gamma = 1.30 \pm 0.05$, which is in good accord with the accepted values (see Table 1). These calculations were performed as a validation of the numerical approach based on Equation (5), which has been used for obtaining unknown values of the exponents on random fractal supports (see Sections 3.2.2 and 3.2.3). The exponent γ is known exactly in 2D, $\gamma = 43/32$ [334], while recent calculations yield the most accurate value in 3D, $\gamma = 1.15695300(95)$ [335]. It is interesting to note that an approximate relation for $\gamma(d)$ has been obtained by relying on analytical calculations of diffusion in fractal structures, such as RWs, SAWs and percolation clusters, i.e., $d > 1$, yielding [336],

$$\gamma(d) = 1 + \frac{4-d}{6}, \quad \text{with } \gamma(d) = 1 \text{ for } d \geq 4, \tag{6}$$

from which one finds $\gamma(2) = 4/3$ and $\gamma(3) = 7/6$, which is in very good agreement with the accepted values.

In contrast to its simple one-dimensional topology, an SAW chain attains a rather complex structure in space in the sense that monomers located at the same Euclidean distance r from a given chosen site (see Figure 29 for an illustration in 2D) can be found at quite a different number of steps, or topological (chemical) distances ℓ [310], from it. The inset in the figure displays the values of ℓ for SAW sites at fixed distance r from the site chosen as the origin. This feature leads to a rather broad distribution function, $\Phi_{\rm SAW}(r,\ell)$, as a function of ℓ (see, e.g., Figure I.5 in [324]), which obeys the scaling form,

$$\Phi_{\rm SAW}(r,\ell) = \frac{1}{r}\,f(x), \quad \text{with } x = \frac{r}{\ell^{\nu}}, \tag{7}$$

where the normalization condition $\int_0^\infty dr\,\Phi_{\rm SAW}(r,\ell) = 1$. This scaling form was originally derived for the end-to-end distance, $r = R_{A,B}(N)$ (Figure 28), where one of the ends, A or B, is taken as the origin, yielding $\Phi_{\rm SAW}(R_{A,B}, \ell_{A,B} = N) \equiv \Phi_{\rm SAW}(r,N)$ [337]). Here, we use the notation $\Phi(r,\ell)$ whenever we refer to the internal structure of a fractal substrate, while when we consider the statistics of SAWs describing the distribution of the end-to-end Euclidean distance on a given substrate (either regular or fractal one), we refer to it as $P(r,N)$, with $x = r/N^{\nu}$. The distribution can be also considered using the ℓ-metric, $P(\ell,N)$, with $x = \ell/N^{\nu}$, where ℓ is the topological distance calculated on the substrate between the end sites. On regular lattices, one has the equivalence, $\Phi_{\rm SAW}(r,\ell = N) \equiv P(r,N) \propto P(\ell,N)$ (cf. Figure 30 below).

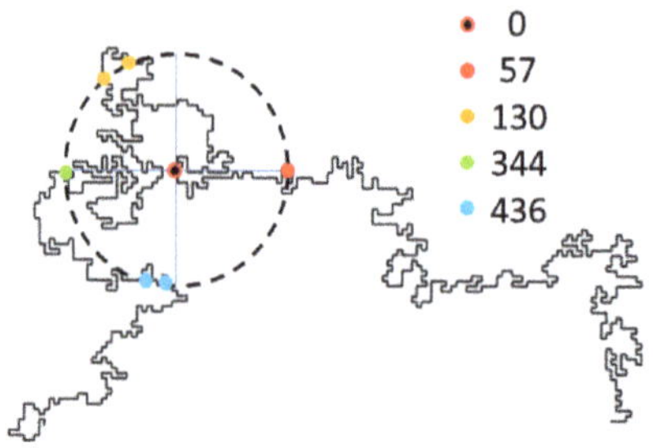

Figure 29. The internal structure of SAWs: The metric ℓ, yielding the distance along the chain between two monomers, compared with the Euclidean distance r in space between them. In the figure, a site is taken as the origin (black–red circle at $r = 0$), from which one can determine the chemical distances ℓ (see the inset for the indicative values of ℓ) of those monomers located on the dashed circle at distance r from the origin. Then, $\Phi_{SAW}(r, \ell)$ is obtained by averaging over all pairs (r,ℓ).

The scaling function $f(x)$ in Equation (7) displays the asymptotic behaviors,

$$f(x) \simeq x^{g_1+d}, \qquad x \ll 1, \tag{8}$$

$$f(x) \simeq x^{g_2+d}\, e^{-c\,x^{\delta}}, \quad x \gg 1, \tag{9}$$

where the exponents (g_1, g_2, δ) are related to (ν, γ, d) according to

$$g_1 = \frac{\gamma - 1}{\nu}, \tag{10}$$

$$g_2 = [d(\nu - 1/2) - (\gamma - 1)]\,\delta, \tag{11}$$

$$\delta = \frac{1}{1-\nu}, \tag{12}$$

yielding $g_1 = g_2 = 0$, and $\delta = 2$ for RWs. The expression for g_1 is due to des Cloizeaux [338], g_2 was obtained by McKennzie and Moore [339], and δ was first derived by Fisher [337]. A summary of the updated values of the exponents is given in Table 1.

Table 1. The exponents describing SAWs in 2D and 3D in regular systems. The values for the coordination numbers $\mu(d)$ are for the square and simple cubic lattices, respectively. For comparison, the exponents in 4D are reported, without taking into account logarithmic corrections, and the coordination was taken as $\mu(4) = 2d - 1 = 7$. The values for (g_1, g_2, δ) have been calculated from the exponents (ν, γ, d) using Equations (10)–(12).

	2D	3D	4D
ν	3/4	0.587597(7)	1/2
γ	43/32 (1.34375)	1.15695300(95)	1
μ	2.63815853032790(3)	4.6840401(50)	7
g_1	11/24 (0.45833...)	0.26711(2)	0
g_2	5/8 (0.625)	0.25664(5)	0
δ	4	2.42481(4)	2

The remarkable accuracy reported for $\mu(2)$ in Table 1 [340], was the last station of a long and fascinating numerical journey triggered by an interesting conjecture by Guttmann in the 1980s. Based on the exact result on the exagonal lattice, $\mu_{ex} = \sqrt{2+\sqrt{2}}$, he suggested that $\mu(2)$ might be given by the positive root of $13t^4 - 7t^2 - 581 = 0$, yielding $\mu(2) = \sqrt{7+\sqrt{30261}/\sqrt{26}} \simeq 2.63815853034174$. In 2001, Guttmann and Conway [341] obtained the value $\mu(2) = 2.638158534(4)$, supporting the conjecture. In 2012, Clisby and Jensen [342] confirmed the previous result, obtaining $\mu(2) = 2.63815853035(2)$.

It was only in 2016 [340] that the final word on this issue was pronounced, yielding $\mu(2) = 2.63815853032790(3)$ (Table 1), thus ruling out the conjecture. On the cubic lattice, the most accurate value obtained so far, $\mu(3) = 4.6840401(50)$, was reported in [343].

To numerically determine the distribution function, $P(r, N)$, one considers the case $\ell = N$ in Equation (7) and performs a configurational average of SAWs on the lattice considered. It is, however, more convenient to work with the end-to-end topological distance, $\ell(N) = |x_1 - x_N| + |y_1 - y_N|$, instead of $r(N) = \sqrt{(x_1 - x_N)^2 + (y_1 - y_N)^2}$, here written for the 2D case, thus eliminating the fluctuations arising from the conversion of the real value r to an integer, $i_r = \mathrm{int}(r \pm \Delta r)$, required to obtain the histogram $H(i_r, N)$ from which the distribution function is derived. This change of metric is possible because on regular lattices, and also on Sierpinski carpets and sponges, both the Euclidean and the chemical distances on the substrate scale the same, in the sense that $\ell \simeq r$. An example of an SAW for $N = 9$ is shown in Figure 30, on the square lattice. It is easy to convince oneself that $r \leq \ell \leq \sqrt{d}r$, where $d = 2$ in this case, suggesting that indeed $\ell \simeq r$.

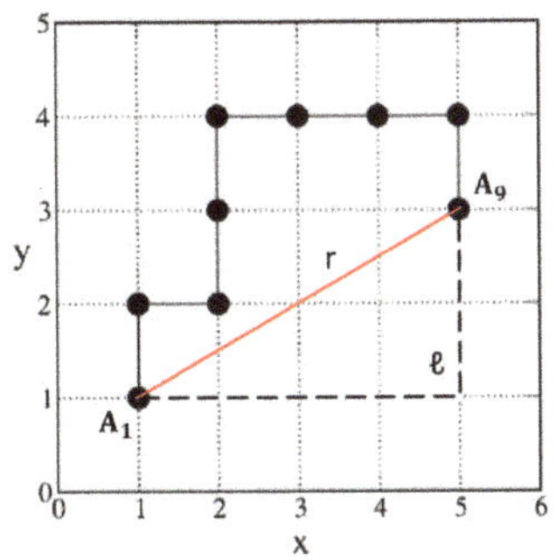

Figure 30. The end-to-end topological distance, $\ell(N)$ (dashed line), and Euclidean one, $r(N)$ (red line), for an SAW with $N = 9$. Here, $\ell(9) = 6$ and $r(9) = \sqrt{20} \simeq 4.472$, in lattice units, which are consistent with the bounds, $4.47 \lesssim \ell \lesssim 6.32$.

The knowledge of exact values of exponents for SAWs is very important for testing numerical approaches aimed at studying cases in which the exponents are not known exactly, as in the fractal structures to be discussed in Sections 3.2.2 and 3.2.3. Indeed, the reptation algorithm (see Section 3.2.4) turns out to be very accurate to deal with long linear polymers in complex environments. A test case is illustrated in Figure 31, where we have determined the structural function, $P(\ell, N)$ (the equivalent to $P(r, N)$ in ℓ space), on square and simple cubic lattices, using the reptation method. The results agree very well with the accepted values reported in Table 1.

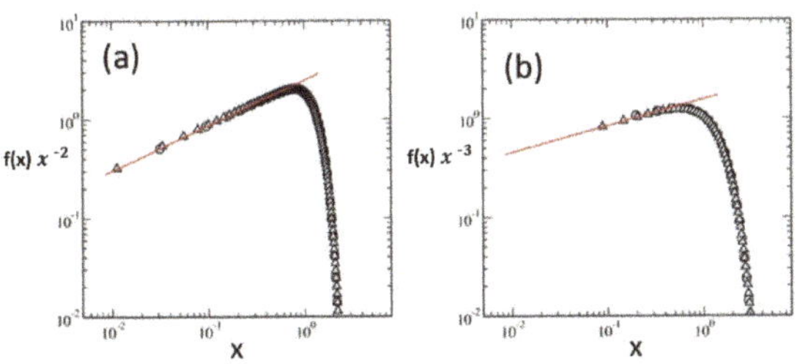

Figure 31. The scaling function for the end-to-end chemical distance (see Equations (8) and (9)), $f(x)/x^d$ vs. $x = \ell/N^\nu$, for SAWs on regular lattices: (**a**) square lattice and (**b**) simple cubic lattice, which were obtained using the reptation method. Here, we have used $\nu = 3/4$ in 2D and $\nu = 0.5876$ in 3D. The circles correspond to the case $N = 100$ steps, and the triangles to $N = 400$ steps. The red straight lines have slopes, $g_1(2) = 0.458(1)$ and $g_1(3) = 0.268(1)$, which are consistent with the results, $11/24 \simeq 0.45833$ and $0.26711(2)$, respectively, as reported in Table 1 (adapted from [344]).

3.2.2. SAWs in Deterministic Fractals

In order to understand the way in which the scaling behaviors of SAWs are modified if they are embedded in self-similar structures, it is convenient to consider first the case of deterministic fractals, where $\ell \simeq r$ holds due to the lack of any source of randomness in the substrate. In fractals in general, the polymer structural functions for the Euclidean and chemical end-to-end distances, $P(r, N)$ and $P(\ell, N)$, respectively, obey a similar scaling form as in regular systems, Equations (7)–(9), with appropriate values of the exponents (ν, γ), and with d replaced by the fractal dimension, d_f, of the embedding fractal medium. However, the present knowledge on the associated exponents is limited, and further theoretical work is needed to attain a more accurate picture.

Deterministic fractals can be classified into two main categories, finitely and infinitely ramified structures (see, e.g., [345,346]). Finite and infinite ramifications refer to the number of cut operations required to fully disconnect any given subset of the structure. The well-known Sierpinski triangle in 2D, and its 3D counterpart the Sierpinski pyramid (known also as Sierpinski gaskets), are examples of finitely ramified objects, for which renormalization group (RG) techniques can be applied to determine the critical exponents (see, e.g., [347]). In contrast, for infinitely ramified structures, such as those shown in Figure 26, the number of cuts required grows with the length scale. As a result, RG methods cannot be applied and the statistical properties of SAWs must be obtained numerically.

Regarding the exponents, the des Cloizeaux relation [338], Equation (10), seems to hold for finitely ramified fractals, in addition to the expression for δ, Equation (12), which seems to remain valid in general. However, analytical expressions for g_2, Equation (11), are not known. For infinitely ramified fractals, the des Cloizeaux relation breaks down, since it is essentially based on an RG type of approach. Results for the Sierpinski carpet and Sierpinski sponge (Figure 26) are summarized in Table 2, indicating that the values $g_1 \neq (\gamma - 1)/\nu$. Also, notice that γ increases with the dimensionality of the embedding space.

Table 2. The exponents describing SAWs embedded in a Sierpinski carpet (2D) and Sierpinski sponge (3D) of fractal dimensions, $d_f \simeq 1.89278$ and $d_f \simeq 2.872683$, respectively. The values of the exponents are from exact enumeration results [348] except for g_1, which have been taken from the MC simulation results based on the reptation algorithm [344]. The values for μ can be estimated quite accurately using the relation, $\mu = (d_f/d)\mu(d)$ [348], where $\mu(d)$ is the coordination number of the regular lattice. This relation yields, 2.50(2) in $d = 2$ and 4.26(2) in $d = 3$, which are in very good agreement with the measured values.

	2D	3D
d_f	$\ln 8/\ln 3$	$\ln 20/\ln 3$
ν	0.75(5)	0.58(3)
γ	1.23(4)	1.36(3)
μ	2.515(15)	4.26(2)
g_1	0.52(5)	0.33(5)
g_2	1.41(8)	0.10(5)
δ	4.00(30)	2.65(50)

A similar behavior for γ has been found for the triangular Sierpinski fractal [349], yielding $\gamma(2) = 1.36(3)$ and $\gamma(3) = 1.42(4)$, suggesting the lack of an upper critical dimension in contrast to the behavior on regular systems. This result is consistent with another family of Sierpinski gaskets in 3D [350], with $d_f = \ln[(b+2)(b+1)b/6]/\ln b$, that can be varied in the range $2 \leq d_f < 3$ when the parameter b is varied in the interval $2 \leq b < \infty$. For $b = 2$ ($d_f = 2$), they find $\gamma = 1.446(17)$, consistent with the above result, $\gamma(3) = 1.42(4)$ [349], and also that $\gamma(b)$ increases with b (see their Figure 3).

As suggested in Figure 26, the problem of polymers deposition on a rough surface could be a possible application of the present model of SAWs on a Sierpinski carpet. One can therefore consider a heterogeneous surface in terms of an energy landscape, on which the monomer units have different interaction energies, ϵ_i, on different locations depending on the type of the local substrate element i. For instance, in the case of the Sierpinski carpet, one can prepare a system in which the available fractal sites have an interaction energy, say ϵ_{f}, while the remaining non-fractal sites possess a higher interaction energy, $\epsilon_{\mathrm{b}} > \epsilon_{\mathrm{f}}$, thus favoring the occupation of the carpet sites by the linear polymer. Indeed, at very low temperatures, $k_{\mathrm{B}}T \ll \epsilon_{\mathrm{b}}$, one expects the polymer to be fully located on the fractal sites. In contrast, at very high temperatures, $k_{\mathrm{B}}T \gg \epsilon_{\mathrm{b}}$, the chain would behave as on a regular (flat) surface. In this way, one could study the temperature dependence of the polymer structure function, say, $P(\ell, N) \propto P(r, N)$, and the way the critical exponents can be modified accordingly. An attempt to answer this question has been suggested in [351].

3.2.3. SAWs in Fractal Percolation Clusters

The behavior of SAWs immersed within a conducting network modeled by percolation clusters, 'sufficiently' close to the critical percolation threshold, is very different from the one discussed above for regular lattices and deterministic fractals. The essential difference relies on the existence of two distinct spatial metrics within the random structure, which can be quantified by the relation between the length of the shortest path, ℓ, and the Euclidean distance, r, between two points on the fractals, i.e.,

$$\ell \simeq r^{d_{\mathrm{min}}}, \tag{13}$$

where d_{min} is the fractal dimension of the shortest path (see Figure 27). The shortest paths are scale-invariant and look similar to SAWs trails (see Figure 28). The relation (13) holds for all length scales exactly at the percolation critical concentration, p_c, and it crosses over to a linear regime, $\ell \simeq r$, above the correlation length (see, e.g., [310,312,313]). In the following, we assume we are dealing with percolation systems in which the fractal correlation length is much larger than the system sizes considered. Therefore, all the quantities discussed are described by the scaling behavior expected at criticality.

There is a second important issue regarding the structure of percolation clusters, which is the presence of singly connected portions of the cluster, like a peninsula connected by a narrow bridge to the main land. Such configurations act as traps for an SAW because the chain becomes stuck if it enters them. Therefore, when discussing percolation, we need to exclude those cluster dangling ends and confine our treatment to the cluster substructure which is able to sustain an 'infinitely' long SAW.

The substructure of a percolation cluster we are talking about is the backbone of the cluster. It is suitably defined as the subset of cluster sites carrying an electric current when a voltage difference is applied between two sites of the cluster. The so-defined backbone depends on the pair of sites selected for the input and output of the current; however, its statistical properties are robust and well defined [352]. As explained above, it is convenient to study the statistical properties of the backbone in chemical space, since its Euclidean distance counterparts can be obtained via the value of d_{min}. For instance, the fractal dimension of the backbone in ℓ-space, d_{ℓ}^{B}, determines the total number of backbone sites, $N_{\mathrm{B}}(\ell)$, within a distance ℓ such that

$$N_{\mathrm{B}}(\ell) \simeq \ell^{d_{\ell}^{\mathrm{B}}}, \tag{14}$$

yielding $d_{\mathrm{f}}^{\mathrm{B}} = d_{\mathrm{min}}\, d_{\ell}^{\mathrm{B}}$ in Euclidean space. The structural function, $\Phi_{\mathrm{B}}(r, \ell)$, giving the probability that two backbone sites, connected by a shortest path of length ℓ, are at spatial distance r from each other, obeys a scaling form similar to those in Equations (7)–(9), i.e.,

$$\Phi_{\mathrm{B}}(r, \ell) = \frac{1}{r} f_{\mathrm{B}}(x), \quad \text{with } x = \frac{r}{\ell^{\tilde{\nu}}}, \tag{15}$$

with $\tilde{\nu} = 1/d_{\mathrm{min}}$, yielding $\delta = 1/(1-\tilde{\nu})$, and the g-exponents are now denoted as $g_{1,2}^{\mathrm{B}}$, but no analytical results are known. A summary of the backbone exponents are reported in Table 3.

Table 3. The exponents describing the structural properties of the backbone of critical percolation clusters in 2D and 3D [352]. Values of d_{min} are from [319]. Above the critical dimension, $d_c = 6$, structural loops are irrelevant, and the backbone behaves like random walks, yielding $d_{\text{min}} = 2$, $\tilde{\nu} = 1/2$, $\delta = 2$, $d_\ell^{\text{B}} = 1$, and $d_{\text{f}}^{\text{B}} = 2$.

	2D	3D
d_{min}	1.13077(2)	1.3756(6)
d_ℓ^{B}	1.45(1)	1.36(2)
d_{f}^{B}	1.64(1)	1.87(2)
$\tilde{\nu}$	0.88435(2)	0.7270(6)
g_1^{B}	1.30(20)	1.02(20)
g_2^{B}	1.97(20)	0.59(20)
δ	8.6468(2)	3.662(6)

Several studies of SAW in percolation clusters, both on all clusters and on the backbone of the incipient infinite cluster, have been performed for site percolation at criticality (see [353] and refs. therein). In [353], the authors study SAWs in 2D percolation clusters near the critical concentration, $p_c \simeq 0.592745(5)$. On the backbone, they find for the end-to-end distance exponent ν the value, $\nu_r^{\text{B}} \simeq 0.77(1)$, using the Euclidean metric r, the enhancement exponent $\gamma_{\text{B}} \simeq 1.31(3)$, close to $\gamma = 1.34375$ valid on regular 2D lattices, and an effective coordination number $\mu_{\text{B}} \simeq 1.46(3)$, which is barely consistent with the suggested value $\mu_{\text{B}} = p_c\, \mu_{\text{reg}} \simeq 1.5637$ (see Table 1).

In [354], the mean distribution function of the end-to-end distance of SAWs in ℓ-space, $P(\ell, N)$, was studied on percolation clusters at criticality, which obeys the scaling form

$$P(\ell, N) = \frac{1}{\ell} f(x), \qquad \text{with } x = \frac{\ell}{N^{\nu_\ell}}, \tag{16}$$

$$f(x) = x^{g_1^\ell + d_\ell^{\text{B}}}, \qquad \text{for } x \ll 1, \tag{17}$$

$$f(x) = x^{g_2^\ell + d_\ell^{\text{B}}} e^{-c x^{\delta_\ell}}, \quad \text{for } x \gg 1, \tag{18}$$

where $\delta_\ell = 1/(1 - \nu_\ell)$. The scaling yields the mean value $\bar{\ell}(N) \simeq N^{\nu_\ell}$, ensuring the normalization condition $\int_{\ell_{\text{min}}}^{\infty} d\ell\, P(\ell, N) = 1$. The function in r-space, $P(r, N)$, where now $x = r/N^{\nu_r}$ and $\bar{r}(N) \simeq N^{\nu_r}$, obeys the same scaling behavior as $P(\ell, N)$, with the corresponding r-exponents. MC simulations were performed for site percolation on the square lattice at p_c [354] to obtain $P(\ell, N)$ and $P(r, N)$. Both distributions are related to each other by the convolution

$$P(r, N) = \int_{\ell_{\text{min}}}^{\infty} d\ell\, \Phi_{\text{iIC}}(r, \ell)\, P(\ell, N), \tag{19}$$

where $\Phi_{\text{iIC}}(r, \ell)$ is the equivalent of $\Phi_{\text{B}}(r, \ell)$ for the whole incipient infinite cluster (iIC) (see [354] for more details). From Equation (19), one can obtain exact relations between the set of exponents in both metrics. The idea is that working in ℓ-space yields more accurate values of the critical exponents, from which one can obtain the results in r-space according to $d_r^{\text{B}} = d_\ell^{\text{B}}\, d_{\text{min}}$, $\nu_r = \nu_\ell / d_{\text{min}}$, $\delta_r = 1/(1 - \nu_r)$, and $g_1^r = g_1^\ell\, d_{\text{min}}$, but $g_2^r \neq g_2^\ell\, d_{\text{min}}$ [354].

It was soon realized that the very existence of structural disorder and their scale-invariant fluctuations yield a multifractal behavior for the enhancement exponent γ and the effective coordination number μ [330]. In other words, the expression in Equation (4) must be generalized to account for the presence of fluctuations, as a function of N, arising from different backbone configurations. To investigate the contributions from the latter, one introduces a tune parameter, denoted as q in the multifractal literature, where $-\infty < q < \infty$, with the help of which one can select different configurations. In the limit $q \to -\infty$, one looks at very rare, almost linear backbones, while in the opposite limit, i.e., for $q \to \infty$, one can select the most compact ones. The multifractal equation describing the total number of SAW configurations can be written as [330],

$$\langle C_N^q \rangle_B^{1/q} = A_q\, \mu_q^N\, N^{\gamma_q - 1}, \tag{20}$$

where $\langle \ldots \rangle_B$ represents averages over different backbone configurations, μ_q are generalized effective coordination numbers and γ_q generalized enhancement exponents. By varying q in (20), one can extract the values (μ_q, γ_q) associated with each different type of backbone structure.

In order to proceed, we write Equation (20) in a form similar to the one considered in Equation (5) for regular systems,

$$\frac{1}{N} \ln\left(\langle C_N^q \rangle_B^{1/q} \right) = \frac{\ln A_q}{N} + \ln \mu_q + \frac{\gamma_q - 1}{N} \ln N. \tag{21}$$

Among all the values of q, just two of them play the most relevant roles. They are the $q = 1$ and $q = 0$ moments of C_N. The former yields the so-called annealed average,

$$\langle C_N \rangle_B = A_1\, \mu_1^N\, N^{\gamma_1 - 1}, \tag{22}$$

which is obtained using exact enumeration (EE) calculations [330], and the second one, the quenched average,

$$\frac{1}{N} \langle \ln C_N \rangle_B = \frac{\ln A_0}{N} + \ln \mu_0 + \frac{\gamma_0 - 1}{N} \ln N, \tag{23}$$

yielding 'typical' mean values accessible in MC calculations. Therefore, the discrepancy found in previous works between EE and MC calculations can be resolved within this multifractal scheme, suggesting that the case $q = 1$ corresponds to the EE results, while $q = 0$ to the MC ones. Indeed, it is found that the relation $\mu_1 = p_c\, \mu$, where μ is the effective coordination number in the regular lattice, holds very well in both 2D and 3D [330]. A summary of the exponents describing SAWs on the backbone of critical percolation clusters in 2D and 3D is reported in Table 4.

The problem with g_1^r in percolation clusters: The results for g_1^r presented in Table 4 are rather inconsistent (even in 2D) with the des Cloizeaux relation, $g = (\gamma - 1)/\nu$, obtained for SAWs on regular systems. Nor does it work in the case of infinitely ramified deterministic fractals (see Section 3.2.2). So, the question arises regarding whether we may gain some new insight yielding a better theoretical estimate for g_1^r in the case of critical percolation clusters. A step along this direction has been attempted in [355], yielding the new relation,

$$g_1^r = \frac{\gamma_1 - 1}{\nu_r} + \frac{\beta}{\nu}, \tag{24}$$

where β describes the probability that an arbitrary site belongs to the incipient infinite cluster, standardly denoted as P_∞, which for concentrations $p \gtrsim p_c$ behaves as

$$P_\infty \simeq (p - p_c)^\beta, \quad \text{for} \quad 0 < (p - p_c)/p_c \ll 1, \tag{25}$$

and ν describes the divergence of the correlation length near p_c, $\xi \simeq |p - p_c|^{-\nu}$.

The second term in (24) describes the additional difficulty that the Nth monomer of the SAW encounters to return close to the first one, taken as the origin, due to the topological constraints caused by the random nature of fractal percolation clusters. Indeed, Equation (24) yields $g_1^r = 0.54(7)$ in 2D, and $g_1^r = 0.92(8)$ in 3D, which is in very good agreement with the numerical results in Table 4. To be noted is that above the critical dimension $d_c = 6$, $\beta = 1$, $\nu = 1/2$ and $\gamma_1 = 1$; therefore, the second term in (24) yields the expected result, $g_1^r = 2$ (see [355] for further details). Finally, a review of the SAWs behavior in fractals, including triangular Sierpinski gaskets, can be found in [356].

Table 4. The exponents describing the structural properties of SAWs embedded in the backbone of critical site percolation clusters in 2D and 3D, obtained using EE calculations [330]. The relation $g_2^r = g_2^\ell d_{\min}$ does not appear to hold in this case, at least within the accuracy of these EE studies, and the values reported here have been obtained directly from the simulation results. Also, the des Cloizeaux relation, $g_1^r = (\gamma_1 - 1)/\nu_r$, shows significant departures from the reported numerical results, in particular in 3D. The numerically obtained values of μ_1 are, however, consistent with the relation, $\mu_1 = p_c\,\mu$ (see Table 1 for μ), with $p_c = 0.5927$ (2D), and 0.3116 (3D).

	2D	3D
γ_1	1.34(5)	1.29(5)
γ_0	1.26(5)	1.19(5)
μ_1	1.565(5)	1.462(5)
μ_0	1.456(5)	1.317(5)
ν_ℓ	0.890(5)	0.910(5)
ν_r	0.778(15)	0.66(1)
$\nu_r = \nu_\ell/d_{\min}$	0.787(10)	0.662(6)
g_1^ℓ	0.45(10)	0.66(15)
g_1^r	0.51(10)	0.91(15)
$g_1^r = g_1^\ell d_{\min}$	0.51(11)	0.91(20)
g_2^ℓ	1.60(16)	1.95(17)
g_2^r	1.26(40)	2.96(18)
δ_ℓ	9.5(5)	12.0(5)
δ_r	4.85(20)	3.10(20)

3.2.4. The Reptation Method

The reptation model [357] aims at describing the dynamics of a whole SAW chain in a melt, thus describing the motion of a single linear polymer in the presence of other similar ones (see also [321,322] and more recent literature [11,12,52,358,359]). Clearly, de Gennes' idea of an SAW attempting to move among 'obstacles' [357] can be easily extended to study SAW diffusion within an either deterministic or disorder fractal.

Here, we extend the method further to deal with static SAW configurations in a lattice. We have already shown that this is indeed possible, yielding very accurate results in the case of regular lattices (see Figure 31). In the following, we discuss the lattice reptation method able to generate suitable SAW configurations in a general domain that is either regular or fractal. We present MC simulations indicating that reptation yields results consistent with the SAW statistical properties in regular lattices (see Section 3.2.1).

The rules for the lattice reptation method are illustrated in Figure 32 in 2D, where the lattice constant equals the intra-polymer bond length. In this example, a chain of $N = 9$ monomers is considered, and the possible motions of the end monomers, denoted with the letters A and B, are indicated. If the selected new site for the end monomer is free, its motions carries the whole chain with it. Notice that we have taken just three possible jumps which point 'outside' the chain. Indeed, a fourth movement, for instance the upward one for A or B are automatically realized when B or A move, respectively. In a standard SAW chain, those movements are forbidden.

To generate an SAW configuration of say, $N = 100$ steps in a square lattice of size $L = 100$, one can start from an initial configuration in which the monomers are just disposed along a straight line. Clearly, there is the issue of the lattice boundary, and one can use PBC to treat cases in which the chain attempts to exit the square domain. However, one soon realizes that the initial configuration can be of arbitrary shape, for instance of an L-shape whose sides are 50 lattice units in length, thus avoiding the problem of touching the boundaries, at least in the first reptation steps. Since the span of an SAW, R_N, grows as $R_N \simeq N^{3/4}$, the chain will, on average, be of size $R_{100} \approx 32$ lattice units, suggesting that the lattice size $L = 100$ should be sufficiently large to admit SAW configurations which do not touch the boundary.

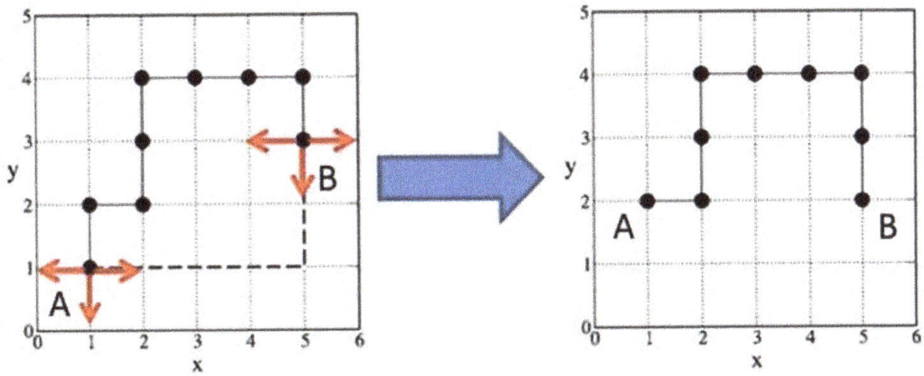

Figure 32. The reptation method on a square lattice. A chain of 9 monomers is shown, where the polymer ends are indicated with the letters A and B. The thick lines between the circles represent the chain links. The left panel illustrates the initial configuration, which can be changed by selecting any of the six possible jumps of the chain starting at any of its ends (red arrows). The right panel depicts the new configuration after end B has moved downward carrying the whole chain with it.

In general, an acceptable SAW configuration will be one in which the trail of the arbitrary initial configuration has been completely left after a sufficiently large number of reptation steps has been performed. This ensures full independence from the initial configuration as required to obtain an SAW. If one takes this final configuration as the new starting one, one should implement additional reptation steps until the new starting configuration has been exited. In this way, a newly indipendent chain configuration can be obtained and used to perform accurate SAW statistics.

We suggest a much simpler method for generating an SAW configuration, consisting of growing a chain from some suitable site, s_0, of the lattice, in such a way that at every reptation step, the chain increases its length by one unit inside the lattice (imagine that the whole chain is confined at the chosen site, gradually unfolding into the lattice), and the growth stops when the length N has been reached. If the leading end of the reptating chain becomes stuck within its self-generated trail, the chain is moved m_{back} steps backwards along the trail, like rewinding it back into the site s_0. This amounts to delete the last m_{back} monomers of the incipient trail, which are generated again.

In some circumstances, two or more m_{back} consecutive deletions may be needed. The parameter m_{back} can be adjusted so that the final chain is created within a maximum number of reptation steps, T_{max}. Here, the total number of steps to fully exit the trail is denoted as the exit time, $T_{\text{Exit}} < T_{\text{max}}$. We have chosen T_{max} so as to minimize the computing time, avoiding the occurrence of rejected configurations which have not reached the N steps within the maximum allowed time. An illustrative example of typical values of the mean exit times, $\langle T_{\text{Exit}}(N)\rangle$, is reported in Figure 33a as a function of chain length N. In Figure 33b, two SAW configurations for $N = 100$ and $N = 1000$ are shown, where one can notice an emerging self-similarity of the chain structures.

In the case of a fractal, two types of averages are required. The first one consists of generating several SAWs starting from a given fractal site, and there is a second one in which different starting locations are considered to explore the self-similarity of the structure. Clearly, for deterministic fractals, one should only consider initial points which are not equivalent by symmetry, while in random structures, any fractal site is a suitable starting point for growing SAWs.

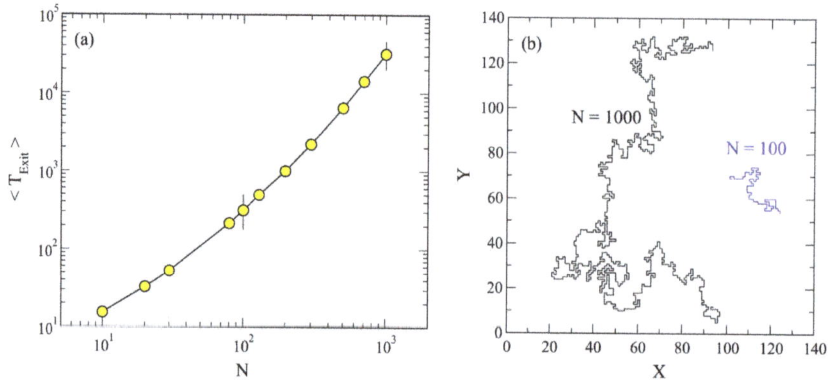

Figure 33. (**a**) Mean exit time steps, $\langle T_{\text{Exit}} \rangle$, to generate a reptation SAW configuration vs. N. The continuous line is a guide to the eyes, and the vertical lines are included for illustration, representing a standard deviation. The numerical data can be well fitted with the form, $\overline{T} = 0.7 \exp(1.92\, N^{1/4})$, for $N > 20$. Averages over 10^4 SAW configurations have been performed for the case $T_{\max} = 10^5$ steps and lattice size $L = 2000$. (**b**) Reptating SAWs on a square lattice illustrated in the cases: $N = 1000$ and $N = 100$ monomers.

The question remains whether these reptating linear structures are truly SAWs. The first test consists in calculating the mean end-to-end distance, $R(N)$, of the chains as a function of N. Results for $R(N)$ are shown in Figure 34, suggesting that the reptating chains are consistent with the expected scaling behavior for SAWs with $\nu = 3/4$. Similarly good results are found in 3D (in simple cubic lattices), which are discussed in Section 5.2 when modeling the attachment of polyethylene glycol (PEG) chains to a plasma-treated surface.

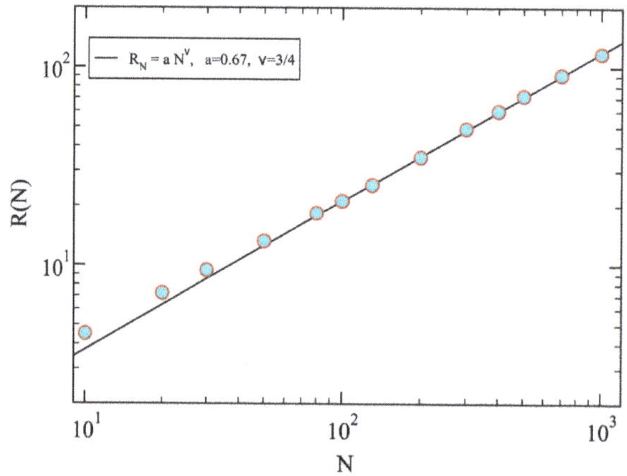

Figure 34. Root mean square of the end-to-end distance, $R(N) = \sqrt{\langle R^2(1,N) \rangle}$, vs. number of monomers, N, on the square lattice (full circles), for chains generated with the reptation method fulfilling the 'exit' condition discussed in the text. The straight line has the slope $\nu = 3/4$, which is in excellent agreement with the expected exact value. The square lattice has size $L = 2000$, and averages over 10^4 SAW configurations have been perfomed.

Another interesting quantity to look at is the distribution function of the number of non-local bridges in SAWs. A non-local bridge between monomers i and j can occur if they are located at a chemical distance $\ell = |i - j| \geq 3$. Illustrative examples of such bridges are shown in Figure 35, in the cases $N = 100$ (upper panel) and $N = 200$ (lower panel). For convenience of visualization, the positions of the non-local 'contacts' are indicated by the red dots, which are located at the middle point between sites (i, j), which are one lattice constant apart in space. Knowledge of the total number of non-local contacts for an SAW is important for estimating values of the collapsed temperature, T_θ, as we briefly discussed in the Introduction to Section 3 (see also [116]). The distribution function of the number of bridges, $P(B)$, is shown in Figure 36, in the cases $N = 100$ and $N = 200$. Notice the asymmetric shape of $P(B)$ displaying a positive skewness. Calculations of T_θ for reptating chains and the comparison with known results in the literature remain to be performed.

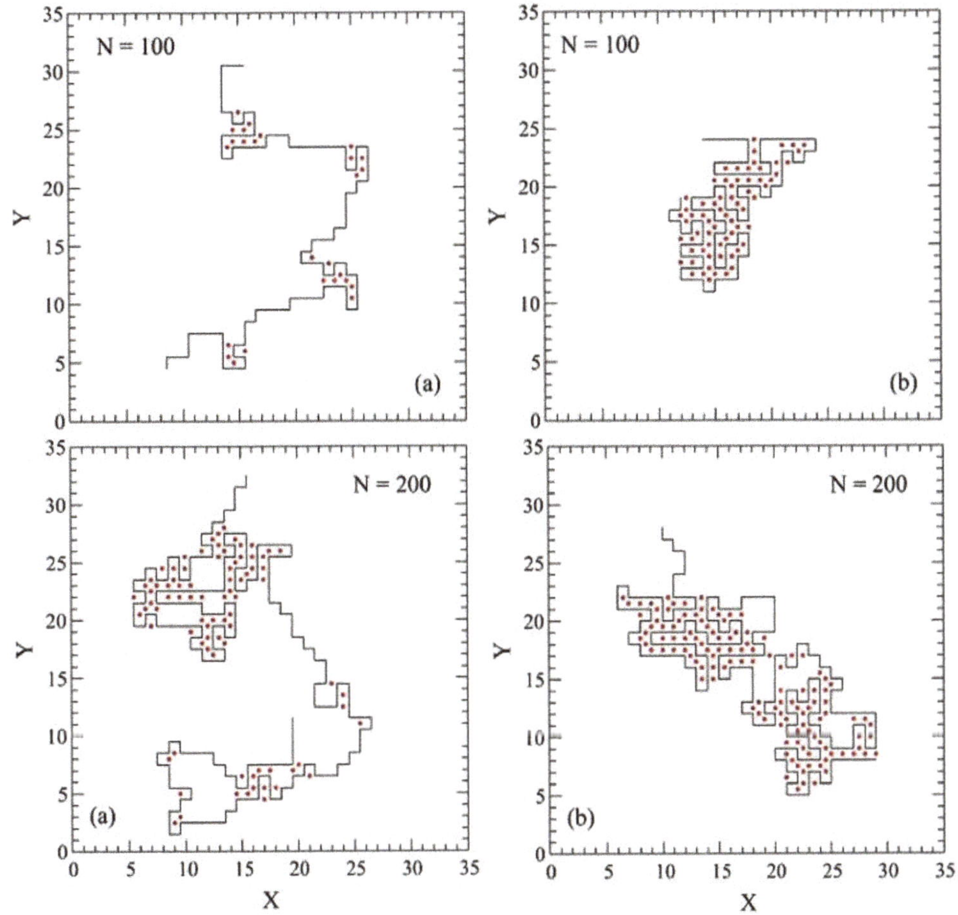

Figure 35. Examples of bridges, B (virtual non-local links between two monomers in space indicated by the red dots), for SAWs generated with the reptation method on a square lattice. (**Upper panel**) $N = 100$ in the cases: (Upper a) Typical configuration with a low number $B = 26$ bridges, and (Upper b) compact one with $B = 67$ bridges. (**Lower panel**) Same as in the upper panel in the case $N = 200$ for: (Lower a) Typical configuration with $B = 83$ bridges, and (Lower b) compact one with $B = 123$ bridges. A bridge is a virtual link connecting two monomers (i, j) which are nearest neighbors (one lattice unit distance) in space, but they are separated by at least three links along the SAW chain, i.e., $|i - j| \geq 3$.

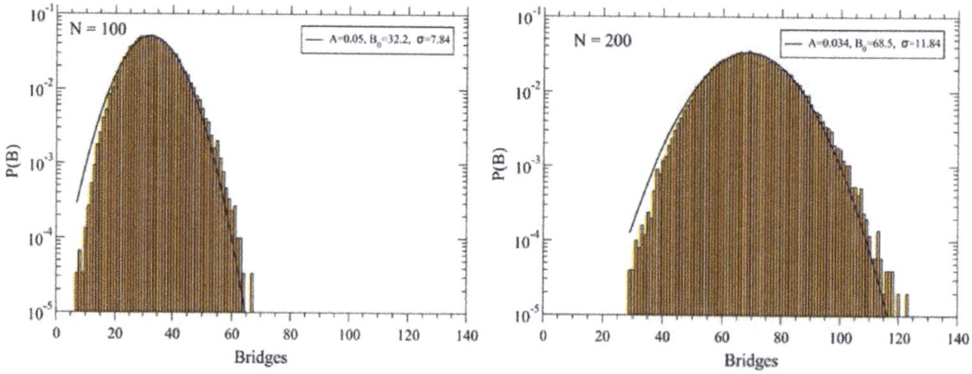

Figure 36. The distribution of the number of bridges, $P(B)$ (bars), vs. number B, for reptating SAWs in the cases: $N = 100$ and $N = 200$, obtained by generating $5\,10^4$ configurations each. The continuous lines represent Gaussian fits, $P(B) = A\ \exp[-(1/2)((B - B_0)/\sigma)^2]$, yielding ($A = 0.05$, $B_0 = 32.2$, $\sigma = 7.84$) for $N = 100$, and ($A = 0.034$, $B_0 = 68.5$, $\sigma = 11.84$) for $N = 200$.

The use of the reptation method is not limited to the problem of constructing single SAW configurations, but it can be extended straightforwardly to the realm of a many-chain ensemble. To generate an initial many-polymer configuration, one can grow reptating chains sequentially, in which the previously created chains remain fixed at their locations. In this case, the newly generated chains experience an ever-increasing difficulty to find available sites for growth, meaning that the chains are treated differently. In some cases, however, it might be necessary to avoid this undesired growth artifact explicitly. To this end, one can also generate reptating chains from single sites, the latter distributed at random within the system. Then, one starts growing chains one after the other, but one monomer at a time. If a chain becomes stuck, monomer deletions are applied to all chains to treat them in a consistent fashion, without giving preference to any particular chain, and also to keep the algorithm as simple as possible. There is, however, a much simpler solution to this problem if one lets the chains move around when dealing with a many-chain problem confined within a finite system, such as a cell (see Section 6.2).

3.3. *SAWs in Biology*

We discuss the application of SAWs in biology with particular focus on DNA molecules (Section 3.3.1), protein structure and folding (Section 3.3.2), and protein sequence evolution (Section 3.3.3).

3.3.1. The Case of DNA

Experimental studies of the behavior of linear polymers on surfaces is essential to test, and eventually validate, the SAW models discussed in the previous sections. For our present purposes, we refer to an experiment in which DNA molecules are kept in contact with a fluid lipid membrane by electrostatic interactions, still allowing the long chains to diffuse on the membrane surface [360]. The work complements observations of reptation dynamics of actin filaments in solution [361]. In [360], the authors were able to vary the concentration of the deposited λ phage DNA molecules by covering a quite broad range of surface densities from a very diluted regime to one characterized by full coverage. Both the structure of DNA chains, as well as their diffusive dynamics, were observed using fluorescence microscopy. For low and medium concentrations, they find that the mean spatial extension of the chains (radius of gyration) obeys $\overline{R}_G = \langle R^2_{DNA}(N)\rangle^{1/2} \simeq N^\nu$, as a function of the number of base pairs N, with $\nu = 0.79 \pm 0.4$, which is consistent with the theoretical value, $\nu_{SAW} = 3/4$ in 2D. They were also able to determine the structure factor in the low–medium concentration regime, yielding the expected fractal scaling, $S(k) \simeq k^{-1/\nu}$.

Remarkably, at high concentrations, the collapse of DNA chains was observed for the first time (cf. Figure 37), in which the molecules segregate and achieve a much smaller size, $\overline{R}_G$, than in the low-concentration SAW regime.

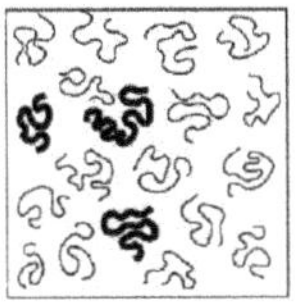

Figure 37. A cartoon of the type of collapsed DNA chain observed in the experiments. The thick lines represent fluorescently labeled DNA molecules, while the thin lines the remaining ones in the sample studied. The former are the actually observed DNA chains (adapted from [360]).

When looking at the distribution of collapsed DNA chains shown in Figure 37, one may wonder how the ensemble of linear polymers would behave if they were confined inside a finite spherical domain of say, radius $R_S < \overline{R}_G$. Actually, this question is relevant to the issue of packing DNA long chains into the chromosomes located inside a cell nucleus. An attempt to shed some light into this quest was put forward in [358], based on MD simulations of chains with $N = 1000$ monomers each, confined within a spherical cavity of radius $R_S < \overline{R}_G \simeq a\, N^{\nu_{\mathrm{SAW}}}$, where a is the monomer size and $\nu_{\mathrm{SAW}} \simeq 3/5$ (Equation (2)). It is found that that the rearrangement of the chain configurations does not cost additional free energy with respect to the case of non-confined chains. From these confinement studies, it is then concluded that within a cell nucleus, it is therefore very likely that a global reorganization of eukaryotic chromosomes can be readily achieved during most of the cell cycle. More importantly, it is then suggested that within a spherical boundary, eukaryotic chromosomes may behave themselves as they would in a semidilute polymer solution.

An extension of the simple SAW model presented here is considered in a recent review by Baschnagel et al. [359], in which the authors discuss numerical and analytical results on semiflexible chains (SFC) focused on static properties. SFCs are more involved than simple SAWs in the sense that they possess a local persistent length in which the monomers keep a preferred type of configuration characterized by a stiffness parameter, the latter responsible for the presence of correlations typically decaying with increasing relative chemical distance between monomers. SFC models admit theoretical approaches of a mean-field type, making them very attractive for analytical manipulations. A great deal of attention is dedicated to the issue of polymer interactions with surfaces, which is relevant to many experiments. Indeed, different variants of the model are discussed in detail to cope with the broad range of applications such as polymers confined in thin layers and within strict two-dimensional (2D) layers.

3.3.2. Protein Structure and Folding Dynamics

Proteins are highly complex molecules that are essential for life. They are directly involved in many chemical processes, in addition of being of nutritional value for all living organisms. Proteins of one species differ from those of another species, which make them species-specific. Within a single species, they differ according to which organ they belong to, thus being also organ-specific.

A protein is a very large molecule consisting of a series of amino acids which are joined together via the so-called peptide bonds, yielding a long chain of connected units at the very moment of its formation in the cytoplasma of a cell. Although there are more than 500 known naturally occurring amino acids, only 22 play a role in the constitution of proteins and are thus encoded in the DNA. Amino acids are organic compounds and all possess the same structure, as illustrated in Figure 38a. An amino acid is built upon the so-called α-carbon, C_α, connected to an H atom, a side-chain residue, and amino and carboxyl groups. Two amino acids can be joined together, forming a dipeptide (see the example in Figure 38b in the case of glycylglycine (GG)).

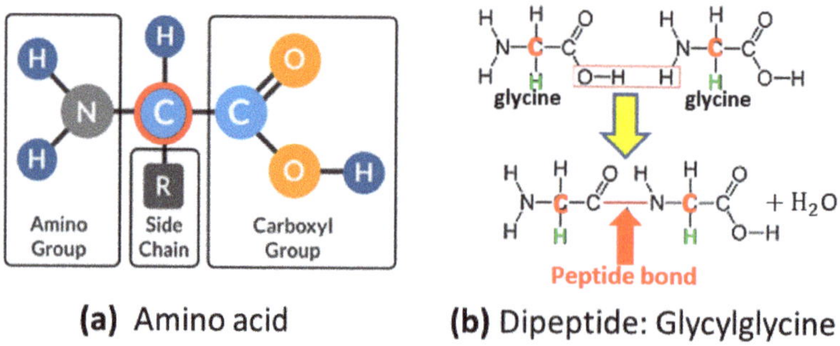

Figure 38. (**a**) The chemical structure of an amino acid. It consists of a central carbon atom (red circle), denoted as C_α, to which a hydrogen atom and a small molecule, or side-chain residue, are attached. The two remaining C_α bonds are occupied by an amino group and a carboxyl one. Amino acids therefore just differ from each other by the type of side chain considered. (**b**) The formation of a dipeptide: the case of glycylglycine. A peptide bond arises when the O-H subgroup of one amino acid (in this case glycine) joins one H unit of the amino group of the second amino acid (the second glycine here), yielding a water molecule. The newly created C-N link is the peptide bond (red line). Note that there are $(N-1)$ peptide bonds within a polypeptide chain of N amino acids. Notice that in the case of 20 different amino acids, the size of the sequence space would be $\mathcal{N}_s(N) = 20^N \simeq 10^{1.30N}$, yielding $\mathcal{N}_s(100) \simeq 10^{130}$.

Peptide bonds require about 18 kcal/mol to be formed. This energy is provided by adenosine triphosphate (ATP) hydrolysis within the cell. They are quite strong bonds remaining unaffected up to very high temperatures (above 1000 °K). However, they can be broken by hydrolysis in a water environment, releasing about 3 kcal/mol. Remarkably, the process is extremely slow with a half-life of several hundred years. In the cell, peptide bond breaking is obtained efficiently under the presence of enzymes, known as peptidases or proteases, which accelerate the transition when required.

Once the polypeptide chain has been released in the cytoplasma from a ribosome, it undergoes a spontaneous process of structural collapse, or folding, yielding a unique 3D structure of the protein in its so-called native state. The latter is very specific to the sequence of amino acids present along the polypeptide chain, and it is characterized by a suitable thermal stability within a relatively wide range of temperatures. Once formed, the protein is ready to enter the life cycle of the host body.

The native state has a rather small size compared to the original long structure of the underlying polypeptide, which becomes the internal backbone of the protein. The issue here is that side chains interact among themselves in 3D space, often via H-bonds or van der Waals links, whose interaction strengths are much weaker than that of peptide bonds, the latter being responsible for the backbone stability. These off-chain weak links remind us of the bridges between monomers discussed in Figure 35 for SAWs. The point is that such local pairwise interactions depend on the type of amino acids in 'contact' yielding a very complex and typically rough energy landscape, thus making the native state configuration extremely difficult to predict. The final proof that there might exist a one-to-one correspondance between the sequence of amino acids and its ground-state configuration remains so far elusive. This is known as the protein-folding problem. Recently, however, very promising results have been achieved based on the use of deep learning algorithms yielding very accurate predictions of actual native state conformations [362].

One way to qualitatively understand the behavior of polypeptide chains, and as a result shed some light into the mechanism of folding, consists of considering a coarse-grained model of a protein defined in a simple cubic lattice (see, e.g., [363]). A maximally compact coarse-grained native structure of a lattice protein, obtained in the simple cubic lattice, is shown in Figure 39.

In order to mimic the folding dynamics toward a target structure (e.g., the one in Figure 39), which occurs within a suitable range of temperatures, a fitting sequence of amino acids needs to be found. Indeed, if one takes a sequence at random and simulates a chain collapse, it will never reach the target in any finite amount of time. The basic idea is that only 'special' sequences of amino acids yielding extremely low values of the total non-local interactions will do the job. This sequence design can be applied very efficiently for any given target structure, since the rare configurations are widely separated from the energy density of states of random sequences, displaying a cospicuous energy gap [363]. The search for optimized sequences can be performed with standard MC simulations by assuming a given typical amino acids composition, in which the non-local contact interactions between, say, 20 amino acid types is taking from a 20×20 matrix, $\varepsilon_{n,m}$, with $n, m \in (1-20)$, which has both positive and negative entries, yielding a vanishing value, $\langle \varepsilon_{n,m} \rangle \simeq 0$, on average. It is indeed found that independently of the initial configuration in space, the chain carrying an optimized sequence, which we denote here as S_{36} in our example of Figure 39, successfully folds into the native configuration, which coincides with its ground state. In this sense, the folding problem admits a rather simple solution in the lattice at least for single domain (globular) types of structures and for not too large numbers of amino acids [363,364].

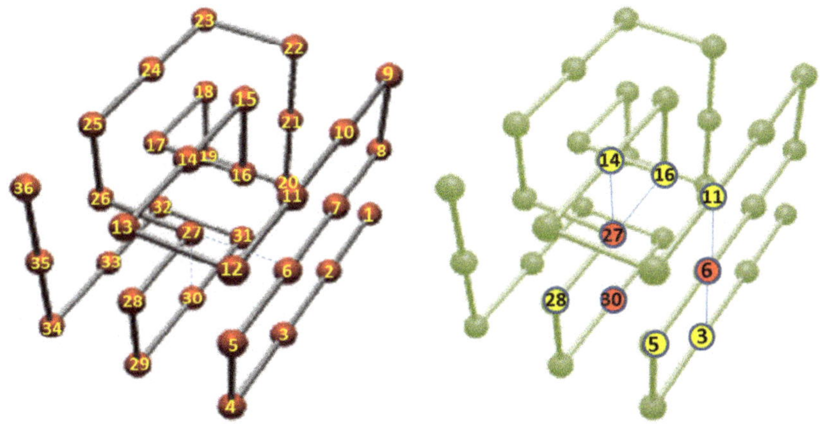

Figure 39. (**Left panel**) A native structure of a lattice protein in a simple cubic lattice having $N = 36$ amino acids (red circles). The structure is chosen to be maximally compact, where the numbers indicate the index of the C_α along the chain. This structure has $n_c = 40$ non-local contacts (or bridges) [363]. The sites (6, 27, 30) (connected by thin dashed lines) represent a sort of 'hydrophobic core' of the protein (see text below). (**Right panel**) The red sites (6,27,30) are denoted as hot sites, and the yellow ones (3,5,11,14,16,28) as warm sites [364]. Hot and warm sites happen to be first neighbors in the native structure either along the chain or as non-local contacts (see text below).

Once the folding issue of a sequence toward its native structure has been dealt with, the question arises of how stable the structure remains if changes in the amino acids sequence are considered. These changes represent amino acids mutations which are known to occur in natural proteins. Considering this question is also important to understand why a single native structure can admit different amino acids sequences [365].

To this end, we consider the following method for dealing with single amino acid mutations [364]. Let us denote by $m(i)$ the amino acid of type m at site i. From the sequence of amino acids $m(i)$, $i \in (1-N)$, one can calculate the total internal energy of the 'wild-type' chain, $E_N(\{m(i)\})$. To characterize quantitatively the behavior of a mutated chain carrying a single amino acid mutation, $m'(i) \neq m(i)$, one can calculate the mean local energy variation of the protein, $\Delta E_{m',m}(i)$, as the difference between the energies of the mutated chain, $E'_N(\{m'(i)\})$, and the wild-type one, averaged over all possible 19 types of different amino acids $m'(i) \neq m(i)$ at site i,

$$\Delta E_{m',m}(i) = \frac{1}{19} \sum_{m'=1}^{19} [E'_N(\{m'(i)\}) - E_N(\{m(i)\})]. \tag{26}$$

The quantity $\Delta E_{m',m}(i)$ contains essential information about the way the different amino acids become arranged in space within the native configuration. Indeed, there is a great deal of information that one can obtain from this simple approach. An example is reported in Figure 40 in the case of mutations on the optimized sequence, S_{36} [364,366].

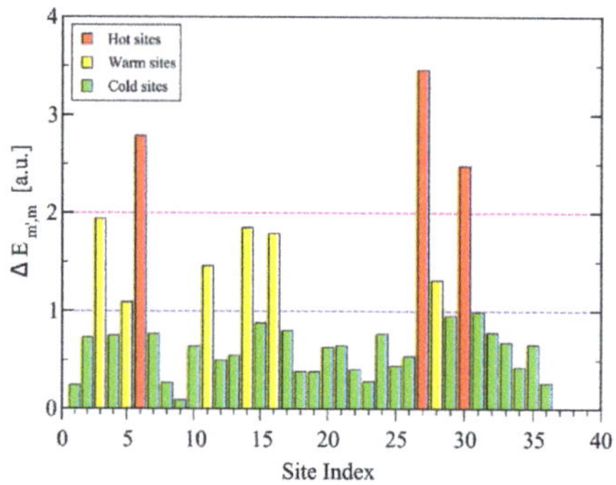

Figure 40. Local energy variations for single mutations at a site i in the native structure of S_{36} as a function of site index i. The colors indicate sites belonging to the same subset, denoted here as hot sites (6,27,30) (red bars) with $\Delta E_{m',m} > 2$, warm sites (3,5,11,14,16,28) (yellow bars) with $1 < \Delta E_{m',m} < 2$, and cold sites (1,2,4,7,8,9,10...) (green bars) with $\Delta E_{m',m} < 1$.

The results shown in Figure 40 suggest that one can classify different sites along the chain into roughly three groups: (1) cold sites (green bars in the figure) characterized by mean values $\Delta E_{m',m} < 1$; (2) warm sites (yellow bars) with $1 < \Delta E_{m',m} < 2$; and (3) hot sites (red bars) where $\Delta E_{m',m} > 2$. Although these numbers can depend on the actual values of interaction energies employed, a more general picture can be drawn. It is found that for the first set (cold sites), the mutated chain always folds into the target. This is very important, since the cold sites are the majority in the sequence, which explains the fact that a 'wild-type' chain may actually admit many different sequences, as observed in natural proteins. In the second group, the mutated chain still folds to a stable configuration, which differs from the target by a small amount, with $Q \lesssim 1$, where Q is the fraction of the same contacts present in the target. In some cases, it still folds to the target, yielding $Q = 1$. The third group is critical in the sense that some chains still fold to a state with $Q \simeq 1$, but many of them just end into a misfolded state with $Q \ll 1$, representing a sort of twilight zone for single mutated chains [364]. In addition to these findings, it has been shown that S_{36} also admits more than one mutation at a time [366], which are allowed when the mutated chain energies fall within the energy gap [363] between the wild-type target energy and the onset of the energy distribution for random sequences.

Another interesting problem that can be studied using lattice protein models is the issue of protein aggregation of similar polypeptides [367]. In the presence of a sufficiently high concentration of (denatured or non yet folded) chains, the folding processes do not follow straightforwardly as for isolated chains, but rather, interactions between particular monomers located on different chains can yield to misfolding. The basic mechanism for the occurrence of protein aggregation can be identified by looking at the hot sites of S_{36} (Figure 39), for which the simulations have been performed. A careful examination of the folding path followed by a pair of closely interacting chains provides a clue about what is happening. One can conclude that if one or more hot sites of a protein come in contact with any of the hot sites of the second chain, then the formation of an interchain link takes place and, as a result, none of the chains can fold to the native conformation. They just remain locked in a misfolded state. Remarkably, this is due to that fact that a protein has a (small) number of local partially folded intermediates that are responsible for the control of both protein folding and aggregation. It turns out that for designed lattice proteins, such intermediates are also the elementary structures constituting their folding nucleus [367]. Similarly, heteropolymer interactions within the cell in the form of intrinsically disordered proteins (IDS) display a quite complex behavior (see, e.g., [368,369]).

Coarse-grained protein models can be studied in the continuum, in which, in addition to the non-local interactions, one needs to specify how to deal with C_α–C_α interactions along the backbone. In lattice models, this question does not arise because there are no dynamical degrees of freedom associated to such 'peptide' bonds; they just keep a constant length all the time. In the continuum, one possibility is to introduce harmonic forces between nearest-neighbor sites along the chain, which must be treated simultaneously with the motion of the chain in space. There is one way to go beyond this traditional approach if one treats the peptide links as rigid without the need for treating any internal dynamics. The price for this simplification consists of rewriting the equations of motion using a Lagrangian approach with holonomic constraints [370]. An illustrative example is reported in Figure 41.

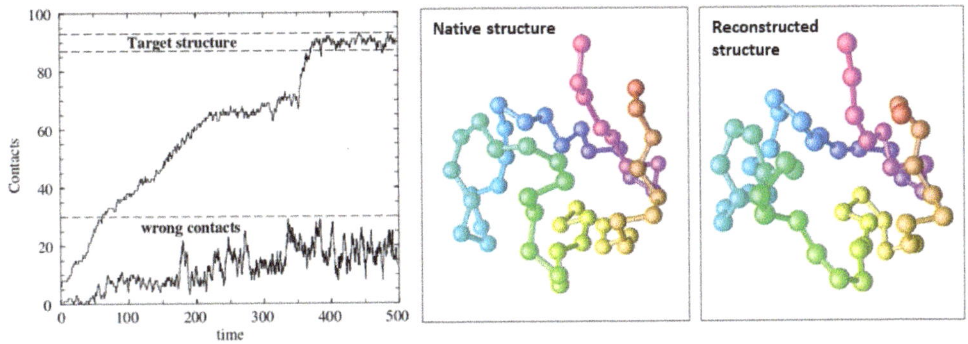

Figure 41. Off-lattice simulation of the reconstruction of Rubredoxin native structure using holonomic constraints for the intrachain bonds. (**Left panel**) The evolution of the number of emerging native contacts as a function of time. For comparison, also the number of new contacts not present in the native structure (denoted as wrong ones) are displayed for illustration. (**Right panel**) Comparison between Rubredoxin native conformation with a reconstructed one. The colors are to facilitate the location of the sites in space (adapted from [370]).

The equations of motion of the chain are solved by assuming a coupling to a thermostat which requires the use of friction coefficients, whose equations of motion are coupled to monomers momenta (for each spatial coordinate) and to the global temperature of the system. No other forces, such as due to the presence of water molecules, are considered in this case [370]. This method allows for reaching the long time-scale dynamics of folding, and of polymer chains in general, which cannot be accessible with traditional, more standard methods. It could also be employed to generate a suitable chain configuration to be used as the starting guess for full-atom MD simulations. The results thus suggest its possible utility in the context of protein structure predictions as well as for dealing with the modeling of polymeric materials at the nanoscale [370].

3.3.3. Protein Evolution in Sequence Space

The lattice protein model discussed in Figure 39 opened the door to the study of the general problem of protein sequence evolution by relying on a solvable model of folding [371]. The basic idea is to search for suitable single amino acids mutations which can be accepted if the resulting new sequence obeys the following three conditions. The first one is that the ground state of the mutated chain must be identical to the reference target structure (conservation of the phenotype). The second one regards its thermodynamic stability, and the third one its kinetic accessibility [371].

These conditions can be monitored in terms of the contact matrix, C_{ij}, with $(i,j) \in (1-N)$, defined as $C_{ij} = 1$ if a (non-local) contact between sites (i,j) is present in the structure and $C_{ij} = 0$ otherwise. To compare two structures, one can rely on the fraction of contacts that both structures have in common (see also the definition of Q above). Specifically, the structural overlap is given by $Q(C', C^0)$, where C' is the contact matrix of the ground state of the mutated sequence and C^0 the reference one, defined as,

$$Q(C', C^0) = \frac{1}{\max(n'_c, n^0_c)} \sum_{i<j} C'_{ij} C^0_{ij}, \tag{27}$$

where (n'_c, n^0_c) are the total number of contacts within each structure, i.e. $0 \leq Q \leq 1$. Notice that with the help of C_{ij}, the configurational energy $E_N(\{m(i)\})$ discussed in Equation (26) can be also written as

$$E_{N,C}(\varepsilon) = \sum_{i<j}^{N} \varepsilon_{m(i),m(j)} C_{i,j}. \tag{28}$$

The first condition amounts to taking $Q(C', C^0) = Q(C^0, C^0) = 1$. The second one requires the stability of C' along the folding path taking place at a fixed temperature T. It is quantified by the similarity between C'_{ij} and C^0_{ij}, which should remain above a predetermined threshold, Q_{thr}, that is,

$$\langle Q(C', C^0) \rangle_T > Q_{\text{thr}}, \tag{29}$$

where the symbol $\langle\rangle$ indicates an average over the different chain configurations at the fixed temperature T. Finally, the third condition imposes a maximum number of attempts to reach the ground state, which is based on the idea that a suitable amino acids sequence should proceed smoothly and swiftly toward its ground state (fast folder), thus mimicking the funnel scenario expected for real protein folding [372].

The most important consequence of these results regards the emergence of 'neutral networks' in sequence space, representing sets of sequences, S' associated with C', having the same ground state, $C' = C^0$ as the original sequence S^0, which is connected by point mutations. Therefore, a typical pair of distant sequences along the neutral network is indistinguishable from the behavior of just randomly chosen sequences. In other words, sequences that originated from the common ancestor, that keep the same ground state, can diverge in sequence space, yielding sequence similarities consistent with those found for random sequences. It is concluded that the thermodynamic stability of the ground state plays a crucial role for the emergence of neutral evolution in sequence space [371].

Based on this structural approach, a new model denoted the structurally constrained neutral evolution (SCNE) model has been developed [373], in which conservation (say $Q_{\text{thr}} \gtrsim 0.95$) of the thermodynamic stability of the wild-type native structure is imposed. Results from the SCNE model help understand the nature of neutral evolution trajectories in sequence space, suggesting new universal features for proteins. A prominent aspect is the existence of a broad distribution of the number of neutral mutations from one sequence to another, which is responsible for the breakdown of the traditional assumption of a Poissonian substitution process. Second, a much less known feature is the presence of strong correlations between the number of neutral mutations along an evolutionary trajectory, which cause the breakdown of self-averaging of the substitution process [373,374]. The correlations can be calculated using a so-called fluctuation analysis introduced within the framework of time series fractal analysis [375,376].

Further insight into the issue of predicting the native conformation from its sequence of amino acids has been achieved by realizing that for globular proteins, the contact map can be fully reconstructed from its principal eigenvector (PE) [377], in keeping with the fact that reconstructed contact maps allow in turn for an accurate representation of the 3D native structure (see, e.g., Figure 41). The results suggest that the single vectorial representation contained in the principal eigenvector of C is equivalent to the protein structure itself. Furthermore, it is also possible to map the sequence of amino acids to an hydrophobicity profile (HP), using a generalized hydrophobicity scale (interactivity scale), from which an optimal HP of a protein fold can be obtained. The latter is predicted to be strongly correlated with the PE of C [378]. This vectorial representation thus provides a sequence (HP)-to-structure (PE) map to be used in bioinformatics algorithms for structure comparison and sequence alignment and, more importantly, as an additional tool for protein structure predictions. Additional results and reference papers related to the issue of protein evolution and structure predictions are available in the literature [379].

4. Experimental Aspects of Polymers

Here, we deal with some experimental characteristics of polymers relevant to our discussions. In Section 4.1, we briefly review the chemical and structural properties of polymers, such as polypropylene (PP), and PTFE, and in Section 4.2, we explore the behavior of polymeric surfaces, such as polycaprolactone (PCL) and PTFE, after plasma treatments.

4.1. Chemical and Structural Aspects of Polymers

During the decade 1955–1965, significant developments to the then rapidly evolving field of polymer chemistry were achieved. They regard the polymerization of ethylene ($H_2C{=}CH_2$) by K. Ziegler [380], yielding polyethylene, and that of propylene ($CH_3CH{=}CH_2$) by G. Natta [381], in the isotactic and atactic forms of PP (see Figure 42). These discoveries lead to the method known as Ziegler–Natta catalysis (see, e.g., [382]), for which they shared the Nobel Prize in Chemistry in 1963. A third form of PP, known as syndiotactic PP, was also discovered [383]. More details on these systems can be found in the book by Roberts and Caserio [384].

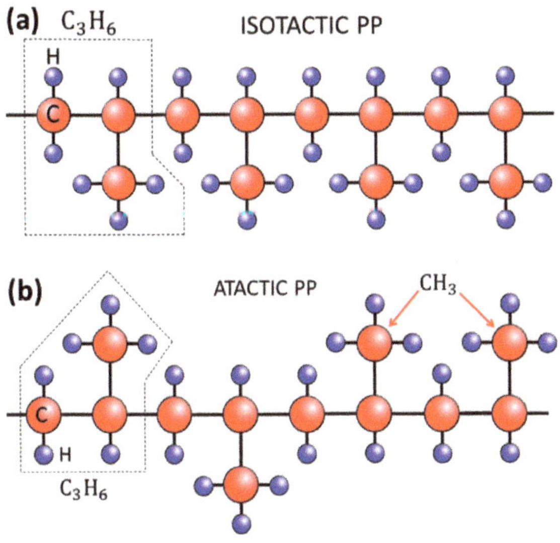

Figure 42. The two main forms of polypropylene of unit [C_3H_6]: (**a**) isotactic PP in which all the chiral carbons (CH_3) have the same configuration; i.e., they are on the same side of the chain. (**b**) Atactic PP, where the chiral carbons have random orientations. In syndiotactic PP, the chiral units alternate regularly along the chain (see, e.g., [384] for more details).

The two PP configurations shown in the figure have very different behaviors when forming a polymeric material. Isotactic PP displays a conspicuous degree of crystallinity, while its atactic counterpart is amorphous (i.e., not crystalline). As a result, the physical properties of these materials are quite different. The isotactic one is a strong crystalline polymer which melts at 175 °C, being also practically insoluble in all organic solvents at room temperature. The atactic one is much softer, elastic, and rather soluble in suitable solvents. In contrast, polyethylene melts around 100 °C. Polypropylene is a very versatile polymer with many applications in almost all of the plastics employed today.

Atactic PP polypropylene is not of commercial use, however under the presence of metallocene catalysts (see, e.g., [385]), it is expected that block polymers can be made containing alternating strings of both isotactic and atactic PP, as illustrated in Figure 43. Such mixed PP chains can eventually condense in such a way that isotactic parts meet each other, yielding little crystals inside the structure, thus creating sort of crosslinks that bring a partial rigidity to the network. In contrast, the atactic regions play the role of the elastic component of the structure, making it softer and amenable for applications as a good elastomer. Typically, plastic containers and bottles are made of a (50–60)% mix of isotactic–atactic PP.

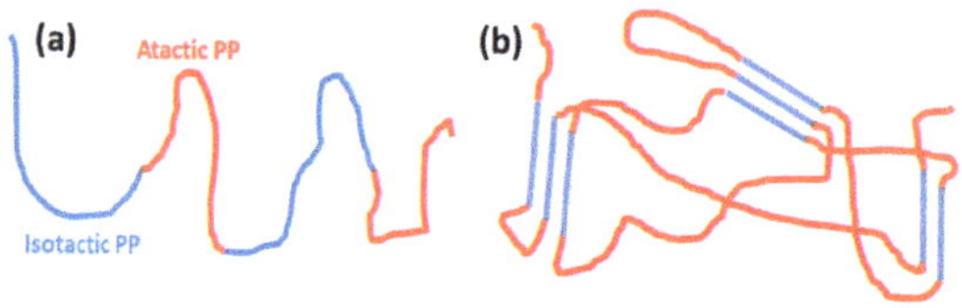

Figure 43. (**a**) Mixed polymerization connecting an isotactic PP string (blue lines) followed by an atactic one (red lines). (**b**) Illustration of a piece of material formed by mixed PP chains shown in (a). The blue straight lines represent isotactic PP crystalline zones, giving some rigidity to the structure, while the chosen proportion of atactic PP pieces provide the newtowrk with a desired elasticity.

To complement this experimental section, it is also useful to briefly review few theoretical papers which deal with polypropylene on both chemical as well as structural features.

Crystallization of isotactic PP via MC simulations: In [386], the authors model the temperature dependence of structural changes of isotactic PP near its melting temperature, using a coarse-grained model in a lattice with up to second nearest-neighbor interactions (see also [387–390]). At equilibrium, two different behaviors are found separated by a temperature $T_{PP} = 450\,\mathrm{K}$. For $T > T_{PP}$, the chains display Gaussian disorder, while for $T < T_{PP}$ helix formation becomes favorable even in the presence of environmental residual disorder. Below a second temperature, $T_{crys} = 385\,\mathrm{K}$, the system crystallizes. It is suggested that the polymeric material behaves similarly to a liquid crystal due to the long-chain helix configurations, which are locally emerging just above T_{crys}.

MC simulations of interlamellar isotactic PP: In [391], MC simulations are reported to investigate the semi-crystalline interphase for isotactic PP, in which the α_2 crystal phase of isotactic PP is used to model the structure of crystalline lamellae, and it is brought to equilibrium together with the remaining disordered domains [392–395]. The interactions between the different groups of the PP molecule are considered in detail via a united atom force field software. The work may shed some light into the quest of how crystal-amorphous interphases play a role in semi-crystalline polymer materials.

MD simulations of solvent effects on polyethylene oxide (PEO) and PP oxide (PPO) chains: In [396], the authors study using MD simulations the structural and dynamical properties of PEO and PPO chains at room temperature, both in a melt and at very high dilutions in different substances (see also [397–400]). The results are in very good accord with experiments, and conclusions can be drawn about the structural units which are formed in the solutions as found in nuclear magnetic resonance (NMR) studies.

MC simulations on effects of chain tacticity on polyethylene/PP blends: In [401], the authors search the mechanism responsible for the demixing of polyethylene/PP at equal concentrations, for the three PP tacticities at its melting temperature based on a coarse-grained chain model using MC simulations [402–405]. The model consists of both short- and long-range interactions, which is able to describe the transport behavior of the chain quite accurately. Evaluation of the degree of blend immiscibility was possible by calculating the interchain pair correlation functions, suggesting different tendencies of demixing according to the type of PP tacticity involved.

Single-chain folding of an isotactic PP chain: In [406], the authors study the folding behavior of a single isotactic PP chain via MD simulations, estimating the local rigidity strength of the folded chain. Care must be exercised in order to keep the underlying tacticity of the PP chain in its folded state. The study supports the expectation that angle and torsional potentials contribute to the speeding up of folding (see [407–410]).

Modified polyvinylidene fluoride (PVDF)–hexafluoropropylene (HFP) membranes: In [411], the authors report experiments on PVDF–HFP flat sheet membranes added with PEG, of different molecular weights and concentrations, which act as a pore former aimed at improving vacuum membrane distillation (VMD). The latter is a hybrid technology possessing several advantages over traditional desalination processes [412–415]. A 3D computational fluid dynamics (CFD) model is proposed for the validation of several properties relevant for comparison with experimental data.

MC simulations for complex polymer systems: In [416], the author reviews the Metropolis MC method to study complex polymer systems, showing that it remains a valid tool when compared with more elaborated full atoms MD simulations. The review describes applications to several systems in soft matter in which nanoscale structures play a central role. Examples include polymer nanocomposites and soft nanostructured materials, confined polymer chains, polymer rings and knots, crystalline systems, hydrogels and networks [417–420]. The extension of the method to non-equilibrium systems is also discussed together with some remaining open questions for future studies.

Fluorinated polymers, such as PTFE (see Figure 44), are very popular due to their low chemical reactivity, suitable wettability and adhesion properties, making them very attractive to be employed in coating. PTFE is a thermoplastic polymer, insoluble in water, solid at room temperature, having a density of 2200 kg/m^3 and 600 K melting temperature. It is produced by the free-radical polymerization of tetrafluoroethylene (TFE), the fluorocarbon C_2F_4. It is the simplest perfluorinated alkene. PTFE is hydrophobic, exhibits low electric polarizability due to fluorine, and has an extremely low friction coefficient suitable to be used as lubricant. PTFE typical chemical inertness comes from the high strength of carbon–fluorine bonds, finding also useful applications in surgery and as a coating on catheters. Initially, PFAS were used in the production of PTFE, but they have been replaced by less toxic substances having a much lower impact on human health.

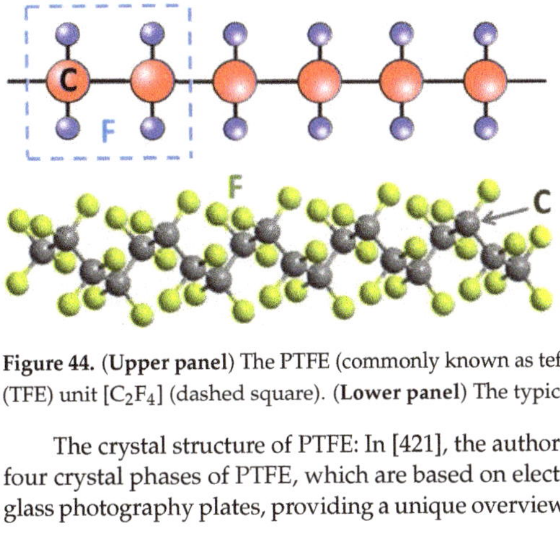

Figure 44. (**Upper panel**) The PTFE (commonly known as teflon) is built upon the tetrafluoroethylene (TFE) unit [C_2F_4] (dashed square). (**Lower panel**) The typical PTFE helix configuration.

The crystal structure of PTFE: In [421], the author reviews experimental studies on the four crystal phases of PTFE, which are based on electron diffraction patterns recorded on glass photography plates, providing a unique overview of reciprocal space that improves on the limitations from X-ray data. The paper is a useful reference for highlighting unexplained features of the PTFE crystal structure, resolving some conflicting issues remaining from previous works (see also [422–425]).

Crystalline lattice strains measurements in phase IV PTFE: In [426], a discussion is presented on neutron diffraction measurements of crystalline lattice strains as a result of applied deformations on PTFE crystals [427–430]. The hexagonal unit cell can be identified from the observed six primary diffraction peaks of the corresponding plane deformations of the cell. It is possible to determine the associated chain compression moduli in different directions, resulting in values which exceed those of the bulk by (1–3) orders of magnitude, in agreement with the theoretical values for the PTFE chains moduli.

Optical and mechanical properties of thin PTFE films from gas phase deposition: In [431], the authors produce PTFE thin films by gas phase deposition, using electron-enhanced vacuum deposition (EEVD), which was also combined with a cold plasma technique. The experimental probes include IR spectroscopy, AFM and X-ray photoelectron spectroscopy, allowing the study of the film structure, morphology and composition. Several quantities are obtained such as the film surface roughness, which depends on gas pressure and plasma parameters, the anisotropy of the refractive and extinction indices measured by spectral ellipsometry, hardness and Young moduli, and the thermal stability of the films [432–435]. The film contact angle equals that of bulk PTFE. It is concluded that PTFE molecules are mostly oriented perpendicularly to the substrate surface.

4.2. Fractal Aspects of Polymeric Surfaces

Polymeric materials surfaces can be modified, or altered in a well-defined fashion, by using chemical methods. This chemistry often requires the use of toxic elements, and a way to alleviate this issue relies on the application of gaseous plasma techniques (see Section 2.2 and Figure 14). Different plasma configurations and plasma setups are used, and the treated surfaces can develop a prevailing wetting character ranging from super-hydrophilic to super-hydrophobic, depending on the type of polymer considered. The resulting chemical composition of the new surface can be studied for instance by means of X-ray photoelectron spectroscopy. In some cases, the hydrophilic behavior may be due to the surface polymer bond breaking induced by the presence of ultraviolet radiation rather than the chemical functionalization of some reactive species (see, e.g., [436,437]).

Plasma techniques have become very popular for their versatility and reproducible results. Some examples are shown in Figure 45 in the case of PCL (upper panel) and PTFE (lower panel) (see [438] for more details). In the following section, we discuss the effects of plasma treatment to the wetting properties of some polymeric materials' surfaces.

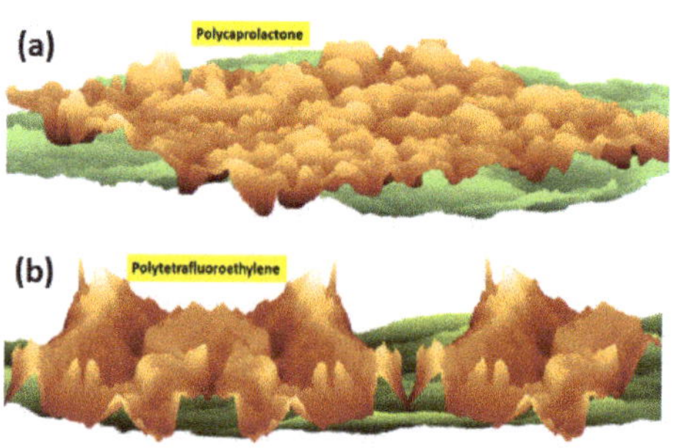

Figure 45. Illustration of non-treated (NT) (green color) and plasma-treated (PT) (red color) samples of polymeric surfaces, depicted roughly on scale. (**a**) PCL, the vertical scale for the NT sample is 60 nm, while for the PT one, it is 120 nm. (**b**) PTFE, the NT scale is 70 nm, and the PT one 790 nm.

5. Applications in Materials Science

The applications to materials science we discuss here regard plasma treatments of polymeric surfaces, examples of which concern PTE, PTFE, PP, and PCL. The effects of plasmas, typically cold ones, are described in terms of modifications in the wettability properties of the surfaces, which find many useful applications in industry (Section 5.1). A second theme of interest is the fabrication of new thin films by polymer attachment to a surface, which has been activated using plasma techniques, allowing for the presence of suitable sites on which a polymer chain, in this case PEG, becomes locked (Section 5.2).

5.1. Plasma Treatment: Wetting Properties and Surface Fractal Scaling

Plasma-treated polymeric materials surfaces are illustrated in Figure 46 in the case of PET, where oxygen gas is used at working powers of a few hundred Watts [439]. Based on these experiments, it is possible to draw conclusions about specific properties of the modified surface, in particular at the nanoscale, regarding wettability and surface new morphology. Typically, AFM and scanning electron microscopy (SEM) techniques are employed to extract relevant structural information on the treated surfaces. Remarkably, in most cases, such modified surfaces present fractal scaling behavior over the (10–100) nm range, which is however found to be already present in the non-treated surface too but having lower amplitude fluctuations extending on a smaller range of length scales [438].

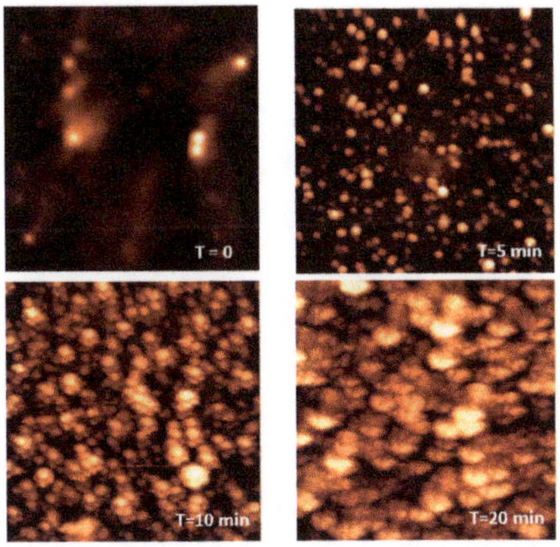

Figure 46. AFM images of PET surfaces after oxygen plasma treatment of different time durations, $T = (0, 5, 10, 20)$ [min] (adapted from [439]).

In the following, we consider the wetting properties of material surfaces by briefly reviewing a fractal surface model aimed at describing the dependence of the contact angle, θ_c, on the plasma treatment duration time, T. In the present notation, T should not be confused with temperature, since the latter does not play a role in our discussions here. The wetting behavior of surfaces is quantified by measuring the contact angle between a little water droplet gently deposited on top of the material. According to the Young equation, the contact angle θ_c between a fluid (L) and a solid (S) interface, immersed in a gas phase (G), can be written as [438,439]

$$\cos\theta_c(T) = A\,(1 - B\,F(T)), \tag{30}$$

with $A = \gamma_{SG}/\gamma_{LG}$, $B = \gamma_{SL}/\gamma_{SG}$, where γ_{ij} represents the interfacial surface tensions between phases i and j. The function $F(T)$ has been introduced to account for the variation of $\cos\theta_c$ with treatment time T. Notice that the original Young equation corresponds to taking $F(T) = 1$. Qualitatively, Equation (30) can be interpreted in terms of chemical effects between the interacting surfaces via the surface tensions γ_{ij} and physical ones due to the varying surface roughness as a function of T. This separation appears reasonable at first sight, but it needs to be carefully validated with experimental data. Illustrative examples of the observed behavior of a water droplet located on the PET-treated surfaces are reported in Figure 47a.

Interestingly, the form of $F(T)$ is rather simple, containing explicit information about the fractal scaling of the surfaces as a function of time T. The continuous lines shown in Figure 47b are obtained if one defines the roughnes function $F(T)$ as follows,

$$F(T) = \frac{1}{1+R(T)}, \quad \text{with} \quad R(T) = \left(\frac{\sigma(T)-\sigma(0)}{\sigma(0)}\right)^{d_S} \simeq \left(\frac{T}{T_0}\right)^{d_S H}, \tag{31}$$

where $\sigma(T)$ is the standard deviation of surface amplitude fluctuations, $d_S = 3 - H$ is the fractal dimension of the surface, H is the Hurst exponent of linear profiles of the surface, yielding the behavior $\sigma(T) \simeq \sigma(0) + \sigma(0)(T/T_0)^H$, and T_0 is a characteristic time for plasma treatment [438,439]. According to its definition, $R(0) = 0$, and $\cos\theta_c(0)$ is given by the surface tension values alone. In applying Equation (30), it is assumed that the latter (i.e., the chemistry of the process) remains unchanged upon plasma treatment. In other words, within this approximation, the γ values do not change with the plasma treatment. This assumption needs to be verified in future experiments, but in the cases discussed so far [438,439], it seems to hold quite accurately.

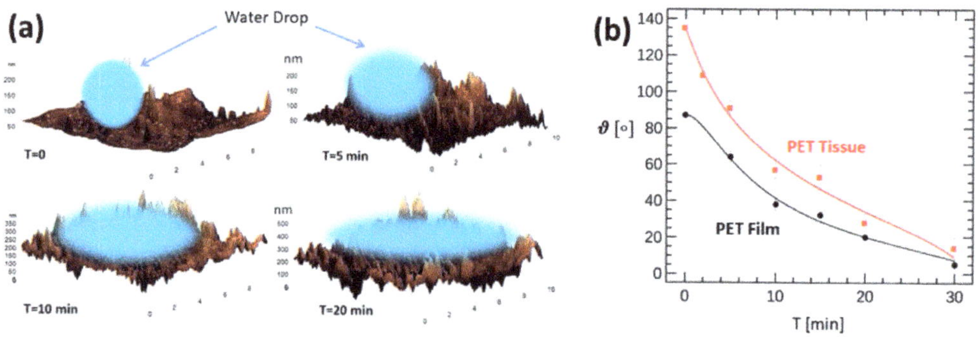

Figure 47. (**a**) The treated PET surfaces shown in Figure 46 illustrating the hydrophilicity effects on a virtual water droplet positioned on top of the material for different treatment times T [min]. The original surface is at $T = 0$, and the vertical scales are: (200, 200, 350, 500) nm. (**b**) The variation of the contact angle with treatment time expected theoretically (lines) and compared with experimental results (symbols). The images correspond to films of size 10 µm × 10 µm. In the case of the PET film, the fractal parameters are $H = 0.82 \pm 0.02$ and $d_S = 2.18 \pm 0.02$, while for the PET tissue, they become $H = 0.46 \pm 0.02$ and $d_S = 2.54 \pm 0.02$ (see Equation (31)) (adapted from [439]).

It is useful to make contact with the standard approach to wetting based on the Cassie–Baxter equation for the contact angle, θ_c, between a liquid droplet and a two-component solid surface, characterized by displaying fractional surface areas f_1 and f_2, with $f_1 + f_2 = 1$. Each surface component has its own contact angle denoted as $\theta_{1,2}$, respectively, yielding the relation,

$$\cos\theta_c = f_1 \cos\theta_1 + f_2 \cos\theta_2. \tag{32}$$

Now, for a fully hydrophobic second component, i.e., when $\theta_2 = 180°$, one has the known result $\cos\theta_c = f_1 \cos\theta_1 + f_1 - 1$. Finally, in the case of a single component, $f_1 = 1$, we arrive at the Wenzel regime, where the surface is characterized by a roughness $r \geq 1$ [440–442], so that $\cos\theta_c = r \cos\theta_1$. The Wenzel limit is appropriate for rough hydrophilic surfaces where θ_c decreases when r increases. For instance, in the examples shown in Figure 47, one has for non-treated samples, i.e., $r = 1$, $\theta_c \simeq 89°$ for

PET films, and $\theta_c \simeq 135°$ for PET tissues (see [439]), suggesting that NT samples are moderately hydrophilic.

To make contact with Equations (30) and (31), notice that the roughness parameter r can be written as the ratio between the surface area of the treated material, $S(T)$, and the non-treated one, $S(0)$, i.e., $r = S(T)/S(0) \equiv 1 + a\,R(T)$, where a is a parameter. Thus, one finds $\cos\theta_c = \cos\theta_1 + a\cos\theta_1\,R(T)$, which is denoted as the fractal-Wenzel model [439]. The model in Equation (30) can be expanded up to the first order in $R(T)$, yielding $\cos\theta_c \simeq A(1-B) + AB\,R(T)$, which coincides with the fractal-Wenzel result when $A(1-B) = \cos\theta_1$, and $AB = a\cos\theta_1$. Thus, a fractal-Wenzel equation describes a hydrophilic homogeneous fractal surface in the limit of low roughness, corresponding to short treatment times in the more general expression Equation (30). Further generalization of the latter to the case of hydrophobic surfaces, as in the case of PTFE, is possible yielding a very good agreement with the experimental results [438].

5.2. PEG: Thin Film Deposition via Reptation Dynamics

Thin films of linear polymers deposited on a surface have many useful applications in the biomedical sector. Among them, PEG plays a prominent role as a biocompatible coating displaying reduced protein adsorption and cell adhesion [443]. The deposition of thin films of PEG chains on a substrate necessitates the presence of active sites on the surface on which one of the ends of a PEG chain can become attached. The active sites are carboxyl (COOH) groups randomly dispersed onto a silicon wafer by means of chemical or physical processes. Plasma techniques are suitable for efficiently achieving this so-called pre-functionalization of the surface. Although this phenomenon was known for quite some time, the questions of how the PEG chains diffuse in solution and become finally attached to a surface have remained unanswered. In [444], thin films of PEG have been created on carboxyl-activated silicon wafers using chains of a fixed number of monomers, N. Different experiments were performed for samples with $N = 68$ and $N = 114$ monomers each. In order to understand the mechanism of PEG chain deposition and thin film formation, MC simulations of reptating SAWs in a simple cubic lattice were performed (see Figure 48), where the substrate has a density of active sites in accord with experiments [444].

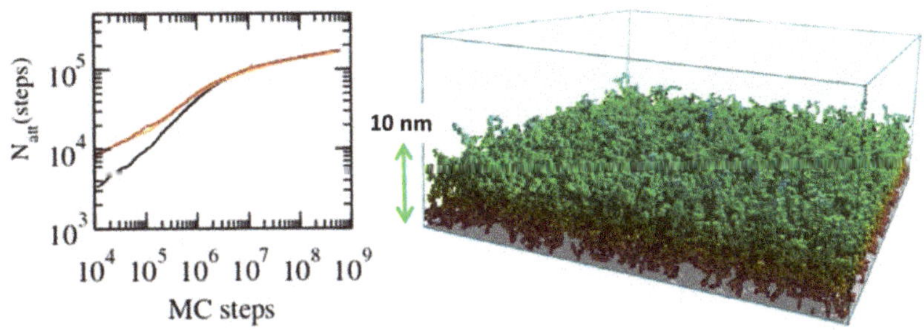

Figure 48. Monte Carlo simulations of SAW chains deposition on a flat plasma-activated substrate. (**Left panel**) Number of attached monomer units, $N_{att} = N\,N_{chain}$, with N_{chain} being the numer of attached chains of N monomers each, as a function of MC steps. Here, $N = 68$ (black line), $N = 114$ (orange line) and $N = 200$ (red line). The data collapse above 10^6 steps suggests that N_{att} becomes independent of N asymptotically. (**Right panel**) An example of a simulated PEG thin film, on a substrate of 200×200 sites and system height 100 sites, with a lattice constant $a_0 = 0.36$ nm, and SAW chains of $N = 200$ monomers. The total number of attached chains is 834 in this example. The color scale is proportional to monomer height (adapted from [444]).

It is found that less than 20% of total active sites ($\simeq 2\,\text{sites/nm}^2$) are effectively occupied by attached chains, corresponding to less than 5% of the total available substrate sites. Scaling arguments suggest a universal power-law dependence of film density, $\rho(N)$, as a function of the number of polymer units N, i.e., $\rho(N) \simeq \rho_0/N^{\nu}$, with $\rho_0 \simeq 5\,\mathrm{g\,cm^{-3}}$ and $\nu \simeq 0.6$, the latter consistent with the Flory exponent for SAWs in 3D (see Section 3.2).

The theoretical predictions of the dependence of film density on PEG length N and the prefactor ρ_0 on the density of carboxyl active sites [444] are important for the development of suited fabrics for specific biomedical applications. Further experimental research is still required to clarify several open issues such as deposited PEG film mass and density, and the polymer attaching dynamics to validate the model predictions of the film growth time scales determining the PEG deposition processes.

PP films on silicon wafers: In [445], the authors consider PP film growth on silicon wafers based on a simple and versatile method favoring a covalent attachment to produce ultrathin films [446–448]. The attachment sites are created by employing perfluorophenyl nitrenes produced by the thermolysis or photolysis of perfluorophenyl azides. A careful development of film–wafer covalent bonds allows the fabrication of both crystalline and amorphous PP films and, as an extension of the method, an elastomeric PP copolymer is also successfully realized.

6. Applications in Biology: DNA Chains Within the Cell and Aminocyanines

We consider applications to the DNA chain structure in particular regarding the occurrence of 'i-motifs' (Section 6.1). The question of DNA transcription is discussed in some detail in Section 6.2, where contact is made with the reptation method for generating SAWs configurations in a lattice discussed in Section 3.2.4. Finally, recent results on the promising use of aminocyanines are discussed in Section 6.3.

6.1. DNA Chains: I-Motifs

The DNA consists of two connected chains in space typically forming a perfect closed double helix. Each chain of DNA consists of a backbone upon which a sequence of nucleotides or bases (A = adenine, T = thymine, and G = guanine, C = cytosine) are attached. The two chains are connected by bonds between pairs of opposite bases, (AT) and (GC) pairs. Thus, the sequence of bases on one chain corresponds exactly to the one of the second chain, in which A is replaced by T and G by C. Remarkably, often C bases pair with each other, i.e., forming a (CC) link instead of a (CG) one, yielding a deformation and partial opening of DNA, thus creating a protrusion denoted as an i-motif. This is likely to occur within cytosine-rich regions of the genome. Therefore, DNA is prone to contain i-motifs or knots along the chain (Figure 49), leading to the suspicion that they could be important regulators of the genome in relation to which genes are on or off [449]. However, both the locations and number of knots were not identified.

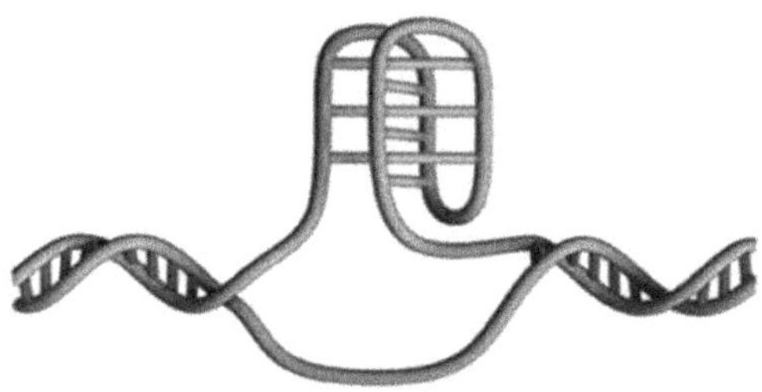

Figure 49. Schematic illustration of an i-motif: a piece of DNA with a knot inside its structure. Such openings make the transcription process possible, providing key regulatory functions (adapted from [449]).

Recently, a thorough examination of DNA led to the conclusion that such i-motifs are much more common than presently believed. Some 50,000 knot-type structures were found along DNA strands [450]. This work is relevant for improving the predictions about the exact locations of i-motifs within a DNA strand, which in turn can help the development of new strategies for the treatments and perhaps even the prevention of diseases.

6.2. DNA Chains: Transcription Process Within the Cell

The problem of DNA transcription has also attracted a great deal of attention among theoreticians. As a good starting point to this problem, one considers coarse-grained polymer models studied by either MC or MD simulations. One popular approach is the so-called facilitated diffusion model, taking place within a closed volume, mimicking the nucleus of a living cell, within which a long chain, representing a strand of DNA, is confined. In facilitated diffusion, one considers a point-like transcriptor factor, representing a protein that binds to DNA in appropriate locations (targets), that mimics the important task of genetic information transcription from DNA to messenger RNA. Such a particle performs, alternatively and randomly, both a 3D random walk in the free cell space and a random walk along the DNA chain until it reaches the target. The latter represents the finite coded sequence to be read. The MC simulations in [451] performed in the continuum suggest that facilitated diffusion is most efficient in the presence of an elastic DNA chain, which appears relevant for in vitro studies and diffusion on yeats chromatin.

An illustration of a long chain inside a cube defined in a simple cubic lattice is shown in Figure 50a. The chain configuration was obtained using the reptation method in 3D by letting a polymer chain grow inside the finite volume consisting of hard walls. When the leading SAW chain end touches the wall, it retreats as explained in Section 3.2.4 and performs another growth attempt. In 3D, the growing of SAW chains is much easier than in 2D due to the less attrition present in the system. Notice also the exceedingly large fraction of cell space left unoccupied by the single SAW chain. Indeed, the method allows for simulating many SAWs at the same time, making the model much more realistic (see example in Figure 50b).

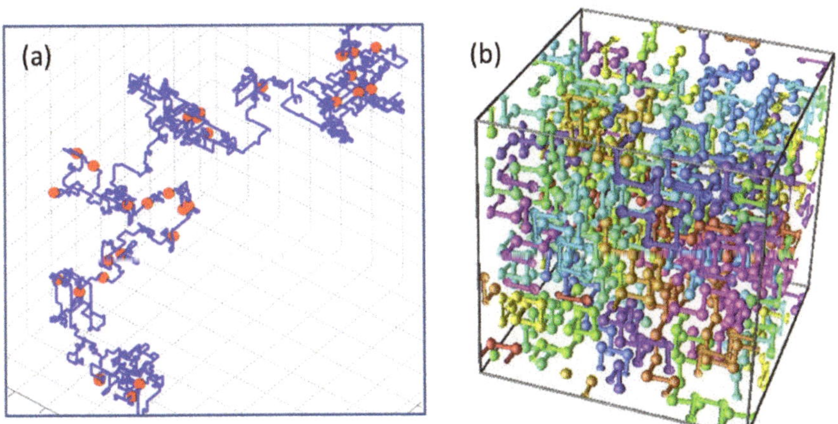

Figure 50. (**a**) Transcription process: Simulation of a long DNA (10,000 units) in a cubic lattice of linear size 100 lattice units, using the reptation method. The red circles represent the transcription targets. A single transcriptor executes a 3D random walk inside the free lattice space and can perform a 1D random walk along the blue path. Once moving along the DNA chain, the transcriptor can leave its 1D walk and move to the 3D free lattice if the target has not been found. Its remains along the DNA for a finite number of steps only, in keeping with the faster diffusion in the free space that accelerates the target-finding process. (**b**) Simulation of 20 SAW chains with $N = 20$ monomers each, using the reptation method, within a cubic lattice of linear size $L = 10$ lattice units with PBC, yielding a total chain occupation of 40% inside the confining cube. Courtesy of Francesco Marini.

In [452], the authors describe the process of gene transcription target, by means of MC simulations in a simple cubic lattice, using many proteins that start their motion from a single lattice position, $\vec{\ell}$, simultaneously. They study first-passage times of proteins touching a single DNA chain and average them over several configurations. The mean first-passage time is then studied as a function of the distance from the starting position $\vec{\ell}$ and a single DNA target position and the resulting concentration of proteins employed. The results suggest the possible relevance and importance of colocalization of the initial protein search inside real biological cells. Additional research has been performed on a simpler 1D model of DNA in which facilitated diffusion occurs in 3D space undergoing anomalous diffusion to mimic the complex transport behavior of a many-protein system [453].

6.3. *Aminocyanines: A Physical Method to Kill Cancer Cells*

Polymethines are compounds made up from an odd number of methine groups (=CH–) bound together by alternating single and double bonds, as, e.g., in piperylene or 1,3-pentadiene CH_3[–CH=CH–CH=]CH_2 indicated within square brakets. A chain with alternating single and double bonds can form a conjugated system, which is characterized by the presence of connected π-orbitals carrying delocalized electrons along the molecule. Methine groups can become closed into a single ring as in benzene (3[=CH–CH=]$\equiv C_6H_6$), having an aromatic character. For even number of methine groups, the compounds are called polyenes. Finally, cyanines (Cyn-Residue) are synthetic dyes and belonging to a polymethine group. In case the residue is the amino group NH_2, the compound is called aminocyanine (see the example of Cy3-NH_2 in Figure 51 and [454] for more details). Such polymethine–carbocyanine dyes, known also as aminocyanine fluorophores, are widely utilized as fluorescence probes specially for obtaining images of biological targets in vivo due to their remarkable photophysical properties.

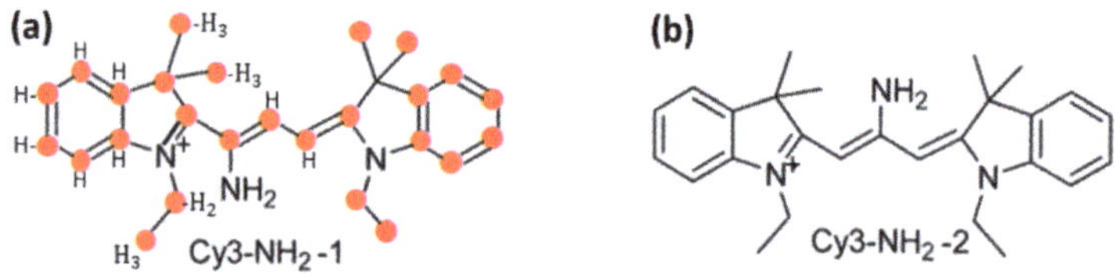

Figure 51. The structure of aminocyanine dyes Cy3-NH_2-(1,2). The red circles represent C atoms, while few H atoms are indicated for illustration. (**a**) Cy3-NH_2-1: The unit NH_2 is close to N^+. Here, a 3C-polymethine chain is transformed into a 2C-polymethine chain in the center of the molecule, and the third C is attached to the amine group NH_2 rather than to a H. (**b**) Cy3-NH_2-2: The unit NH_2 is at the center of the polymethine chain (adapted from [454]).

In [455], the authors study a remarkable application of aminocyanines to kill cancer cells in vitro using near-infrared light (NIR) in the 650–900 nm wavelength range. The basic idea is to exploit the significantly high electron–phonon couplings typical of these dye molecules. The NIR light produces the excitation of plasmon modes, which are collective electronic states extending over the whole molecule [456]. In the case of aminocyanines, the strong coupling between plasmons and collective atomic motions called phonons result in a mechanical response of the molecule which is used to destroy the membrane protecting a cancer cell. The authors called this effect vibronic-driven action (VDA), and the mechanical disruptive behavior of the aminocyanine is referred to as a molecular jackhammer. It is expected that a cell cannot avoid necrosis by developing resistance to such jackhammers, suggesting that they can be used with success to treat cancer cells which have developed deep inside the skin, perhaps reaching several cm in depth. In addition to experiments in vitro, they obtained a 50% tumor-free efficacy in mouse with melanoma, thus opening the door for future studies aimed at treating cancer cells in vivo by exciting aminocyanines with NIR light.

Recently, a new discovery on unexpected double bonding in carbon composites has been reported on a class of geometrically distorted ϕ-bonded molecules [457]. These are known as anti-Bredt olefins (ABOs), the latter were theoretically known since 1924 but expected to be difficult or even impossible to realize. The authors discuss a new strategy to deal with these elusive compounds displaying significant distortion due to the presence of geometrically constrained ϕ-bonds, offering a method to produce geometrically distorted alkenes to be used in future synthetic applications.

7. Concluding Remarks

We have reviewed theoretical aspects of linear polymers modeled by SAWs in regular lattices, and in both deterministic and random fractal structures, which should provide the reader with the state of the art on SAWs in those environments. The use of square and simple cubic lattices simplifies the numerical studies to a large extent, in keeping with the universality class of SAWs in 2D and 3D cases. Fractal structures are of central interest here, since they may play a prominent role in many important applications, in particular when some type of complex structural disorder yields an unusual behavior of the polymeric system. By studying simple fractal models, one can understand how the statistical properties of SAWs are modified in the presence of scale invariance. Clearly, the easiest cases to consider are deterministic fractals which lack structural disorder before entering the realm of fractal disordered structures as in the case of critical percolation clusters. These studies provide a benchmark for the fractal behavior of polymers, which can be used as guidelines to build realistic models aimed at describing experimental observations of real polymeric systems.

Specifically, we have discussed in detail the geometrical aspects of SAWs in regular and fractal lattices with emphasis on the two types of metrics required to determine the structural functions of single chains, such as the distribution functions of the end-to-end distance of N-steps SAWs, both in Euclidean space, $P(r, N)$, and in topological or chemical space, $P(\ell, N)$, where ℓ is the topological distance between the chain ends (see, e.g., Figure 30). While in regular lattices, and deterministic fractals modeled by Sierpinski fractal carpets in 2D and the fractal sponges in 3D (see Figure 26), both metrics are equivalent, i.e., $r \simeq \ell$, in critical percolation clusters they scale differently, $r \simeq \ell^{d_{\min}}$, where $d_{\min} > 1$ is the fractal dimension of the shortest path. The fact that $d_{\min} > 1$ yields a multifractal behavior of SAWs in such disordered systems. However, one can study the statistical properties of the chains in ℓ-space, which offers the possibility of reducing considerably the fluctuations of the measured quantites of interest, while the behavior of SAWs in real space can in most cases be obtained from their counterparts in chemical space.

The question arises regarding whether the deterministic fractals discussed here (as those in Figure 26) are just of theoretical interest, or they may have some experimental realization. We have suggested that their structures may become experimentally accessible as in the case of the 2D carpet, which may be created by deposition techniques using a fractal mask. The 3D sponge could be built using 3D-printing techniques, the latter representing for instance a regular type of porous material. From a theoretical point of view, they offer the possibility of studying statistical properties of linear polymers in scale-invariant systems accurately, thus allowing for the determination of critical exponents which may differ from their standard values in regular lattices. In particular, much effort is still needed to improve on the values of the exponents in the case of percolation clusters (see Section 3.2.3). Furthermore, future work is envisaged to have a complete picture on the transport properties of diffusing chains within fractals. To this end, applications of the reptation method (Section 3.2.4) should be explored, as it may yield accurate and interesting results. In Figure 31, we have provided an example demonstrating the excellent performance of the method in regular lattices regarding static properties. We have used reptation also on deterministic fractals [344] (Sierpinski carpet (2D) and Sierpinski sponge (3D) in Section 3.2.2) and the numerical results were summarized in Table 2.

The reptation method is suitable for generating SAWs configurations in arbitrary environments. Yet it can be easily extended to a many-chain system, also in the presence of geometric constraints inside the host's structure. The problem of creating an ensemble of reptating chains, briefly discussed in Section 3.2.4, has not been studied systematically, and future work on this issue is certainly welcome. The applications are countless due to the many different types of polymer melts one can consider, including heteropolymers where different types of monomers are present, yielding a complex interaction-energy landscape. The method can be applied to study a melt at thermal equilibrium using MC rules and the outcome compared with more standard simulations methods of a many-chain system (see Section 2 for examples). The method can be used to find a suitable initial configuration of a many-polymer system, as we briefly discussed in Section 6.2 in the case of periodic boundary conditions. Indeed, homopolymer chains can be taken as the backbone of more complex polymers and use their location to start integrating the equations of motion of the more realistic polymers within an MD scheme.

Protein folding and protein sequence evolution have been discussed in Section 3.3 based on simple SAWs heteropolymer models of proteins in the simple cubic lattice, where the basic concepts of folding and the characteristic structures of proteins are illustrated. The last two sections of the work are dedicated to applications in materials science and biology. The former deals with a discussion of the important role that fractal features play for determining the wetting properties of plasma-treated polymeric materials surfaces. The basic concepts underlying the main equations used for describing wetting are explained, and contact is made with experimental data. Further experimental investigations are welcome to extend or modify the range of validity of these theoretical predictions. In addition, the interesting issue of modeling the growth of thin films by polymer deposition is mentioned, but many questions remain open such as a detailed numerical study on the scaling effects of the density of active sites created on the substrate on which the chains become attached. Experiments should be designed to investigate other polymers by exploring possible modifications of the reptation approach used in the case of PEG. Finally, applications in biology are discussed dealing with long linear polymers, such as DNA, confined within a finite domain, such as a cell. The material examined here is just a short introduction to the subject, and much more work can be conducted to establish a full picture on the issue of a high-density polymer ensemble confined within closed boundaries. We hope that the present review can be of help to explore new avenues in polymer science in different areas of physics, chemistry and biology.

Funding: This research received no external funding.

Acknowledgments: I would like to kindly thank my colleagues M. Porto, A. Ordemann, F. Marini, A. Bunde, S. Havlin, J. Dräger, Dietrich Stauffer, Ricardo A. Broglia, G. Tiana, E. Vigezzi, S. Pasquali, E. Shakhnovich, A. Bulgac, U. Bastolla, M. Vendruscolo, C. Riccardi and R. Ziano, for their inestimable collaboration over the years resulting in many new research results. I dedicate this review to the memory of Marek Cieplak (1950–2021) [458], Ricardo A. Broglia (1939–2022) [459] and Dietrich Stauffer (1943–2019) [460].

Conflicts of Interest: The author declares no conflicts of interest.

Abbreviations

ABS	acrylonitrile butadiene styrene
ACN	acrylonitrile
AI	artificial intelligence
AIMD	ab initio molecular dynamics
AFM	atomic force microscopy
AFMIR	infrared (IR) laser-based AFM
ABO	anti-Bredt olefin
AMSilica	acrylamide functionalized silica
APS	ammonium persulfate

ATP	adenosine triphosphate
BB	Bergström–Boyce model
BCP	block copolymer
BNNS	boron nitride nanosheet
BPADA	4,4′-bisphenol A dianhydride
BAPB	4,4′-bis(4-aminophenoxy)biphenyl
CFD	computational fluid dynamics
CNT	carbon nanotube
CFD	computational fluid dynamics
cPAM	crosslinked polyacrylamides
CROW	chemical retrieval on the web
CRR	constraint release Rouse
DNA	deoxyribonucleic acid
DDS	drug delivery system
DFT	density functional theory
DPD	dissipative particle dynamics
EE	exact enumeration
EEVD	electron-enhanced vacuum deposition
FCC	face-centered cubic
FDM	fused deposition modeling
FP	frontal polymerization
FRP	fiber-reinforced polymer
GG	glycylglycine
GIP	glucose-dependent insulinotropic polypeptide
GLP	glucagon-like peptide
HCP	hexagonal closed packed
HDDMA	1,6-hexanediol dimethacrylate
HFP	hexafluoropropylene
HiG	hierarchical grid
HP	hydrophobicity profile
HSP	Hansen solubility parameter
IC	interaction chromatography
iIC	incipient infinite cluster
IDS	intrinsically disordered proteins
IoT	Internet of Things
ISP	in situ polymerization
KMC	kinetic Monte Carlo
LC	liquid chromatography
LG	liquid wood
LIB	lithium-ion battery
LL37	trifluoroacetate salt
LPI	liquid phase infiltration
MBB	molecular bottlebrush
MC	Monte Carlo
MD	molecular dynamics
MMM	mixed-matrix membrane
MPCD	multiparticle collision dynamics
MPD	m-phenylenediamine
mRNA	messenger RNA
MWD	molecular weight distribution
NBR	nitrile butadiene rubber
NIR	near infrared
NMR	nuclear magnetic resonance
NP	nanoparticles
NT	non-treated
PA	phthalic anhydride
PAM	polyacrylamide
PAMAM	polyamidoamine

PANI	polyaniline
PASA	poly(aspartic acid)
PASJD	plasma-assisted supersonic jet deposition
PBI	polybenzimidazole
PBC	periodic boundary conditions
PCF	pair correlation function
PCL	polycaprolactone
PCSAFT	perturbed-chain statistical associating fluid theory
PDB	Protein Data Bank
PDMS	poly(dimethylsiloxane)
PE	principal eingenvector
PEEK	polyetheretherketone
PEG	polyethylene glycol
PEI	polyetherimide
PEM	proton exchange membrane
PEMFC	polymer electrolyte membrane fuel cell
PEO	polyethylene oxide
PES	polyethersulfone
PET	polyethylene terephthalate
PETG	polyethylene terephthalate glycol
PEV	polyethylene vanillate
PFAS	per- and polyfluoroalkyl substances
PGA	poly(glutamic acid)
PHA	polyhydroxyalkanoate
PI	polyimide
PLA	polylactic acid
PLL	polylysine
PMM	polymeric metamaterials
PMIA	poly-m-phenyleneisophthalamide
PP	polypropylene
PPI	polypropylene imine
PPC	polypropylene carbonate
PPM	porous polymeric material
PPO	polypropylene oxide
PRISM	polymer reference interaction site model
PS	polystyrene
TFE	tetrafluoroethylene
PT	plasma treated
PTFE	polytetrafluoroethylene
P3HT	poly(3-hexylthiopene)
PVDF	polyvinylidene fluoride
PVDF-HFP	polyvinylidene fluoride–hexafluoropropylene
rHCP	random hexagonal close packed
RISC	RNA-induced silencing complex
RNA	ribonucleic acid
RW	random walk
SAW	self-avoiding walk
SCFT	self-consistent field theory
SEC	size-exclusion chromatography
SFC	semiflexible chains
SEM	scanning electron microscopy
Silica	SiO_2
siRNA	small interfering RNA
SLS	selective laser sintering
SMP	shape-memory polymer
SCNE	structurally constrained neutral evolution
SPAES	sulfonated polystyrene grafted poly(arylene ether sulfone)
SWCNT	single-walled carbon nanotube

TAPE	tris[4-(4-aminophenoxy)phenyl] ethane
TENG	triboelectric nanogenerator
TGA	thermogravimetric analysis
TN	three-network model
TNV	three-network viscoplastic model
TPMS	triply periodic minimal surfaces
VDA	vibronic-driven action
VMD	vacuum membrane distillation
XLPE	crosslinked polyethylene

References

1. Schiemann, R.; Bachmann, M.; Janke, W. Exact enumeration of three-dimensional lattice proteins. *Comput. Phys. Commun.* **2005**, *166*, 8–16. [CrossRef]
2. Lee, J.H.; Kim, S.Y.; Lee, J. Exact partition function zeros of a polymer on a simple cubic lattice. *Phys. Rev. E* **2012**, *86*, 011802. . [CrossRef]
3. Chen, C.N.; Hsieh, Y.H.; Hu, C.K. Heat capacity decomposition by partition function zeros for interacting self-avoiding walks. *Europhys. Lett.* **2013**, *104*, 20005. [CrossRef]
4. Hsieh, Y.H.; Chen, C.N.; Hu, C.K. Efficient algorithm for computing exact partition functions of lattice polymer models. *Comput. Phys. Commun.* **2016**, *209*, 27–33. . [CrossRef]
5. Huang, S.S.; Hsieh, Y.H.; Chen, C.N. Exact Enumeration Approach to Estimate the Theta Temperature of Interacting Self-Avoiding Walks on the Simple Cubic Lattice. *Polymers* **2022**, *14*, 4536. [CrossRef]
6. Grest, G.S.; Kremer, K. Molecular Dynamics Simulation for Polymers in the Presence of a Heat Bath. *Phys. Rev. A* **1986**, *33*, 3628–3631. [CrossRef]
7. Graessley, W.W. *Polymeric Liquids & Networks: Dynamics and Rheology*; Taylor & Francis: New York, NY, USA, 2008. [CrossRef]
8. Tsolou, G.; Stratikis, N.; Baig, C.; Stephanou, P.S.; Mavrantzas, V.G. Melt structure and dynamics of unentangled polyethylene rings: Rouse theory, atomistic molecular dynamics simulation, and comparison with the linear analogues. *Macromolecules* **2010**, *43*, 10692–10713. [CrossRef]
9. Colmenero, J. Non-exponential Rouse correlators and generalized magnitudes probing chain dynamics. *J. Non Cryst. Solids* **2015**, *407*, 302–308. [CrossRef]
10. Tsalikis, D.G.; Koukoulas, T.; Mavrantzas, V.G.; Pasquino, R.; Vlassopoulos, D.; Pyckhout-Hintzen, W.; Wischnewski, A.; Monkenbusch, M.; Richter, D. Microscopic structure, conformation, and dynamics of ring and linear poly (ethylene oxide) melts from detailed atomistic molecular dynamics simulations: Dependence on chain length and direct comparison with experimental data. *Macromolecules* **2017**, *50*, 2565–2584. [CrossRef]
11. Phillies, G.D.J. Simulational Tests of the Rouse Model. *Polymers* **2023**, *15*, 2615. [CrossRef]
12. Phillies, G.D.J. Kirkwood–Riseman Model in Non-Dilute Polymeric Fluids. *Polymers* **2023**, *15*, 3216. [CrossRef]
13. Volpe, S.C.; Leporini, D.; Puosi, F. Structure and Mechanical Properties of a Porous Polymer Material via Molecular Dynamics Simulations. *Polymers* **2023**, *15*, 358. [CrossRef]
14. Araki, T.; Tanaka, H. Three-dimensional numerical simulations of viscoelastic phase separation: Morphological characteristics. *Macromolecules* **2001**, *34*, 1953–1963. [CrossRef]
15. Testard, V.; Berthier, L.; Kob, W. Intermittent dynamics and logarithmic domain growth during the spinodal decomposition of a glass-forming liquid. *J. Chem. Phys.* **2014**, *140*, 164502. [CrossRef]
16. Sun, Y.; Andrews, W.B.; Thornton, K.; Voorhees, P.W. Self-similarity and the dynamics of coarsening in materials. *Sci. Rep.* **2018**, *8*, 17940. [CrossRef]
17. Tateno, M.; Tanaka, H. Power-law coarsening in network-forming phase separation governed by mechanical relaxation. *Nat. Commun.* **2021**, *12*, 912. [CrossRef]
18. Rzayev, J. Molecular bottlebrushes: New opportunities in nanomaterials fabrication. *ACS Macro Lett.* **2012**, *1*, 1146–1149. [CrossRef]
19. Paturej, J.; Sheiko, S.S.; Panyukov, S.; Rubinstein, M. Molecular structure of bottlebrush polymers in melts. *Sci. Adv.* **2016**, *2*, e1601478. [CrossRef]
20. Xie, G.; Martinez, M.R.; Olszewski, M.; Sheiko, S.S.; Matyjaszewski, K. Molecular bottlebrushes as novel materials. *Biomacromolecules* **2018**, *20*, 27–54. [CrossRef]
21. Tu, S.; Choudhury, C.K.; Luzinov, I.; Kuksenok, O. Recent advances towards applications of molecular bottlebrushes and their conjugates. *Curr. Opin. Solid State Mater. Sci.* **2019**, *23*, 50–61. [CrossRef]
22. Tu, S.; Choudhury, C.K.; Giltner, M.; Luzinov, I.; Kuksenok, O. Mesoscale Modeling of Agglomeration of Molecular Bottlebrushes: Focus on Conformations and Clustering Criteria. *Polymers* **2022**, *14*, 2339. [CrossRef]
23. Gohy, J.F.; Zhao, Y. Photo-responsive block copolymer micelles: Design and behavior. *Chem. Soc. Rev.* **2013**, *42*, 7117–7129. [CrossRef]
24. Mai, Y.; Eisenberg, A. Self-assembly of block copolymers. *Chem. Soc. Rev.* **2012**, *41*, 5969–5985. [CrossRef]

25. Rud, O.V.; Mercurieva, A.A.; Leermakers, F.A.M.; Birshtein, T.M. Collapse of polyelectrolyte star: Theory and modeling. *Macromolecules* **2012**, *45*, 2145–2160. [CrossRef]
26. Zhulina, E.B.; Borisov, O.V. Theory of block polymer micelles: Recent advances and current challenges. *Macromolecules* **2012**, *45*, 4429–4440. [CrossRef]
27. Shah, S.; Leon, L. Structural dynamics, phase behavior, and applications of polyelectrolyte complex micelles. *Curr. Opin. Colloid Interface Sci.* **2021**, *53*, 101424. [CrossRef]
28. Kuperkar, K.; Patel, D.; Atanase, L.I.; Bahadur, P. Amphiphilic block copolymers: Their structures, and self-assembly to polymeric micelles and polymersomes as drug delivery vehicles. *Polymers* **2022**, *14*, 4702. [CrossRef]
29. Procházka, K.; Limpouchová, Z.; Štěpánek, M.; Šindelka, K.; Lísal, M. DPD Modelling of the Self- and Co-Assembly of Polymers and Polyelectrolytes in Aqueous Media: Impact on Polymer Science. *Polymers* **2022**, *14*, 404. [CrossRef]
30. Gaddam, A.; Garg, M.; Agrawal, A.; Joshi, S.S. Modeling of liquid–gas meniscus for textured surfaces: Effects of curvature and local slip length. *J. Micromechan. Microeng.* **2015**, *25*, 125002. [CrossRef]
31. Rezaei, H.; Modarress, H. Dissipative particle dynamics (DPD) study of hydrocarbon–water interfacial tension (IFT). *Chem. Phys. Lett.* **2015**, *620*, 114–122. [CrossRef]
32. Tang, X.; Xiao, S.; Lei, Q.; Yuan, L.; Peng, B.; He, L.; Luo, J.; Pei, Y. Molecular dynamics simulation of surfactant flooding driven oil-detachment in nano-silica channels. *J. Phys. Chem. B* **2018**, *123*, 277–288. [CrossRef]
33. Nguyen, T.X.; Vu, T.V.; Razavi, S.; Papavassiliou, D.V. Coarse Grained Modeling of Multiphase Flows with Surfactants. *Polymers* **2022**, *14*, 543. [CrossRef]
34. Erdemir, D.; Lee, A.Y.; Myerson, A.S. Nucleation of crystals from solution: Classical and two-step models. *Accounts Chem. Res.* **2009**, *42*, 621–629. [CrossRef]
35. Russo, J.; Tanaka, H. The microscopic pathway to crystallization in supercooled liquids. *Sci. Rep.* **2012**, *2*, 505. [CrossRef]
36. Palberg, T. Crystallization kinetics of colloidal model suspensions: Recent achievements and new perspectives. *J. Phys. Condens. Matter* **2014**, *26*, 333101. [CrossRef]
37. Manoharan, V.N. Colloidal matter: Packing, geometry, and entropy. *Science* **2015**, *349*, 1253751. [CrossRef]
38. Herranz, M.; Foteinopoulou, K.; Karayiannis, N.C.; Laso, M. Polymorphism and Perfection in Crystallization of Hard Sphere Polymers. *Polymers* **2022**, *14*, 4435. [CrossRef]
39. Martínez-Fernández, D.; Herranz, M.; Foteinopoulou, K.; Karayiannis, N.C.; Laso, M. Local and Global Order in Dense Packings of Semi-flexible Polymers of Hard Spheres. *Polymers* **2023**, *15*, 551. [CrossRef]
40. Landau, D.P.; Binder, K. *A Guide to Monte Carlo Simulations in Statistical Physics*, 5th ed.; Cambridge University Press: Cambridge, MA, USA, 2015.
41. Espinosa, J.R.; Vega, C.; Valeriani, C.; Frenkel, D.; Sanz, E. Heterogeneous versus homogeneous crystal nucleation of hard spheres. *Soft Matter* **2019**, *15*, 9625–9631. [CrossRef]
42. Lam, M.A.; Khusid, B.; Kondic, L.; Meyer, W.V. Role of diffusion in crystallization of hard-sphere colloids. *Phys. Rev. E* **2021**, *104*, 054607. [CrossRef]
43. Venu, L.B.S.; Shim, E.; Anantharamaiah, N.; Pourdeyhimi, B. Three-dimensional structural characterization of nonwoven fabrics. *Microsc. Microanal.* **2012**, *18*, 1368–1379. [CrossRef]
44. Soto, M.; Esteva, M.; Martínez-Romero, O.; Baez, J.; Elías-Zúñiga, A. Modeling percolation in polymer nanocomposites by stochastic microstructuring. *Materials* **2015**, *8*, 6697–6718. [CrossRef]
45. Luu, H.T.; Perrot, C.; Monchiet, V.; Panneton, R. Three-dimensional reconstruction of a random fibrous medium: Geometry, transport, and sound absorbing properties. *J. Acoust. Soc. Am.* **2017**, *141*, 4768–4780. [CrossRef]
46. Mostaço-Guidolin, L.B.; Smith, M.S.; Hewko, M.; Schattka, B.; Sowa, M.G.; Major, A.; Ko, A.C.T. Fractal dimension and directional analysis of elastic and collagen fiber arrangement in unsectioned arterial tissues affected by atherosclerosis and aging. *J. Appl. Physiol.* **2019**, *126*, 638–646. [CrossRef]
47. Filla, N.; Zhao, Y.; Wang, X. A Tractable, Transferable, and Empirically Consistent Fibrous Biomaterial Model. *Polymers* **2022**, *14*, 4437. [CrossRef]
48. Schweizer, K.S.; Curro, J.G. Integral-Equation Theory of the Structure of Polymer Melts. *Phys. Rev. Lett.* **1987**, *58*, 246–249. [CrossRef]
49. Douglas, J.F.; Hubbard, J.B. Semiempirical theory of relaxation: Concentrated polymer solution dynamics. *Macromolecules* **1991**, *24*, 3163–3177. [CrossRef]
50. Yethiraj, A.; Schweizer, K.S. Self-consistent polymer integral equation theory: Comparisons with Monte Carlo simulations and alternative closure approximations. *J. Chem. Phys.* **1992**, *97*, 1455–1464. [CrossRef]
51. Jaramillo, E.; Wu, D.T.; Grest, G.S.; Curro, J.G. Anomalous mixing behavior of polyisobutylene/polypropylene blends: Molecular dynamics simulation study. *J. Chem. Phys.* **2004**, *120*, 8883–8886. [CrossRef]
52. Guenza, M.G. Anomalous Dynamics in Macromolecular Liquids. *Polymers* **2022**, *14*, 856. [CrossRef]
53. Sofos, F.; Karakasidis, T.E.; Liakopoulos, A. How wall properties control diffusion in grooved nanochannels: A molecular dynamics study. *Heat Mass Transf.* **2013**, *49*, 1081–1088. [CrossRef]
54. Sofos, F.; Karakasidis, T.E.; Liakopoulos, A. Fluid flow at the nanoscale: How fluid properties deviate from the bulk. *Nanosci. Nanotechnol. Lett.* **2013**, *5*, 457–460. [CrossRef]

55. Sofos, F.; Karakasidis, T.E.; Spetsiotis, D. Molecular Dynamics simulations of ion separation in nano-channel water flows using an electric field. *Mol. Simul.* **2019**, *45*, 1395–1402. [CrossRef]
56. Paun, M.A.; Paun, V.A.; Paun, V.P. Polymer Translocation through Nanometer Pores. *Polymers* **2022**, *14*, 1166. [CrossRef]
57. Monteferrante, M.; Succi, S.; Pisignano, D.; Lauricella, M. Simulating Polymerization by Boltzmann Inversion Force Field Approach and Dynamical Nonequilibrium Reactive Molecular Dynamics. *Polymers* **2022**, *14*, 4529. [CrossRef]
58. Duval-Terrié, C.; Lebrun, L. Polymerization and characterization of PMMA. Polymer chemistry laboratory experiments for undergraduate students. *J. Chem. Educ.* **2006**, *83*, 443. [CrossRef]
59. Ansong, O.E.; Jansen, S.; Wei, Y.; Pomrink, G.; Lu, H.; Patel, A.; Li, S. Accelerated controlled radical polymerization of methacrylates. *Polym. Int.* **2009**, *58*, 54–65. [CrossRef]
60. Monteferrante, M.; Tiribocchi, A.; Succi, S.; Pisignano, D.; Lauricella, M. Capturing free-radical polymerization by synergetic ab initio calculations and topological reactive molecular dynamics. *Macromolecules* **2022**, *55*, 1474–1486. [CrossRef]
61. Wang, X.J.; Zhao, X.Y.; Li, Q.G.; Chan, T.W.; Wu, S.Z. Artificial neural network modeling and mechanism study for relaxation of deformed rubber. *Ind. Eng. Chem. Res.* **2016**, *55*, 4059–4070. [CrossRef]
62. Wang, W.; Tanaka, Y.; Takada, T.; Iwata, S.; Uehara, H.; Li, S. Influence of oxidation on the dynamics in amorphous ethylene-propylene-diene-monomer copolymer: A molecular dynamics simulation. *Polym. Degrad. Stab.* **2018**, *147*, 187–196. [CrossRef]
63. Celina, M.; Linde, E.; Brunson, D.; Quintana, A.; Giron, N. Overview of accelerated aging and polymer degradation kinetics for combined radiation-thermal environments. *Polym. Degrad. Stab.* **2019**, *166*, 353–378. [CrossRef]
64. Liu, Q.; Huang, W.; Liu, B.; Wang, P.C.; Chen, H.B. Gamma radiation chemistry of polydimethylsiloxane foam in radiation-thermal environments: Experiments and simulations. *ACS Appl. Mater. Interfaces* **2021**, *13*, 41287–41302. [CrossRef] [PubMed]
65. Yang, J.; Lou, W. Molecule Dynamics Simulation of the Effect of Oxidative Aging on Properties of Nitrile Rubber. *Polymers* **2022**, *14*, 226. [CrossRef] [PubMed]
66. Sun, H.; Ren, P.; Fried, J.R. The COMPASS Force Field: Parameterization and Validation for Phosphazenes. *Comput. Theor. Polym. Sci.* **1998**, *8*, 229–246. [CrossRef]
67. Lemkul, J.A.; Huang, J.; Roux, B.; MacKerell, J.A.D. An empirical polarizable force field based on the classical drude oscillator model: Development history and recent applications. *Chem. Rev.* **2016**, *116*, 4983–5013. [CrossRef]
68. John, M.J.; Thomas, S. Biofibres and biocomposites. *Carbohydr. Polym.* **2008**, *71*, 343–364. [CrossRef]
69. Faruk, O.; Bledzki, A.K.; Fink, H.P.; Sain, M. Biocomposites reinforced with natural fibers: 2000–2010. *Prog. Polym. Sci.* **2012**, *37*, 1552–1596. [CrossRef]
70. Bharath, K.N.; Basavarajappa, S. Applications of biocomposite materials based on natural fibers from renewable resources: A review. *Sci. Eng. Compos. Mater.* **2016**, *23*, 123–133. [CrossRef]
71. Petrescu, T.C.; Mihai, P.; Voordijk, J.T.; Nedeff, V.; Văideanu, D.; Nedeff, F.; Babor, T.D.; Vasincu, D.; Agop, M. Complex Behavior in the Dynamics of a Polymeric Biocomposite Material–"Liquid Wood": Experimental and Theoretical Aspects. *Polymers* **2022**, *14*, 64. [CrossRef]
72. Bourbigot, S.; Flambard, X. Heat resistance and flammability of high performance fibres: A review. *Fire Mater.* **2002**, *26*, 155–168. [CrossRef]
73. Tian, W.; Qiu, T.; Shi, Y.; He, L.; Tuo, X. The facile preparation of aramid insulation paper from the bottom-up nanofiber synthesis. *Mater. Lett.* **2017**, *202*, 158–161. [CrossRef]
74. Liu, B.; Lv, F.; Fan, X.; Li, Y.; Jiang, B. Molecular Dynamics Study of the Influence of Nano SiO_2 on the Thermodynamic Properties of PMIA Composites. *Polymers* **2022**, *14*, 3134. [CrossRef] [PubMed]
75. Yang, X.; Wang, X.; Wang, W.; Fu, Y.; Xie, Q. Atomic-scale insights into interface thermal resistance between epoxy and boron nitride in nanocomposites. *Int. J. Heat Mass Transf.* **2020**, *159*, 120105. [CrossRef]
76. Allahyarov, E.; Taylor, P.L. Simulation study of the equilibrium morphology in ionomers with different architectures. *J. Polym. Sci. Part Polym. Phys.* **2011**, *49*, 368–376. [CrossRef]
77. Bolintineanu, D.S.; Stevens, M.J.; Frischknecht, A.L. Atomistic simulations predict a surprising variety of morphologies in precise ionomers. *ACS Macro Lett.* **2013**, *2*, 206–210. [CrossRef]
78. Frischknecht, A.L.; Winey, K.I. The evolution of acidic and ionic aggregates in ionomers during microsecond simulations. *J. Chem. Phys.* **2019**, *150*, 064901.. [CrossRef]
79. Karatrantos, A.V.; Khantaveramongkol, J.; Kröger, M. Structure and Diffusion of Ionic PDMS Melts. *Polymers* **2022**, *14*, 3070. [CrossRef]
80. Maass, P.; Petersen, J.; Bunde, A.; Dieterich, W.; Roman, H.E. Non-Debye relaxation in structurally disordered ionic conductors: Effect of Coulomb interaction. *Phys. Rev. Lett.* **1991**, *66*, 52. [CrossRef]
81. Sheavly, S.B.; Register, K.M. Marine debris & plastics: Environmental concerns, sources, impacts and solutions. *J. Polym. Environ.* **2007**, *15*, 301–305. [CrossRef]
82. Smith, M.; Love, D.C.; Rochman, C.M.; Neff, R.A. Microplastics in seafood and the implications for human health. *Curr. Environ. Health Rep.* **2018**, *5*, 375–386. [CrossRef]
83. Schwarz, A.E.; Ligthart, T.N.; Boukris, E.; Van Harmelen, T. Sources, transport, and accumulation of different types of plastic litter in aquatic environments: A review study. *Mar. Pollut. Bull.* **2019**, *143*, 92–100. [CrossRef] [PubMed]
84. Hong, S.-H.; Gorce, J.B.; Punzmann, H.; Francois, N.; Xia, H. Surface waves control bacterial attachment and formation of biofilms in thin layers. *Sci. Adv.* **2020**, *6*, aaz9386.. [CrossRef] [PubMed]

85. Sangkhawasi, M.; Remsungnen, T.; Vangnai, A.S.; Poo-Arporn, R.P.; Rungrotmongkol, T. All-atom Molecular Dynamics Simulations on a Single Chain of PET and PEV Polymers. *Polymers* **2022**, *14*, 1161. [CrossRef] [PubMed]
86. Lyu, J.; Gao, Y.; Zhang, Z.; Greiser, U.; Tai, H.; Wang, W. Can Flory-Stockmayer theory be applied to predict conventional free radical polymerization of multivinyl monomers? A study via Monte Carlo simulations. *Sci. China Chem.* **2018**, *61*, 319–327. [CrossRef]
87. Lyu, J.; Gao, Y.; Zhang, Z.; Greiser, U.; Polanowski, P.; Jeszka, J.K.; Matyjaszewski, K.; Tai, H.; Wang, W. Monte Carlo simulations of atom transfer radical (homo) polymerization of divinyl monomers: Applicability of Flory–Stockmayer theory. *Macromolecules* **2018**, *51*, 6673–6681. [CrossRef]
88. Schamboeck, V.; Iedema, P.D.; Kryven, I. Dynamic networks that drive the process of irreversible step-growth polymerization. *Sci. Rep.* **2019**, *9*, 2276. [CrossRef]
89. Wen, C.; Odle, R.; Cheng, S. Molecular Weight Distribution of Branched Polymers: Comparison between Monte Carlo Simulation and Flory-Stockmayer Theory. *Polymers* **2023**, *15*, 1791. [CrossRef]
90. Barbero, C.A. Functional Materials made by Combining Hydrogels (cross-linked polyacrylamides) and Conducting Polymers (polyanilines): A Critical Review. *Polymers* **2023**, *15*, 2240. [CrossRef]
91. Green, M.R.; Sambrook, J. Polyacrylamide gel electrophoresis. *Cold Spring Harb. Protoc.* **2020**. [CrossRef]
92. Zhang, H.P.; Cao, J.J.; Jiang, W.B.; Yang, Y.Q.; Zhu, B.Y.; Liu, X.Y.; Wu, Y.; Sun, X.; Essouma, A.F.B.E.; Liu, J.; et al. Synthesis and mechanical properties of polyacrylamide gel doped with graphene oxide. *Energies* **2022**, *15*, 5714. [CrossRef]
93. Lv, Y.; Zhang, S.; Zhang, Y.; Yin, H.; Feng, Y. Hydrophobically Associating Polyacrylamide "Water-in-Water" Emulsion Prepared by Aqueous Dispersion Polymerization: Synthesis, Characterization and Rheological Behavior. *Molecules* **2023**, *28*, 2698. [CrossRef] [PubMed]
94. Ferrarese Lupi, F.; Giammaria, T.J.; Miti, A.; Zuccheri, G.; Carignano, S.; Sparnacci, K.; Seguini, G.; De Leo, N.; Boarino, L.; Perego, M. and Laus, M. Hierarchical order in dewetted block copolymer thin films on chemically patterned surfaces. *ACS Nano* **2018**, *12*, 7076–7085. [CrossRef]
95. Liu, C.C.; Franke, E.; Mignot, Y.; Xie, R.; Yeung, C.W.; Zhang, J.; Chi, C.; Zhang, C.; Farrell, R.; Lai, K.; et al. Directed self-assembly of block copolymers for 7 nanometre FinFET technology and beyond. *Nat. Electron.* **2018**, *1*, 562–569. [CrossRef]
96. Ferrarese Lupi, F.; Murataj, I.; Celegato, F.; Angelini, A.; Frascella, F.; Chiarcos, R.; Antonioli, D.; Gianotti, V.; Tiberto, P.; Pirri, C.F.; et al. Tailored and Guided Dewetting of Block Copolymer/Homopolymer Blends. *Macromolecules* **2020**, *53*, 7207–7217. [CrossRef]
97. Murataj, I.; Cara, E.; Baglieri, N.; Pirri, C.F.; De Leo, N.; Ferrarese Lupi, F. Liquid Phase Infiltration of Block Copolymers. *Polymers* **2022**, *14*, 4317. [CrossRef]
98. Mustonen, K.; Susi, T.; Kaskela, A.; Laiho, P.; Tian, Y.; Nasibulin, A.G.; Kauppinen, E.I. Influence of the diameter of single-walled carbon nanotube bundles on the optoelectronic performance of dry-deposited thin films. *Beilstein J. Nanotechnol.* **2012**, *3*, 692–702. [CrossRef]
99. Escárcega-Bobadilla, M.V.; Zelada-Guillén, G.A.; Pyrlin, S.V.; Wegrzyn, M.; Ramos, M.M.; Giménez, E.; Stewart, A.; Maier, G.; Kleij, A.W. Nanorings and rods interconnected by self-assembly mimicking an artificial network of neurons. *Nat. Commun.* **2013**, *4*, 2648. [CrossRef]
100. Wang, A.; Ye, J.; Humphrey, M.G.; Zhang, C. Graphene and Carbon-Nanotube Nanohybrids Covalently Functionalized by Porphyrins and Phthalocyanines for Optoelectronic Properties. *Adv. Mater.* **2018**, *40*, 1705704. . [CrossRef]
101. Pyrlin, S.; Lenzi, V.; Silva, A.; Ramos, M.; Marques, L. Adhesion of Bis-Salphen-Based Coordination Polymers to Graphene: Insights from Free Energy Perturbation Study. *Polymers* **2022**, *14*, 4525. [CrossRef]
102. Levchenko, I.; Baranov, O.; Riccardi, C.; Roman, H.E.; Cvelbar, U.; Ivanova, E.P.; Mohandas, M.; Ščajev, P.; Malinauskas, T.; Xu, S.; Bazaka, K. Nanoengineered Carbon-Based Interfaces for Advanced Energy and Photonics Applications: A Recent Progress and Innovations. *Adv. Mater. Interfaces* **2023**, *10*, 2201739. [CrossRef]
103. Morita, M.; Oya, Y.; Kato, N.; Mori, K.; Koyanagi, J. Effect of Electrostatic Interactions on the Interfacial Energy between Thermoplastic Polymers and Graphene Oxide: A Molecular Dynamics Study. *Polymers* **2022**, *14*, 2579. . [CrossRef] [PubMed]
104. Karatasos, K.; Kritikos, G. Characterization of a graphene oxide/poly (acrylic acid) nanocomposite by means of molecular dynamics simulations. *RSC Adv.* **2016**, *6*, 109267–109277. [CrossRef]
105. Duan, K.; Li, L.; Wang, F.; Meng, W.; Hu, Y.; Wang, X. Importance of interface in the coarse-grained model of CNT/epoxy nanocomposites. *Nanomaterials* **2019**, *9*, 1479. [CrossRef] [PubMed]
106. Zhang, X.; Nguyen, H.; Daly, M.; Nguyen, S.T.; Espinosa, H.D. Nanoscale toughening of ultrathin graphene oxide-polymer composites: Mechanochemical insights into hydrogen-bonding/van der Waals interactions, polymer chain alignment, and steric parameters. *Nanoscale* **2019**, *11*, 12305–12316. [CrossRef]
107. Wen, X.; Wang, D.; Tang, J.; Yang, Z. A Trefoil Knot Polymer Chain Translocates through a Funnel-like Channel: A Multi-Particle Collision Dynamics Study. *Polymers* **2022**, *14*, 1164. [CrossRef]
108. Guo, J.; Li, X.; Liu, Y.; Liang, H. Flow-induced translocation of polymers through a fluidic channel: A dissipative particle dynamics simulation study. *J. Chem. Phys.* **2011**, *134*, 134906. [CrossRef]
109. Chelakkot, R.; Winkler, R.G.; Gompper, G. Flow-induced helical coiling of semiflexible polymers in structured microchannels. *Phys. Rev. Lett.* **2012**, *109*, 178101. [CrossRef]
110. Bazilevskii, A.B.; Rozhkov, A.N. Dynamics of the Capillary Breakup of a Bridge in an Elastic Fluid. *Fluid Dyn.* **2015**, *50*, 800–811. [CrossRef]

111. Chakrapani, T.H.; den Otter, W.K. Capillary imbibition of binary fluid mixtures in nanochannels. *Langmuir* **2020**, *36*, 12712–12722. [CrossRef]
112. Yeou, S.; Lee, N.K. Single-molecule methods for investigating the double-stranded DNA bendability. *Mol. Cells* **2022**, *45*, 33–40. [CrossRef]
113. Pang, Z.; Sun, H.; Guo, Y.; Du, J.; Li, L.; Li, Q.; Yang, J.; Zhang, J.; Wu, W.; Yang, S. Research Advances of Porous Polyimide-Based Composites with Low Dielectric Constant. *Polymers* **2023**, *15*, 3341. [CrossRef] [PubMed]
114. Min, C.K.; Wu, T.B.; Yang, W.T.; Chen, C.L. Functionalized mesoporous silica/polyimide nanocomposite thin films with improved mechanical properties and low dielectric constant. *Compos. Sci. Technol.* **2008**, *68*, 1570–1578. [CrossRef]
115. Lee, T.; Park, S.S.; Jung, Y.; Han, S.; Han, D.; Kim, I.; Ha, C.S. Preparation and characterization of polyimide/mesoporous silica-hybrid nanocomposites based on water-soluble poly (amic acid) ammonium salt. *Eur. Polym. J.* **2009**, *45*, 19–29. [CrossRef]
116. Huang, B.; Li, K.; Peng, M.; Cheng, J. Polyimide/fluorinated silica composite films with low dielectric constant and low water absorption. *High Perform. Polym.* **2022**, *34*, 434–443. [CrossRef]
117. Levchenko, I.; Baranov, O.; Keidar, M.; Riccardi, C.; Roman, H.E.; Xu, S.; Alexander, K. Additive Technologies and Materials for the Next-Generation CubeSats and Small Satellites. *Adv. Funct. Mater.* **2024**, 2407602. [CrossRef]
118. Cai, W.; Shalaev, V. *Optical Metamaterials: Fundamentals and Applications*; Springer: New York, NY, USA, 2010. [CrossRef]
119. Cummer, S.A.; Christensen, J.; Alù, A. Controlling sound with acoustic metamaterials. *Nat. Rev. Mater.* **2016**, *1*, 1–13. [CrossRef]
120. Li, Y.; Li, W.; Han, T.; Zheng, X.; Li, J.; Li, B.; Fan, S.; Qiu, C.W. Transforming heat transfer with thermal metamaterials and devices. *Nat. Rev. Mater.* **2021**, *6*, 488–507. [CrossRef]
121. Almesmari, A.; Baghous, N.; Ejeh, C.J.; Barsoum, I.; Abu Al-Rub, R.K. Review of Additively Manufactured Polymeric Metamaterials: Design, Fabrication, Testing and Modeling. *Polymers* **2023**, *15*, 3858. [CrossRef]
122. Khim, D.; Luzio, A.; Bonacchini, G.E.; Pace, G.; Lee, M.J.; Noh, Y.Y.; Caironi, M. Uniaxial alignment of conjugated polymer films for high-performance organic field-effect transistors. *Adv. Mater.* **2018**, *30*, 1705463. [CrossRef]
123. Fratini, S.; Nikolka, M.; Salleo, A.; Schweicher, G.; Sirringhaus, H. Charge transport in high-mobility conjugated polymers and molecular semiconductors. *Nat. Mater.* **2020**, *19*, 491–502. [CrossRef]
124. Xiao, M.; Sadhanala, A.; Abdi-Jalebi, M.; Thomas, T.H.; Ren, X.; Zhang, T.; Chen, H.; Carey, R.L.; Wang, Q.; Senanayak, S.P.; et al. Linking Glass-Transition Behavior to Photophysical and Charge Transport Properties of High-Mobility Conjugated Polymers. *Adv. Funct. Mater.* **2021**, *31*, 2007359. [CrossRef]
125. Qu, T.; Nan, G.; Ouyang, Y.; Bieketuerxun, B.; Yan, X.; Qi, Y.; Zhang, Y. Structure-Property Relationship, Glass Transition, and Crystallization Behaviors of Conjugated Polymers. *Polymers* **2023**, *15*, 4268. [CrossRef] [PubMed]
126. Matsen, M.W.; Bates, F.S. Unifying weak-and strong-segregation block copolymer theories. *Macromolecules* **1996**, *29*, 1091–1098. [CrossRef]
127. Gompper, G.; Schick, M. *Soft Matter: Polymer Melts and Mixtures*; Wiley-VCH Verlag GmbH & Co. KGaA: Weinheim, Germany, 2006. [CrossRef]
128. Vigil, D.L.; Quah, T.; Sun, D.; Delaney, K.T.; Fredrickson, G.H. Self-Consistent Field Theory Predicts Universal Phase Behavior for Linear, Comb, and Bottlebrush Diblock Copolymers. *Macromolecules* **2022**, *55*, 4237–4244. [CrossRef]
129. Qiao, L.; Giannakou, M.; Schmid, F. An Efficient and Accurate SCF Algorithm for Block Copolymer Films and Brushes Using Adaptive Discretizations. *Polymers* **2024**, *16*, 1228. [CrossRef]
130. Pater, J.T.; Weickert, G.; Swaaij, W.P.V. Polymerization of liquid propylene with a fourth-generation Ziegler–Natta catalyst: Influence of temperature, hydrogen, monomer concentration, and prepolymerization method on powder morphology. *J. Appl. Polym. Sci.* **2003**, *87*, 1421–1435. [CrossRef]
131. Kasih, T.P.; Kuroda, S.I.; Kubota, H. Poly (methyl methacrylate) Films Deposited via Non-Equilibrium Atmospheric Pressure Plasma Polymerization Using Argon as Working Gas. *Plasma Process. Polym.* **2007**, *4*, 648–653. [CrossRef]
132. Amberg, M.; Höhener, M.; Rupper, P.; Hanselmann, B.; Hufenus, R.; Lehner, S.; Perret, E.; Hegemann, D. Surface modification of recycled polymers in comparison to virgin polymers using Ar/O_2 plasma etching. *Plasma Process. Polym.* **2022**, *19*, 2200068. [CrossRef]
133. Rusu, B.G.; Ursu, C.; Olaru, M.; Barboiu, M. Frequency-tuned porous polyethylene glycol films obtained in atmospheric-pressure dielectric barrier discharge (DBD) plasma. *Appl. Sci.* **2023**, *13*, 1785. [CrossRef]
134. Hegemann, D. Plasma activation mechanisms governed by specific energy input: Potential and perspectives. *Plasma Process. Polym.* **2023**, *20*, 2300010. [CrossRef]
135. Dufour, T. From Basics to Frontiers: A Comprehensive Review of Plasma-Modified and Plasma-Synthesized Polymer Films. *Polymers* **2023**, *15*, 3607. [CrossRef] [PubMed]
136. Riccardi, C.; Roman, H.E. (Eds.). *Plasma Applications in Gases, Liquids and Solids: Technology and Methods*; World Scientific: Singapore, 2023. [CrossRef]
137. Zhong, S.; Liu, C.; Na, H. Preparation and properties of UV irradiation-induced crosslinked sulfonated poly (ether ether ketone) proton exchange membranes. *J. Membr. Sci.* **2009**, *326*, 400–407. [CrossRef]
138. Zhang, Z.; Wu, L.; Xu, T. Synthesis and properties of side-chain-type sulfonated poly (phenylene oxide) for proton exchange membranes. *J. Membr. Sci.* **2011**, *373*, 160–166. [CrossRef]
139. Kim, D.J.; Choi, D.H.; Park, C.H.; Nam, S.Y. Characterization of the sulfonated PEEK/sulfonated nanoparticles composite membrane for the fuel cell application. *Int. J. Hydrogen Energy* **2016**, *41*, 5793–5802. [CrossRef]

140. Pan, M.; Pan, C.; Li, C.; Zhao, J. A review of membranes in proton exchange membrane fuel cells: Transport phenomena, performance and durability. *Renew. Sustain. Energy Rev.* **2021**, *141*, 110771. [CrossRef]
141. Li, X.; Zhang, H.; Lin, C.; Tian, R.; Zheng, P.; Hu, C. Morphological Effect of Side Chain Length in Sulfonated Poly(arylene ether sulfone)s Polymer Electrolyte Membranes via Molecular Dynamics Simulation. *Polymers* **2022**, *14*, 5499. [CrossRef]
142. Ammineni, S.P.; Nagaraju, C.; Lingaraju, D. Thermal degradation of naturally aged NBR with time and temperature. *Mater. Res. Express* **2022**, *9*, 065305. [CrossRef]
143. Xie, X.; Yang, D. Simultaneously improved thermal conductivity and dielectric properties of NBR composites by constructing 3D hybrid filler networks. *Macromol. Mater. Eng.* **2023**, *308*, 2300016. [CrossRef]
144. da Silva Ruiz, N.M.; de Sousa, A.M.F.; Calderari, M.R.D.C.M.; de Figueiredo, M.A.G.; Lima, A.P.; de Azevedo, E.R. Time domain NMR evaluation of thermal and thermochemical aging of nitrile rubber on crosslinking and mechanical properties. *Polym. Degrad. Stab.* **2024**, *224*, 110727. [CrossRef]
145. Jing, Y.; Liu, G. Systematic Investigation on the Swelling Response and Oil Resistance of NBR Using the Prediction Models Determined by the Modified Flory-Huggins Interaction Parameter. *Polymer* **2024**, *16*, 2696. [CrossRef]
146. Seo, S.E.; Hawker, C.J. The Beauty of Branching in Polymer Science. *Macromolecules* **2020**, *53*, 3257–3261. [CrossRef]
147. Tsukamoto, T.; Imaoka, T.; Yamamoto, K. Unique functions and applications of rigid dendrimers featuring radial aromatic chains. *Accounts Chem. Res.* **2021**, *54*, 4486–4497. [CrossRef] [PubMed]
148. Nikzamir, M.; Hanifehpour, Y.; Akbarzadeh, A.; Panahi, Y. Applications of dendrimers in nanomedicine and drug delivery: A review. *J. Inorg. Organomet. Polym. Mater.* **2021**, *31*, 2246–2261. [CrossRef]
149. Patel, P.; Patel, V.; Patel, P.M. Synthetic strategy of dendrimers: A review. *J. Indian Chem. Soc.* **2022**, *99*, 100514. [CrossRef]
150. Sapra, R.; Verma, R.P.; Maurya, G.P.; Dhawan, S.; Babu, J.; Haridas, V. Designer Peptide and Protein Dendrimers: A Cross-sectional Analysis. *Chem. Rev.* **2019**, *119*, 11391–11441. [CrossRef]
151. Pérez-Ferreiro, M.; Abelairas, A.M.; Criado, A.; Gómez, I.J.; Mosquera, J. Dendrimers: Exploring Their Wide Structural Variety and Applications. *Polymers* **2023**, *15*, 4369. [CrossRef]
152. Karaduman, A. Pyrolysis of polystyrene plastic wastes with some organic compounds for enhancing styrene yield. *Energy Sources* **2002**, *24*, 667–674. [CrossRef]
153. Çelikgöğüs, Ç.; Karaduman, A. Thermal-catalytic pyrolysis of polystyrene waste foams in a semi-batch reactor. *Energy Sources Part Recover. Util. Environ. Eff.* **2015**, *37*, 2507–2513. [CrossRef]
154. Ordaz-Quintero, A.; Monroy-Alonso, A.; Saldívar-Guerra, E. Thermal Pyrolysis of Polystyrene Aided by a Nitroxide End-Functionality: Experiments and Modeling. *Processes* **2020**, *8*, 432. [CrossRef]
155. Monroy-Alonso, A.; Ordaz-Quintero, A.; Ramirez, J.C.; Saldívar-Guerra, E. Thermal Pyrolysis of Polystyrene Aided by a Nitroxide End-Functionality Improved Process and Modeling of the Full Molecular Weight Distribution. *Polymers* **2022**, *14*, 160. [CrossRef]
156. Olajire, A.A. A review of oilfield scale management technology for oil and gas production. *J. Pet. Sci. Eng.* **2015**, *135*, 723–737. [CrossRef]
157. Wang, Q.; Liang, F.; Al-Nasser, W.; Al-Dawood, F.; Al-Shafai, T.; Al-Badairy, H.; Shen, S.; Al-Ajwad, H. Laboratory study on efficiency of three calcium carbonate scale inhibitors in the presence of EOR chemicals. *Petroleum* **2018**, *4*, 375–384. [CrossRef]
158. Xu, Z.; Zhao, Y.; He, J.; Qu, H.; Wang, Y.; Wang, B. Fouling characterization of calcium carbonate on heat transfer surfaces with sodium carboxymethyl cellulose as an inhibitor. *Int. J. Therm. Sci.* **2021**, *162*, 106790. [CrossRef]
159. Onawole, A.T.; Hussein, I.A.; Saad, M.A.; Ismail, N.; Alshami, A.; Nasser, M.S. Theoretical Studies of a Silica Functionalized Acrylamide for Calcium Scale Inhibition. *Polymers* **2022**, *14*, 2333. [CrossRef] [PubMed]
160. Weitsman, Y.J.; Elahi, M. Effects of fluids on the deformation, strength and durability of polymeric composites—An overview. *Mech. -Time-Depend. Mater.* **2000**, *4*, 107–126. [CrossRef]
161. Xu, C.L.; Wang, Y.Z. Durability, anti-corrosion and self-clean in air/oil of a transparent superhydrophobic polyimide film. *Appl. Mater. Today* **2018**, *10*, 18–23. [CrossRef]
162. Huang, X.; Ding, Z.A.; Wang, J.; Wang, J.; Li, Q. The impacts of chemical modification on the initial surface creepage discharge behaviors of polyimide insulating film in power electronics. *ACS Appl. Electron. Mater.* **2020**, *2*, 3418–3425. [CrossRef]
163. Ren, L.; Li, H.; Xie, Z.; Ai, D.; Zhou, Y.; Liu, Y.; Zhang, S.; Yang, L.; Zhao, X.; Peng, Z.; et al. High-temperature high-energy-density dielectric polymer nanocomposites utilizing inorganic core-shell nanostructured nanofillers. *Adv. Energy Mater.* **2021**, *11*, 2101297. [CrossRef]
164. Wang, X.; Zhao, T.; Wang, Y.; Zhang, L.; Zou, L. Microscopic Pyrolytic and Electric Decomposition Mechanism of Insulating Polyimide/boron Nitride Nanosheet Composites based on ReaxFF. *Polymers* **2022**, *14*, 1169. [CrossRef]
165. Gondal, I.A.; Sahir, M.H. Prospects of natural gas pipeline infrastructure in hydrogen transportation. *Int. J. Energy Res.* **2012**, *36*, 1338–1345. [CrossRef]
166. Zheng, Y.; Tan, Y.; Zhou, C.; Chen, G.; Li, J.; Liu, Y.; Liao, B.; Zhang, G. A review on effect of hydrogen on rubber seals used in the high-pressure hydrogen infrastructure. *Int. J. Hydrogen Energy* **2020**, *45*, 23721–23738. [CrossRef]
167. Jung, J.K.; Kim, I.G.; Kim, K.T.; Ryu, K.S.; Chung, K.S. Evaluation techniques of hydrogen permeation in sealing rubber materials. *Polym. Test.* **2021**, *93*, 107016. [CrossRef]
168. Jung, J.K.; Lee, J.H.; Park, J.Y.; Jeon, S.K. Modeling of the Time-Dependent H_2 Emission and Equilibrium Time in H_2-Enriched Polymers with Cylindrical, Spherical and Sheet Shapes and Comparisons with Experimental Investigations. *Polymers* **2024**, *16*, 2158. [CrossRef] [PubMed]

169. Emadian, S.M.; Onay, T.T.; Demirel, B. Biodegradation of bioplastics in natural environments. *Waste Manag.* **2017**, *59*, 526–536. [CrossRef]
170. Sanjay, M.R.; Madhu, P.; Jawaid, M.; Senthamaraikannan, P.; Senthil, S.; Pradeep, S. Characterization and properties of natural fiber polymer composites: A comprehensive review. *J. Clean. Prod.* **2018**, *172*, 566–581. [CrossRef]
171. Kalpana, S.; Priyadarshini, S.R.; Leena, M.M.; Moses, J.A.; Anandharamakrishnan, C. Intelligent packaging: Trends and applications in food systems. *Trends Food Sci. Technol.* **2019**, *93*, 145–157. [CrossRef]
172. Reis, R.S.; Souza, D.D.H.S.; Marques, M.D.F.V.; da Luz, F.S.; Monteiro, S.N. Novel bionanocomposite of polycaprolactone reinforced with steam-exploded microfibrillated cellulose modified with ZnO. *J. Mater. Res. Technol.* **2021**, *13*, 1324–1335. [CrossRef]
173. Ilyas, R.A.; Zuhri, M.Y.M.; Norrrahim, M.N.F.; Misenan, M.S.M.; Jenol, M.A.; Samsudin, S.A.; Nurazzi, N.M.; Asyraf, M.R.M.; Supian, A.B.M.; Bangar, S.P.; et al. Natural Fiber-reinforced Polycaprolactone Green and Hybrid Biocomposites for various Advanced Applications. *Polymers* **2022**, *14*, 182. [CrossRef]
174. Hopewell, J.; Dvorak, R.; Kosior, E. Plastics recycling: Challenges and opportunities. *Philos. Trans. R. Soc. Biol. Sci.* **2009**, *364*, 2115–2126. [CrossRef]
175. Xanthos, D.; Walker, T.R. International policies to reduce plastic marine pollution from single-use plastics (plastic bags and microbeads): A review. *Mar. Pollut. Bull.* **2017**, *118*, 17–26. [CrossRef]
176. Rhodes, C.J. Plastic pollution and potential solutions. *Sci. Prog.* **2018**, *101*, 207–260. [CrossRef] [PubMed]
177. Akinmulewo, A.B.; Nwinyi, O.C. Polyhydroxyalkanoate: A biodegradable polymer (a mini review). *J. Phys. Conf. Ser.* **2019**, *1378*, 042007. [CrossRef]
178. Bejagam, K.K.; Gupta, N.S.; Lee, K.S.; Iverson, C.N.; Marrone, B.L.; Pilania, G. Predicting the Mechanical Response of Polyhydroxyalkanoate Biopolymers using Molecular Dynamics Simulations. *Polymers* **2022**, *14*, 345. [CrossRef]
179. Bernardo, P.; Drioli, E.; Golemme, G. Membrane gas separation: A review/state of the art. *Ind. Eng. Chem. Res.* **2009**, *48*, 4638–4663. [CrossRef]
180. Park, H.B.; Kamcev, J.; Robeson, L.M.; Elimelech, M.; Freeman, B.D. Maximizing the right stuff: The trade-off between membrane permeability and selectivity. *Science* **2017**, *356*, eaab0530. [CrossRef]
181. Aburabie, J.; Peinemann, K.V. Crosslinked poly (ether block amide) composite membranes for organic solvent nanofiltration applications. *J. Membr. Sci.* **2017**, *523*, 264–272. [CrossRef]
182. Kardani, R.; Asghari, M.; Mohammadi, T.; Afsari, M. Effects of nanofillers on the characteristics and performance of PEBA-based mixed matrix membranes. *Rev. Chem. Eng.* **2018**, *34*, 797–836. [CrossRef]
183. Clarizia, G.; Bernardo, P. A Review of the Recent Progress in the Development of Nanocomposites Based on Poly (ether-block-amide) Copolymers as Membranes for CO_2 Separation. *Polymers* **2022**, *14*, 10. [CrossRef]
184. Roman, H.E.; Bunde, A.; Dieterich, W. Conductivity of dispersed ionic conductors: A percolation model with two critical points. *Phys. Rev. B* **1986**, *34*, 3439. [CrossRef]
185. Barzegar, T.; Hassanajili, S. Fabrication and characterization of dual layer PEBAX-SiO2/polyethersulfone nanocomposite membranes for separation of CO2/CH4 gases. *J. Appl. Polym. Sci.* **2022**, *139*, 51624. [CrossRef]
186. Gao, H.; Dong, C.; Wang, Q.; Zhu, H.; Meng, X.; Cong, C.; Zhou, Q. Improving the proton conductivity of proton exchange membranes via incorporation of HPW-functionalized mesoporous silica nanospheres into SPEEK. *Int. J. Hydrogen Energy* **2018**, *43*, 21940–21948. [CrossRef]
187. Hasan, M.M.; Islam, M.D.; Rashid, T.U. Biopolymer-based electrolytes for dye-sensitized solar cells: A critical review. *Energy Fuels* **2020**, *34*, 15634–15671. [CrossRef]
188. Al Munsur, A.Z.; Goo, B.H.; Kim, Y.; Kwon, O.J.; Paek, S.Y.; Lee, S.Y.; Kim, H.J.; Kim, T.H. Nafion-based proton-exchange membranes built on cross-linked semi-interpenetrating polymer networks between poly (acrylic acid) and poly (vinyl alcohol). *ACS Appl. Mater. Interfaces* **2021**, *13*, 28188–28200. [CrossRef] [PubMed]
189. Huang, H.; Zeng, X.; Zhang, X.; Li, H. Proton exchange membrane with excellent proton conductivity and superior stability for application at high operating temperatures. *Energy Fuels* **2023**, *37*, 17516–17525. [CrossRef]
190. Rani, M.S.A.; Norrrahim, M.N.F.; Knight, V.F.; Nurazzi, N.M.; Abdan, K.; Lee, S.H. A Review of Solid-State Proton-Polymer Batteries: Materials and Characterizations. *Polymers* **2023**, *15*, 4032. [CrossRef] [PubMed]
191. Ali, A.M.M.; Subban, R.H.Y.; Bahron, H.; Yahya, M.Z.A.; Kamisan, A.S. Investigation on modified natural rubber gel polymer electrolytes for lithium polymer battery. *J. Power Sources* **2013**, *244*, 636–640. [CrossRef]
192. Yu, X.; Bi, Z.; Zhao, F.; Manthiram, A. Hybrid lithium-sulfur batteries with a solid electrolyte membrane and lithium polysulfide catholyte. *ACS Appl. Mater. Interfaces* **2015**, *7*, 16625–16631. [CrossRef]
193. Zhu, Y.S.; Xiao, S.Y.; Li, M.X.; Chang, Z.; Wang, F.X.; Gao, J.; Wu, Y.P. Natural macromolecule based carboxymethyl cellulose as a gel polymer electrolyte with adjustable porosity for lithium ion batteries. *J. Power Sources* **2015**, *288*, 368–375. [CrossRef]
194. Manthiram, A.; Yu, X.; Wang, S. Lithium battery chemistries enabled by solid-state electrolytes. *Nat. Rev. Mater.* **2017**, *2*, 1–16. [CrossRef]
195. Chattopadhyay, J.; Pathak, T.S.; Santos, D.M. Applications of Polymer Electrolytes in Lithium-Ion Batteries: A Review. *Polymers* **2023**, *15*, 3907. [CrossRef]
196. Dallaev, R.; Pisarenko, T.; Papež, N.; Sadovský, P.; Holcman, V. A Brief Overview on Epoxies in Electronics: Properties, Applications, and Modifications. *Polymers* **2023**, *15*, 3964. [CrossRef] [PubMed]

197. Gu, H.; Ma, C.; Gu, J.; Guo, J.; Yan, X.; Huang, J.; Zhang, Q.; Guo, Z. An overview of multifunctional epoxy nanocomposites. *J. Mater. Chem. C* **2016**, *4*, 5890–5906. [CrossRef]
198. Ahmadi, Z. Epoxy in nanotechnology: A short review. *Prog. Org. Coatings* **2019**, *132*, 445–448. [CrossRef]
199. Xiang, Q.; Xiao, F. Applications of epoxy materials in pavement engineering. *Constr. Build. Mater.* **2020**, *235*, 117529. [CrossRef]
200. Tian, Z.S.; Wang, Y.Q.; Hou, X.L. Review of chemical recycling and reuse of carbon fiber reinforced epoxy resin composites. *New Carbon Mater.* **2022**, *37*, 1021–1041. [CrossRef]
201. Li, S.; Wang, K.W. Plant-inspired adaptive structures and materials for morphing and actuation: A review. *Bioinspiration Biomim.* **2016**, *12*, 011001. [CrossRef]
202. Walsh, C. Human-in-the-loop development of soft wearable robots. *Nat. Rev. Mater.* **2018**, *3*, 78–80. [CrossRef]
203. Li, W.; Matsuhisa, N.; Liu, Z.; Wang, M.; Luo, Y.; Cai, P.; Chen, G.; Zhang, F.; Li, C.; Liu, Z.; et al. An on-demand plant-based actuator created using conformable electrodes. *Nat. Electron.* **2021**, *4*, 134–142. [CrossRef]
204. Chen, Y.; Yang, J.; Zhang, X.; Feng, Y.; Zeng, H.; Wang, L.; Feng, W. Light-driven bimorph soft actuators: Design, fabrication, and properties. *Mater. Horiz.* **2021**, *8*, 728–757. [CrossRef]
205. Razzaq, M.Y.; Balk, M.; Mazurek-Budzyńska, M.; Schadewald, A. From Nature to Technology: Exploring Bioinspired Polymer Actuators via Electrospinning. *Polymers* **2023**, *15*, 4029. [CrossRef]
206. Lee, H.; Nguyen, N.H.; Hwang, S.I.; Lee, H.J.; Hong, S.K.; Byun, S.S. Personalized 3D kidney model produced by rapid prototyping method and its usefulness in clinical applications. *Int. Braz. J. Urol.* **2018**, *44*, 952–957. [CrossRef] [PubMed]
207. Lupulescu, C.; Sun, Z. A systematic review of the clinical value and applications of three-dimensional printing in renal surgery. *J. Clin. Med.* **2019**, *8*, 990. [CrossRef] [PubMed]
208. Moldovanu, C.G.; Lebovici, A.; Buruian, M.M. A systematic review of the clinical value and applications of three-dimensional virtual reconstructions in renal tumors. *Med. Pharm. Rep.* **2022**, *95*, 11. [CrossRef] [PubMed]
209. Li, N.; Khan, S.B.; Chen, S.; Aiyiti, W.; Zhou, J.; Lu, B. Promising New Horizons in Medicine: Medical Advancements with Nanocomposite Manufacturing via 3D Printing. *Polymers* **2023**, *15*, 4122. [CrossRef]
210. Vasquez, P.A.; McKinley, G.H.; Cook, L.P. A network scission model for wormlike micellar solutions: I. Model formulation and viscometric flow predictions. *J. Non Newton. Fluid Mech.* **2007**, *144*, 122–139. [CrossRef]
211. Pipe, C.J.; Kim, N.J.; Vasquez, P.A.; Cook, L.P.; McKinley, G.H. Wormlike micellar solutions: II. Comparison between experimental data and scission model predictions. *J. Rheol.* **2010**, *54*, 881–913. [CrossRef]
212. de Souza Mendes, P.R. Thixotropic elasto-viscoplastic model for structured fluids. *Soft Matter* **2011**, *7*, 2471–2483. [CrossRef]
213. Castillo-Sánchez, H.A.; de Souza, L.F.; Castelo, A. Numerical Simulation of Rheological Models for Complex Fluids Using Hierarchical Grids. *Polymers* **2022**, *14*, 4958. [CrossRef]
214. Komuraiah, A.; Kumar, N.S.; Prasad, B.D. Chemical composition of natural fibers and its influence on their mechanical properties. *Mech. Compos. Mater.* **2014**, *50*, 359–376. [CrossRef]
215. Venkatarajan, S.; Subbu, C.; Athijayamani, A.; Muthuraja, R. Mechanical properties of natural cellulose fibers reinforced polymer composites—2015–2020: A review. *Mater. Today Proc.* **2021**, *47*, 1017–1024. [CrossRef]
216. Santha, R.D.; Gopala Krishna, M.O. Evaluation of mechanical and micro structural properties of natural fiber reinforced polymer composites. *Mater. Sci. Forum* **2022**, *1065*, 69–77. [CrossRef]
217. Farazin, A.; Khan, A. An extensive study on strain dependence of glass fiber-reinforced polymer-based composites. *J. Strain Anal. Eng. Des.* **2022**, *57*, 411–432. [CrossRef]
218. Ilyas, R.A.; Zuhri, M.Y.M.; Aisyah, H.A.; Asyraf, M.R.M.; Hassan, S.A.; Zainudin, E.S.; Sapuan, S.M.; Sharma, S.; Bangar, S.P.; Jumaidin, R.; et al. Natural Fiber-reinforced Polylactic Acid, Polylactic Acid Blends and their Composites for Advanced Applications. *Polymers* **2022**, *14*, 202. [CrossRef] [PubMed]
219. Rodríguez Soto, A.A.; Valín Rivera, J.L.; Sanabio Alves Borges, L.M.; Palomares Ruiz, J.E. Parameter Identification of Fractional Index Viscoelastic Model for Vegetable-Fiber Reinforced Composite. *Polymers* **2022**, *14*, 4634. [CrossRef] [PubMed]
220. Bhattacharjee, P.K.; Nguyen, D.A.; McKinley, G.H.; Sridhar, T. Extensional stress growth and stress relaxation in entangled polymer solutions. *J. Rheol.* **2003**, *47*, 269–290. [CrossRef]
221. Huang, Q.; Hengeller, L.; Alvarez, N.J.; Hassager, O. Bridging the gap between polymer melts and solutions in extensional rheology. *Macromolecules* **2015**, *48*, 4158–4163. [CrossRef]
222. Stephanou, P.S.; Ch. Tsimouri, I.; G. Mavrantzas, V. Simple, accurate and user-friendly differential constitutive model for the rheology of entangled polymer melts and solutions from nonequilibrium thermodynamics. *Materials* **2020**, *13*, 2867. [CrossRef]
223. Masubuchi, Y.; Yang, L.; Uneyama, T.; Doi, Y. Analysis of Elongational Viscosity of Entangled Poly(propylene carbonate) Melts by Primitive Chain Network Simulations. *Polymers* **2022**, *14*, 741. [CrossRef]
224. Masubuchi, Y.; Takimoto, J.I.; Koyama, K.; Ianniruberto, G.; Marrucci, G.; Greco, F. Brownian simulations of a network of reptating primitive chains. *J. Chem. Phys.* **2001**, *115*, 4387–4394. [CrossRef]
225. Ibar, J.P. The great myths of polymer melt rheology, Part I: Comparison of experiment and current theory. *J. Macromol. Sci. Part B* **2009**, *48*, 1143–1189. [CrossRef]
226. Ibar, J.P. The great myths of rheology, Part II: Transient and steady-state melt deformation: The question of melt entanglement stability. *J. Macromol. Sci. Part B* **2010**, *49*, 1148–1258. [CrossRef]
227. Ibar, J.P. Trouble with Polymer Physics: Development of "Sustained Orientation" Contradicts the Current Understanding of the Liquid State of Polymers. *J. Macromol. Sci. Part B* **2015**, *54*, 722–748. [CrossRef]

228. Ibar, J.P. Shear-Thinning of Polymeric Melts: The Failure of the Reptation Model. In *The Physics of Polymer Interactions—A Novel Approach to Rheology and Processing*; Hanser Publishers: Munich, Germany, 2019.
229. Ibar, J.P. The Challenges Facing the Current Paradigm Describing Viscoelastic Interactions in Polymer Melts. *Polymers* **2023**, *15*, 4309. [CrossRef] [PubMed]
230. Xie, X.L.; Mai, Y.W.; Zhou, X.P. Dispersion and alignment of carbon nanotubes in polymer matrix: A review. *Mater. Sci. Eng. Rep.* **2005**, *49*, 89–112. [CrossRef]
231. Rong, C.; Ma, G.; Zhang, S.; Song, L.; Chen, Z.; Wang, G.; Ajayan, P.M. Effect of carbon nanotubes on the mechanical properties and crystallization behavior of poly (ether ether ketone). *Compos. Sci. Technol.* **2010**, *70*, 380–386. [CrossRef]
232. Mittal, V. Functional polymer nanocomposites with graphene: A review. *Macromol. Mater. Eng.* **2014**, *299*, 906–931. [CrossRef]
233. Nazarychev, V.M.; Vaganov, G.V.; Larin, S.V.; Didenko, A.L.; Elokhovskiy, V.Y.; Svetlichnyi, V.M.; Yudin, V.E.; Lyulin, S.V. Rheological and Mechanical Properties of Thermoplastic Crystallizable Polyimide-Based Nanocomposites Filled with Carbon Nanotubes: Computer Simulations and Experiments. *Polymers* **2022**, *14*, 3154. [CrossRef]
234. Likhtman, A.E.; McLeish, T.C. Quantitative theory for linear dynamics of linear entangled polymers. *Macromolecules* **2002**, *35*, 6332–6343. [CrossRef]
235. Shahid, T.; Huang, Q.; Oosterlinck, F.; Clasen, C.; Van Ruymbeke, E. Dynamic dilution exponent in monodisperse entangled polymer solutions. *Soft Matter* **2017**, *13*, 269–282. [CrossRef]
236. Lentzakis, H.; Costanzo, S.; Vlassopoulos, D.; Colby, R.H.; Read, D.J.; Lee, H.; Chang, T.; Van Ruymbeke, E. Constraint release mechanisms for H-polymers moving in linear matrices of varying molar masses. *Macromolecules* **2019**, *52*, 3010–3028. [CrossRef]
237. Hannecart, C.; Clasen, C.; Van Ruymbeke, E. Constraint Release Rouse Mechanisms in Bidisperse Linear Polymers: Investigation of the Release Time of a Short-Long Entanglement. *Polymers* **2023**, *15*, 1569. [CrossRef] [PubMed]
238. Ouardad, S.; Bakleh, M.E.; Kostjuk, S.V.; Ganachaud, F.; Puskas, J.E.; Deffieux, A.; Peruch, F. Bio-inspired cationic polymerization of isoprene and analogues: State-of-the-art. *Polym. Int.* **2012**, *61*, 149–156. [CrossRef]
239. Marquet, F.; Stojceski, F.; Grasso, G.; Patrulea, V.; Danani, A.; Borchard, G. Characterization of the Interaction of Polymeric Micelles with siRNA: A Combined Experimental and Molecular Dynamics Study. *Polymers* **2022**, *14*, 4409. [CrossRef] [PubMed]
240. Limbach, H.J.; Arnold, A.; Mann, B.A.; Holm, C. ESPResSo—An extensible simulation package for research on soft matter systems. *Comput. Phys. Commun.* **2006**, *174*, 704–727. [CrossRef]
241. Abraham, M.J.; Murtola, T.; Schulz, R.; Páll, S.; Smith, J.C.; Hess, B.; Lindahl, E. GROMACS: High performance molecular simulations through multi-level parallelism from laptops to supercomputers. *SoftwareX* **2015**, *1*, 19–25. [CrossRef]
242. Kwok, A.; McCarthy, D.; Hart, S.L.; Tagalakis, A.D. Systematic Comparisons of Formulations of Linear Oligolysine Peptides with si RNA and Plasmid DNA. *Chem. Biol. Drug Des.* **2016**, *87*, 747–763. [CrossRef]
243. Cong, Y.; Topf, M.; Sali, A.; Matsudaira, P.; Dougherty, M.; Chiu, W.; Schmid, M.F. Crystallographic conformers of actin in a biologically active bundle of filaments. *J. Mol. Biol.* **2008**, *375*, 331–336. [CrossRef]
244. Arii, Y.; Hatori, K. Relationship between the flexibility and the motility of actin filaments: Effects of pH. *Biochem. Biophys. Res. Commun.* **2008**, *371*, 772–776. [CrossRef]
245. Tassieri, M. Dynamics of semiflexible polymer solutions in the tightly entangled concentration regime. *Macromolecules* **2017**, *50*, 5611–5618. [CrossRef]
246. Alva, E.; George, A.; Brancaleon, L.; Marucho, M. Hydrodynamic and Polyelectrolyte Properties of Actin Filaments: Theory and Experiments. *Polymers* **2022**, *14*, 2438. [CrossRef]
247. Venneti, K.C.; Hewage, C.M. Conformational and molecular interaction studies of glucagon-like peptide-2 with its N-terminal extracellular receptor domain. *FEBS Lett.* **2011**, *585*, 346–352. [CrossRef] [PubMed]
248. Li, Y.; Sun, J.; Li, D.; Lin, J. Activation and conformational dynamics of a class B G-protein-coupled glucagon receptor. *Phys. Chem. Chem. Phys.* **2016**, *18*, 12642–12650. [CrossRef] [PubMed]
249. Song, G.; Yang, D.; Wang, Y.; De Graaf, C.; Zhou, Q.; Jiang, S.; Liu, K.; Cai, X.; Dai, A.; Lin, G.; et al. Human GLP-1 receptor transmembrane domain structure in complex with allosteric modulators. *Nature* **2017**, *546*, 312–315. [CrossRef] [PubMed]
250. Su, J.; Sun, T.; Wang, Y.; Shen, Y. Conformational Dynamics of Glucagon-like Peptide-2 with Different Electric Field. *Polymers* **2022**, *14*, 2722. [CrossRef]
251. Antonietti, M.; Förster, S. Vesicles and liposomes: A self-assembly principle beyond lipids. *Adv. Mater.* **2003**, *15*, 1323–1333. [CrossRef]
252. Meng, F.; Engbers, G.H.; Feijen, J. Biodegradable polymersomes as a basis for artificial cells: Encapsulation, release and targeting. *J. Control. Release* **2005**, *101*, 187–198. [CrossRef]
253. Li, W.; Luo, T.; Yang, Y.; Tan, X.; Liu, L. Formation of controllable hydrophilic/hydrophobic drug delivery systems by electrospinning of vesicles. *Langmuir* **2015**, *31*, 5141–5146. [CrossRef]
254. Li, P.; Kang, N.; Chai, A.; Lu, D.; Luo, S.; Yang, Z. Vesicle Geometries Enabled by Semiflexible Polymer. *Polymers* **2022**, *14*, 757. [CrossRef]
255. Li, Y.; Zhang, H.; Wang, Z.; Bao, M. Micelle-vesicle transitions in catanionic mixtures of SDS/DTAB induced by salt, temperature, and selective solvents: A dissipative particle dynamics simulation study. *Colloid Polym. Sci.* **2014**, *292*, 2349–2360. [CrossRef]
256. Li, Z.; Wu, J. Density functional theory for polyelectrolytes near oppositely charged surfaces. *Phys. Rev. Lett.* **2006**, *96*, 048302. [CrossRef]

257. Knychała, P.; Timachova, K.; Banaszak, M.; Balsara, N.P. 50th anniversary perspective: Phase behavior of polymer solutions and blends. *Macromolecules* **2017**, *50*, 3051–3065. [CrossRef]
258. Batys, P.; Kivistö, S.; Lalwani, S.M.; Lutkenhaus, J.L.; Sammalkorpi, M. Comparing water-mediated hydrogen-bonding in different polyelectrolyte complexes. *Soft Matter* **2019**, *15*, 7823–7831. [CrossRef] [PubMed]
259. Xu, X.; Shi, H.; Wang, F. Near-critical phase behavior in polyelectrolyte solutions: Effect of charge fluctuations. *J. Phys. Chem. B* **2020**, *124*, 4203–4210. [CrossRef]
260. Wang, Y.; Qiu, Q.; Yedilbayeva, A.; Kairula, D.; Dai, L. Phase Behavior of Ion-Containing Polymers in Polar Solvents: Predictions from a Liquid-State Theory with Local Short-Range Interactions. *Polymers* **2022**, *14*, 4421. [CrossRef]
261. Micheletti, C.; Marenduzzo, D.; Orlandini, E.; Summers, D.W. Knotting of random ring polymers in confined spaces. *J. Chem. Phys.* **2006**, *124*, 064903. [CrossRef]
262. Dabrowski-Tumanski, P.; Sulkowska, J.I. Topological knots and links in proteins. *Proc. Natl. Acad. Sci. USA* **2017**, *114*, 3415–3420. [CrossRef]
263. Jackson, S.E.; Suma, A.; Micheletti, C. How to fold intricately: Using theory and experiments to unravel the properties of knotted proteins. *Curr. Opin. Struct. Biol.* **2017**, *42*, 6–14. [CrossRef]
264. Rusková, R.; Račko, D. Knot Formation on DNA Pushed Inside Chiral Nanochannels. *Polymers* **2023**, *15*, 4185. [CrossRef]
265. Lang, P.; Frey, E. Disentangling entanglements in biopolymer solutions. *Nat. Commun.* **2018**, *9*, 494. [CrossRef]
266. Lorenz, J.S.; Schnauß, J.; Glaser, M.; Sajfutdinow, M.; Schuldt, C.; Käs, J.A.; Smith, D.M. Synthetic Transient Crosslinks Program the Mechanics of Soft, Biopolymer-Based Materials. *Adv. Mater.* **2018**, *30*, 1706092. [CrossRef]
267. Händler, T.; Tutmarc, C.; Glaser, M.; Freitag, J.S.; Smith, D.M.; Schnauß, J. Measuring structural parameters of crosslinked and entangled semiflexible polymer networks with single-filament tracing. *Phys. Rev. E* **2021**, *103*, 062501. [CrossRef] [PubMed]
268. Elbalasy, I.; Mollenkopf, P.; Tutmarc, C.; Herrmann, H.; Schnauß, J. Keratins determine network stress responsiveness in reconstituted actin-keratin filament systems. *Soft Matter* **2021**, *17*, 3954–3962. [CrossRef] [PubMed]
269. Händler, T.; Tutmarc, C.; Freitag, J.S.; Smith, D.M.; Schnauß, J. Constraint Release for Reptating Filaments in Semiflexible Networks Depends on Background Fluctuations. *Polymers* **2022**, *14*, 707. [CrossRef] [PubMed]
270. Kim, S.; Lee, N.K.; Chae, M.K.; Johner, A.; Park, J.M. Translocation of Hydrophobic Polyelectrolytes under Electrical Field: Molecular Dynamics Study. *Polymers* **2023**, *15*, 2550. [CrossRef]
271. Jeon, J.; Dobrynin, A.V. Molecular Dynamics Simulations of Polyelectrolyte–Polyampholyte Complexes: Effect of Solvent Quality and Salt Concentration. *J. Phys. Chem. B* **2006**, *110*, 24652–24665. [CrossRef]
272. Palyulin, V.V.; Ala-Nissila, T.; Metzler, R. Polymer translocation: The first two decades and the recent diversification. *Soft Matter* **2014**, *10*, 9016–9037. [CrossRef]
273. Payet, L.; Martinho, M.; Merstorf, C.; Pastoriza-Gallego, M.; Pelta, J.; Viasnoff, V.; Auvray, L.; Muthukumar, M.; Mathé, J. Temperature effect on ionic current and ssDNA transport through nanopores. *Biophys. J.* **2015**, *109*, 1600–1607. [CrossRef]
274. Lee, N.K.; Jung, Y.; Johner, A.; Joanny, J.F. Globular polyampholytes: Structure and translocation. *Macromolecules* **2021**, *54*, 2394–2411. [CrossRef]
275. Lukasheva, N.; Tolmachev, D.; Martinez-Seara, H.; Karttunen, M. Changes in the Local Conformational States Caused by Simple Na^+ and K^+ Ions in Polyelectrolyte Simulations: Comparison of Seven Force Fields with and without NBFIX and ECC Corrections. *Polymers* **2022**, *14*, 252. [CrossRef]
276. Malkin, A.; Ilyin, S.; Vasilyev, G.; Arinina, M.; Kulichikhin, V. Pressure losses in flow of viscoelastic polymeric fluids through short channels. *J. Rheol.* **2014**, *58*, 433–448. [CrossRef]
277. Galata, A.A.; Kröger, M. Globular Proteins and Where to Find Them within a Polymer Brush: A Case Study. *Polymers* **2023**, *15*, 2407. [CrossRef] [PubMed]
278. Senaratne, W.; Andruzzi, L.; Ober, C.K. Self-assembled monolayers and polymer brushes in biotechnology: Current applications and future perspectives. *Biomacromolecules* **2005**, *6*, 2427–2448. [CrossRef] [PubMed]
279. Barbey, R.; Lavanant, L.; Paripovic, D.; Schuwer, N.; Sugnaux, C.; Tugulu, S.; Klok, H.A.. Polymer brushes via surface-initiated controlled radical polymerization: Synthesis, characterization, properties, and applications. *Chem. Rev.* **2009**, *109*, 5437–5527. [CrossRef] [PubMed]
280. Wei, S.; Knotts, T.A. A coarse grain model for protein-surface interactions. *J. Chem. Phys.* **2013**, *139*, 095102. [CrossRef]
281. Han, Y.; Cui, J.; Jin, J.; Jiang, W. Hydrogen bonding induced protein adsorption on polymer brushes: A Monte Carlo study. *J. Mater. Chem. B* **2017**, *5*, 8479–8486. [CrossRef]
282. Bloomfield, V.A. DNA Condensation by Multivalent Cations. *Biopolymers* **1997**, *44*, 269–282. [CrossRef]
283. Grosberg, A.Y.; Nguyen, T.T.; Shklovskii, B.I. Colloquium: The Physics of Charge Inversion in Chemical and Biological Systems. *Rev. Mod. Phys.* **2002**, *74*, 329. [CrossRef]
284. McGinty, R.K.; Tan, S. Nucleosome structure and function. *Chem. Rev.* **2015**, *115*, 2255–2273. [CrossRef]
285. Kornberg, R.D.; Lorch, Y. Primary role of the nucleosome. *Mol. Cell* **2020**, *79*, 371–375. [CrossRef]
286. Nodelman, I.M.; Bowman, G.D. Biophysics of chromatin remodeling. *Annu. Rev. Biophys.* **2021**, *50*, 73–93. [CrossRef]
287. Shi, D.; Huang, Y.; Bai, C. Studies of the Mechanism of Nucleosome Dynamics: A Review on Multifactorial Regulation from Computational and Experimental Cases. *Polymers* **2023**, *15*, 1763. [CrossRef] [PubMed]
288. Škrbić, T.; Giacometti, A.; Hoang, T.X.; Maritan, A.; Banavar, J.R. A Tale of Two Chains: Geometries of a Chain Model and Protein Native State Structures. *Polymers* **2024**, *16*, 502. [CrossRef] [PubMed]

289. Auhl, R.; Everaers, R.; Grest, G.S.; Kremer, K.; Plimpton, S.J. Equilibration of long chain polymer melts in computer simulations. *J. Chem. Phys.* **2003**, *119*, 12718–12728. [CrossRef]
290. Rossi, G.; Monticelli, L.; Puisto, S.R.; Vattulainen, I.; Ala-Nissila, T. Coarse-graining polymers with the MARTINI force-field: Polystyrene as a benchmark case. *Soft Matter* **2011**, *7*, 698–708. [CrossRef]
291. Newberry, R.W.; Raines, R.T. Secondary forces in protein folding. *ACS Chem. Biol.* **2019**, *14*, 1677–1686. [CrossRef]
292. Ramakrishna, S.; Mayer, J.; Wintermantel, E.; Leong, K.W. Biomedical applications of polymer-composite materials: A review. *Compos. Sci. Technol.* **2001**, *61*, 1189–1224. [CrossRef]
293. Senra, M.R.; Marques, M.D.F.V. Synthetic polymeric materials for bone replacement. *J. Compos. Sci.* **2020**, *4*, 191. [CrossRef]
294. Agrawal, R.; Kumar, A.; Mohammed, M.K.; Singh, S. Biomaterial types, properties, medical applications, and other factors: A recent review. *J. Zhejiang-Univ.-Sci. A* **2023**, *24*, 1027–1042. [CrossRef]
295. Ornaghi, H.L., Jr.; Monticeli, F.M.; Agnol, L.D. A Review on Polymers for Biomedical Applications on Hard and Soft Tissues and Prosthetic Limbs. *Polymers* **2023**, *15*, 4034. [CrossRef]
296. Wang, Z.; Li, H.; Tang, Z.; Liu, Z.; Ruan, Z.; Ma, L.; Yang, Q.; Wang, D.; Zhi, C. Hydrogel electrolytes for flexible aqueous energy storage devices. *Adv. Funct. Mater.* **2018**, *28*, 1804560. [CrossRef]
297. Chan, C.Y.; Wang, Z.; Jia, H.; Ng, P.F.; Chow, L.; Fei, B. Recent advances of hydrogel electrolytes in flexible energy storage devices. *J. Mater. Chem. A* **2021**, *9*, 2043–2069. [CrossRef]
298. Correa, S.; Grosskopf, A.K.; Lopez Hernandez, H.; Chan, D.; Yu, A.C.; Stapleton, L.M.; Appel, E.A. Translational applications of hydrogels. *Chem. Rev.* **2021**, *121*, 11385–11457. [CrossRef] [PubMed]
299. Mariani, A.; Malucelli, G. Polymer Hydrogels and Frontal Polymerization: A Winning Coupling. *Polymers* **2023**, *15*, 4242. [CrossRef] [PubMed]
300. Aitcheson, S.M.; Frentiu, F.D.; Hurn, S.E.; Edwards, K.; Murray, R.Z. Skin wound healing: Normal macrophage function and macrophage dysfunction in diabetic wounds. *Molecules* **2021**, *26*, 4917. [CrossRef]
301. An, Y.; Lin, S.; Tan, X.; Zhu, S.; Nie, F.; Zhen, Y.; Gu, L.; Zhang, C.; Wang, B.; Wei, W.; et al. Exosomes from adipose-derived stem cells and application to skin wound healing. *Cell Prolif.* **2021**, *54*, e12993. [CrossRef]
302. Walker, M. Human skin through the ages. *Int. J. Pharm.* **2022**, *622*, 121850. [CrossRef]
303. Liang, W.; Ni, N.; Huang, Y.; Lin, C. An Advanced Review: Polyurethane-Related Dressings for Skin Wound Repair. *Polymers* **2023**, *15*, 4301. [CrossRef]
304. Sepahi, M.T.; Abusalma, H.; Jovanovic, V.; Eisazadeh, H. Mechanical properties of 3D-printed parts made of polyethylene terephthalate glycol. *J. Mater. Eng. Perform.* **2021**, *30*, 6851–6861. [CrossRef]
305. Raj, R.; Dixit, A.R.; Łukaszewski, K.; Wichniarek, R.; Rybarczyk, J.; Kuczko, W.; Górski, F. Numerical and experimental mechanical analysis of additively manufactured ankle–foot orthoses. *Materials* **2022**, *15*, 6130. [CrossRef]
306. Zhang, P.; Li, J.; Ashour, E.A.; Chung, S.; Wang, H.; Vemula, S.K.; Repka, M.A. Development of multiple structured extended release tablets via hot melt extrusion and dual-nozzle fused deposition modeling 3D printing. *Int. J. Pharm.* **2024**, *653*, 123905. [CrossRef]
307. Mian, S.H.; Umer, U.; Moiduddin, K.; Alkhalefah, H. Predicting Mechanical Properties of Polymer Materials Using Rate-Dependent Material Models: Finite Element Analysis of Bespoke Upper Limb Orthoses. *Polymers* **2024**, *16*, 1220. [CrossRef] [PubMed]
308. By Niabot—Own Work, CC BY 3.0. Available online: https://commons.wikimedia.org/w/index.php?curid=7793731 (accessed on 26 November 2024)
309. Mandelbrot, B.B. *The Fractal Geometry of Nature*; Freemann: San Francisco, CA, USA, 1982.
310. Havlin, S.; Ben-Avraham, D. Diffusion in Disordered Media. *Adv. Phys.* **1987**, *36*, 695–798. [CrossRef]
311. Feder, J. *Fractals*; Springer: New York, NY, USA, 1988. [CrossRef]
312. Bunde, A.; Havlin, S. (Eds.) *Fractals and Disordered Systems*; Springer Science & Business Media: Berlin/Heidelberg, Germany, 1996. Available online: https://link.springer.com/book/10.1007/978-3-642-84868-1 (accessed on 26 November 2024).
313. Stauffer, D.; Aharony, A. *Introduction to Percolation Theory*; Taylor & Francis: London, UK, 2017. [CrossRef]
314. Hughes, B.D. Conduction and Diffusion in Percolating Systems. In *Complex Media and Percolation Theory*; Sahimi, M., Hunt, A.G., Meyers, R.A., Eds.; Encyclopedia of Complexity and Systems Science Series; Springer: New York, NY, USA, 2020. [CrossRef]
315. Leath, P.L. Cluster size and boundary distribution near percolation threshold. *Phys. Rev. B* **1976**, *14*, 5046. [CrossRef]
316. Alexandrowicz, Z. Critically branched chains and percolation clusters. *Phys. Lett. A* **1980**, *80*, 284–286. [CrossRef]
317. Ordemann, A.; Roman, H.E.; Bunde, A. Cluster growth at the percolation threshold with a finite lifetime of growth sites. *Phys. A Stat. Mech. Its Appl.* **1999**, *266*, 92–95. [CrossRef]
318. Jan, N.; Stauffer, D. Random site percolation in three dimensions. *Int. J. Mod. Phys. C* **1998**, *9*, 341–347. [CrossRef]
319. Zhou, Z.; Yang, J.; Deng, Y.; Ziff, R.M. Shortest-path fractal dimension for percolation in two and three dimensions. *Phys. Rev. E* **2012**, *86*, 061101. [CrossRef]
320. Flory, P.J. *Principles of Polymer Chemistry*; Cornell University Press: Ithaca, NY, USA; London, UK, 1953.
321. de Gennes, P.G. *Scaling Concepts in Polymer Physics*; Cornell University Press: Ithaca, NY, USA; London, UK, 1979.
322. Doi, M.; Edwards, S.F. *The Theory of Polymer Dynamics*; Oxford University Press: Oxford, UK, 1986; Volume 73.
323. Vanderzande, C. *Lattice Models of Polymers. Cambridge Lecture Notes in Physics 11*; Cambridge University Press: Cambridge, UK, 1998. [CrossRef]

324. Roman, H.E. Static and Transport Properties of Disordered Systems: Random Fractal Structures and Composite Materials. Habilitation Thesis, University of Hamburg, Hamburg, Germany, 1990. [CrossRef]
325. Madras, N.; Sokal, A.D. The pivot algorithm: A highly efficient Monte Carlo method for the self-avoiding walk. *J. Stat. Phys.* **1988**, *50*, 109–186. [CrossRef]
326. Eizenberg, N.; Klafter, J. Self-avoiding walks on a simple cubic lattice. *J. Chem. Phys.* **1993**, *99*, 3976–3982. [CrossRef]
327. Li, B.; Madras, N.; Sokal, A.D. Critical exponents, hyperscaling, and universal amplitude ratios for two-and three-dimensional self-avoiding walks. *J. Stat. Phys.* **1995**, *80*, 661–754. [CrossRef]
328. Van Rensburg, E.J. Monte Carlo methods for the self-avoiding walk. *J. Phys. A Math. Theor.* **2009**, *42*, 323001. [CrossRef]
329. Clisby, N. Accurate estimate of the critical exponent ν for self-avoiding walks via a fast implementation of the pivot algorithm. *Phys. Rev. Lett.* **2010**, *104*, 055702. [CrossRef] [PubMed]
330. Ordemann, A.; Porto, M.; Roman, H.E.; Havlin, S.; Bunde, A. Multifractal Behavior of Linear Polymers in Disordered Media. *Phys. Rev. E* **2000**, *61*, 6858. [CrossRef] [PubMed]
331. Guttmann, A.J.; Wang, J. The extension of self-avoiding random walk series in two dimensions. *J. Phys. A Math. Gen.* **1991**, *24*, 3107. [CrossRef]
332. Masand, B.; Wilensky, U.; Massar, J.P.; Redner, S. An extension of the two-dimensional self-avoiding walk series on the square lattice. *J. Phys. A Math. Gen.* **1992**, *25*, L365. [CrossRef]
333. Conway, A.R.; Guttmann, A.J. Square lattice self-avoiding walks and corrections to scaling. *Phys. Rev. Lett.* **1996**, *77*, 5284. [CrossRef]
334. Nienhuis, B. Exact critical point and critical exponents of $O(n)$ models in two dimensions. *Phys. Rev. Lett.* **1982**, *49*, 1062. [CrossRef]
335. Clisby, N. Scale–free Monte Carlo method for calculating the critical exponent γ of self-avoiding walks. *J. Phys. A Math. Theor.* **2017**, *50*, 264003. [CrossRef]
336. Roman, H.E. Structure of random fractals and the probability distribution of random walks. *Phys. Rev. E* **1995**, *51*, 5422. [CrossRef]
337. Fisher, M.E. Shape of a self-avoiding walk or polymer chain. *J. Chem. Phys* **1966**, *44*, 616–622. [CrossRef]
338. Des Cloizeaux, J. Lagrangian theory for a self-avoiding random chain. *Phys. Rev. A* **1974**, *10*, 1665. [CrossRef]
339. McKenzie, D.S.; Moore, M.A. Shape of self-avoiding walk or polymer chain. *J. Phys. A Gen. Phys.* **1971**, *4*, L82. [CrossRef]
340. Jacobsen, J.L.; Scullard, C.R.; Guttmann, A.J. On the growth constant for square-lattice self-avoiding walks. *J. Phys. A Math. Theor.* **2016**, *49*, 494004. [CrossRef]
341. Guttmann, A.J.; Conway, A.R. Square lattice self-avoiding walks and polygons. *Ann. Comb.* **2001**, *5*, 319–345. [CrossRef]
342. Clisby, N.; Jensen, I. A new transfer-matrix algorithm for exact enumerations: Self-avoiding polygons on the square lattice. *J. Phys. A Math. Theor.* **2012**, *45*, 115202. [CrossRef]
343. Schram, R.D.; Barkema, G.T.; Bisseling, R.H. Exact enumeration of self-avoiding walks. *J. Stat. Mech. Theory Exp.* **2011**, *2011*, P06019. [CrossRef]
344. Marini, F.; Ordemann, A.; Porto, M.; Roman, H.E. Violation of the des Cloizeaux Relation for Self-avoiding Walks on Sierpinski Square Lattices. *Phys. Rev. E* **2006**, *74*, 051102. [CrossRef]
345. Gefen, Y.; Aharony, A.; Mandelbrot, B.B. Phase transitions on fractals. III. Infinitely ramified lattices. *J. Phys. A Math. Gen.* **1984**, *17*, 1277. [CrossRef]
346. Reis, F.D.A.A.; Riera, R. Critical behavior of self-avoiding walks on fractals. *J. Stat. Phys.* **1993**, *71*, 453–470. [CrossRef]
347. Chalub, F.A.C.C.; Reis, F.A.; Riera, R. Connective constant of SAWs on the Sierpinski gasket family. *J. Phys. A Math. Gen.* **1997**, *30*, 4151. [CrossRef]
348. Ordemann, A.; Porto, M.; Roman, H.E. Self-avoiding Walks on Self-similar Structures: Finite versus Infinite Ramification. *J. Phys. A Math. Gen.* **2002**, *35*, 8029. [CrossRef]
349. Ordemann, A.; Porto, M.; Roman, H.E. Self-avoiding Walks on Sierpinski Lattices in Two and Three Dimensions. *Phys. Rev. E* **2002**, *65*, 021107. [CrossRef] [PubMed]
350. Živić, I.; Milošević, S.; Djordjević, B. On the total number of distinct self-interacting self-avoiding walks on three-dimensional fractal structures. *J. Phys. A Math. Gen.* **2005**, *38*, 555. [CrossRef]
351. Fritsche, M.; Roman, H.E.; Porto, M. Temperature-dependent Structural Behavior of Self-avoiding Walks on Sierpinski Carpets. *Phys. Rev. E* **2007**, *76*, 061101. [CrossRef]
352. Porto, M.; Bunde, A.; Havlin, S.; Roman, H.E. Structural and dynamical properties of the percolation backbone in two and three dimensions. *Phys. Rev. E* **1997**, *56*, 1667. [CrossRef]
353. Woo, K.Y.; Lee, S.B. Monte Carlo study of self-avoiding walks on a percolation cluster. *Phys. Rev. A* **1991**, *44*, 999. [CrossRef]
354. Roman, H.E.; Dräger, J.; Bunde, A.; Havlin, S.; Stauffer, D. Distributions of Polymers in Disordered Structures. *Phys. Rev. E* **1995**, *52*, 6303. [CrossRef]
355. Ordemann, A.; Porto, M.; Roman, H.E.; Havlin, S. "Generalized des Cloizeaux" Exponent for Self-avoiding Walks on the Incipient Percolation Cluster. *Phys. Rev. E* **2001**, *63*, 020104. [CrossRef]
356. Ordemann, A.; Porto, M.; Roman, H.E. Self-avoiding Walks on Deterministic and Random Fractals: Numerical Results. In *Statistics of Linear Polymers in Disordered Media*; Chakrabarti, B.K., Ed.; Elsevier: Amsterdam, The Netherlands, 2005; pp. 195–233. [CrossRef]
357. de Gennes, P.G. Reptation of a polymer chain in the presence of fixed obstacles. *J. Chem. Phys.* **1971**, *55*, 572–579. [CrossRef]

358. Jun, S.; Arnold, A.; Ha, B.Y. Confined space and effective interactions of multiple self-avoiding chains. *Phys. Rev. Lett.* **2007**, *98*, 128303. [CrossRef]
359. Baschnagel, J.; Meyer, H.; Wittmer, J.; Kulić, I.; Mohrbach, H.; Ziebert, F.; Nam, G.M.; Lee, N.K.; Johner, A. Semiflexible chains at surfaces: Worm–like chains and beyond. *Polymers* **2016**, *8*, 286. [CrossRef] [PubMed]
360. Maier, B.; Rädler, J.O. Conformation and self-diffusion of single DNA molecules confined to two dimensions. *Phys. Rev. Lett.* **1999**, *82*, 1911. [CrossRef]
361. Käs, J.; Strey, H.; Sackmann, E. Direct imaging of reptation for semiflexible actin filaments. *Nature* **1994**, *368*, 226–229. [CrossRef] [PubMed]
362. Jumper, J.; Evans, R.; Pritzel, A.; Green, T.; Figurnov, M.; Ronneberger, O.; Tunyasuvunakool, K.; Bates, R.; Žídek, A.; Potapenko, A.; et al. Highly accurate protein structure prediction with AlphaFold. *Nature* **2021**, *596*, 583–589. [CrossRef]
363. Shakhnovich, E.I. Proteins with selected sequences fold into unique native conformation. *Phys. Rev. Lett.* **1994**, *72*, 3907. [CrossRef]
364. Tiana, G.; Broglia, R.A.; Roman, H.E.; Vigezzi, E.; Shakhnovich, E. Folding and Misfolding of Designed Protein-like Chains with Mutations. *J. Chem. Phys.* **1998**, *108*, 757–761. [CrossRef]
365. Rost, B. Twilight zone of protein sequence alignments. *Protein Eng.* **1999**, *12*, 85–94. [CrossRef]
366. Broglia, R.A.; Tiana, G.; Roman, H.E.; Vigezzi, E.; Shakhnovich, E. Stability of Designed Proteins against Mutations. *Phys. Rev. Lett.* **1999**, *82*, 4727. [CrossRef]
367. Broglia, R.A.; Tiana, G.; Pasquali, S.; Roman, H.E.; Vigezzi, E. Folding and Aggregation of Designed Proteins. *Proc. Natl. Acad. Sci. USA* **1998**, *95*, 12930–12933. [CrossRef]
368. Uversky, V.N. Intrinsically disordered proteins in overcrowded milieu: Membrane-less organelles, phase separation, and intrinsic disorder. *Curr. Opin. Struct. Biol.* **2017**, *44*, 18–30. [CrossRef]
369. Adachi, K.; Kawaguchi, K. Predicting heteropolymer interactions: Demixing and hypermixing of disordered protein sequences. *Phys. Rev.* **2024**, *14*, 031011. [CrossRef]
370. Bulgac, A.; Roman, H.E. Equations of Motion for Polymer Chains in a Thermostat. *New J. Phys.* **2005**, *7*, 2. [CrossRef]
371. Bastolla, U.; Roman, H.E.; Vendruscolo, M. Neutral evolution of model proteins: Diffusion in sequence space and overdispersion. *J. Theor. Biol.* **1999**, *200*, 49–64. [CrossRef]
372. Dill, K.A. Polymer principles and protein folding. *Protein Sci.* **1999**, *8*, 1166–1180. [CrossRef]
373. Bastolla, U.; Porto, M.; Roman, H.E.; Vendruscolo, M. Lack of self-averaging in neutral evolution of proteins. *Phys. Rev. Lett.* **2002**, *89*, 208101. [CrossRef]
374. Bastolla, U.; Porto, M.; Roman, H.E.; Vendruscolo, M. The Structurally Constrained Neutral Model of Protein Evolution. In *Structural Approaches to Sequence Evolution*; Bastolla, U., Porto, M., Roman, H.E., Vendruscolo, M., Eds.; Biological and Medical Physics, Biomedical Engineering; Springer: Berlin/Heidelberg, Germany, 2007.
375. Kantelhardt, J.W.; Roman, H.E.; Greiner, M. Discrete Wavelet Approach to Multifractality. *Phys. Stat. Mech. Its Appl.* **1995**, *220*, 219–238. [CrossRef]
376. Koscielny-Bunde, E.; Bunde, A.; Havlin, S.; Roman, H.E.; Goldreich, Y.; Schellnhuber, H.-J. Indication of a Universal Persistence Law Governing Atmospheric Variability. *Phys. Rev. Lett.* **1998**, *81*, 729. [CrossRef]
377. Porto, M.; Bastolla, U.; Roman, H.E.; Vendruscolo, M. Reconstruction of protein structures from a vectorial representation. *Phys. Rev. Lett.* **2004**, *92*, 218101. [CrossRef]
378. Bastolla, U.; Porto, M.; Roman, H.E.; Vendruscolo, M. Principal Eigenvector of Contact Matrices and Hydrophobicity Profiles in Proteins. *Proteins* **2005**, *58*, 22–30. [CrossRef]
379. Bastolla, U.; Porto, M.; Roman, H.E.; Vendruscolo, M. (Eds.). *Structural Approaches to Sequence Evolution: Molecules, Networks, Populations*; Book Series: Biological and Medical Physics, Biomedical Engineering (BIOMEDICAL); Springer Science & Business Media: Berlin, Germany, 2007.
380. Ziegler, K.; Holzkamp, E.; Breil, H.; Martin, H. Das Mülheimer normaldruck-polyäthylen-verfahren. *Angew. Chem.* **1955**, *67*, 541–547. [CrossRef]
381. Natta, G.; Corradini, P. Structure and properties of isotactic polypropylene. *Nuovo Cim.* **1960**, *15* (Suppl. 1), 40–51. [CrossRef]
382. Cossee, P. Ziegler–Natta catalysis I. Mechanism of polymerization of α-olefins with Ziegler-Natta catalysts. *J. Catal.* **1964**, *3*, 80–88. [CrossRef]
383. Shimanouchi, T. Stable conformations of syndiotactic and isotactic vinyl polymers and their two-unit and three–unit model molecules. *Pure Appl. Chem.* **1966**, *12*, 287–310. [CrossRef]
384. Roberts, J.D.; Caserio, M.C. *Basic Principles of Organic Chemistry*, 2nd ed.; W.A. Benjamin, Inc.: Menlo Park, CA, USA , 1977.
385. Kaminsky, W.; Laban, A. Metallocene Catalysis. *Appl. Catal. Gen.* **2001**, *222*, 47–61. [CrossRef]
386. Chen, X.; Kumar, S.K.; Ozisik, R. Monte Carlo simulations of the crystallization of isotactic polypropylene. *J. Polym. Sci. Part Polym. Phys.* **2006**, *44*, 3453–3460. [CrossRef]
387. Grasruck, M.; Strobl, G. Crystallization of s-polypropylene from the glassy state: Indications for a multistage process. *Macromolecules* **2003**, *36*, 86–91. [CrossRef]
388. Li, L.; de Jeu, W.H. Shear-induced smectic ordering in the melt of isotactic polypropylene. *Phys. Rev. Lett.* **2004**, *92*, 075506. [CrossRef]

389. Kaji, K.; Nishida, K.; Kanaya, T.; Matsuba, G.; Konishi, T.; Imai, M. Spinodal crystallization of polymers: Crystallization from the unstable melt. In: Interphases and Mesophases in Polymer Crystallization III; Springer: Berlin/Heidelberg, Germany, 2005; pp. 187–240. [CrossRef]
390. Gee, R.; Lacevic, N.; Fried, L. Atomistic simulations of spinodal phase separation preceding polymer crystallization. *Nat. Mater* **2006**, *5*, 39–43. [CrossRef]
391. Kuppa, V.K.; in't Veld, P.J.; Rutledge, G.C. Monte Carlo simulation of interlamellar isotactic polypropylene. *Macromolecules* **2007**, *40*, 5187–5195. [CrossRef]
392. Ranganathan, R.; Kumar, V.; Brayton, A.L.; Kroger, M.; Rutledge, G.C. Atomistic modeling of plastic deformation in semicrystalline polyethylene: Role of interphase topology, entanglements, and chain dynamics. *Macromolecules* **2020**, *53*, 4605–4617. [CrossRef]
393. Pervaje, A.K.; Pasquinelli, M.A.; Khan, S.A.; Santiso, E.E. Multiscale constitutive modeling of the mechanical properties of polypropylene fibers from molecular simulation data. *Macromolecules* **2022**, *55*, 728–744. [CrossRef]
394. Sigalas, N.I.; Van Kraaij, S.A.; Lyulin, A.V. Effect of Temperature on Flow-Induced Crystallization of Isotactic Polypropylene: A Molecular-Dynamics Study. *Macromolecules* **2023**, *56*, 8417–8427. [CrossRef]
395. Ojedeji, D.; Doxastakis, M. Impact of Chain Orientation on the Melting of the α-Form of Isotactic Polypropylene. *Macromolecules* **2024**, *57*, 8643–8655. [CrossRef]
396. Hezaveh, S.; Samanta, S.; Milano, G.; Roccatano, D. Molecular dynamics simulation study of solvent effects on conformation and dynamics of polyethylene oxide and polypropylene oxide chains in water and in common organic solvents. *J. Chem. Phys.* **2012**, *136*, 124901. [CrossRef] [PubMed]
397. Smith, G.D.; Borodin, O.; Bedrov, D. Quantum chemistry based force field for simulations of poly (propylene oxide) and its oligomers. *J. Phys. Chem.* **1998**, *102*, 10318–10323. [CrossRef]
398. Hess, B.; Kutzner, C.; Van Der Spoel, D.; Lindahl, E. GROMACS 4: Algorithms for highly efficient, load-balanced, and scalable molecular simulation. *J. Chem. Theory Comput.* **2008**, *4*, 435–447. [CrossRef]
399. Shephard, J.J.; Bremer, P.J.; McQuillan, A.J. Structure and conformation in mixtures of methyl-terminated poly (ethylene oxide) and water: Principal component analysis and band fitting of infrared absorptions. *J. Phys. Chem. B* **2009**, *113*, 14229–14238. [CrossRef]
400. Becker, N.B.; Rosa, A.; Everaers, R. The radial distribution function of worm-like chains. *Eur. Phys. J. E* **2010**, *32*, 53–69. [CrossRef]
401. Takhulee, A.; Vao-Soongnern, V. Monte Carlo simulation study of the effect of chain tacticity on demixing of polyethylene/polypropylene blends. *Polym. Sci. Ser. A* **2014**, *56*, 936–944. [CrossRef]
402. Heine, D.; Wu, D.T.; Curro, J.G.; Grest, G.S. Role of intramolecular energy on polyolefin miscibility: Isotactic polypropylene/polyethylene blends. *J. Chem. Phys.* **2003**, *118*, 914–924. [CrossRef]
403. Choi, P.; Mattice, W.L. Molecular origin of demixing, prior to crystallization, of atactic polypropylene/isotactic polypropylene blends upon cooling from the melt. *J. Chem. Phys.* **2004**, *121*, 8647–8651. [CrossRef] [PubMed]
404. Nwabunma, D.; Kyu, T. (Eds.) *Polyolefin Blends*; John Wiley & Sons: Hoboken, NJ, USA, 2008.
405. Pinijmontree, T.; Vao-soongnern, V. Dynamics of polypropylene chains in their binary blends of different stereochemical sequences studied by Monte Carlo simulations. *Chin. J. Polym. Sci.* **2014**, *32*, 640–649. [CrossRef]
406. Hagita, K.; Fujiwara, S. Single-chain folding of a quenched isotactic polypropylene chain through united atom molecular dynamics simulations. *Polymer* **2019**, *183*, 121861. [CrossRef]
407. Naiki, M.; Kikkawa, T.; Endo, Y.; Nozaki, K.; Yamamoto, T.; Hara, T. Crystal ordering of α phase isotactic polypropylene. *Polymer* **2001**, *42*, 5471–5477. [CrossRef]
408. Zhang, X.B.; Li, Z.S.; Lu, Z.Y.; Sun, C.C. The crystallization of low-density polyethylene: A molecular dynamics simulation. *Polymer* **2002**, *43*, 3223–3227. [CrossRef]
409. Yamamoto, T. Computer modeling of polymer crystallization – Toward computer-assisted materials' design. *Polymer* **2009**, *50*, 1975–1985. [CrossRef]
410. Khan, P.; Khan, M.N.; Khan, N.A. Shear and cooling induced regulation of mechanical properties and glass transition temperature of isotactic polypropylene. *Mater. Today: Proc.* **2023**, *72*, 2705–2712. [CrossRef]
411. Yadav, A.; Labhasetwar, P.K.; Shahi, V.K. Fabrication and optimization of tunable pore size poly (ethylene glycol) modified poly (vinylidene-co-hexafluoropropylene) membranes in vacuum membrane distillation for desalination. *Sep. Purif. Technol.* **2021**, *271*, 118840. [CrossRef]
412. Luo, A.; Lior, N.; 2016. Critical review of membrane distillation performance criteria. *Desalin. Water Treat.* **2016**, *57*, 20093–20140. [CrossRef]
413. Feng, C.; Wang, R.; Shi, B.; Li, G.; Wu, Y. Factors affecting pore structure and performance of poly (vinylidene fluoride-co-hexafluoro propylene) asymmetric porous membrane. *J. Membr. Sci.* **2006**, *277*, 55–64. 009. [CrossRef]
414. Lalia, B.S.; Guillen-Burrieza, E.; Arafat, H.A.; Hashaikeh, R. Fabrication and characterization of polyvinylidenefluoride-co-hexafluoropropylene (PVDF-HFP) electrospun membranes for direct contact membrane distillation. *J. Membr. Sci.* **2013**, *428*, 104–115. [CrossRef]
415. Wongchitphimon, S.; Wang, R.; Jiraratananon, R. Surface modification of polyvinylidene fluoride-co-hexafluoropropylene (PVDF–HFP) hollow fiber membrane for membrane gas absorption. *J. Membr. Sci.* **2011**, *381*, 183–191. [CrossRef]
416. Mavrantzas, V.G. Using Monte Carlo to simulate complex polymer systems: Recent progress and outlook. *Front. Phys.* **2021**, *9*, 661367. [CrossRef]

417. Theodorou, D.N. Progress and outlook in Monte Carlo simulations. *Ind. Eng. Chem. Res.* **2010**, *49*, 3047–3058. [CrossRef]
418. Martin, M.G. MCCCS Towhee: A tool for Monte Carlo molecular simulation. *Mol. Simul.* **2013**, *39*, 1212–1222. [CrossRef]
419. Moorthi, K.; Kamio, K.; Ramos, J.; Theodorou, D.N. Monte Carlo simulations of structure and entanglements in polymer melts. *Mol. Simul.* **2015**, *41*, 993–995. [CrossRef]
420. Jeong, S.H.; Kim, J.M.; Yoon, J.; Tzoumanekas, C.; Kröger, M.; Baig, C. Influence of molecular architecture on the entanglement network: Topological analysis of linear, long-and short-chain branched polyethylene melts via Monte Carlo simulations. *Soft Matter* **2016**, *12*, 3770–3786. [CrossRef]
421. Clark, E.S. The crystal structure of polytetrafluoroethylene: Forms I and IV. *J. Macromol. Sci. Part Phys.* **2006**, *45*, 201–213. [CrossRef]
422. Yamamoto, T. Monte Carlo simulation of the crystal structure of the low-temperature phase of polytetrafluoroethylene. *J. Polym. Sci. Polym. Phys. Ed.* **1985**, *23*, 771–786. [CrossRef]
423. Clark, E.S. The molecular conformations of polytetrafluoroethylene: Forms II and IV. *Polymer* **1999**, *40*, 4659–4665. [CrossRef]
424. Perepelkin, K.E. Fluoropolymer fibres: Physicochemical nature and structural dependence of their unique properties, fabrication, and use: A Review. *Fibre Chem.* **2004**, *36*, 43–58. [CrossRef]
425. Xu, Q.; Tang, X.; Zhang, J.; Hu, Y.; Ma, T. Unraveling Tribochemistry and Self-Lubrication Mechanism of Polytetrafluoroethylene by Reactive Coarse-Grained Molecular Dynamics Simulations. *Acs Appl. Mater. Interfaces* **2023**, *15*, 45506–45515. [CrossRef] [PubMed]
426. Brown, E.N.; Clausen, B.; Brown, D.W. In situ measurement of crystalline lattice strains in phase IV polytetrafluoroethylene. *J. Neutron Res.* **2007**, *15*, 139–146. [CrossRef]
427. Rae, P.J.; Dattelbaum, D.M. The properties of poly (tetrafluoroethylene)(PTFE) in compression. *Polymer* **2004**, *45*, 7615–7625. [CrossRef]
428. Rae, P.J.; Brown, E.N.; Clements, B.E.; Dattelbaum, D.M. Pressure-induced phase change in poly (tetrafluoroethylene) at modest impact velocities. *J. Appl. Phys.* **2005**, *98*, 063521. [CrossRef]
429. Rae, P.J.; Brown, E.N. The properties of poly (tetrafluoroethylene)(PTFE) in tension. *Polymer* **2005**, *46*, 8128–8140. [CrossRef]
430. Brown, E.N.; Rae, P.J.; Orler, E.B.; Gray III, G.T.; Dattelbaum, D.M. The effect of crystallinity on the fracture of polytetrafluoroethylene (PTFE). *Mater. Sci. Eng.* **2006**, *26*, 1338–1343. [CrossRef]
431. Grytsenko, K.; Kolomzarov, Y.; Lytvyn, P.; Kondratenko, O.; Sopinskyy, M.; Lebedyeva, I.; Niemczyk, A.; Baranovska, J.; Moszyński, D.; Villringer, C.; et al. Optical and mechanical properties of thin PTFE films, deposited from a gas phase. *Macromol. Mater. Eng.* **2023**, *308*, 2200617. [CrossRef]
432. Li, C.; Jiang, H. Fabrication and characterization of flexible electrowetting on dielectrics (EWOD) microlens. *Micromachines* **2014**, *5*, 432–441. [CrossRef]
433. Mak, T.; Westerwaal, R.J.; Slaman, M.; Schreuders, H.; Van Vugt, A.W.; Victoria, M.; Boelsma, C.; Dam, B. Optical fiber sensor for the continuous monitoring of hydrogen in oil. *Sens. Actuators Chem.* **2014**, *190*, 982–989. [CrossRef]
434. Price, J.S.; Wang, B.; Kim, T.; Grede, A.J.; Sandoval, J.M.; Xie, R.; Shen, Y.; Adams, D.R.; Eller, M.J.; Sokolov, A.; et al. Fluoropolymer-diluted small molecule organic semiconductors with extreme thermal stability. *Appl. Phys. Lett.* **2018**, *113*, 263302. [CrossRef]
435. Wang, B.; Ruud, C.J.; Price, J.S.; Kim, H.; Giebink, N.C. Graded-index fluoropolymer antireflection coatings for invisible plastic optics. *Nano Lett.* **2019**, *19*, 787–792. [CrossRef]
436. Zhou, G.; Gou, X.; Jin, X.; Wang, S.; He, W.; Wang, S.; Deng, L.; Chen, S.; Liu, G. Low fractal dimension modified drilling-hole wall for PTFE high-frequency board copper plating with plasma treatment. *J. Appl. Polym. Sci.* **2019**, *136*, 48052. [CrossRef]
437. Primc, G. Recent advances in surface activation of polytetrafluoroethylene (PTFE) by gaseous plasma treatments. *Polymers* **2020**, *12*, 2295. [CrossRef] [PubMed]
438. Roman, H.E.; Cesura, F.; Maryam, R.; Levchenko, I.; Alexander, K.; Riccardi, C. The fractal Geometry of Polymeric Materials Surfaces: Surface Area and Fractal Length Scales. *Soft Matter* **2024**, *20*, 3082–3096. [CrossRef] [PubMed]
439. Piferi, C.; Bazaka, K.; D'Aversa, D.L.; Di Girolamo, R.; De Rosa, C.; Roman, H.E.; Riccardi, C.; Levchenko, I. Hydrophilicity and Hydrophobicity Control of Plasma-Treated Surfaces via Fractal Parameters. *Adv. Mater. Interfaces* **2021**, *8*, 2100724. [CrossRef]
440. Giacomello, A.; Meloni, S.; Chinappi, M.; Casciola, C.M. Cassie-Baxter and Wenzel states on a nanostructured surface: Phase diagram, metastabilities, and transition mechanism by atomistic free energy calculations. *Langmuir* **2012**, *28*, 10764–10772. [CrossRef]
441. MacGregor-Ramiasa, M.N.; Vasilev, K. Questions and answers on the wettability of nano-engineered surfaces. *Adv. Mater. Interfaces* **2017**, *4*, 1700381. [CrossRef]
442. Cheeseman, S.; Owen, S.; Truong, V.K.; Meyer, D.; Ng, S.H.; Vongsvivut, J.; Linklater, D.; Tobin, M.J.; Werner, M.; Baulin, V.A.; et al. Pillars of life: Is there a relationship between lifestyle factors and the surface characteristics of dragonfly wings? *Acs Omega* **2018**, *3*, 6039–6046. [CrossRef]
443. Förch, R.; Chifen, A.N.; Bousquet, A.; Khor, H.L.; Jungblut, M.; Chu, L.Q.; Zhang, Z.; Osey-Mensah, I.; Sinner, E.K.; Knoll, W. Recent and expected roles of plasma-polymerized films for biomedical applications. *Chem. Vap. Depos.* **2007**, *13*, 280–294. [CrossRef]
444. Riccardi, C.; Roman, H.E.; Ziano, R. Attachment of Polymer Chains on Plasma-treated Surfaces: Experiments and Modeling. *New J. Phys.* **2010**, *12*, 073008. [CrossRef]
445. Yan, M.; Ren, J. Covalent immobilization of polypropylene thin films. *J. Mater. Chem.* **2005**, *15*, 523–527. [CrossRef]

446. Wang, X.; Ramström, O.; Yan, M. Glyconanomaterials: Synthesis, characterization, and ligand presentation. *Adv. Mater.* **2010**, *22*, 1946–1953. [CrossRef] [PubMed]
447. Murugan, P.; Krishnamurthy, M.; Jaisankar, S.N.; Samanta, D.; Mandal, A.B. Controlled decoration of the surface with macromolecules: Polymerization on a self-assembled monolayer (SAM). *Chem. Soc. Rev.* **2015**, *44*, 3212–3243. [CrossRef] [PubMed]
448. Seidi, F.; Zhao, W.F.; Xiao, H.N.; Jin, Y.C.; Saeb, M.R.; Zhao, C.S. Advanced surfaces by anchoring thin hydrogel layers of functional polymers. *Chin. J. Polym. Sci.* **2021**, *39*, 14–34. [CrossRef]
449. Zeraati, M.; Langley, D.B.; Schofield, P.; Moye, A.L.; Rouet, R.; Hughes, W.E.; Bryan, T.M.; Dinger, M.E.; Christ, D. I-motif DNA structures are formed in the nuclei of human cells. *Nat. Chem.* **2018**, *10*, 631–637. [CrossRef]
450. Peña Martinez, C.D.; Zeraati, M.; Rouet, R.; Mazigi, O.; Henry, J.Y.; Gloss, B.; Kretzmann, J.A.; Evans, C.W.; Ruggiero, E.; Zanin, I.; et al. Human genomic DNA is widely interspersed with i-motif structures. *EMBO J.* **2024**, *43*, 4786–4804. [CrossRef]
451. Foffano, G.; Marenduzzo, D.; Orlandini, E. Facilitated diffusion on confined DNA. *Phys. Rev. E* **2012**, *85*, 021919. [CrossRef]
452. Kar, P.; Cherstvy, A.G.; Metzler, R. Acceleration of bursty multiprotein target search kinetics on DNA by colocalisation. *Phys. Chem. Chem. Phys.* **2018**, *20*, 7931–7946. [CrossRef]
453. Liu, L.; Cherstvy, A.G.; Metzler, R. Facilitated diffusion of transcription factor proteins with anomalous bulk diffusion. *J. Phys. Chem.* **2017**, *121*, 1284–1289. [CrossRef]
454. Cao, J.; Fan, J.; Sun, W.; Guo, Y.; Wu, H.; Peng, X. The photoprocess effects of an amino group located at different positions along the polymethine chain in indodicarbocyanine dyes. *R. Soc. Chem. Adv.* **2017**, *7*, 30740–30746. [CrossRef]
455. Ayala-Orozco, C.; Galvez-Aranda, D.; Corona, A.; Seminario, J.M.; Rangel, R.; Myers, J.N.; Tour, J.M. Molecular jackhammers eradicate cancer cells by vibronic-driven action. *Nat. Chem.* **2024**, *16*, 456–465. [CrossRef]
456. Broglia R.A.; Colò G.; Onida G.; Roman H.E. *Solid State Physics of Finite Systems: Metal Clusters, Fullerenes, Atomic Wires, Advanced Texts in Physics;* Springer: Berlin, Germany, 2004. [CrossRef]
457. McDermott, L.; Walters, Z.G.; French, S.A.; Clark, A.M.; Ding, J.; Kelleghan, A.V.; Houk, K.N.; Garg, N.K. A solution to the anti-Bredt olefin synthesis problem. *Science* **2024**, *386*, eadq3519. [CrossRef]
458. Banavar, J.; Szymczak, P.; Turski, Ł.A. Marek Cieplak (1950–2021). *Phys. Today* **2023**, *6*, 16a. [CrossRef]
459. In Memoriam: *Ricardo Americo Broglia* (1939–2022). Italian Physical Society (SIF). Available online: https://www.sif.it/riviste/sif/sag/ricordo/broglia (accessed on 26 November 2024).
460. Herrmann, H.J.; de Oliveira, P.M.C.; Schadschneider, A. Dietrich Stauffer (1943–2019). *Int. Mod. Phys.* **2020**, *31*, 2001001. [CrossRef]

MDPI AG
Grosspeteranlage 5
4052 Basel
Switzerland
Tel.: +41 61 683 77 34

Polymers Editorial Office
E-mail: polymers@mdpi.com
www.mdpi.com/journal/polymers

www.ingramcontent.com/pod-product-compliance
Lightning Source LLC
Chambersburg PA
CBHW061807210926
51850CB00006B/633
* 9 7 8 3 7 2 5 8 5 1 1 5 7 *